Separation Techniques in Clinical Chemistry

Separation Techniques in Clinical Chemistry

edited by
Hassan Y. Aboul-Enein
King Faisal Specialist Hospital and Research Centre
Riyadh, Saudi Arabia

Marcel Dekker, Inc. New York • Basel

ISBN: 0-8247-4013-0

This book is printed on acid-free paper.

Headquarters
Marcel Dekker, Inc.
270 Madison Avenue, New York, NY 10016
tel: 212-696-9000; fax: 212-685-4540

Eastern Hemisphere Distribution
Marcel Dekker AG
Hutgasse 4, Postfach 812, CH 4001 Basel, Switzerland
tel: 41-61-260-6300; fax: 41-61-260-6333

World Wide Web
http://www.dekker.com

The publisher offers discounts on this book when ordered in bulk quantities. For more information, write to Special Sales/Professional Marketing at the headquarters address above.

Current printing (last digit):
10 9 8 7 6 5 4 3 2 1

PRINTED IN THE UNITED STATES OF AMERICA

Preface

With the development of scientific areas such as a combinatorial chemistry, high-throughput screening, and robotics, it has become essential to identify and conduct analytical measurements on thousands of new molecular entities. Separation science has, in turn, become the backbone of analytical chemistry in all its areas of application, including clinical chemistry. Accordingly, it is crucial that separation techniques be reliable, reproducible, and easily applied.

The objective of this volume is to provide comprehensive, up-to-date information for several analytical separation methods used in clinical chemistry. The book begins with an extensive review of sample pretreatment (Chapter 1), followed by a discussion of the use of separation science in therapeutic drug monitoring (Chapter 2). Enantioseparation, a unique and essential field of separation science, is used in the analysis and purification of chiral drugs. Chapter 3 presents the impact of chirality in pharmacokinetics and therapeutic drug monitoring. Several chapters are devoted to reflecting the importance of high-performance liquid chromatography (HPLC). These chapters examine HPLC in bioavailability studies and therapeutic drug monitoring (Chapters 4 and 5), analysis of illicit drugs (Chapter 6), and analysis of biogenic amine neurotransmitters (Chapter 7). Chapters 8 and 9 address the application of affinity chromatography and immunochromatography in clinical analysis. The importance of tandem mass spectrometry in the analysis of inherited metabolic diseases is discussed in Chapter 10. Chapter 11 covers thin-layer chromatography, a very simple and cost-effective technique that is still in use in clinical and pharmaceutical analysis. Finally, Chapter 12 focuses on the use of capillary electrophoresis as a potential separation technique in clinical chemistry and drug analysis in biological fluids.

I would like to thank the contributors for providing us with invaluable knowledge from their respective fields; without their participation, this book would not have been possible. I also extend my thanks to Marcel Dekker, Inc., and, in particular, to Mr. Russell Dekker for his encouragement. Finally, Ms. Shelly Lynde deserves a special thanks for her secretarial assistance and keeping up with all the logistics during the preparation of the book.

I am confident that this volume will be particularly beneficial to analytical, clinical, and medicinal chemists.

Hassan Y. Aboul-Enein

Contents

Contributors

Hassan Y. Aboul-Enein Pharmaceutical Analysis Laboratory, Department of Biological and Medical Research, King Faisal Specialist Hospital and Research Centre, Riyadh, Saudi Arabia

Jan Bladek Institute of Chemistry, Military University of Technology, Warsaw, Poland

M. J. Bogusz Department of Pathology and Laboratory Medicine, King Faisal Specialist Hospital and Research Centre, Riyadh, Saudi Arabia

Jacek Bojarski Department of Organic Chemistry, College of Medicine, Jagiellonian University, Kraków, Poland

Iwona Cendrowska Pharmaceutical Research Institute, Warsaw, Poland

Zdzisław Chilmonczyk National Institute of Public Health, Warsaw, and University of Białystok, Białystok, Poland

William Clarke Department of Pathology, Johns Hopkins School of Medicine, Baltimore, Maryland, U.S.A.

David S. Hage Department of Chemistry, University of Nebraska, Lincoln, Nebraska, U.S.A.

Kamila Kobylińska Pharmaceutical Research Institute, Warsaw, Poland

Adarsh M. Kumar Department of Psychiatry and Behavioral Sciences, University of Miami School of Medicine, Miami, Florida, U.S.A.

Barry Levine Toxicology Laboratory, Office of the Chief Medical Examiner, Baltimore, Maryland, U.S.A.

Karla A. Moore Toxicology Laboratory, Office of the Chief Medical Examiner, Baltimore, Maryland, U.S.A.

Slawomir Neffe Institute of Chemistry, Military University of Technology, Warsaw, Poland

I. N. Papadoyannis Laboratory of Analytical Science, Department of Chemistry, Aristotle University of Thessaloniki, Thessaloniki, Greece

Mohamed S. Rashed Metabolic Screening Laboratory, Department of Genetics, King Faisal Specialist Hospital and Research Centre, Riyadh, Saudi Arabia

V. F. Samanidou Laboratory of Analytical Science, Department of Chemistry, Aristotle University of Thessaloniki, Thessaloniki, Greece

Zak K. Shihabi Department of Pathology, Wake Forest University School of Medicine, Winston-Salem, North Carolina, U.S.A.

Joanna Szymura-Oleksiak Department of Pharmacokinetics and Physical Pharmacy, College of Medicine, Jagiellonian University, Kraków, Poland

1

Sample Pretreatment in Clinical Chemistry

I. N. Papadoyannis and V. F. Samanidou
Aristotle University of Thessaloniki, Thessaloniki, Greece

1 INTRODUCTION

The development of any analytical method includes a number of steps from sample collection to the final report of the results. Intermediate stages involve sample storage, sample preparation, matrix modification, isolation of analytes, identification, and finally quantification. Among theses steps the most tedious and time-consuming is the preanalytical phase, known as *sample pretreatment*, which takes almost 60% of the time distribution of the whole procedure of analysis and determination. It is also the most error-prone part of the process, as it contributes 30% of the sources of errors, impacting on precision and accuracy of the overall analysis [1–3].

In clinical chemistry, sample preparation is crucial, as the determination and identification of endogenous compounds is important for the diagnosis and prevention of disorders. Analytical chemistry is continuously used in the research and development of new drugs, e.g., for pharmacokinetics, drug monitoring, metabolism studies, and toxicological studies.

The pretreatment of samples of biological origin is aimed mainly at the adequate reduction in matrix interference. The successful extraction of drugs from biological matrices such as urine, blood, saliva, and hair presents several challenges, as their composition is variable and complex, and the concentrations of analytes are very low. Biological matrices often contain proteins, salts, acids, bases, and various organic compounds with similar chemistries to the target analytes. Selective analysis of drugs from a variety of biological matrices can be performed by both traditional and modern techniques of sample pretreatment. The objectives of sample pretreatment are listed in Table 1. The term sample pretreatment may refer to various stages of the analysis procedure as shown in Fig. 1 [4–11].

Several sample pretreatment techniques are reported in the literature depending on the sample volume and required quantities for measurement. These techniques can also be classified in terms of sample nature, liquid or solid, as shown in Table 2 [10].

An ideal sample pretreatment technique should have the following characteristics:

- Simplicity and rapidity
- High extraction efficiency with quantitative and reproducible analyte recoveries
- Specificity for the analytes
- High sample throughput; fewer manipulation steps to minimize analyte losses
- Amenability to automation
- Use of the minimum amount of solvent, compatible with many analytical techniques
- Low cost, regarding reagents and equipment [12]

Although there is no single ideal technique for sample preparation, there are several, each with its own advantages and disadvantages. The most commonly used techniques are discussed in the following paragraphs of this

TABLE 1 Objective of Sample Preparation

1. Matrix modification
   - Preparation for introduction (injection) onto chromatographic column
   - Rendering of the solvent suitable for the analytical technique to be used
   - Prolongation of instrument lifetime, e.g., column lifetime
2. Cleanup–purification
   - Removal of impurities in order to obtain the required analytical performance and selectivity
3. Analyte enrichment (preconcentration)
   - Improvement of method sensitivity (reduction of limits of detection and quantification).

Sources: Refs. 7, 8.

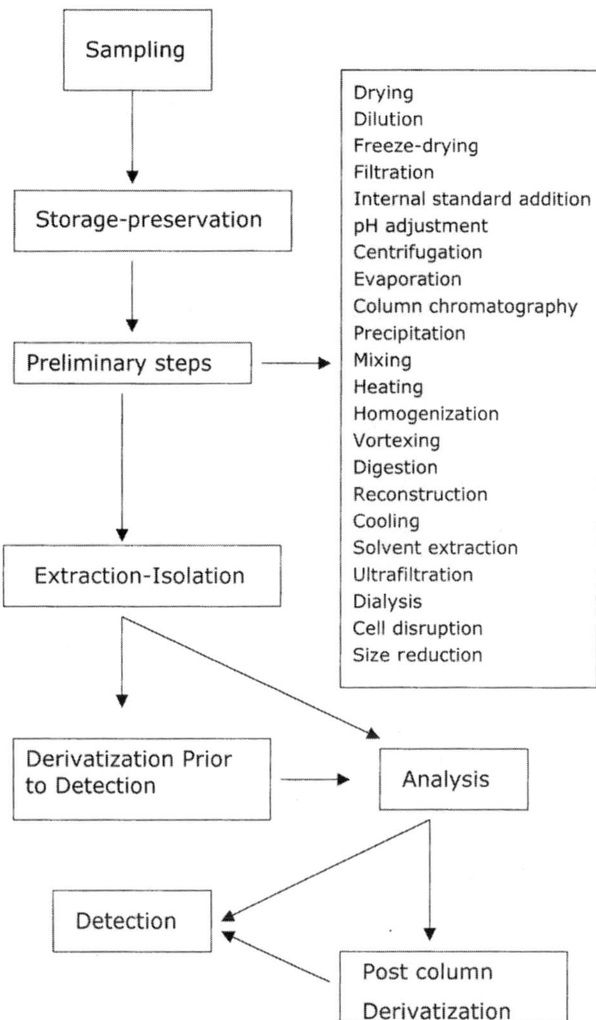

FIGURE 1 Stages of sample preparation. (From Refs. 9–11.)

chapter with the emphasis on clinical samples. The analyst may choose the most appropriate technique, when developing a method depending on several parameters such as sample matrix, nature, physical and chemical properties of analytes, concentration, and analytical technique that is to be applied.

As the biggest problem with sample preparation is time, an increasingly important consideration when developing techniques is the possibility of

TABLE 2 Typical Preparation Techniques for Liquid and Solid Samples

Liquid samples	Solid samples
Dilution	Solid–liquid extraction (shake filter)
Evaporation	Forced flow leaching
Distillation	Soxhlet extraction and automated Soxhlet extraction
Microdialysis	Homogenization
Lyophilization	Sonication
Liquid–liquid extraction (LLE)	Dissolution
Automated LLE	Matrix solid phase dispersion
Solid phase extraction (SPE)	Accelerated solvent extraction
Automated SPE	Supercritical fluid extraction (SFE)
SPE disc technology	Microwave-assisted extraction
SPME	Gas-phase extraction
Direct analysis by column switching techniques (on-line techniques)	Thermal desorption
Stir bar sorptive extraction (SBSE)	

Source: Ref. 2.

automating the entire analytical process. The benefits of automation, apart from the obvious economic ones, the reduced number of manual operations and amount of laboratory equipment required, include increased sample throughput and productivity per instrument and higher precision. Moreover, automation may improve contamination control in trace analysis, as handling of contagious or radioactive samples can be performed in closed analytical systems. Thus high-throughput automated analyzers are the workhorses in the central hospital laboratory [7,10].

2 SAMPLES OF CLINICAL INTEREST

Clinical analysis deals mainly with the determination of drugs, metabolites, poisons, chemicals of environmental exposure, and endogenous substances in body fluids and tissues. Additionally, the quantitative and qualitative analysis of drugs and metabolites is applied to pharmacokinetic studies. Variables such as time to maximal concentration in plasma, clearance, and bioavailability have to be identified for the approval of a new drug. Therapeutic drug monitoring (TDM) may be a useful tool for the improvement of drug therapy. Drugs of abuse and illicit drugs are analyzed in clinical and forensic toxicology. Finally, as part of environmental medicine, a wide variety of chemicals such as dioxins are analyzed in human body fluids for the investigation of environmental and occupational exposure.

The nature and typical characteristics of the most common biological samples are discussed in this section.

2.1 Urine

This is one of the most commonly studied biological matrices for drug determination, and its analysis is of paramount importance in clinical chemistry, as it is relatively easy to collect. It can be used for drug screening, for forensic purposes, and to monitor workplace exposure to chemicals. It is a universal means of excretion of both parent drug compound and diagnostic metabolites. As a matrix, it has moderate complexity and a relatively high variability, and it typically contains both organic and inorganic constituents. Drug extraction efficiency can vary for several reasons such as variations in pH value and ionic strength and the presence of additional components [13,14].

2.2 Blood: Serum—Plasma—Whole Blood

Serum is the straw-colored liquid that separates from the clot that is formed in whole blood. *Plasma* is prepared from whole blood that has been treated with an anticlotting substance such as heparin. It is the supernatant that results when the cellular components of blood are removed by centrifugation. As matrices for drug analysis, the significant difference between the two is that serum does not contain fibrinogen and certain clotting factors (2.5–5% of proteins). Both serum and plasma are routinely used for drug analysis. A method developed for plasma can normally be applied to serum without modification.

Plasma samples contain significant amounts of salt and protein. Major constituents of normal human serum are albumins 35–45 mg/mL, globulins 30–35 mg/mL, lipids 4–7 mg/mL, salts 7 mg/mL, and carbohydrates 1.34–2.0 mg/mL. Drugs will bind to plasma proteins to varying degrees depending on their individual physicochemical properties. In general, acid and natural drugs bind primarily to albumin, and basic drugs primarily to a-acid glycoprotein. Although it is general practice to report total drug concentration (free plus protein-bound) in serum, there are isolation techniques, like solid-phase microextraction (SPME), that can measure the concentration of free drug, which is therapeutically relevant.

In most cases a method developed for the determination of drugs in plasma could also apply to the analysis of *whole blood*, which is particularly useful in postmortem analyses. The determination of blood constituents can be performed either by employing headspace analysis, as blood is considered a "dirty" sample, or by means of direct immersion extraction, e.g., a range of local anesthetics [13,14].

2.3 Saliva

The determination of drugs in saliva is convenient, as sampling is noninvasive. A sample is easy to collect, and quantitative measurements may reflect the nonpolar protein bound, i.e., the therapeutically relevant fraction of the drug in plasma. Compared to other biological samples, levels of protein and lipids are quite low, making it amenable to microextraction. Extraction efficiency from saliva may vary from 5 to 9% relative to 100% for extraction from pure water, owing to a combination of reduced mass transfer kinetics in the sample due to its viscous nature, and binding of the compounds to proteins present in the sample. By acidifying the samples with acetic acid, extraction efficiencies are improved as the samples are clarified, and proteinaceous material and cellular debris are precipitated and removed prior to extraction [13,14].

2.4 Tissue

Tissue samples are treated as solid samples and can originate from liver, myocardium, muscle, kidney cortex, cerebellum, and brain stem. Tissue samples are treated by an extraction technique, or headspace sampling is applied for volatile compounds.

2.5 Hair

Hair is a popular target for drug analysis as collection is relatively noninvasive; therefore its analysis can be used for forensic purposes, and to monitor drug compliance and abuse. As with biological fluids, drugs and their metabolites are expressed in hair. Measurements along a strand of hair can provide a record of drug usage or exposure. Before analysis the hair matrix must be either digested enzymatically (e.g., with a protease) or more usually with strong alkali, e.g., 1 M NaOH. Though it suffers from variability in drug concentration by hair type, due to drug affinity variations, an additional advantage is that it is a fairly nonpolar matrix, so it tends to absorb parent drug molecules, which are typically less polar than metabolites. Because of this, it is ideal for extraction of analytes to nonpolar extraction phases, especially when the parent drug is extensively metabolized and often nondetectable in other tissues [13–15].

2.6 Human Milk

The main constituents of human milk are water (88%), proteins (3%), lipids (3%), and carbohydrate in the form of lactose (6%). The lipids are in the form of fat droplets suspended in the aqueous matrix. The nonwater constituents are present in different physical forms: dissolved (lactose), colloidally dispersed (protein), and in water (lipids). Human milk can serve as a means of exposure of a newborn to compounds the mother has been previously exposed to and provides a relatively convenient means of biomonitoring for toxicant exposure. It is quite constant in composition from the third week on, while collostrum from early lactation, differs significantly as it contains less lactose and practically no fat [13,14].

2.7 Breath

Analysis of compounds in breath is a new area mainly of microextraction investigation. Expired breath can be collected in Tedlar bags, and the gas sample can be further treated, e.g., by SPME, where the fiber is subsequently exposed directly to the contents of the bag. The determination of compounds in breath is of great importance for clinical diagnosis and toxicological purposes.

3 PRELIMINARY SAMPLE PREPARATION PROCEDURES

Frequently, intermediate stages may be required prior to extraction techniques for sample preparation. The primary goal is to minimize analytes' solubility in the matrix and maximize selective isolation. The sample has to be prepared to optimize conditions for quantitative, rapid extraction of the analyte of interest. This step may include dilution, changing of pH or ionic strength, filtration, centrifugation, and addition of an organic solvent and/or internal standard.

In the presence of proteins these must be removed prior to extraction by denaturation with organic solvents or chaotropic agents, precipitation with acid (pH modification), adding a compound that competes for binding sites, or using restricted access media. Conjugated components may need to be hydrolyzed to release the free compound. Lipids may be extracted into a nonpolar organic solvent [16].

Once transferred in a liquid phase, the sample can either be directly injected or treated further as described in Table 3.

3.1 Distillation

Distillation is a widely used method for separating the components of liquid mixtures based on the distribution of mixture constituents between liquid and vapor. There are several distillation techniques that can be applied for purification (sample cleanup) or preparation of mixtures prior to analysis.

TABLE 3 Further Treatment Requirements for a Sample in Liquid Phase Before Analysis

Dilution to the appropriate concentration ranges with a compatible solvent in case it is too concentrated
Concentration by LLE, SPE, evaporation, or lyophilization
Derivatization to stabilize, freeze-drying, storage at 4°C, far from light or air, in case it is reactive or thermally or hydrolytically unstable
Removal of unwanted high molecular weight substances by size exclusion chromatography, dialysis, ultrafiltration, precipitation, use of supported liquid membrane
Removal of particulate matter by filtration, centrifugation, or sedimentation
Solvent exchange if the solvent is not compatible with the analytical method
Solvent removal by evaporation, lyophilization or distillation, and analyte reconstitution in a solvent and at a concentration suitable for the technique that will be used for analysis
Addition of internal standard

Source: Ref. 16.

The main purpose of distillation is the separation of volatile components in a mixture from either semivolatile components or each other. The variation in the saturation vapor pressure of a material versus its temperature is the basis of a distillation separation. Two phases, the liquid and the vapor, can be recovered separately with the more volatile compounds enriched in the vapor phase and the less volatile compounds concentrated in the liquid phase. Simple, fractional, vacuum, steam, membrane, and azeotropic distillation approaches can be applied depending on the purification and/or sample preparation needs (Figs. 2a, 2b).

Fractional distillation (countercurrent or rectification) in a batch process can be more often performed or a continuous process occasionally for the separation of volatile mixtures. In this multistage approach the distillate product (i.e., the more volatile components) is removed as the distillation proceeds. Vapor and liquid move countercurrently through the various stages. The vapors condense and redistill as they move up the distillation column. As the process goes on, the less volatile components remain behind and become enriched in the liquid phase. The degree of separation in a typical distillation depends on the physical properties of the components in the liquid mixture, the type and design of the still used, and the method of distillation. Another important parameter is distillate throughput, which is more important when large amounts of samples are needed.

Performance and productivity of distillation can be improved by automation, including the incorporation of microprocessors into control

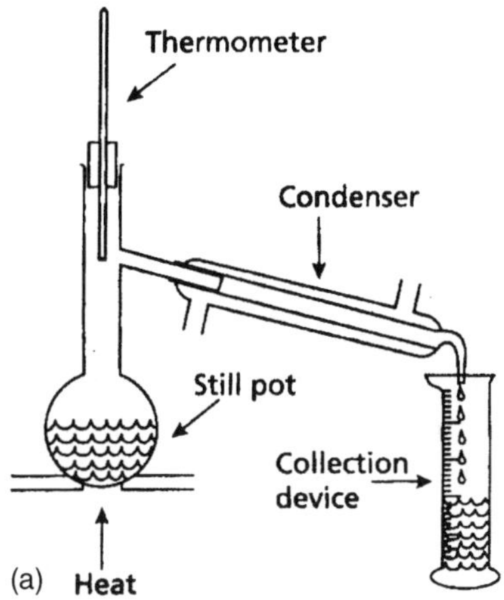

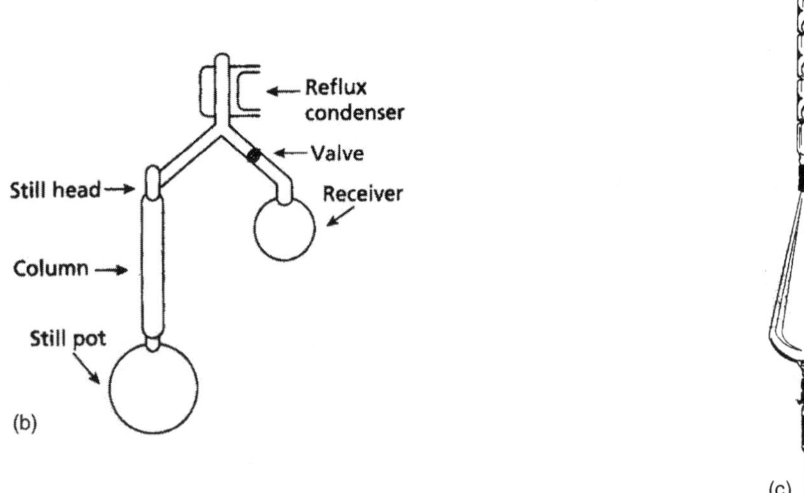

FIGURE 2 (a) Simple distillation apparatus. (From Ref. 17. Reprinted with permission of LCGC Europe.) (b) Fractional distillation apparatus. (From Ref. 17. Reprinted with permission of LCGC Europe.) (c) Kuderna–Danish concentrator. (From Ref. 19. Reprinted with permission of Supelco, Bellefonte, Pennsylvania, U.S.A.)

circuits, automated fraction collectors, and automated vacuum regulation for distillation at reduced pressures. However, automation features render the distillation equipment more expensive. Distillation of essential oils in drugs is an interesting application example of this technique [17].

3.2 Evaporation

When organic solvents are involved to isolate, elute, or extract an analyte of interest from a sample matrix, the analytes are usually suspended in a larger volume of this solvent than is desirable for the subsequent analysis. Thus the sample has to be concentrated so that the volume injected into the chromatograph contains detectable amounts of the individual solutes of interest. Analytes' integrity must be maintained during this process of reducing the solvent amount. For best quantification, no losses should occur during concentration, and no change in chemical composition of the sample should take place. In addition, the solvent in which the analyte is isolated must be compatible with the analytical, most often chromatographic, method. Consequently, at one time or another, most chemists face the problem of partial or total reduction of solvent volume used in analyte isolation. Techniques used for this purpose include evaporation, drying, and lyophilization.

Evaporation is a similar process to distillation as they both constitute the transfer of liquid into a vapor state, but the result is different. While in evaporation and drying the less volatile are the retained constituents and the more volatile constituents are discarded, in distillation the more volatile constituent is obtained in a pure form. Thus distillation is the removal of the solvent or mixtures of volatile compounds to purify and reclaim them, while evaporation is the removal of solvent so that the analyte is in a concentrated form. For evaporation or distillation to continue rapidly, the gas vapor must be removed as fast as it forms. For this reason, evaporation is accelerated by blowing an inert gas across the surface of the liquid or by removing vapors by vacuum.

The rate of evaporation is one factor that influences the removal of vapor (solvent) from the surface of a liquid. Another factor is the boiling point of the solvent or the latent heat of vaporization (ΔH), which is the amount of heat absorbed or released per mass unit during evaporation. Water is the slowest solvent to evaporate, as it has a much higher latent heat of vaporization than any other solvent. Methanol, though volatile, evaporates relatively slowly as it has a high ΔH. The more nonvolatile solvents are hard to evaporate because they may be overheated by the amount of heat necessary to volatilize them. Volatile solvents such as methylene chloride evaporate more quickly because they have a low ΔH [18].

Various types of microglassware are commercially available for small volumes of organic samples. One of the most popular concentration devices is the Kuderna–Danish (K–D) concentrator (Fig. 2c). Clean solvent is added on the top of the Snyder column and sample is concentrated until a volume of 3–4 mL remains in the receiver. Then the Kuderna–Danish concentrator is removed from the heat source and it is allowed to cool. The extract is transferred to a volumetric flask for final concentration under a gentle stream of nitrogen. Some K–D systems can concentrate volumes of 500 mL to 1 mL. Extraction to volumes lower to 1 mL might cause analyte loss. K–D concentrators can also be combined with other devices such as continuous liquid–liquid extractors to improve productivity and solvent saving [19].

3.2.1 Rotary Evaporators

When an evaporation step is required, time is an important matter of concern. Devices for evaporation can be manual or automatic. Rotary evaporators can have capacities for volumes ranging from a few mL to as large as 100 L. Sophisticated models with functional utilities can also be used. Rotary evaporators can be used for a single sample at a time. For simultaneous evaporation or concentration of multiple samples, solvent or nitrogen evaporators are used. The simplest multiple solvent evaporator is an aluminum heating block with holes drilled or machined into it to accommodate vials, test tubes, cuvettes, or 96-well plates. The heater block can provide temperatures up to 200°C. The speed of evaporation can be enhanced by delivering a flow of nitrogen gas from a height-adjustable hypodermic needle placed over each evaporation container. Systems with up to 100 evaporation positions are available. Other constructions that can be used are

1. *Microwave evaporation* systems, where organic solvents are evaporated under closely controlled vacuum and reduced temperature microwave heating conditions to retain volatile compounds, have been also introduced.
2. *Centrifugal evaporators/concentrators* comprise a centrifuge with a heater that can operate under vacuum and a vacuum pump with a vapor condenser. The centrifuge compartment can be heated from ambient temperature to 60°C to provide higher solvent evaporation rates. Such a concentrator can be used in a molecular biology laboratory, where nucleic acids are dried in a vacuum concentrator after precipitation by isopropanol and ethanol and centrifugation. Centrifugal concentrators have been also adapted to the 96-well plate format to be used in clinical laboratories or in any lab with combinatorial chemistry needs [18].

3.3 Lyophilization

Freeze-drying or lyophilization is used to reduce water solvent from easily degradable biological samples that cannot be heated. Its principle is based on rapidly freezing the material and subsequently removing the solvent using sublimation under vacuum. The solvent in a solid state is changed directly to a vapor, avoiding the liquid phase since it can cause change in a product. Lyophilization can be used to prepare tissue samples for analyses that require maintenance of the analyte's structure integrity. A lyophilizer system includes a refrigerated chamber with a manifold setup that holds various sizes of sample containers ranging from high to low volumes, a condensing module to trap the evaporated solvent, and a reliable vacuum source.

3.4 Storage of Biological Samples

Biological samples are often stored in common freezers at 4°C or deep-freezers. The main purpose is to freeze down a specific sample as fast as possible and keep it at a constant temperature below its critical point, which is often around −65 or −70°C. The standard freezers of today can perform a reasonable heat removal according to the convection principle and can freeze down to −87°C [20].

4 LIQUID SAMPLE PRETREATMENT

The choice of the proper sample preparation technique is sample dependent. The methods used for treating biological samples prior to their introduction into the chromatographic system include isolation of analytes by means of extraction and direct injection. Though the direct injection technique of biological samples is the simplest and most rapid, proteins may be precipitated in the column. A protein-free sample can be obtained by precipitation and filtration or centrifugation, but samples are not yet sufficiently clean, as a large fraction of them remain soluble and can cause interference with the analytes. Also, strongly protein-bound analytes may coprecipitate, resulting in low irreproducible recoveries, while low-molecular-weight compounds, for example lipids, stay in the sample. To overcome this drawback, an extraction technique, such as liquid–liquid extraction (LLE) and solid-phase extraction (SPE), is frequently used so that the compounds of interest are removed from the biological matrix under conditions that selectively isolate (extract) the desired components and leave behind unwanted materials. The solvent is then removed by gentle evaporation and the dried residue reconstituted in a small volume of the eluent solvent for injection onto the chromatographic column [15,21].

4.1 Liquid–Liquid Extraction (LLE)

For many years, liquid–liquid extraction was the classical technique for sample preparation of liquids in organic analysis. In spite of several drawbacks, it is widely used in all fields of analysis and until recently, it was exclusively applied. It usually involves mixing of an aqueous sample solution with an equal volume of immiscible organic solvent for a period of time and then allowing these two liquid phases to interact so that the analytes of interest are extracted from the aqueous layer into the organic layer, as the organic solvent has a larger affinity for the analyte. The selectivity and efficiency of the extraction process is governed by the choice of the two immiscible liquids. The more hydrophilic compounds prefer the aqueous phase, and the more hydrophobic compounds will be found in the organic solvent, as like dissolves like. Organic solvents are preferred, as they can be easily removed, so analytes can be concentrated by evaporation. These include aliphatic hydrocarbons, diethyl and other ethers, methylene chloride, chloroform, ethyl acetate and other esters, toluene, xylenes, aliphatic ketones C6+, and aliphatic alcohols C6+. Chelating and other complexing agents, ion-pairing, and chiral may also be used [7,22,23].

A limitation of this technique is that polar water miscible solvents cannot be used for the extraction. Because of this, LLE works best for more nonpolar analytes, while it is not suitable for very polar analytes. Since many samples also have very polar metabolites, this is a significant drawback. Typically, after the immiscible liquids are separated, the organic layer containing the extracted analytes is removed, concentrated to dryness, and reconstituted in an appropriate solvent compatible with the analytical system, e.g., the high-pressure liquid chromatography (HPLC) mobile phase.

Other drawbacks of this technique are

1. The use of large quantities of organic solvents that are required, leading to a considerable cost for their acquisition and disposal, especially when chlorinated and/or fluorinated solvents are involved with a potential risk for human health and environmental pollution, though microextraction techniques using reduced solvent volumes are also introduced.
2. The formation of emulsions during the mixing procedure.
3. Time-consuming evaporation procedures.
4. Coextraction of proteins and other matrix components.

5. Furthermore, LLE in its classical form (using a separation funnel or something similar) is difficult to automate and to connect in-line with analytical instruments. However, a number of flow-system LLE approaches have been presented in a tube: the aqueous and organic phases in a tube coil are mixed and then separated. Continuous liquid–liquid extraction can be performed as illustrated in Fig. 3 [7,15,22].

4.1.1 Liquid Back-Extraction (LLLE)

For many compound classes, it is possible to apply a two-step LLE. As an example, amines can be extracted from a basic aqueous sample into an organic solvent and then reextracted (or back extracted) into a second, acidic aqueous phase.

Two very significant advantages of liquid back-extraction (LLLE) are, firstly, that solvents are completely eliminated from the injection, so the solvent peak does not obscure early eluting peaks. Secondly, the entire sample extracted is injected, with better sensitivity. The most significant problem encountered is that any factor that alters the partition or absorption coefficient of analyte, or competes with the extraction phase for absorption or absorption of the analyte, will change the recovery. Applications are in the field of drug abuse analysis and in the analysis of legitimate pharmaceuticals at therapeutic concentration as well [7,22,23].

4.1.2 Countercurrent Distribution

A countercurrent distribution apparatus can provide 1000 or more equilibration steps for more efficient liquid–liquid extraction, but it requires more time and effort. Countercurrent distribution allows the recovery of analytes that have extremely small K_D values.

In principle, countercurrent distribution could be performed in a series of separatory funnels, each of which would contain an identical lower phase. The mixture could be introduced in the upper phase of the first separatory funnel and transferred to the second funnel after equilibration of the upper phase containing the substance of interest. Next, a new portion of upper-phase solvent would be introduced into the first separatory funnel. The equilibration process could be repeated many times, with each upper phase transferred to the next separatory funnel and new solvent introduced. With automated countercurrent distribution equipment, this process can be performed with several hundred transfers.

A newer countercurrent technique called *centrifugal partition chromatography* is used for extracting analytes from aqueous samples. The liquid

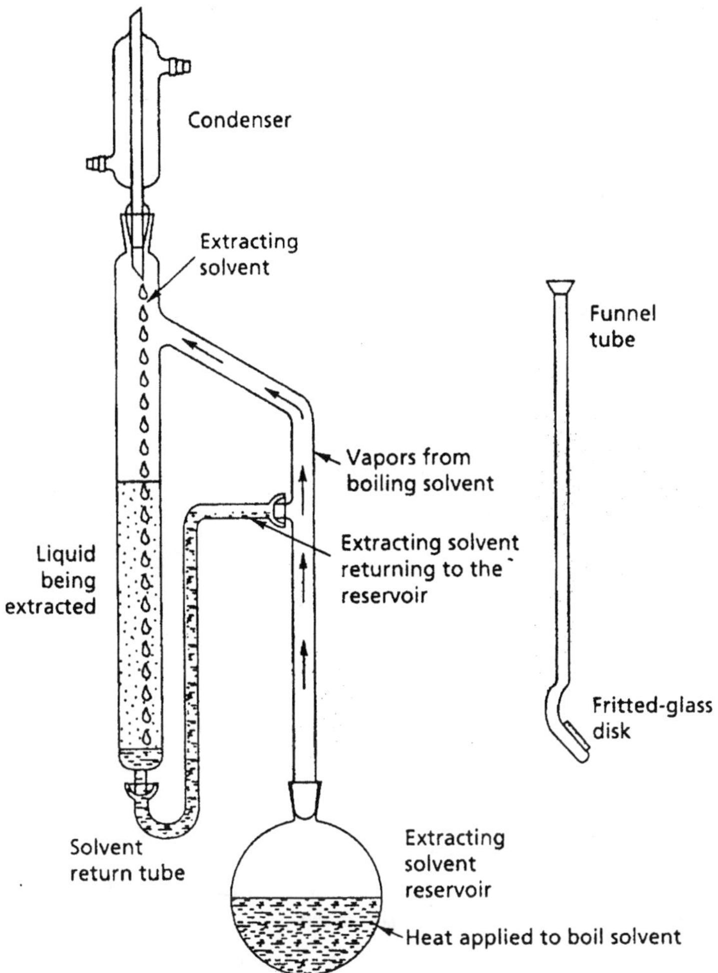

FIGURE 3 Design of a continuous liquid–liquid extraction apparatus. On the left, the apparatus is configured for use with an extracting solvent denser than the liquid to be extracted. For lighter solvent extractions, the solvent return tube is removed and the fritted-glass disk funnel on the right is placed in the extractor. In this case, the extracting solvent returns to the reservoir by overflowing in the extractor sidearm. (From Ref. 22. Reprinted with permission of LCGC Europe.)

stationary phase is retained by centrifugal force within discrete and interconnected partition channels of a series of discs. The mobile phase is pumped continuously through the liquid phase in the form of microdroplets. Sample components are partitioned between the mobile and stationary phases and are separated on the basis of their partition coefficients. This technique, compared to conventional liquid–liquid extraction, provides improvements in solvent reduction, good enrichment factors, and good recovery [24].

4.1.3 On-Line Extractions

A dynamic on-line LLE system applies the principle of flow injection analysis (FIA). Basically, the practice of flow injection analysis is similar to liquid chromatography, except that an open narrow-bore tube (extraction coil) is used instead of a packed column.

In the simplest form of LLE, the introduction of aqueous or organic sample solutions continuously or in discrete volumes into a continuous aqueous stream that contains a reactive or a complexing reagent, performs flow injection analysis. A chemical reaction or complexation takes place in a mixing-reaction coil, resulting in an extractable component segmented with an organic immiscible solvent steam at the phase segmenter, where small reproducible droplets of one phase are formed in the other. As the segmented solvent sections move through the extraction coil, the organic or aqueous solvents tend to wet the inner walls, depending on the type of tubing used for the coil, and they create a thin film that serves as an interface where the analyte partitioning process occurs. The extraction efficiency of the process is a function of the residence time in the extraction coil and the size of the extraction coil. The segments of the aqueous and organic phases are separated in the phase segmenter (Fig. 4) [22].

The phase containing the analytes of interest can be monitored by a flow-through detector, and the unwanted phase is directed to waste. More complex on-line systems that involve the introduction of membrane separation units, water-absorbing microcolumn separators, and preconcentration devices for analyte enrichment can also be used.

The advantages of this technique include the ability to use very small volumes of samples, small amounts of reagent and organic solvent, a closed system (less exposure of sample to the atmosphere and of users to potentially toxic or flammable solvents), high sample extraction throughput, full automation, and interface of the flowing system direct to the analytical measurement technique. Disadvantages include more complex hardware requirements (pumps, phase segmenters, and phase separators) and lower sensitivity than batch extractions unless preconcentration techniques are used.

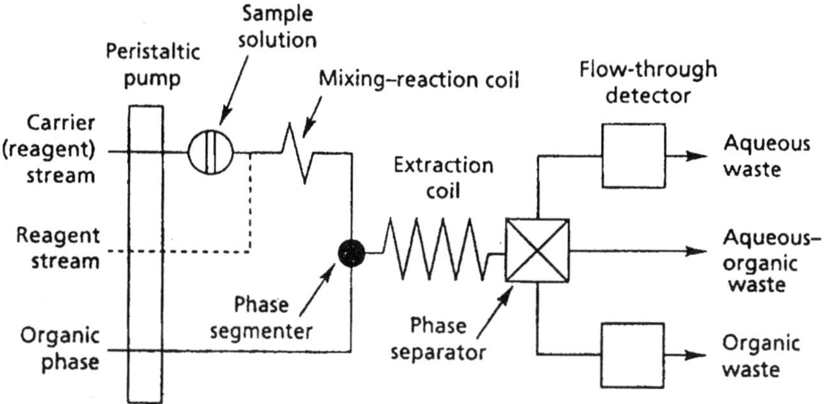

FIGURE 4 Liquid–liquid extraction flow injection manifold. (From Ref. 22. Reprinted with permission of LCGC Europe.)

4.1.4 Automated Liquid–Liquid Extraction

Classical LLE requires a great deal of manual handling, so when the sample load increases beyond a reasonable level, analysts frequently turn to automation. Many instrument manufacturers have developed instrumentation that can automate all or part of the extraction and concentration process. A number of autosamplers and workstations for HPLC and gas chromatography (GC) can perform LLE. Most of these use their liquid dispensing and mixing capabilities to perform LLE in sample vials or in tapered microvials. Some systems mix the layers by alternately shipping the solvents into the autosampler needle and dispensing the contents back into the sample vial. Units can also use vortex mixing to spin the vial at a high rate of speed. After the mixing is complete, the autosampler waits for a specified period of time until the phases separate. By controlling the depth of the needle, either the top layer or the bottom layer can be removed, for injection or further sample preparation. Most of these devices use a few milliliters of the sample and have limited capability to perform LLE of very large volumes. Robotic systems can be used to handle larger volume LLE [25].

4.1.5 Column Extraction LLE

Column extraction based on the theory of liquid–liquid partition is a widely used technique in biochemistry, toxicology, pharmaceutical analysis, and other fields. Extrelut by Merck (Darmstadt, Germany) and Extube by Varian (Harbor City, CA, USA) are commercially available prepacked columns used in these fields. The column extraction method can be easily

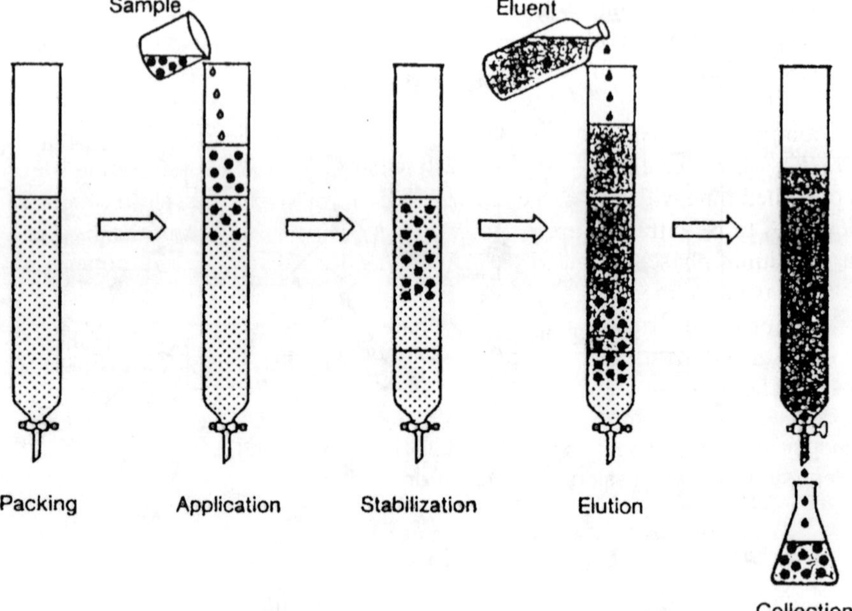

FIGURE 5 General procedure for column LLE. (From Ref. 26. Reprinted with permission of Merck KGaA, Darmstadt, Germany.)

applied to organic experiments by following the procedure illustrated in Fig. 5. Extraction is done by fixing an aqueous solution or suspension to a supporting material (stationary phase) and allowing the other immiscible solvent (mobile phase) to pass over it. The two phases are in contact, thus permitting continuous and multistep extraction to occur. This technique can replace the conventional LLE in a separating funnel, and thus it becomes more efficient and practical, as no emulsions are formed, lesser solvent volumes are used, and time is reduced as no drying is necessary before evaporation. Body fluids such as urine are the best example for application of LLE in column construction [26].

Commercially available supporting materials include Kieselguhr, Celite 545, synthetic and natural aluminum silicates, cellulose, silica gel aluminum oxide, and diatomaceous earth. Ethyl acetate, chloroform, diethyl ether, and benzene can be used as eluents, at an optimum flow rate of 3–5 mL/min.

4.2 Solid-Phase Extraction (SPE)

Solid-phase extraction, introduced in the early 1970s, is a particularly attractive, fast, and effective technique for isolation, and preconcentration of target analytes that avoids or eliminates the disadvantages of LLE; therefore it tends to replace it in many applications. The advantages of SPE versus LLE are listed in Table 4 [8,12].

The principle of SPE is similar to that of LLE, involving a partitioning of compounds between two phases. It is also similar to *liquid–solid extraction* (*LSE*), where the extraction procedure consists of shaking the sample for a controlled time with a suitable absorbent material (solid phase). In SPE, the analytes to be extracted are partitioned between a solid phase, the sorbent, and a liquid phase, the matrix that contains possible interference. Analytes must have a greater affinity for the solid phase than for the sample matrix (retention or adsorption step), and they are subsequently removed by eluting with a solvent that has a greater affinity for the analytes (elution or desorption step). The different mechanisms of retention or elution are due to intermolecular forces among three components: the analyte, the active sites on the surface of the absorbent, and the liquid phase or matrix.

SPE objectives include

1. Removal of interfering compounds
2. Preconcentration of the sample, i.e., the selective enrichment by factors of 100 to 5000
3. Fractionation of the sample into different compounds or groups of compounds as in classical column chromatography
4. Storage of analytes that are unstable in a liquid medium or with relatively high volatility
5. Derivatization reactions between the reactive groups of the analyte(s) and those on the adsorbent surface

TABLE 4 Advantages of SPE over LLE

Improved selectivity and specificity
Higher extraction recoveries (almost 100% recovery with suitable conditions)
Cleaner extracts than those obtained with liquid–liquid extraction
Reduced sample volume required
Decrease in solvent consumption
Reduced exposure of lab personnel to toxic and/or flammable solvents
Elimination of emulsion formation
Saving of time
Applicability to a wide variety of matrices
Simultaneous processing of a large number of samples with manifolds
Automation
Versatility owing to the wide variety of extraction conditions

Sources: Refs. 8, 12.

The major types of commercially available SPE product configurations are

1. Syringe barrel columns
2. Cartridges
3. 96-well plates
4. Disc membranes
5. SPE vacuum manifolds for simultaneous processing of 10, 12, 20, and 24 SPE tubes are provided from various manufacturers [6].

Most current SPE sample preparations are performed using SPE cartridges available in a wide variety of sizes, with packing capacities ranging from 20 mg to 10 g and with reservoir volumes as large as 30 mL [27].

An SPE cartridge/minicolumn is similar in shape to a disposable a medical-grade plastic syringe without plunger and without injection needle. The column housing is of polypropylene, a material that is resistant to a series of organic solvents. Sorbents with various functional groups of different particle size allow the use of small pressure to force the sample and wash solutions through the column. The packing is contained between bottom and top fritted discs, usually constructed of porous PTFE, polyethylene, or metal. The pore size of 20 μm allows macromolecules such as plasma proteins to pass. The average particle size of the sorbent is 40 μm. Above the sorbent is the sample reservoir. At the bottom the column is provided with a regular Luer fitting, by which it can be connected to the extraction systems via a vacuum valve with control manometer and a sidearm vacuum flask to a vacuum source, such as a water aspirator, which is readily available in any laboratory. Although gravity can facilitate the flow of most organic solvents through the columns, samples and more viscous solvents must be drawn by vacuum applied to the column outlet, by positive pressure applied to the column inlet (gas pressure from a syringe), or by centrifugation (Fig. 6). Typical flow rates range from 0.2 to 1.5 mL/s [27]. The SPE process can be carried out either on-line or off-line. The experimental procedure using the SPE cartridge described above is known as off-line SPE.

4.2.1 SPE Steps

Solid phase extraction consists of the following steps:

1. Preparation of the sorbent: this is solvation (activation of functional groups) and conditioning: removal of the excess of solvation solvent and washing with a solvent suitable for analyte retention of the sorbent. This is an extremely important step that prepares the SPE sorbent to interact with the sample. It has to be carried out very carefully, otherwise a negative effect on the extraction recovery and the reproducibility is anticipated.

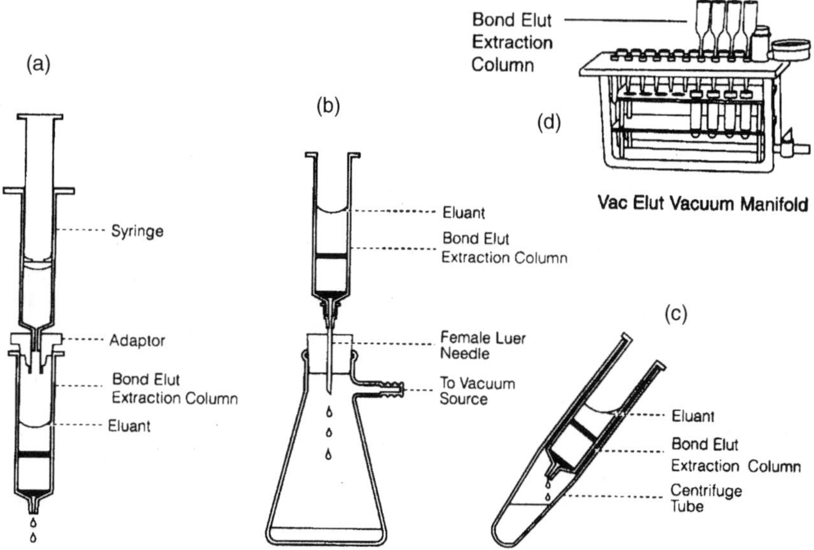

Figure 6 Different ways of SPE process. (a) Positive pressure. (b) Vacuum. (c) Centrifugation. (d) Vacuum manifold. (From Ref. 27. Reprinted with permission of Varian, Harbor City, California, U.S.A.)

In the case that C_{18} is the sorbent of choice, C_{18} groups can exert interaction with the solvent. As a result they will rotate freely and adopt the so-called brush-type configuration, which is energetically the most favorable one. The C_{18} groups are directed toward the solvent. Prior to conditioning, these groups are in the droplet configuration. The choice of the solvent is important. The solvent must have affinity with the non-polar C_{18} groups, but it also has to be miscible with the polar matrix of the sample. The most common choice is methanol. The excess of solvent is removed by washing with water or a buffer solution to suppress the analyte's ionization, while a thin layer of solvent resides on the C_{18} groups, forming a transitional layer between the hydrophobic surface of the sorbent and the polar aqueous matrix. At this point the intrusion of air into the sorbent must be avoided, as this will bring the sorbent back into its original inactive droplet configuration; in this case the solvation step should be repeated [28–29].

2. Sample loading. The next step in the SPE process is the sample, optionally containing the internal standard, percolation through the sorbent. As the sample is forced to pass through the cartridge, the components of interest and some undesired compounds are adsorbed by the sorbent. Quantitative retention of the analyte on the column is required, while the matrix passes unretained. Sample volumes in bioanalysis are between 20 μL and 1 mL.

The interaction between analyte and sorbent can be affected by selecting the most effective column. To facilitate the retention of the analyte, the sample can be diluted with water, so that the polarity of the environment of the analyte is increased.

3. Washing with an appropriate solvent: Rinsing of cartridge removes undesired polar matrix constituents. A solvent of low elution strength is chosen as a washing solvent, so that the analyte is not eluted. Usually, water or the same buffer that was used for the column pretreatment is preferred; a washing volume up to 20 times the bed volume of the column is needed. For a C_{18} column, which contains 100 mg of sorbent, the bed volume is about 120 μL. So 2.5 mL of water is a sufficient washing volume. At this point the analytes are in a purified form on the column. Once this step is completed, some air may pass through the column.

4. Sorbent drying may be necessary to remove water if elution solvent is immiscible with it. Drying can be performed by vacuum aspiration, nitrogen flow, or centrifugation, especially when analytes are volatile. Drying time can range from a few seconds to a few minutes depending on sorbent characteristics, elution solvent, and drying method.

5. Elution of the analyte is achieved by selective desorbing the compounds of interest from the sorbent with a suitable elution solvent. The interaction between analyte and elution solvent should be so strong that the interaction between analyte and sorbent will be overcome by using a relatively weak eluent. A quantitative elution can be effected with a solvent volume of about 5 times the bed volume of the column; so 600 μL of methanol is a sufficient volume. An effective tool in the development of an SPE procedure is the construction of the elution profile of an analyte. An elution profile can be obtained by eluting in small consecutive fractions of the elution solvent. Each fraction is collected and analyzed separately.

6. Solvent evaporation. The extract is directly injected or evaporated under a gentle stream of nitrogen and reconstituted in the mobile phase or after addition of an internal standard.

Optimization of SPE must be executed in each of these steps for best performance. Two different approaches can be chosen:

1. The analyte is retained on the sorbent while components pass through the waste. The analyte will be eluted later from the sorbent with a suitable solvent to be analyzed.
2. The matrix components are adsorbed while the analyte is evacuated.

The first approach is generally preferred, as less sorbent is required and isolate preconcentration is possible. The second approach is selected when interferences are present, but concentration of analytes is not required (Fig. 7) [30].

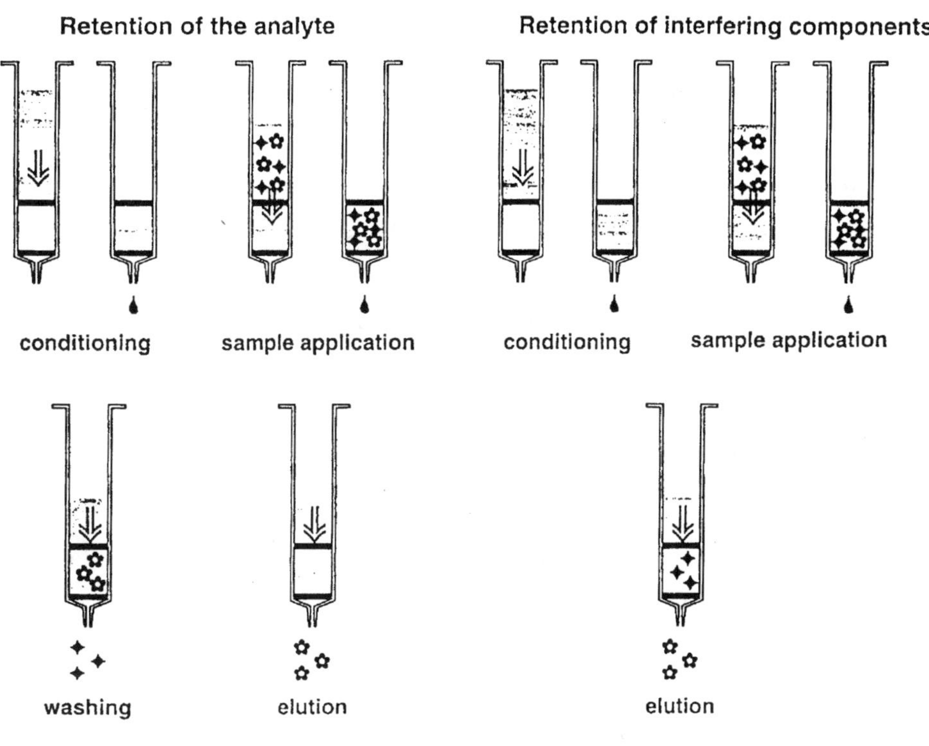

FIGURE 7 The steps of solid-phase extraction technique. (From Ref. 30. Reprinted with permission of Macherey-Nagel GmbH & Co. KG, Düren, Germany.)

4.2.2 SPE Sorbents

In recent years, several sorbents have been developed. The sorbents can be arranged according to the primary interaction possibilities of the functional groups of the sorbents. The majority of applications can be found in the group of sorbents with nonpolar or polar functional groups. A second group consists of cation-exchange and anion-exchange sorbents and contains functional moieties, which can act as ion exchangers. Silica based sorbents are summarized in Table 5 [6,29,14,31].

Besides the silica-based sorbents there are several others with a wide variety of specifications:

Polymeric sorbents are polystyrene-divinylbenzene (PS-DVB) resins that overcome many of the limitations of silica-based solid phases. The

TABLE 5 Silica-Based Sorbents

Reversed phase	High-hydrophobic octadecyl (C_{18}), octadecyl (C_{18}), octyl (C_8), ethyl (C_2), cyclohexyl, phenyl
Wide-pore reversed phase	Butyl (C_4)
Normal phase	Silica modified by cyano (–CN), amino (–NH_2), diols (–COHCOH)
Adsorption	Silica gel (–SiOH), florisil (Mg_2SiO_3), alumina (Al_2O_3)
Ion exchangers	Amino (–NH_2), quaternary amine (N^+), carboxylic acid (–COOH), aromatic sulfonic acid ($ArSO_2OH$)
Wide-pore ion exchangers	Carboxylic acid (–COOH), polyethylene-imine[–$(CH_2CH_2NH)_n$–]

Sources: Refs. 14, 29.

broader range of pH stability increases method development flexibility, typically providing greater analyte retention, which is particularly important for the recovery of polar compounds, than C_{18} sorbents. However, a conditioning step with a wetting solvent is still required. In general, polymeric sorbents retain analytes more strongly than silica-based materials [4].

Those who have introduced hydrophilic–lipophilic polymeric sorbents claim that no conditioning step is required, as the sorbent preserves analyte retention even if the bed dries out. Commercially available products include OASIS by WATERS and Abselut Nexus by Varian. These are copolymers of polyvinyl pyrrolidone and a cross-linked polystyrene divinylbenzene (PS-DVB) resin designed to extract an extensive spectrum of analytes: lipophilic, hydrophilic, acidic, basic, and neutral, where no conditioning is required. The hydrophilic *N*-vinylpyrrolidone, which acts as a hydrogen acceptor, increases the water wettability of the polymer, and the lipophilic di-vinylbenzene provides the reversed-phase retention necessary to retain analytes. These sorbents have greater pH stability and enhanced retention than C_{18}-bonded silica sorbents [8].

Mixed mode sorbents contain both nonpolar and strong ion (cation and/or anion) exchange functional groups targeted for the extraction of basic drugs [32].

Affinity sorbents such as restricted access matrix sorbents, immunosorbents, or molecularly imprinted polymers, have been introduced as well. These are based upon molecular recognition using antibodies with a high degree of selectivity or with molecular imprinted polymer, and they are discussed under Sec. 4.6.

Graphitized carbon: Porous graphitic carbon (PGC) similar to the liquid chromatography (LC)-grade Hypercarb is available in SPE cartridges (Hypersep PGC). This is characterized by a highly homogenous and ordered structure made of large graphitic sheets and by a specific area around 120 m^2/g. Compounds are retained by both hydrophobic and electronic interactions so that nonpolar analytes, but also very polar and water-soluble analytes, can be retained from aqueous matrices. Owing to the different retention mechanisms, acetonitrile and methanol can be inefficient, and it is preferable to use methylene chloride or tetrahydrofuran [6].

4.2.3 Sorbent—Isolate Interaction

The energetics of different interactions upon which SPE is based are as presented in Table 6. The most common chemical interactions between sorbent and isolates are nonpolar, polar (hydrogen bonding, dipole–dipole, induced dipole–dipole ion exchange [ionic]), and covalent interactions.

Based on energetics, the ideal interactions for a highly selective extraction are covalent and ionic, but elution is more difficult. Hydrogen bonding is a weaker interaction but plays an important role in bonded phase extraction due to the large number of possible bonding sites on silica [14,29].

Non-polar interactions are the weakest interactions, the so-called *van der Waals forces* or *dispersion forces*, which are exerted between the C—H bonds of the analyte and those of the sorbent, usually octadecylsilanized silica C_{18}. Since these are facilitated by a polar solvent environment, they play a crucial role in bioanalysis, as blood, plasma, serum, and urine are of polar nature.

Isolated compounds include alkyl, aromatic, alicyclic, or functional groups with significant hydrocarbon structure. All isolates have a potential for nonpolar interaction (except inorganic ions and compounds whose hydrocarbon structure is masked by polar groups [e.g., carbohydrates]). Retention is facilitated by a polar solvent, while elution is facilitated by an

TABLE 6 Bonding Forces in SPE

Interaction	Attractive force	Bond energy (kcal)
Covalent		100–300
Ionic	Electrostatic	50–200
Polar	Hydrogen bonding	3–10
Polar	Induced dipole–dipole	2–6
Nonpolar	van der Waals dispersion (hydrophobic interaction)	1–5

Sources: Refs. 14, 29.

organic solvent or a solvent mixture with sufficient nonpolar character (CH_3OH, CH_3CN, $CHCl_3$, etc). A wash solvent should be more polar than that used for isolate elution.

However, nonpolar interactions are nonselective and allow extraction of groups of compounds with different structures, as they can be found between almost every organic compound and almost every sorbent (Fig. 8a).

Polar interactions occur between a polar group of the sorbent and a polar group of the isolate. Silica is the typical adsorbent. Examples of groups that exhibit polar interactions are carbonyl, carboxyl, hydroxyl, sulfhydril, and amine groups and rings with heteroatoms, with unequal electron distributions, as well as groups with π-electrons, such as aromatic rings and double or triple bonds. Hydrogen bonding is one of the most significant polar interactions. Hydroxyl groups and amines are hydrogen bond donors. Polar interactions are stronger than nonpolar. Though they provide greater selectivity, they have the disadvantage of being facilitated by a nonpolar environment. High polar solvent facilitates elution, while wash solvents should be less polar than those used for isolate elution (Fig. 8b).

Ionic interactions are exhibited between two groups (of analyte and sorbent) with opposite charge. These are the strongest interactions that can be seen between an analyte and a sorbent. Examples of cationic groups are amine groups at pH values lower than the pK_a of the respective groups. However, for the same reason, the applicability is limited. Ionic interactions can only be effected in a polar environment (Fig. 8c). For an actual ionic interaction it is necessary that the respective functional groups of sorbent and analyte be both ionized. Table 7 summarizes the performance characteristics of the ion exchange mechanism in SPE.

Examples of anionic groups are carboxylic acids at pH values higher than the pK_a of the respective groups. Since few compounds possess either a cationic or an anionic group, the selectivity is high. All bonded silicas exhibit cationic secondary interactions in polar solvents due to residual silanols, particularly with amines. Changing the pH to neutralize the silanols or the amine, or adding diethyl or triethylamine as modifier, can suppress these interactions [12,27].

4.2.4 SPE Protocol Development

The method development should take into account the already discussed interactions between isolate and sorbent, sorbent and matrix, and isolate and

---

FIGURE 8 Mechanisms of SPE. (a) Nonpolar. (b) Polar. (c) Ion-exchange. (From Ref. 27. Reprinted with permission of Varian, Harbor City, California, U.S.A.)

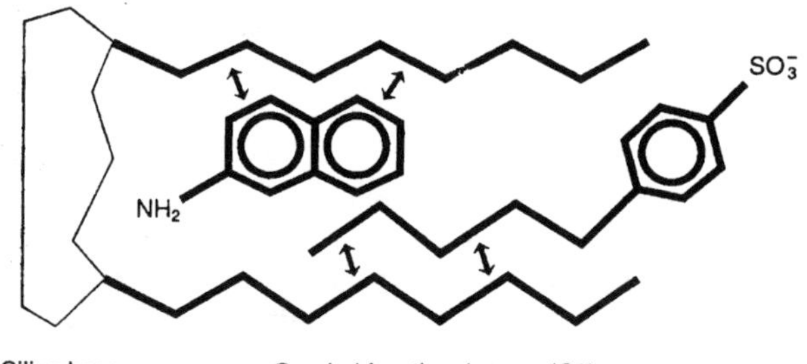

Silica base

Bonded functional group (C8)

⟷ Van der Waals forces

Interactions on nonpolar phases

(a)

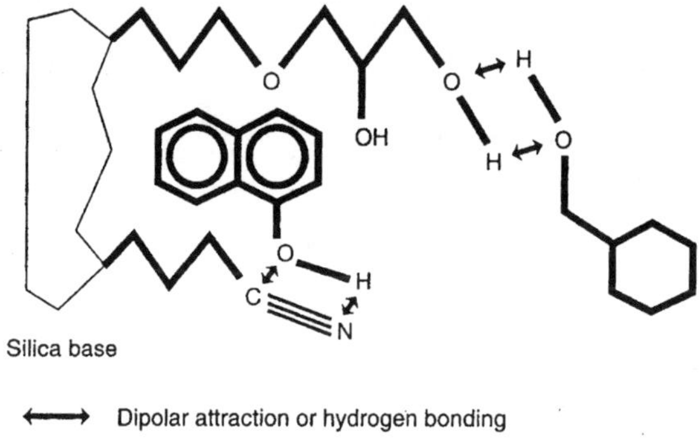

O
OH
O
H
H
O
O
H
C
N
Silica base

⟷ Dipolar attraction or hydrogen bonding

Interactions on polar phases

(b)

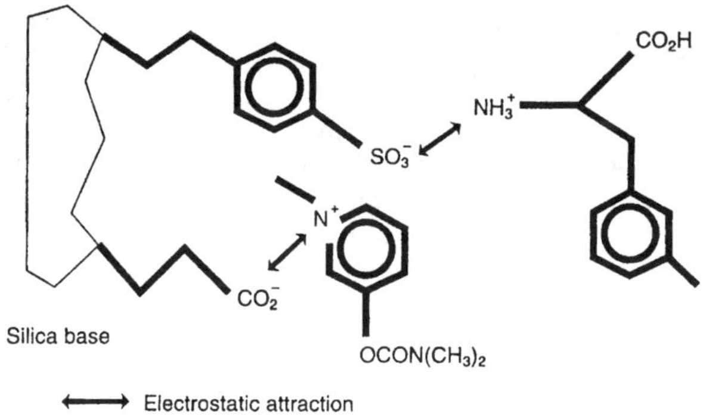

CO_2H
NH_3^+
SO_3^-
N^+
CO_2^-
Silica base
$OCON(CH_3)_2$

⟷ Electrostatic attraction

Interactions on ion-exchange sorbents

(c)

TABLE 7 Performance Characteristics of Ion-Exchange Mechanism in SPE

	Cation exchange	Anion exchange
Retention	Retention is facilitated by solvents with low ionic strength, containing low-selectivity cations, at a pH where both sorbent and isolate are charged: pH = isolate pKa − 2; pH = sorbent pKa + 2.	Retention is facilitated by solvents with low ionic strength, containing high selectivity anions, at a pH where both isolate and sorbent are ionized. pH = isolate pKa + 2; pH = sorbent pKa − 2.
Elution mode 1	By using an elution solvent of such a pH that the respective functional groups of either the sorbent—at a low pH value or the analyte—at a high pH—are neutralized. pH = isolate pKa + 2; pH = sorbent pKa − 2.	By using an elution solvent of such a pH that the respective functional groups of either the sorbent—at high pH values—or the analyte—at low pH values—are neutralized. pH = isolate pKa − 2; pH = sorbent pKa + 2.
Elution mode 2	By using an elution buffer solvent with a high ionic strength, as the high concentration of a counterion will compete with the interaction of the analyte with the sorbent. A counterion is an ion of a charge opposite to that of the functional group of the sorbent, e.g., in the case of cation exchange, a cation (e.g., Na^+) should be used as a counterion. Evidently, a combination of both elution ways is possible.	
Wash solvent	An aqueous solution with a lower ionic strength, a less selective counterion than for isolate elution, and a pH such that both the isolate and the sorbent are charged and can be used as wash solvents.	

Sources: Refs. 12, 27, 33.

matrix. Matrix general characteristics, pH, polarity, ionic strength, which impacts isolate retention or protein binding or adsorption to particular matter, must be also considered [33].

Aspects that influence column selection include

a. Nature of analytes
b. Nature of sample matrix
c. Degree of purity required
d. Nature of major contaminants
e. Analytical technique applied

Method development covers

1. Selection of retention mechanism
2. Selection of sorbent
3. Determination of sample preliminary procedure
4. Optimization of column conditioning
5. Selection of wash solvent for effective interference removal
6. Selection of appropriate elution solvent
7. Format of sorbent (cartridge or disc)
8. Automation

Stacked column configuration may be used for separating species by two different mechanisms, e.g., ion exchange and nonpolar, or by two different solvent systems, depending on the technique that will be used.

Though this technique is useful it is not amenable to automation, in contrast to *layered column configuration*, which can be easily automated. Layered columns contain two sorbents in the same cartridge. The use of layered columns enables the retention of analytes by different types or degree of interaction in the same cartridge. There are two approaches of application of this technique. In the first, analytes and interferences having different properties are retained by the two different layers. For example, interferences can be retained on the top layer and the analytes on the bottom layer. By selecting the proper solvent, interferences are retained, while analytes are eluted. In the second approach, two different sorbents are selected for retaining two different classes of interferences [32,34].

4.2.5 Applications in Clinical Chemistry

Several applications appear in the literature regarding SPE isolation of several analytes such as anthranilic acid derivatives (mefenamic, tolfenamic, and flufenamic acid), theophylline, morphine, codeine, bamifylline, atropine, scopolamine, lamotrigine, water- and fat-soluble vitamins, hydrochlorothiazide, ciprofloxacin, nitrite and nitrate, thyroid gland hormones triiodothyronine (T_3) and thyroxine (T_4), caffeine and its metabolites (methyl xanthines and methyluric acids), glycine betaine and its metabolite *N,N*-dimethylglycine, and analgesics in pharmaceuticals and clinical samples [35–52].

Application of SPE to Isolation of Analgesics (Paracetamol, Codeine, and Caffeine) from Biological Fluids: Blood Serum and Urine. In order to investigate the recovery efficiency of several SPE sorbents from different manufacturers and in different formats several extraction protocols were established for further optimization. The studied adsorbents are RP-18 (Merck-Darmstadt, Germany, Discovery by Supelco-Bellefonte, USA, Varian Harbor City, USA), RP-8 (by Alltech-Deerfield, USA, and Adsorbex by Merck), hydrophilic–lipophilic (Oasis by Waters-Milford, USA, and Abselut Nexus by Varian), and Disc-format C_{18} by SPEC, Ansys Technologies Inc. Recovery was calculated by comparing the peak area ratio of the analytes to the internal standard from the processed sample, to the corresponding peak area ratio of the nonprocessed sample. Table 8 illustrates the examined protocols for each sorbent. The protocols under assay gave various recovery results, as listed in Table 9. Isopropanol was found to be the best eluting solvent for paracetamol and codeine, while OASIS was found to be the best sorbent for the three compounds. Aceto- nitrile was the best eluting solvent for the three compounds when using OASIS, though conditioning was unavoidable. Methanol was shown to be the best solvent to elute the three analytes from nonconditioned NEXUS sorbent, while activation of sorbent improved performance, especially when acetonitrile and isopropanol were used as eluting solvents [52].

The five protocols that gave higher recovery rates for standard solutions were subsequently used for further investigation, when blood serum and urine were to be analyzed.

Human blood serum. Aliquots of 40 μL of pooled human blood serum were spiked with 200 μL of standard solution, at different concentration levels. Each sample was treated with 200 μL of CH_3CN in order to precipitate proteins. After vortex mixing for two minutes, the sample was centrifuged at 3500 rpm for 10 min and the supernatant diluted with 1.5 mL water. Subsequently it was slowly applied to the solid-phase cartridge and treated according to the five protocols that gave the highest recovery rates, as shown in Table 9.

Urine. Aliquots of 100 μL of pooled urine were spiked with 200 μL of standard solution, at different concentration levels. Each sample was treated with 200 μL of CH_3CN in order to precipitate proteins. After vortex mixing for 2 min, the sample was centrifuged at 3500 rpm for 10 min and the supernatant diluted with 1.5 mL water. Subsequently, the sample was slowly

TABLE 8 Setup Parameters for Solid-Phase Extraction Optimization

Protocols	A	B	C	D[a]	E[b]
Conditioning	2 mL CH_3OH 2 mL H_2O	2 mL CH_3OH 2 mL H_2O	2 mL CH_3OH 2 mL H_2O	No conditioning	500 μL CH_3OH 500 μL H_2O
Sample addition	200 μL standard solution: 0.5–1.0 and 3.0 ng/μL				100 μL standard solution
Washing	2 mL H_2O				500 μL H_2O
Drying	The sorbent is dried under vacuum				
Elution	2 mL CH_3OH	2 mL CH_3CN	2 mL $CH_3CH(OH)CH_3$	2 mL $CH_3OH/CH_3CN/CH_3CH(OH)CH_3$	1 mL $CH_3OH/CH_3CN/CH_3CH(OH)CH_3$
Evaporation to dryness under nitrogen 45°C					
Reconstitution to 200 μL internal standard solution					

[a] Only applicable to OASIS and NEXUS.
[b] Applicable to Disc Format C-18.
Source: Ref. 52.

TABLE 9 Recovery Results % of Analgesic Compounds Using the Examined SPE Protocols (Mean Values of Six Measurements at Three Concentration Levels)

	Paracetamol R % ± SD			Caffeine R % ± SD			Codeine R % ± SD		
	CH_3OH	CH_3CN	2-propanol	CH_3OH	CH_3CN	2-propanol	CH_3OH	CH_3CN	2-propanol
Varian C_{18}	85.8 ± 0.5	79.5 ± 0.8	**89.9**[a] ± 4.0	**103.3** ± 5.3	51.5 ± 1.5	91.0 ± 2.7	40.1 ± 1.3	71.8 ± 2.2	**87.7** ± 5.2
Ansys-Disc C_{18}	67.1 ± 0.7	69.1 ± 3.4	**72.2** ± 2.6	**84.1** ± 0.6	65.1 ± 0.7	72.4 ± 3.0	65.9 ± 2.2	60.0 ± 1.3	**84.0** ± 3.5
Supelco C_{18}	64.3 ± 2.3	63.4 ± 1.2	**86.2** ± 0.4	**81.3** ± 2.0	50.3 ± 1.3	72.8 ± 2.1	42.6 ± 1.3	72.6 ± 4.0	**84.8** ± 3.4
Merck C_{18}	62.4 ± 4.1	**79.8** ± 0.1	68.0 ± 1.4	74.2 ± 3.5	**84.9** ± 1.9	58.3 ± 4.3	**89.8** ± 3.0	55.1 ± 0.7	76.4 ± 1.1
Alltech C_8	**76.3** ± 1.0	62.0 ± 1.6	67.9 ± 1.0	**86.9** ± 0.2	45.9 ± 1.9	**86.5** ± 0.3	44.9 ± 0.3	64.7 ± 2.3	**65.1** ± 0.1
Merck C_8	85.6 ± 4.0	59.6 ± 3.1	58.4 ± 2.4	58.2 ± 2.4	**73.1 ± 1.7**	**74.4** ± 1.8	**76.1** ± 5.6	50.1 ± 0.7	66.8 ± 1.5
Oasis	54.7 ± 3.4	**104.6** ± 4.2	29.3 ± 0.6	57.8 ± 2.5	**102.7** ± 3.6	66.04 ± 3.6	44.1 ± 3.0	**100.3** ± 6.8	40.5 ± 2.8
Nexus	80.2 ± 5.7	**86.4** ± 0.3	**85.6** ± 1.0	**73.2** ± 3.8	67.3 ± 4.8	58.7 ± 2.9	40.6 ± 0.6	**74.8** ± 1.0	70.8 ± 3.8
Nexus nc	**101.6** ± 6.4	79.0 ± 3.2	72.8 ± 4.1	**92.9** ± 2.6	75.6 ± 2.0	60.1 ± 0.2	**79.0** ± 1.3	72.6 ± 1.3	59.1 ± 1.4

nc = Nonconditioned.

[a] Higher percentage recovery yielded from different eluent solvents, for each analyte in bold type.

Source: Ref. 52.

TABLE 10 Absolute Recovery of Analgesic Compounds from Human Blood Serum and Urine

SPE sorbent	Elution solvent	Paracetamol R% ± SD[a]	Caffeine R% ± SD[a]	Codeine R% ± SD[a]
		Serum		
C_{18}-Varian	isopropanol	98.8 ± 7.0	98.4 ± 3.1	90.1 ± 0.5
Supelco-Discovery C_{18}	isopropanol	88.1 ± 3.1	96.6 ± 2.0	93.0 ± 2.6
Disc format C_{18}	isopropanol	76.4 ± 1.7	90.0 ± 0.6	86.9 ± 0.7
OASIS (cond.)	CH_3CN	95.5 ± 1.3	99.0 ± 1.7	99.4 ± 0.4
NEXUS Varian n.c.	CH_3OH	86.2 ± 3.1	99.8 ± 3.8	84.3 ± 3.0
		Urine		
C18-Varian	isopropanol	79.5 ± 0.6	82.0 ± 3.2	91.1 ± 0.6
Supelco-Discovery C_{18}	isopropanol	84.4 ± 1.1	79.3 ± 0.3	94.0 ± 1.5
Disc format C_{18}	isopropanol	93.8 ± 1.7	67.8 ± 1.0	73.8 ± 0.4
OASIS (cond.)	CH_3CN	94.2 ± 1.8	97.1 ± 1.9	93.2 ± 0.1
NEXUS Varian n.c.	CH_3OH	84.5 ± 2.8	90.5 ± 2.4	83.3 ± 1.3

[a] Mean value of six measurements at three concentration levels

Source: Ref. 52.

applied to the solid-phase cartridge and extracted according to the five selected protocols. After elution with the appropriate solvent each sample was evaporated to dryness under a gentle stream of nitrogen. Reconstitution was performed with an internal standard solution. Absolute recovery was calculated by comparing the concentrations for spiked serum and urine samples with those for direct injection of pure compounds. The absolute recoveries are shown in Table 10. No interference from matrix compounds was observed in either case. High performance liquid chromatograms of analgesic compounds extracted from serum and urine are shown in Figs. 9a and 9b [52].

4.2.6 SPE Disc Technology

SPE in disc format has been designed to overcome the limitations of conventional packed-bed SPE columns, such as the large void volume (partly in the frits), the tendency to channel, and a substantial binding capacity for unwanted sample components. Discs exhibit minimal channeling, have small void volumes, do not require frits, have low capacity for interference (cleaner extracts are provided), and capture analytes of interest very effectively. The SPE disc format provides reduced bed mass, faster results, and less solvent used. Disc products include rigid discs, flexible discs, and thin packed beds of small particles between two retained screens. It is applicable to many fields, e.g., biomedical drugs of abuse. The low void volumes of the disc reduce the reagent volume requirements for sample processing. Additionally, less solvent means faster sample processing and thus increased throughput.

Disc technology has gained acceptance for processing large sample volumes, and small diameter discs for processing small samples.

The advantages of disc format SPE are reported in Table 11. A typical extraction sequence (shown in Fig. 10) is similar to the traditional SPE, though significantly lower volumes of organic solvents are required:

Disc conditioning with no more than 500 μL of an organic solvent followed by same volume solvent exchange to match the sample matrix.

Sample application after matrix modification, if necessary, such as pH buffer concentration. Careful addition of prepared samples at a

FIGURE 9 (a) High-performance liquid chromatogram of determination of paracetamol 3.359 min, caffeine 4.148 min, codeine 5.669 min, in spiked human blood serum samples with internal standard 1 ng/μL lamotrigine 8.699 min., after SPE using preconditioned OASIS. (b) High-performance liquid chromatogram of determination of paracetamol 3.710 min, caffeine 4.142 min, codeine 5.374 min in blank urine sample with internal standard 1 ng/μL lamotrigine min. after SPE using preconditioned OASIS. (From Ref. 52.)

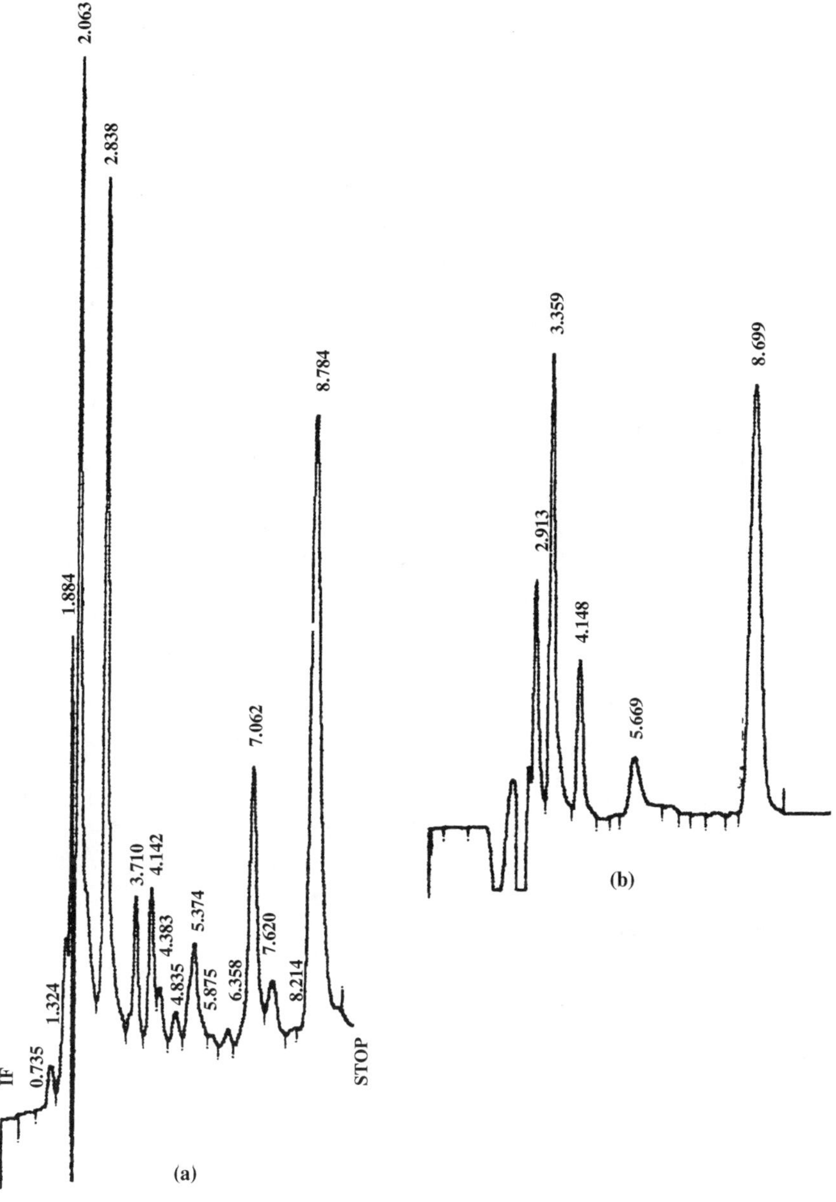
IF
0.735
1.324
1.884
2.063
2.838
3.710
4.142
4.383
4.835
5.374
5.875
6.358
7.062
7.620
8.214
8.784
STOP
(a)
2.913
3.359
4.148
5.669
8.699
(b)

TABLE 11 Advantages of SPEC Technology

High reproducibility (lot-to-lot and within lot)
Selective analyte retention
No channeling, that causes no uniform flow, reducing reproducibility
Cleanliness of sample without interferences
More efficient processing with fewer steps
Faster throughput
Reduced solvent usage

Source: Ref. 53.

vacuum < 17 kPa. Drawing the sample through the discs too rapidly may result in the loss of analyte, and this is a major cause of nonreproducibility.

Washing off the interferences by using the proper solvent in which the analyte is insoluble.

Analyte elution by using the most effective solvent at low flow rates (3–6 kPa) to allow time for efficient desorption of the analyte [53–56].

A wide variety of different chemically modified surfaces ensure maximum extraction selectivity, so that high extraction can be achieved. The extraction device serves as both a selective chemical sorbent and a physical filter. Several types of disc extraction media are commercially available. The most prevalent are paper and glass-fiber-based products and PTFE-based products [55–60].

SPE discs closely resemble membrane filters; they are flat, usually 1 mm or less in thickness, with diameters ranging from 4 to 96 mm. Some discs are

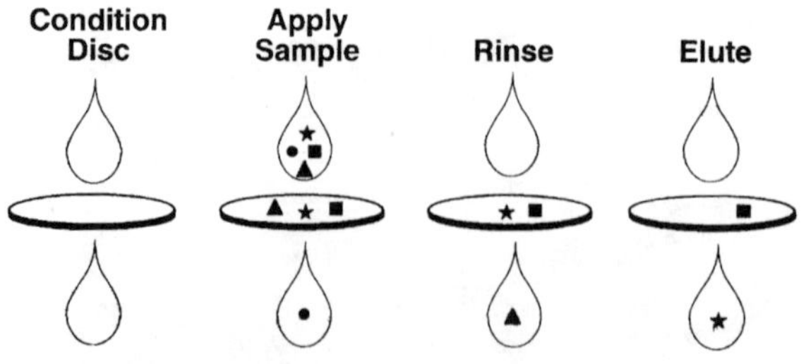

FIGURE 10 Typical extraction sequence using SPE discs. (From Ref. 55. Reprinted with permission of Ansys Technologies Inc.)

sold loose, and users must install them in a filter holder. Other are sold preloaded in disposable holders with Luer fittings.

Speediscs, manufactured by Baker, are commercially available products consisting of a thin sandwich of 10 μm sorbent particles held between two glass-fiber filters with a screen to hold the filters in place. Other disc products consist of flexible PTFE networks (Empore 3M, and Novo Clean Alltech), which are embedded or impregnated with a stationary phase (e.g., silica or bonded silica). The Empore contains approximately 90% silica based packing and 10% PTFE, and the Novo Clean disc contains approximately 60% cation exchange resin. The solid phase extraction concentrator, SPEC, by Ansys Technologies Inc., consists of a rigid fiberglass disc that contains a supporting matrix in which silica and bonded silica particles are embedded. The resulting rigid disc can be considered as an array of interconnected closely packed porous particles of small diameter.

Two major characteristics distinguish SPE discs from SPE cartridges: relatively large cross-sectional area and thinness. These enable higher flow rates, thereby increasing extraction speed. Small SPE discs are typically used for biological sample analysis, requiring smaller capacities and elution volumes. Most applications of such discs that contain 1, 5, 15, and 30 mg of sorbent mass in 3 and 10 mL reservoirs are in clinical chemistry and toxicology. Discs 25, 47, and 90 mm in diameter can be inserted in a disc manifold, while smaller 4, 8, and 12 mm diameter membranes are available in traditional SPE columns and 96-well plates. Ansys cartridges have built-in glass fiber prefilters. Alltech's Novo Clean membrane discs are available individually for replacement into standard 13 and 25 mm filter holders as well as in medical grade polypropylene housing. The PTFE-based disc may contain an impregnated sulfonated PS-DVD cation-exchange resin in hydrogen, silver, and barium forms, in order to be used for sample cleanup in ion chromatography [54,61–63].

4.2.7 Automated Solid-Phase Extraction Systems

SPE involves many steps that require manual intervention, so as the number of samples increases, many analysts search for instruments that automate some or all of these manual procedures. The two main approaches to automate SPE are the use of dedicated instruments, i.e., so-called SPE workstations, and precolumn techniques for on-line coupling of SPE to HPLC [7]. In other words, SPE automation is accomplished off-line and on-line. In off-line automation, SPE cartridges are not directly connected to the high-pressure flow stream of the analytical instrument GC or HPLC. In on-line SPE, the solid-phase device is inserted into and becomes part of the liquid or gas flow stream. Obviously the hardware interfacing requirements for the two approaches differ. In off-line automated SPE, cartridges tops must be sealed to allow transfer of the liquids and gases needed to perform the various SPE steps. In the manual mode, a simple vacuum is used to pull liquids through the cartridge. However, in the automated mode positive gas or liquid pressure is more commonly used to push liquids through the cartridge. Off-line automated SPE usually involves some form of robotic manipulation and has been achieved using flexible robotic arms and semi-flexible modified x-y-z liquid handling devices. Most of these systems are controlled by personal computers and may optionally have bar code readers for sample tracking and identification.

Robot manufacturers and system integrators have built sophisticated systems (robotic workstations) that automate sample preparation procedures and interface them with analytical instruments for unattended analyses. Dedicated SPE devices handle multiple cartridges with the rinsing and elution steps totally automated [25,64].

Some of the major potential advantages and limitations of automated SPE are reported in Table 12. During the course of the automated solid-phase extraction, analysts can redirect their time to other tasks. Additionally, automated systems can provide higher sample throughput with higher accuracy and precision due to minimized systematic errors, than do manual systems, by utilizing the concept of automated parallel sample processing [65].

TABLE 12 Advantages and Disadvantages of Automated Techniques

Advantages
Time savings.
Higher throughput, the use of a parallel processing algorithm.
Improved precision and accuracy.
Automated method development is possible.
Safety: Automation minimizes exposure to pathogenic or otherwise hazardous samples.
Reduced assay tedium.
Disadvantages
Carryover can limit performance.
Systematic errors can occur undetected and error recovery is sometimes a problem.
Precision is worse with systematic errrors.
Sample stability (physical or chemical) is occasionally a problem when sequential processing is used.

Source: Ref. 65.

Early automated systems processed individual samples in series. The next sample in the series was not started until the preceding sample had been completed or was well on its way. This approach, however, was slower than manual systems; this limitation was overcome only because they could operate continuously 24 hours per day so that time saving was still achieved.

Automated parallel processing solid-phase extraction was introduced and demonstrated to be practical, where numerous samples are extracted simultaneously with significantly improved throughput. Today the most common automated systems used are parallel processing systems that can achieve speeds of up to 400 samples per hour. Automation is possible for fast throughput for plasma, serum, and urine samples, using the standard vacuum manifolds and automated SPE systems. Examples of commercially available parallel processing systems include the ASPEC by Gilson (Villiers-le-Bel, France), Rapid Trace by Zymark Corporation (Hopkinton, MA, U.S.A.), and the Spe-ed Wiz from Applied Separations (Pennsylvania, U.S.A.). Most of these are compact, stand-alone systems that automate all of the liquid handling in SPE. They use positive air pressurization, rather than a vacuum, to move liquid through the extraction columns.

On-line automated SPE was introduced in the early 1980s. The AASP (Varian Sample Preparation, Palo Alto, CA, U.S.A.) was the first commercial unit to perform on-line SPE; it has been recently withdrawn from the market. A similar system called Prospekt was introduced by Spark Holland (Emmen, the Netherlands) in the mid-1980s, and a second-generation version is now available. This device incorporates custom solid-phase extraction cartridges into an analytical-scale HPLC separation using three electrically actuated switching valves [65–67].

On-line SPE-LC utilizes precolumns with small dimensions. A typical on-line arrangement is performed using simple switching valves and commercial precolumns and holders. Two automated devices commercially available are OSP-2 (On-line Sample Preparation unit, Merck, Darmstadt, Germany) and Prospect (Programmable Online Solid Phase Extraction Technique, Spark, Holland), which have the capability of using a fresh disposable precolumn for every sample; extraction is carried out in a similar way to the off-line sequence using a solvent delivery unit, which provides the solvent necessary to purge, wash, and activate the precolumn and apply the required volume of sample. The main difference is at the process of desorption, since the analytes from the precolumn are transferred into the analytical column by a suitable mobile phase.

Besides the complex instrumentation, a significant disadvantage of this on-line approach is the single use of the extraction cartridge.

SPE can be also integrated into an HPLC system by coupled column switching as described in Sec. 6. For this purpose, a small (typically 5–30 mm × 3–4.6 mm I.D.) precolumn is connected to a conventional analytical HPLC column via an electrically or pneumatically driven (six-port) valve. In the most common case, the pretreated biofluid is injected onto the cartridge or precolumn, which retains the target molecules. Interfering sample constituents are flushed into waste. The analytes retained on the bonded phase of the cartridge or precolumn are eluted on-line via the switching valve onto the series-connected analytical column. Simultaneously with the analytical separation, an exchange of the cartridge or reconditioning of the precolumn takes place (Fig. 11) [68].

Autosamplers offer at some level sample preparation functions such as dilution, internal standard addition, and preparation of working standards. In addition, some have sample preparation devices at the next level of sophistication, such as the automated liquid handling devices, and some have built-in heating, cooling, SPE, and other functions for performing automated sample preparation.

Typical sample preparation functions that can be combined into sample preparation methods include internal standard addition, preparation of calibration standards, linear or serial dilution, LLE, SPE, filtration, heating and evaporation, digestion, derivatization, and concentration. These operations are controlled by a computer system. Other automated systems incorporate some sample preparation modules to an autosampler such as

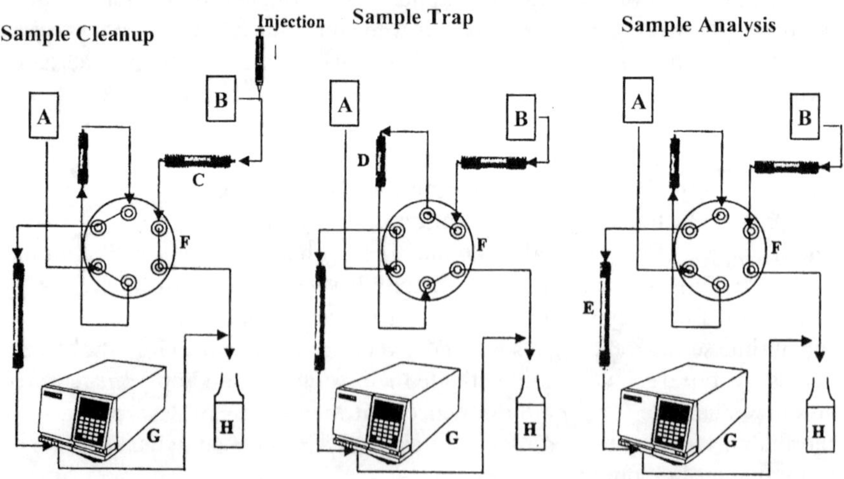

FIGURE 11 Schematic diagram of on-line SPE analysis. (A) Binary pump A. (B) Binary pump B. (C) Cleanup column. (D) Trap column. (E) Analytical column. (F) Switching valve. (G) Detector. (H) Waste. (From Ref. 68.)

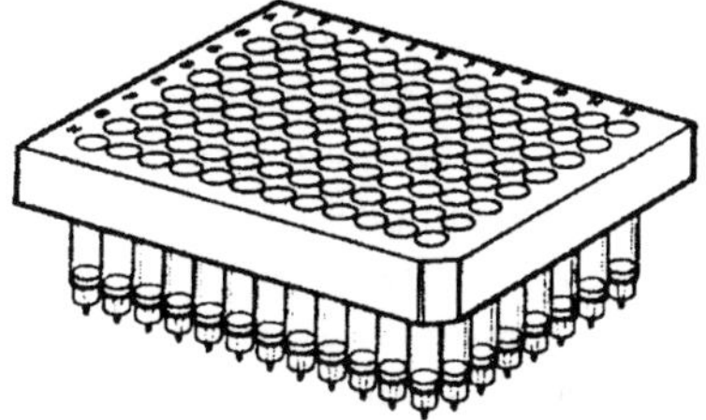

FIGURE 12 Typical 96-well extraction plate. (From Ref. 69. Reprinted with permission of LCGC Europe.)

the autosampler-based SPE Instrument HP 7886 PrepStation system (Palo Alto, California, USA) [63].

96-Well Workstations. The 96-well format workstations reflect the recent explosion of automated solid-phase extraction equipment owing to the commercial availability of SPE materials in 96-well format (Fig. 12), for parallel processing systems. These systems were designed as liquid handlers and as such have more flexibility and capability at the expense of efficiency. They allow for almost completely automated approaches [69].

4.3 Solid-Phase Microextraction (SPME)

Solvent-free sample preparation techniques attract widespread attention because they reduce the use of toxic organic solvents. These can be classified according to the separation medium. Most techniques fall into one of three categories: gas-phase, membrane, and sorbent extraction.

Solid-phase microextraction (*SPME*) is a solvent-free technique for sample preparation that can integrate sampling, extraction, concentration, and sample introduction into a single step, resulting in high sample throughput.

Generally, the term *microextraction* refers to an extraction technique in which the volume of the extracting phase is very small in relation to the volume of the sample, and the extraction of analytes is not exhaustive. In many cases, only a small fraction of the initial analyte is extracted for analysis, depending on the partitioning of analyte between the sample matrix and the extraction phase. The higher the affinity the analyte has for the extraction phase relative to the sample matrix, the greater the amount of analyte extracted. Partitioning is controlled by the physiochemical properties of the analyte, the sample matrix, and the extraction phase. Once sufficient extraction time has been reached and equilibrium is established, further increases in extraction time do not affect the amount of analyte extracted. Thus extraction technique is simplified and precision is improved [13,70].

Arthur and Pawliszyn, investigators at the University of Waterloo (Ontario, Canada), developed solid-phase microextraction in the late 1980s. SPME is an adsorption/desorption technique used to analyze volatile and nonvolatile compounds in both liquid and gaseous samples. It is an alternative to the sample preparation of volatiles that usually involves concentrating the analytes of interest by using headspace, purge and trap, solid phase extraction, or simultaneous distillation/extraction techniques. Furthermore, it allows organic compounds to be directly introduced into any gas chromatograph or GC/MS system, as well as in an HPLC system with the proper interface [70]. The SPME device consists of a 1 cm length of narrow diameter fused-silica optical fiber, coated on the outer surface with a thin film of stationary phase and bonded to a stainless steel plunger, and a holder like a modified microliter syringe (Fig. 13a). The fused silica fiber can be drawn into a hollow needle by using the plunger on the fiber holder. The fiber assembly

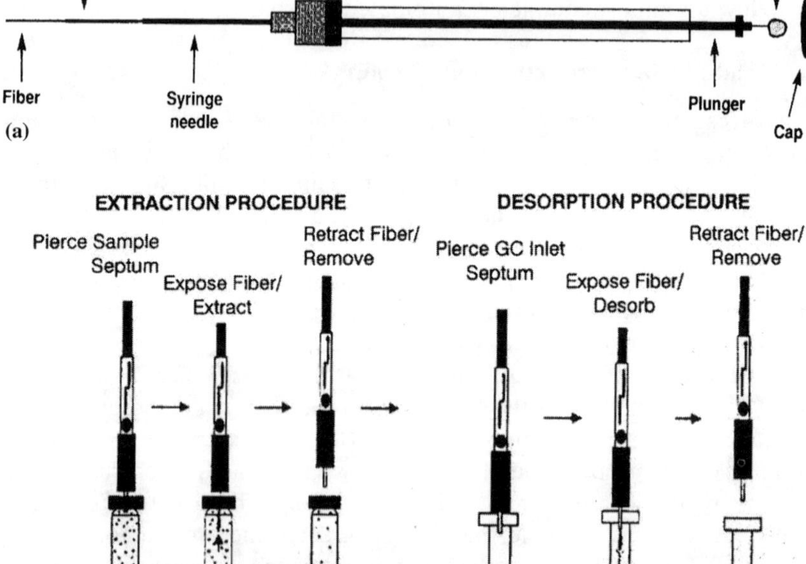

FIGURE 13 (a) Solid-phase microextraction (SPME) device. (From Ref. 71. Reprinted with permission of LCGC North America.) (b) Solid-phase microextraction (SPME) procedure. (From Ref. 19. Reprinted with permission of Supelco, Bellefonte, Pennsylvania, U.S.A.)

consists of an outer protective septum-piercing needle and an inner fiber attachment needle to which the fiber is epoxied. The septum on the piercing needle flange prevents GC/HPLC mobile phase from escaping when injecting. A threaded retaining nut ensures that the fiber assembly remains in the holder. When not in use or during transfer, the fiber retracts into the needle [71].

The SPME processing steps include

Penetration of the syringe through the septum of sample vial.

Extension of the fiber into the sample or in the headspace. Organic analytes partition into the stationary phase of the fiber until equilibrium is reached.

Retraction of the fiber into the syringe needle, and after equilibrium is reached, withdrawal of the syringe from the sample vial.

Insertion/introduction of the needle into the GC port, depression of the plunger and thermal desorption of analytes. Alternatively, the analytes are washed out of the fiber by the mobile phase to be analyzed by HPLC, via a modified HPLC valve.

Removal of the syringe from the injection port after desorption. It is now ready for resampling. The thermal or solvent treatment of the coated fiber in the injection port ensures that the fiber is free from interferences or residual compounds (Fig. 13b) [15,19].

With dirty matrices such as sludge and biological fluids, or when using solid samples, the technique can be operated in the headspace mode as the fiber is exposed to the gas. Agitation by stirring or sonication of the sample matrix improves transport of analytes from the bulk sample phase to the vicinity of the fiber. Other parameters that influence partition equilibrium are sample matrix, temperature, and properties of the coating and analyte. An equilibrium time of 30–60 min is usually enough for direct immersion SPME. Extraction time can be shorter than the equilibrium time, but this results in lower extraction yields and therefore higher detection limits, which means that there must be a compromise between extraction time and yield [72].

In SPME, equilibria are established among the concentrations of analyte in the liquid or solid sample, in the headspace above the sample, and in the phase on the fused silica fiber. The amount of analyte adsorbed by the fiber depends on the phase type, the thickness of the stationary phase coating, and the distribution constant for the analyte. Extraction time is determined by the highest distribution constant. The distribution constant generally increases with increasing molecular weight and boiling point of the analyte. Volatile compounds require a thick phase coating, while for semivolatile analytes a thin coat is most effective. Full equilibrium is not necessary for high accuracy and precision from SPME, but consistent sampling time and other sampling parameters are essential. The vial size, the sample volume and, when using liquid samples, the depth that the fiber is immersed in the sample should be consistent.

Desorption of an analyte from SPME fiber depends on the boiling point of the analyte, the thickness of the coating on the fiber, and the temperature of the injection port. Some analytes can take up to 30 seconds to desorb, and cryogenic cooling might be required to focus these compounds at the inlet of the capillary column. Use of an inlet liner with a narrow internal diameter (e.g., 1 mm) generally provides sharp peaks and can eliminate the need for cooling.

As with any other extraction/concentration technique, it is best to use multiple internal standards, and to treat the standards and the analytes in an identical manner.

There are two different implementations of the SPME technique explored extensively to date. One is associated with a tube design and the other with fiber design. The tube design can use very similar arrangements as SPE; the primary difference, in addition to volume of the extracting phase, is that the objective of SPME is, as already mentioned, not an exhaustive extraction. This substantially simplifies the design of systems. For example, in tube SPME, for analysis of liquids one uses 0.25 mm I.D. tubes and about 0.1 μL of extraction phase, because concern about break-through is not relevant. In fact, the objective is to produce full breakthrough as soon as possible, since this indicates that equilibrium extraction has been reached [73].

A more traditional approach to SPME involves coated fibers. The transport of analytes from the matrix into the coating begins as soon as the coated fiber has been placed in contact with the sample [72].

Typically, SPME extraction is considered to be complete when the analyte concentration has reached distribution equilibrium between the sample matrix and the fiber coating. The higher the distribution constant of a compound, the higher is the affinity of the compound for the SPME fiber coating. The equilibrium conditions can be described as

$$n = \frac{K_{fs} V_f V_s C_0}{K_{fs} V_f + V_s} \tag{1.1}$$

where n is the number of moles extracted by the coating, K_{fs} is a fiber coating/sample matrix distribution constant, V_f is the fiber coating volume, V_s is the sample volume, and C_0 is the initial concentration of a given analyte in the sample.

As indicated by Eq. (1.1), after equilibrium has been reached there is a direct proportional relationship between sample concentration and the amount of analyte extracted, and this is the basis for quantifying analytes.

Coated fibers can be used to extract analytes from very small samples. For example, the use of submicrometer diameter fibers permits the investigation of single cells. Since SPME does not extract target analytes exhaustively, its presence in a living system should not result in significant disturbance. Complete extraction can be achieved for small sample volumes when distribution constants are reasonably high. This observation can be used to advantage if exhaustive extraction is required, since it is very difficult to work with small sample volumes using conventional sample preparation techniques. Also, SPME allows rapid extraction and transfer to an analytical instrument. In addition, when sample volume is very large, Eq. (1.2) can be simplified to

$$n = K_{fs} V_f C_0 \tag{1.2}$$

In this equation, the amount of extracted analyte is independent of the volume of the sample. In practice, there is no need to collect a defined sample prior to analysis, as the fiber can be exposed directly to the ambient air, water, production steam, etc. The amount of extracted analyte will correspond directly to its concentration in the matrix, without being dependent on the sample volume. As the sampling step is eliminated, the whole analytical process is accelerated, and errors associated with analyte losses through decomposition or adsorption on the sampling container walls is prevented [13,70,73].

4.3.1 SPME Fibers

The commercially available fibers include polydimethylsiloxane (PDMS, 100 μm, 30 μm, 7 μm), polydimethylsiloxane-divinylbenzene (PDMS-DVB, 65 μm), polyacrylate (PA, 85 μm), carboxen-polydimethylsiloxane (CAR-PDMS, 75 and 85 μm), carbowax-divinylbenzene (CW-DVB, 65 and 75 μm), carbowax-templated resin (CW-TPR, 50 μm), and divinylbenzene-carboxen-polydimethylsiloxane (DVB-CAR-PDMS, 50/30 μm). The type of fiber used affects the selectivity of extraction. In general, polar fibers are used for polar analytes and nonpolar types for nonpolar analytes, as with conventional GC stationary phases. For example, the dipolar porous PDMS-carboxen fiber is designed to retain highly volatile solvents and gases [74–76].

The preferred coating is PDMS, which has the advantage of sorptive interaction between the coating and the extracted compounds. In this case, the analytes partition onto the extraction phase and the molecules are solvated by the coating molecules. This interaction is much weaker, as adsorption on an active surface and the degradation of unstable analytes is significantly less than adsorption. Furthermore, the retaining capacity of the PDMS material is not influenced by other analytes because each analyte has its own partition equilibrium in the PDMS phase.

Coating volume determines method sensitivity as well, but thicker coatings result in longer extraction times, so it is important to use the appropriate coating for a given application. In addition to liquid polymeric coatings such as PDMS for general applications, other more specialized materials have been developed. For example, ion-exchange coatings were used to remove metal ions and proteins from aqueous solutions, carbowax for polar analytes, and Naflon coatings to extract polar compounds from non-polar matrices.

Selectivity toward target analytes and interferences can be achieved by using surfaces common to affinity chromatography. Polypyrrole coatings, for example, have been recently developed to extract polar or even ionic analytes and possibly to explore the conductive polymer properties of the polymer during extraction in order selectively to extract analytes of interest; then the charge is reversed to facilitate desorption.

Polymers with molecular/analyte recognition as described in Sec. 4.6.2 have also been used, coated on the inner surface of a fuse-silica GC capillary by the chemical polymerization method [15].

Different methods (bonded, nonbonded, cross-linked) are used to attach the coating to the fused silica core. Most polymer films are coated directly (nonbonded types) or partially cross-linked. These can be damaged if exposed to high levels of organic analyte or strong acid or alkali. All fibers require initial conditioning prior to use and have a maximum desorption temperature, similar to GC stationary phases. High-purity carrier gases are essential, as trace levels of oxygen can easily oxidize some phases.

Fibers can be reused in up to 100 analyses or more depending on the sample matrix, on the care of the analyst, and on the applications used for as well.

4.3.2 SPME Method Development and Optimization

When developing SPME methods for drug analysis, the analyst has to face numerous choices: fiber and extraction method choice, adjustments to the matrix, derivatization, and desorption conditions. Often, the matrix must be adjusted to enhance the extraction recovery of analyte into the fiber; this involves pH adjustment or salt content of an aqueous matrix, such as urine, blood, or saliva. Recoveries of basic analytes are enhanced by basic pH; recoveries of acidic analytes are enhanced by acid pH. Addition of salt, such as sodium chloride, often increases recovery due to salting out effects. Careful adjustment of the extraction conditions concerning agitation, temperature, and time-significant enhancements in sensitivity can be achieved to enable the detection of even semivolatile analytes.

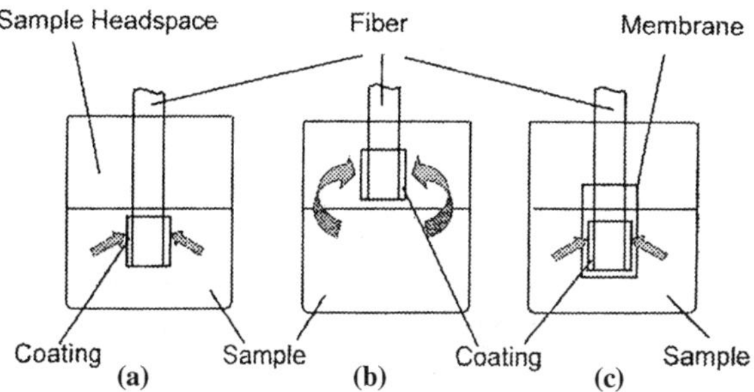

Figure 14 Modes of SPME operation: (a) Direct extraction. (b) Headspace SPME. (c) Membrane-protected SPME. (From Ref. 73. Reprinted with permission of Elsevier Science.)

SPME desorption is perceived to be a slow process, although it can be made more rapid by using a low-thickness fiber and a narrow-bore inlet liner. The inlet conditions become more important as the analyte becomes more volatile; obtaining the narrowest possible injection bandwidth is critical [77].

Three basic types of extractions can be performed using SPME: *direct extraction, headspace configuration, and a membrane protection approach.* Figure 14 illustrates the differences among these modes. In the direct extraction mode, coated fiber is inserted into the sample and the analytes are transported directly from the sample matrix to the extracting phase. Figure 14c shows the principle of indirect SPME extraction through a membrane. The main purpose of the membrane barrier is to protect the fiber against damage, as with the use of headspace SPME when very dirty samples are analyzed. However, membrane protection is advantageous for determination of analytes having volatilities too low for the headspace approach. In addition, a membrane made from appropriate material can add a certain degree of selectivity to the extraction process [73].

4.3.3 Solid-Phase Microextraction in Biomedical Analysis—Direct and Headspace

Several SPME methods are reported in literature with satisfactory sensitivity, linearity, precision, and accuracy, and they can, in many cases, be used instead of a classical sample preparation method. For example, they can be applied for the analysis of drugs, metabolites, and endogenous substances in body fluids. However, the trace analysis in plasma may be limited to highly volatile compounds and only in samples with high concentrations. Additionally, as already mentioned, the qualitative and quantitative analysis of biological fluids is hampered by the presence of dissolved biopolymers. An advantage of

SPME over other sample preparation methods is that a direct assay of free concentration can be performed without the separation of phases. This is possible because the binding of analyte to proteins is not impaired by SPME, and because SPME is an equilibrium extraction.

HS-SPME is ideal for the analysis of biological specimens for investigations in clinical, forensic, and toxicology laboratories. The outstanding advantage of SPME used in the headspace approach, in biomedical analysis, is the prevention of direct contact of the fiber with the sample such as urine or blood matrices and therefore prevention of contamination of the surface of the fiber with organic polymers. No diffusion barrier of clotted proteins is formed, no burning-in of adsorbed organic material is possible during desorption in the hot injector, and the lifetime of the fibers is considerably increased. On the other hand, HS-SPME is limited to special analytes because of the requirement of a high vapor pressure of the analyte. Furthermore, the transfer of fibers to the gas chromatograph and desorption should be performed immediately after extraction because of the high vapor pressure of analytes also in the coating and the risk of loss of analytes during storage of the loaded fiber [77].

Analysis of drugs by SPME has significantly benefited pharmaceutical, clinical, and forensic analysis. Applications include alcohol and volatiles in whole blood and urine, amphetamines, methadone and amphetamines, analgesics from urine, anesthetics from human blood, antidepressants and their metabolites, benzodiazepines, cannabinoids, cocaine, proteins, and steroids and other drugs from expired breath and saliva [15,78–99].

4.3.4 SPME Coupled to HPLC

Though at the beginning, SPME was developed to be used solely in GC, it was adapted to LC by specially designed interfaces or by desorption of the extracted analytes into a small amount of solvent placed in a LC autosampler vial [13].

Analytes extracted by SPME must be desorbed into a suitable receiving solvent prior to analysis. There are two modes by which SPME may be coupled to LC, and there are some minor variations in the way interfaces are incorporated, depending on the LC manufacturer. Either conventional fiber coupling or the newer in-tube SPME may be incorporated by placing the interface in the position of either the sample loop or a transfer line. Fiber extractions for LC applications are completely analogous to those for GC applications. With conventional fiber coupling, analysts are currently limited to performing manual extraction and desorption.

The interface shown in Fig. 15 consists of a six-port injection valve and a desorption chamber that replaces the injection loop in the HPLC system. The SPME fiber is introduced into the desorption chamber under ambient pressure, when the injection valve is in the load position. As soon as the

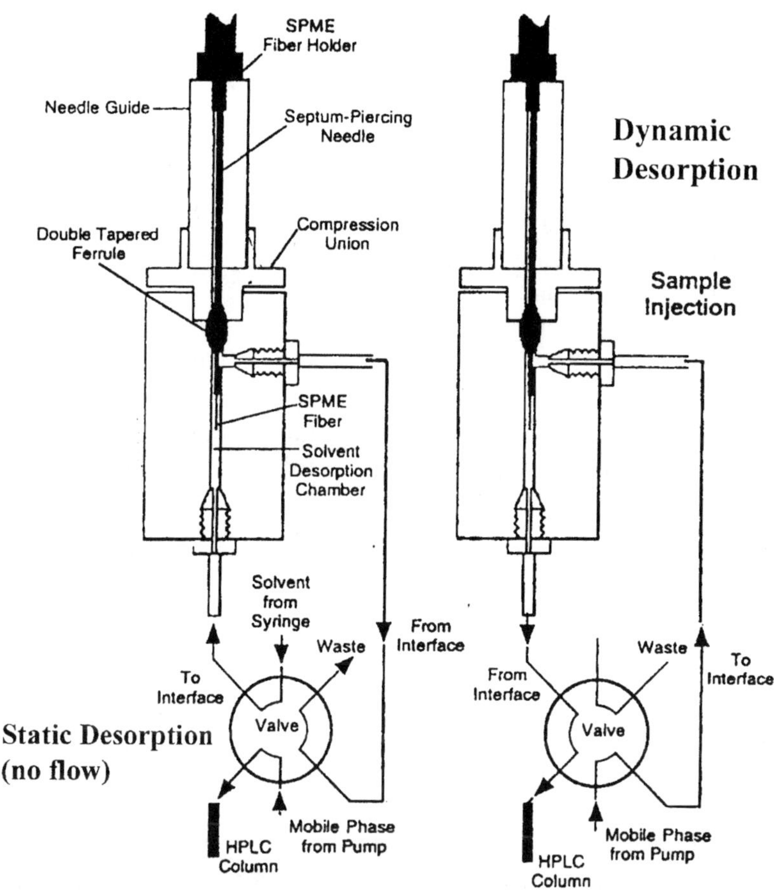

FIGURE 15 SPME–HPLC interface. (From Ref. 101. Reprinted with permission of Supelco, Bellefonte, Pennsylvania, U.S.A.)

fiber is inserted through the ferrule, the unit is made leak tight by closing the clamp and compressing the ferrule against the SPME needle. The SPME-HPLC interface enables the mobile phase to contact the SPME fiber, remove the adsorbed analytes, and deliver them to the separation column. Analytes can be removed via a stream of mobile phase (dynamic desorption), or, when analytes are more strongly adsorbed to the fiber, the fiber can be soaked in mobile phase or another stronger solvent for a specific period of time (e.g., 1 minute) before the material is injected onto the column (static desorption) [75,100–101].

For automated extraction and analysis, in-tube SPME is relatively simple to implement. A schematic of this interface is shown in Fig. 16 [100].

4.3.5 Derivatization in SPME

Derivatization may be necessary for many drugs of interest that are not volatile, polar, or strongly bound to the matrix. Various derivatization techniques that can be implemented in combination with SPME include

1. Direct derivatization in sample matrix
2. Derivatization in GC injection port
3. Derivatization in SPME fiber coating
   A. Simultaneous derivatization and extraction
   B. Derivatization following extraction

Some of the techniques, such as direct derivatization in the sample matrix, are analogous to well-established approaches used in solvent extraction as described in Sec. 7.

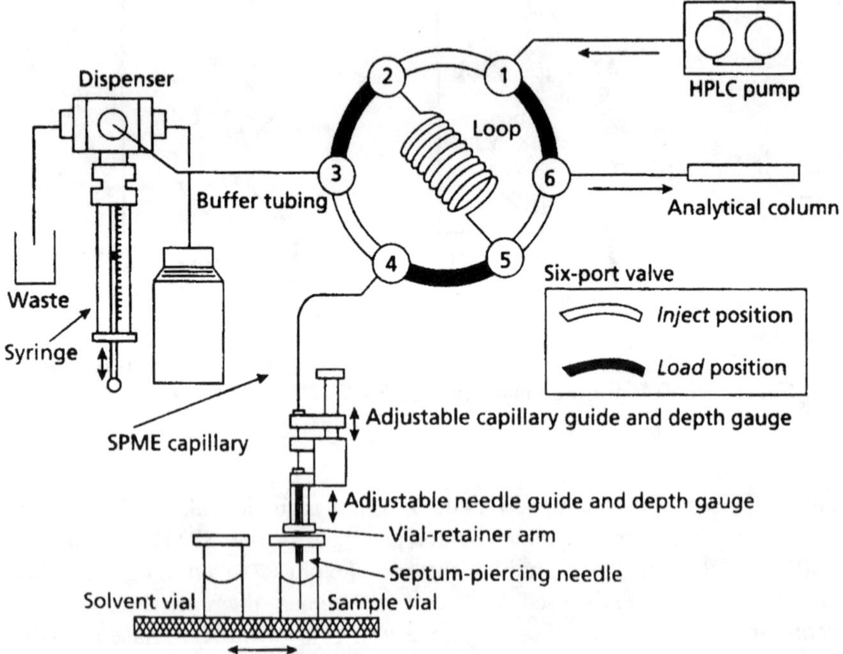

FIGURE 16 In-tube SPME interface for HPLC. (From Ref. 100. Reprinted with permission of LCGC Europe.)

In the direct technique the derivatization reagent is first added into the vial containing the sample matrix; the SPME fiber then extracts the derivatized analytes either in solution or in the headspace and delivers them to the GC. For example, trimethyloxonium tetrafluoroborate has been used to form methyl esters of urinary organic acids, and propyl chloroformate to derivatize the amino group on amphetamines in urine. The derivatives are then extracted by SPME and introduced into the analytical instrument. In addition, selective derivatization to analogues containing high detector response groups will result in enhancement in sensitivity and selectivity of detection.

Derivatization in the GC injector is an analogous approach, but it is performed at high injection port temperatures. For example, long-chain carboxylic acids can be extracted onto the coating as ion pairs when tetramethylammonium hydrogen sulfite is added to the sample.

The most interesting and potentially useful technique is simultaneous derivatization and extraction, performed directly in the coating, allowing high efficiencies. The simplest way to execute the process is to dope the fiber with a derivatization reagent and subsequently expose it to the sample. Subsequently the analytes are extracted and simultaneously converted to analogues that have higher affinity for the coating. This is no longer an equilibrium process, as derivatized analytes are collected in the coating as long as extraction continues. This approach, which is used for low molecular mass carboxylic acids, results in exhaustive extraction of gaseous samples [73].

On-fiber derivatization (e.g., with diazomethane) can be employed after the extraction procedure. Extracted compounds on the fiber are exposed (e.g., in a heated sealed vial) to the derivatizing reagent in the vapor phase for a given time. Damage to the coating is prevented by headspace derivatization. This has been employed for serum steroids, urinary organic acids, and urinary hydroxyl metabolites of polycyclic aromatic hydrocarbons (naphthalene, phenanthrene, pyrene).

4.3.6 Advantages and Disadvantages of Solid-Phase Microextraction

Several advantages are claimed for SPME. However, this is true in some areas of biomedical analysis. The combination of low volatility of analyte and a complex matrix with polymer components, e.g., plasma or cell cultures, considerably limits the application of SPME. The extraction is very slow in contrast to LLE and SPE with packed bed columns. Another disadvantage of SPME is that the sensitivity is low with regard to the target concentrations of analyte; the therapeutic concentration of a drug in plasma, for example,

TABLE 13 Advantages and Disadvantages of SPME

Advantages
No use of solvents.
Easy handling.
Little equipment necessary.
Fast method.
Ease of automation.
Good linearity and high sensitivity.
Direct assay possible without protein removal.
Disadvantages
Longer GC programs owing to low initial temperature. Because of desorption times of at least 1 min, cryofocusing of the analyte is needed.
The desorption needs more time than the injection of fluid extracts after LLE and SPE.
Potential carryover because of the repeated use of one fiber.
Multiple step LLE provides cleaner extracts than SPME.
Because SPME is a nonexhaustive sample preparation, the methods cannot equally compensate for changes of the composition of the matrix, as in the case of LLE.
Quantitation is more prone to errors due to changes of the matrix even if internal standard methods are applied. Thus matrix effects must be extensively investigated during method validation.

Source: Ref. 102.

should be in the range of 1 to 100 μg/mL. Table 13 summarizes the advantages and disadvantages of SPME [15,73,102].

4.4 Stir Bar Sorptive Extraction (SBSE)

Stir bar sorptive extraction (*SBSE*) is a new technique that uses stir bars (stirrers) 10 and 40 mm long, coated with 50–300 μL PDMS. This technique enables the effective extraction and sensitive determination of volatile and semivolatile organic compounds from aqueous samples [74].

The recovery of an analyte from the sample can be estimated by the equation

$$\frac{m_s}{m_0} = \frac{K_{o/w}/\beta}{1 + K_{o/w}/\beta} \tag{1.3}$$

where m_s is the mass of the analyte in the PDMS phase, m_0 the total amount of the analyte originally present in the water sample, $K_{o/w}$ the octanol–water partition coefficient, β the V_w/V_s phase ratio, and V_w and V_s are the volumes of water and PDMS phase of the stir bar, respectively. Sensitivity can be increased by a factor of 100 to 1000. Complete recovery is possible for solutes with $K_{o/w}$ larger than 500. Solutes with $K_{o/w}$ from 10 to 500 can also be analyzed using calibration as is done in SPME. While high recoveries ($>$ 50%) are only obtained for solutes with $K_{o/w} > 10000$ using SPME, the recovery obtained by stir bar sorptive extraction is already higher than 50% for solutes with $K_{o/w} > 100$. PDMS coated stir bars are commercially available under the trade name Twister, by Gerstel GmbH, Germany. Magnetic stirring rods are incorporated in a glass jacket and coated with a 1 mm layer of PDMS. Two Twisters are available: 10 mm $\times$ 3.2 mm o.d. and 40 mm $\times$ 3.2 mm o.d. Typically the shorter bars are used for 1–50 mL sample volumes and the 40 mm stir bars are used for 100–250 mL sample volumes.

Sample extraction is performed by placing a suitable sample amount (typically 10–25 mL) in a vial, adding a stir bar, and stirring for 30–120 min. After extraction, the stir bar is introduced into a glass thermal desorption tube (4 mm i.d. $\times$ 187 mm), placed in a thermal desorption unit, and thermally desorbed. The desorption temperature is application dependent and ranges between 150 and 130°C for 5–15 min. Alternatively, liquid desorption can be performed by ultrasonic treatment in autosampler vials with glass flat bottom inserts using 150 μL of organic solvent or organic solvent and water mixture. Extraction time is about 1 h while desorption time is 10 min [103].

4.5 Membrane-Based Separations

Membrane-based separations have been applied as a useful tool for simplifying the sample preparation step for complicated matrix samples. The analytical applications of membranes include filtration, extraction, and preconcentration of analytes prior to chromatographic and electrophoretic separations.

In simple terms, a membrane can be thought of as a selective barrier between two phases. Mass transfer occurs from the donor or feeding phase to the receiver known as the acceptor or permeate phase (Fig. 17). Although synthetic membranes may be of very different properties, they can be classified into four groups: porous, nonporous, symmetric, and asymmetric. The forces able to generate migration through membranes are directly related to differences in pressure, electrical potential, and concentration. The mass flow brought about by differences in concentration is determined by the Fick law; differences in pressure produce a volume flow related to the Hagen–Poiseuille equation; and the charge flow related to differences in electrical potentials is governed by Ohm's law. Table 14 summarizes the characteristics of membrane based separation processes [104–106].

There are a number of different membrane techniques, which have been suggested as alternatives to the SPE and LLE techniques. It is necessary to distinguish between porous and nonporous membranes, as these have different characteristics and fields of application. In porous membrane techniques,

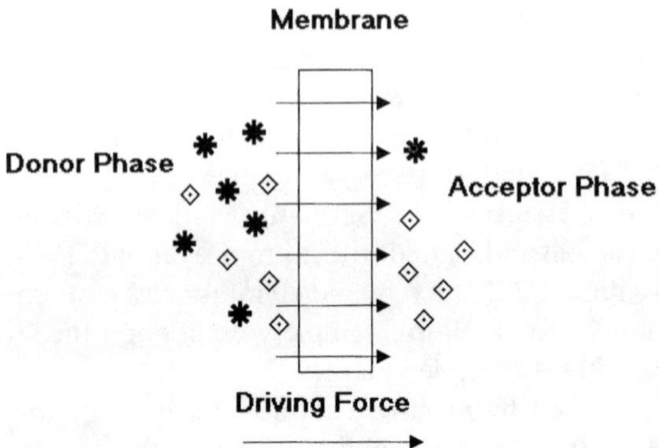

FIGURE 17 Schematic presentation of a membrane system. (From Refs. 104–106.)

the liquids on each side of the membrane are physically connected through the pores. These membranes are used in Donnan dialysis to separate low molecular mass analytes from high molecular mass matrix components, leading to an efficient cleanup, but no discrimination between different small molecules. No enrichment of the small molecules is possible if the mass transfer process is a simple concentration difference over the membrane. These techniques can provide some distinctive advantages over more commonly used alternatives, especially regarding selectivity, enrichment power, and automation potential. Nonporous membranes are used for extraction techniques [7].

Membranes have some drawbacks compared to other sample preparation methods. For example, porous membranes are prone to fouling by particulates or macromolecules, so that flow rates decrease and effectiveness is diminished. In some cases, samples must be pretreated before they can be dialyzed or cleaned up using other membrane techniques.

The introduction of membranes in the field of sample preparation contributes to minimal organic solvent use, minimal contamination and exposure to toxic or dangerous samples, automation, and effective cleanup and analyte isolation.

4.5.1 Membranes in Sample Filtration

Membrane filters are used to remove particulates from samples and solvents before chromatographic analysis and also for the preparation of gaseous and liquid samples, especially for the reason that no solvent is used. Typical materials of construction for membrane filters are usually synthetic polymeric materials, although natural substances such as cellulose and inorganic materials such as glass fibers are also used: acrylic

TABLE 14 Characteristics of Membrane-Based Separation Processes

Separation process	Driving force	Separation mechanism	Membrane structure
Dialysis	Concentration gradient	Difference in diffusion rate	Symmetric, porous/ nonporous
Osmosis	Concentration gradient	Difference in diffusion rate	Symmetric, porous
Hyperfiltration	Pressure gradient (1–10 Mpa)	Difference in solubility and diffusion rate	Asymmetric, nonporous (0.1–1 nm)
Ultrafiltration	Pressure gradient (50 kPa–1 Mpa)	Difference in membrane permeation (sieving)	Asymmetric, porous (1–100 nm)
Microfiltration	Pressure gradient (10–100 kPa)	Sieving	Symmetric, porous (100–1000 nm)
Electrodialysis	Electrical potential	Selective ion transport	Symmetric ionic
Electro-osmosis	Electrical potential	Difference in diffusion rate	Symmetric ionic
Gas separation	Pressure gradient	Difference in solubility and diffusion rate	Asymmetric, nonporous
Pervaporation	Pressure gradient	Difference in solubility and diffusion rate	Asymmetric, nonporous

Source: Ref. 105.

copolymer, aluminum oxide, cellulose acetate, glass fiber, mixed cellulose esters, nitrocellulose, nylon, polycarbonate, polyester, polyether sulfone, polypropylene polysulfone, polytetrafluoroethylene (PTFE), polyvinylchloride (PVC), polyvinylidene fluoride (PVDF) [106,107].

Membranes are available in sheet, roll, disc, capsule, cartridge, and hollow-fiber formats. A wide variety of synthetic polymers can be purchased from various manufacturers. Compatibility of the polymeric material with the solvents used must be a great concern of their different chemical properties. Manufacturers usually provide tables with information about compatibility of filters and solvents.

For sample filtration, disc format membranes are the most popular devices. Discs are sold in loose form or packed in disposable syringe filters or cartridges; common diameters commercially available are 3, 7, 13, 25, 47, and 96 mm or even larger. The disposable filters use membranes enclosed in plastic housings that have Luer-type fittings. Samples are filtered manually, applying positive pressure, or in a vacuum manifold.

Important characteristics for membrane performance is pore size, pore size distribution, filter thickness, amount of extractable material, hydrophobic–hydrophilic character, nonspecific binding properties, pyrogenicity, gas and liquid flow rate, burst strength, autoclavability, and absolute and nominal particulate retention.

If the sample has a hydrophobic character, the membrane itself can interact with or adsorb some of the hydrophobic components and interfere with sample recovery. Adsorption may occur when filtering protein-containing samples through a hydrophobic PTFE filter. Hydrophilic membranes, available with wide solvent compatibility, generally provide low protein binding.

4.5.2 Dialysis and Microdialysis

Dialysis is a membrane barrier separation process in which differential concentration forces one or more sample solutes to transfer from one fluid to another through a membrane. In dialysis, the solution containing the solute of interest (whose concentration is depleted) is called the feed, and the fluid receiving the solute is called the dialyzate. Dialysis is used to remove salts and low molecular weight substances from solutions or alternatively to remove high molecular weight interferences, e.g., proteins, and allow the measurement of small molecules [108].

Other variations of porous membrane techniques include microdialysis, which is extensively used in neuroscience research for in vivo sampling, and electrodialysis, in which a membrane promotes selective transport of charged compounds. A number of micro- and nanofiltration techniques belong to the field of porous membrane techniques; however, they are not considered extraction techniques [109].

Microdialysis is a specialized application of dialysis used for the dynamic monitoring of extracellular chemical events in living tissues. The most important element of microdialysis is the probe containing the semipermeable membrane. The membrane is placed at the end of a small piece of small-diameter fused silica tubing. In practice, the probe is inserted into the tissue of an animal, and it is in direct contact with the interstitial space. A flowing fluid that can be void of substances of interest or include physiologically or pharmacologically active exogenous or endogenous substances is used on the other side of the membrane. Thus a concentration gradient is set up across the membrane that enables the diffusion of substances from interstitial space into the dialysis probe. Usually, the probe is held in place mechanically so that the animal can move freely. The inlet of the probe is connected via a short length of small-diameter tubing to a microsyringe pump that can provide flow rates of 0.5–10 μL/min. The other end of the probe is connected to a collection vial or, in some cases, an HPLC microvalve that enables on-line analysis using a microbore column. With the on-line connections, analysts can obtain dynamic measurements of chemicals in the extracellular tissue. Different shapes, sizes, and designs of the probes are used depending on the type of tissue and the region being investigated. In order to increase the membrane area available for diffusion, hollow-fiber membranes are used instead of planar ones.

Microdialysis can be applied in biotechnology and biomedical analysis. It has the advantages of easy operation, rapid isolation of components of interest from complicated and dirty matrixes, and less or no use of organic solvents. Serum, plasma, and muscle tissue can be analyzed after sampling by means of on-line dialysis. Pharmacokinetic studies can also be executed.

Microdialysis has proven its usefulness for in vivo studies of brain catecholamines in laboratory rats. Continuous sampling of the neuronal microenvironment is important to understand the brain's function and metabolism. The natural level of catecholamines in brain tissue is in the subpicogram range. Studies of these biogenic amines and metabolites require measurement at or near these levels. However, the catecholamine levels dramatically increase in animals under stress, thereby obliterating the value of any observations. Using microdialysis with inserted brain probes allows studies to be performed on rats in undisturbed states. Coupling the HPLC system to an electrochemical detector enables sensitive detection of catecholamines and metabolites [101,108].

Dialysis can also be used as an on-line sample preparation technique for the deproteinization of biological samples before HPLC analysis. Selecting the appropriate semipermeable membrane for the dialyzer can prevent interference from large macromolecules. Samples are introduced into the feed (or donor) chamber, and solvent is pumped through the lower acceptor (or recipient) chamber. The smaller molecules diffuse through the membrane to the acceptor chamber and are directed to an HPLC valve for injection. In case the concentrations of compounds of interest in the dialyzate are undetectable, a trace-enrichment step is required in which a column is placed downstream of the dialyzer that will retain the analyte until the concentration is sufficient for detection. After this step the analyte can be backflushed into the HPLC system. The technique is useful for blood studies, as sampling can be achieved continuously without blood withdrawal. On-line dialysis has been widely used in on-line bioprocess monitoring. A commercial on-line system, such as ASTED from Gilson, is used for both cleanup and enrichment by a combination of dialysis with SPE; it is shown in Fig. 18. Typical applications involve drug analysis in blood plasma and on-line sampling of bovine serum, muscle tissue, and plasma. When analytes are bound to the protein in the plasma, additional analytical steps can replace the bound analytes before dialysis [64,108,110].

4.5.3 Ultrafiltration

Similar to dialysis, *capillary ultrafiltration* is a sample preparation technique in which the pores of a semipermeable membrane prevent passage of macromolecules with masses greater than 30 kDa. Unlike microdialysis, diffusion across the membrane caused by concentration differences is not the driving force. A vacuum is applied to the probe for ultrafiltration, using vacuum or peristaltic pumps or even simple evacuated containers, which make the technique simple and portable.

During ultrafiltration sampling, the analyte molecules are carried along (pulled) with the flow of water and other electrolytes. The ultrafiltration process is unidirectional and results in a net loss of fluid volume from the sample environment. The concentration recovery tends to be higher for ultrafiltration than for dialysis because no external perfusion fluid dilutes the collected analyte. However, the outer surface of the membrane can become obstructed by large molecules that can reduce the flow rate, so ultrafiltration must use larger probes than those used for microdialysis.

It is considerably easier to determine the absolute concentration of an analyte in sampled tissue with an ultrafiltration probe. The probe delivers filtered, undiluted extracellular fluid, so the concentration in the ultrafiltrate is identical to the concentration in the tissue. Ultrafiltration applications include the monitoring of drugs in saliva and the in vivo monitoring of pharmaceuticals in subcutaneous tissues [105,111].

4.5.4 Membranes in Extraction

Nonporous membranes are used as already mentioned in membrane extraction techniques. A nonporous membrane is a liquid or a solid (e.g., polymeric) phase that is placed between two other phases, usually liquid but sometimes gaseous. One of these phases is the sample to be processed, the donor (or feed) phase. On the other side of the membrane is the acceptor (or strip) phase, where the extracted analytes are collected and transferred to the analytical instrument. This arrangement permits the versatile chemistry of LLE to be used and extended, thus providing a highly effective cleanup as well as high enrichment factors, and technical realizations can easily be automated. In most cases there is little or no use of organic solvents [108].

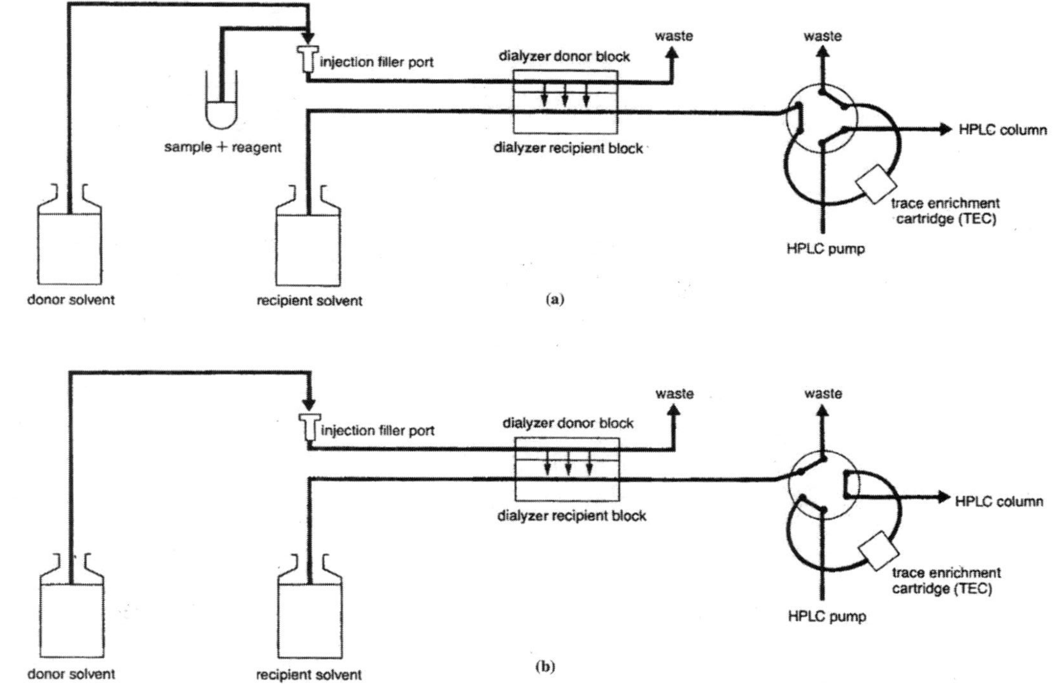

FIGURE 18 On-line sample preparation by dialysis. (a) Sample preparation. (b) Injection and purging. (From Ref. 110. Reprinted with permission of Gilson Medical Electronics S.A., Villiers-le-Bel, France.)

SPE methods have also integrated membrane technology as already mentioned in previous paragraphs. Membranes or discs consist of a 0.5 mm membrane, whereas the adsorbent particles 5 to 10 μm in diameter are immobilized in a web of microfibrils, which allows higher flow rates than SPE cartridges. The extractant particles constitute approximately 90% of the mass, and the remaining 10% are fibrils that hold the membrane together. Liquid samples flow smoothly through the membranes with only a small amount of applied pressure or suction. An example of commercially available membrane is Empore from 3M (St. Paul, MN, USA) [8,61].

Supported Liquid Membrane Separation—Impregnated and Functionalized Membranes. Concerning the *impregnated and functionalized membranes*, a specific functionality is imposed on them by a chemical bonding reaction or by incorporating a silica- or polymer-based packing material within the membrane. The functionality can be very specific, such as an immobilized affinity phase, or very general, such as a reversed phase C_{18} silica [107].

The *supported liquid membrane* (*SLM*) enrichment involves the use of a porous PTFE membrane that separates two aqueous solutions. The membrane is treated and impregnated with an organic solvent, which is held by capillary forces in the pores of a hydrophobic porous membrane (support) and the liquid membrane is the liquid in these pores, between and in contact with the two aqueous phases, mounted between two flat blocks of inert material, e.g., PTFE, with a machined groove in each. The blocks are clamped together with a membrane between them. The entire device is connected to a flow system consisting of two independent flow channels on each side of the membrane. Aqueous solutions are independently pumped through each of the channels. For sample preparation use, channel volumes are typically in the range 10 to 1000 μL. When the solutions are selected properly, compounds can be selectively extracted into the acceptor side.

Enrichment factors of several hundred can be obtained by using the supported membrane technique. Placing a trap or a precolumn between the membrane device and an HPLC instrument can concentrate the analyte further. A switching valve enables users to backflush the concentrated analytes into an HPLC injector.

The organic solvent should have very low water solubility, low viscosity, high partition coefficients for the compounds of interest, and low partition coefficients for possible interferences. Typical solvents are long-chain hydrocarbons like *n*-undecane or kerosene and more polar compounds like dihexyl ether, dioctyl phosphate, and others.

The extraction efficiency can be increased by lowering the donor flow rate, which increases the contact time between the donor phase and the organic membrane. However, the slower the flow rate, the less sample mass is transferred to the acceptor per time unit, but this approach is limited. Another way to increase the extraction efficiency is to increase the channel length, but keep the flow rate constant. The distribution coefficients for this extraction process can be influenced by adding salt to the aqueous donor solution, as this increases the analyte's distribution coefficient in the organic membrane. In the supported-liquid-membrane method, neutral compounds tend to distribute themselves freely throughout all phases. Replacing the donor liquid with a clean washing solution will remove these neutral compounds from the stagnant acceptor solution [7].

Summarizing the principles of SLM extraction: neutral, extractable species should be formed in the donor phase (or at the donor-membrane interface), these species should be transported through the membrane and in the acceptor phase become transformed to another, nonextractable species. Chemically this is similar to liquid–liquid extraction into a second aqueous phase [7,104,111].

As an application example, the extraction of basic compounds, e.g., amines from plasma, is considered. The pH of the sample is adjusted to a value that is high enough for the amines (B) to be uncharged, and therefore they can be extracted into the organic membrane phase, when the sample is pumped through the donor channel. The acceptor channel on the other side of the membrane becomes protonated (B^+) at the membrane–acceptor interface and therefore is prevented from reentering the membrane. This leads to a transport of amine molecules from the donor to the acceptor phase. After the extraction, the acceptor phase is transferred to an analytical instrument, either manually or on-line by a flow system. As the extract is aqueous, the technique is compatible with either reversed-phase liquid chromatography or ion chromatography.

Acidic compounds (HA) will be completely excluded from the membrane as they are charged in the alkaline donor phase. The same is the case for permanently charged compounds. Neutral compounds (N) may be partitioned between the three phases, but not enriched, as their concentration (strictly, their activity) in the acceptor phase will never exceed that in the donor. The extraction rate of uncharged macromolecules will be very low owing to their low diffusion coefficients. Macromolecules, as proteins, will typically be charged and therefore rejected. Considering all this, the SLM extraction will be highly selective for small, basic compounds under the conditions mentioned [111].

4.5.5 Microporous Membrane Liquid–Liquid Extraction (MMLLE)

In the technique of *microporous membrane liquid–liquid extraction (MMLLE)*, the acceptor is an organic solvent, and the same solvent forms the liquid membrane by filling the pores in the porous hydrophobic membrane. MMLLE is more suitable than SLM for highly hydrophobic compounds (e.g., hydrocarbons). These compounds are easily extracted from water to an organic solvent, but unless they can be efficiently trapped, they cannot be back-extracted into a second water phase as required by the SLM approach. It can be considered chemically the same principle as for conventional LLE, performed in a flow system, thus permitting automation and interfacing to analytical instruments. The technique is more easily interfaced to gas chromatography (GC) or to normal phase high-pressure liquid chromatography (NP-HPLC), as the extract ends up in an organic phase. In principle, the membrane could also be hydrophilic, which would lead to an aqueous phase in the membrane pores.

In LLE in a flow system (in the form of FIA) the organic and aqueous phases are mixed in the same flow channel and later separated. The practical problems with the phase separation seem to have prevented this technique from being widely used. In MMLLE, the phases are never mixed and all mass transfer between the phases takes place at the membrane surface.

In MMLLE, as in classical LLE, the extraction efficiency is limited by the partition coefficient. If it is very high, it is possible to work with a stagnant acceptor phase to move with a slow flow rate in order to remove successively the extracted analyte and maintain the diffusion through the membrane. This yields a lesser degree of enrichment. The situation is similar to that for dialysis, and various focusing approaches can be applied to improve it, such as an SPE column or a retention gap [111].

Polymeric Membrane Extraction. A number of applications with a polymeric membrane have been described. The most commonly used membrane material is silicone rubber. There are possibilities for both aqueous-polymer-aqueous extraction, including trapping in the acceptor in a way similar to SLM extraction, and also, e.g., aqueous-polymer-organic extraction, which is similar to MMLLE. As diffusion coefficients in polymers are lower than in liquids, the mass transfer is slower, leading to slower extractions. On the other hand, as the membrane is virtually insoluble, any combination of aqueous and organic liquids can be used [111].

4.5.6 Membrane Extraction with a Sorbent Interface (MESI)

The previously mentioned techniques are all characterized by liquid donor and acceptor phases. However, for best compatibility with gas chromatography, gas acceptor phases are the most convenient, and this is achieved with the *membrane extraction with sorbent interface* (*MESI*) technique. MESI can be used for either gaseous or aqueous samples. The equipment consists of a membrane module with a (usually) silicone rubber hollow

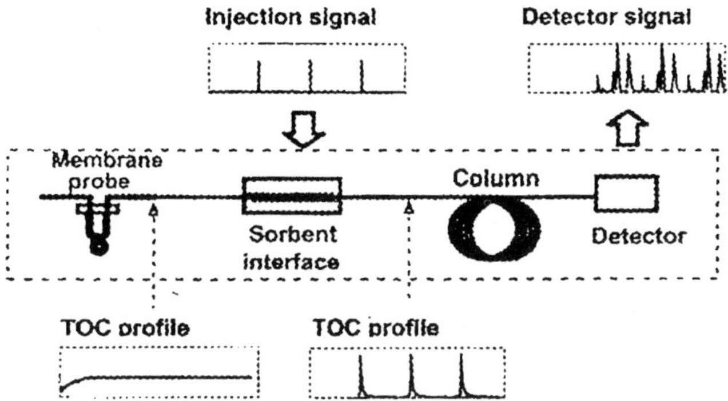

FIGURE 19 MESI system coupled to a GC. (From Ref. 111. Reprinted with permission of Elsevier Science.)

fiber, into which the analytes are extracted from the surrounding liquid or gaseous sample. A gas flows inside the fiber and transports the analyte molecules from the membrane into a cooled sorbent trap where they are trapped. The analytes are desorbed from the sorbent trap by heating, and then they are transferred to GC analysis. In fact, the MESI principle can be seen as a gas phase analogy to the on-line dialysis system principle for liquid chromatography (Fig. 18).

In Fig. 19 a typical MESI set-up is shown. All components are connected on-line so that the carrier gas for the GC passes through the membrane fiber and the sorbent trap. Sampling can also be made off-line with the extraction module and sorbent trap in e.g. field sampling and the sorbent trap can later be connected to the GC and desorbed in a separate step [111].

4.5.7 Automation

An advantage of the membrane techniques is that they are amenable to automation and connection to chromatographic instruments. This is known as hyphenation. With this approach, fully automated analytical systems can be built so that they perform a complete analysis of an untreated sample, even one as complex as urine or blood plasma, up to a finished chromatographic analysis. Flow systems can easily be built up around peristaltic pumps and pneumatically (or electrically) actuated valves, controlled by electronic timers, integrators, or computer systems. These types of automated systems are similar to FIA systems. The membrane extraction unit can in this context be considered as an accessory to FIA in the same way as dialysis cells and gas permeation cells, which are commonly used in FIA practice. The FIA system can be used to handle and treat various types of samples up to chromatographic, spectrometric, or other analytical instrument.

Membrane techniques have been applied to the determination of various compounds, such as drugs, in biological fluids (blood plasma and urine). In these cases, less enrichment is usually obtained, as the sample volumes generally are smaller, but the selectivity is essential.

4.6 Affinity Techniques

4.6.1 Immunoaffinity Extraction

A high degree of molecular selectivity can be achieved with affinity chromatography and affinity extractions. These techniques are based on *molecular recognition* (antigen–antibody interactions). Since antibodies are highly selective toward the analyte used to initiate the immune response, the corresponding immunosorbent may extract and isolate this analyte from complex matrices in a single step, thus eliminating matrix interference [23,112].

Immunosorbents are used in medical and biological fields because they are available for large molecules and easily obtained, while obtaining selective antibodies for small molecules is more difficult.

The first step in making an immunosorbent is to develop antibodies with the ability to recognize the desired analytes. Then immunosorbents are obtained by immobilizing antibodies on solid supports by covalent bonding, adsorption or encapsulation. The sorbent must be chemically and biologically inert, easily activated, and hydrophilic, to avoid nonspecific interactions. The common choices are activated silica, sepharose, agarose gel, etc.

Immunoextraction procedures may be off-line or on-line. In the off-line approach the immunosorbent is packed into a disposable cartridge (Fig. 20). A typical SPE sequence is followed. The sorbent is first conditioned, and then the sample is applied and washed to eliminate interference. Then analytes of interest are desorbed by the appropriate eluent system, which may be a displacer agent, a chaotropic agent, an aqueous–organic solvent mixture, or a solution that alters pH value. With the on-line approach, the immunosorbent is packed into a precolumn incorporated into a six-port switching valve, where the immunoextraction is performed at the load position, while desorption is achieved at the inject position [112].

4.6.2 Molecular Imprinted Polymers (MIP)

The high selectivity provided by immunoextraction led to attempts to synthesize antibody mimics. One approach has been the development of *molecularly imprinted polymers* (*MIP*). These are polymers with specific recognition sites for certain molecules. The desired affinity can be introduced by adding an amount of the compound of interest to the polymerization reaction. This "pattern" (template) chemical may be removed after polymer-

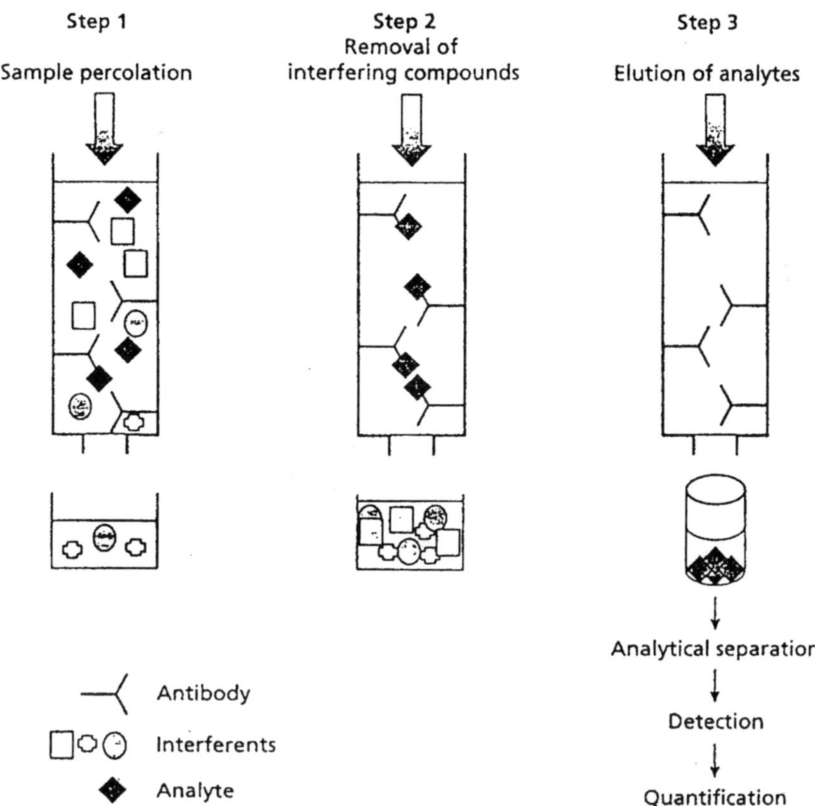

FIGURE 20 Schematic of an off-line immunoextraction procedure. (From Ref. 112. Reprinted with permission of LCGC Europe.)

ization, leaving vacant sites of a specific size and shape, suitable for binding the same chemical again, from an unknown sample. Like immunosorbent, the recognition is due to shape and a mixture of hydrogen, hydrophobic, and electronic interactions. However, they have the advantage of being prepared more rapidly and easily, using well-defined methods, and they are stable at high temperatures in a large pH range and in organic solvents. They have found applications in liquid chromatography, as normal and chiral stationary phases, and in areas where they can be substitutes for natural antibodies, i.e., immunoassays, sensors, and SPE. Noncovalent imprinting protocols are the most commonly used for preparing MIPs using acrylic or methacrylic monomers—often methacrylic acid and ethylene glycol dimethacrylate as cross-linkers.

The molecular imprinting is accomplished through noncovalent interactions, and the solvent, so-called porogenic solvent, has been shown to be one important factor for the determination of the effective molecular recognition.

Problems related to the use of MIP in SPE are the difficulty in removing all the template analyte, even after extensive washing, and the difficulty in establishing quantitative and rapid desorption owing to the high avidity of the MIP for the analyte [113].

4.6.3 Restricted Access Materials (RAM)

Special sorbents possessing restricted access properties have been developed to allow the direct injection of biological matrices into on-line SPE-LC systems. These sorbents, so called *restricted access materials* (*RAMs*), combine size exclusion of proteins and of other high molecular mass matrix components that are prohibited from entering the pores of the packing, and they are not well retained by the column. Therapeutic drugs and other small molecules permeate the pores of the column packing material, where they partition and are retained. The low molecular mass analytes are retained by conventional retention mechanics, such as hydrophobic, ionic, or affinity interactions at the inner pore surface. They are suitable for handling biological samples, since they prevent the access of matrix compounds, like proteins, while by retaining the analytes of interest in the interior of the sorbents they allow extraction and cleanup of samples in the same step (Fig. 21) [65].

The topographical restriction is either obtained by a physical diffusion barrier, i.e., by choosing an appropriate pore diameter (< 8 nm), or by a chemical diffusion barrier, that is, a polymer network at the surface of the particulate. Besides size exclusion, which plays the dominant role in restricted retention of proteins, irreversible protein binding and accumulation is prevented by the generation of a protective, nonabsorptive, hydrophilic outer packing surface, by protein or polymer coating, special binding chemistry, and enzyme or polymer catalyzed hydrolysis [66].

Restricted access materials preferentially have been designed as packing for analytical columns (150×4.6 mm I.D.), where sample cleanup and analyte separation take place simultaneously.

Internal surface reversed-phase (*ISRP*) *supports* are the most popular RAMs. The bonded reversed phase covers the internal pore surface of a glyceryl-modified silica, the ligand being a C_4, C_8, or C_{18} moiety. These sorbents are known as alkyl-diol silica (ADS), and they allow the extraction of a wide range of analytes. Such a precolumn has been used for the analysis of local anesthetics in human plasma, because of its ability to separate rapidly

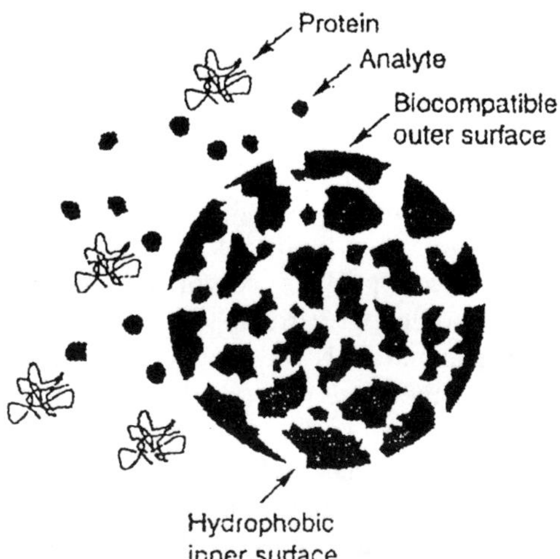

FIGURE 21 Schematic representation of a sorbent particle of restricted access media (RAM) allowing proteins and macromolecules to be excluded and eluted in the solvent front, while small analyte molecules enter the pores and are retained. (From Ref. 65. Reprinted with permission of Elsevier Science.)

the analytes from the proteins and polar endogenous compounds. It has also shown an efficient on-line cleanup and trace enrichment for the determination of several drugs and their metabolites in various biological fluids: serum, urine, intestinal aspirate, supernatants of cell cultures, and supernatants of protein denaturation [114,115].

4.6.4 On-Line Sample Preparation Technique—Direct Analysis of Small Molecules in Serum and Plasma

Although it is not a sample preparation technique, a direct-injection technique can affect sample preparation and save hours of sample-pretreatment time. ISRP supports were first described by Pinkerton and Hagestam. They were designed to enable large biomolecules to be eluted quickly at or near the void volume of the column so that small molecules such as drugs and drug metabolites could be retained beyond the void volume. The early Pinkerton GFF packings consisted of a porous particle that had glycyl-L-phenylalanyl-L-phenylalanine bonded to porous silica packing. By exposing the bonded particles to a large enzyme such as carboxypeptidase A, the phenylalanine is removed from the outer surface, creating a hydrophilic surface (diol-glycine). Because the enzyme could not penetrate inside the pore, the packing retained its inner hydrophobic surface characteristics. When a serum sample is injected, proteins and other large biomolecules would be excluded from the packing by repulsion from the hydrophilic surface group, and small molecules would diffuse into the pores and interact with the hydrophobic surface by a reversed-phase mechanism [59,114–115].

Many variations on this theme have appeared since the original Pinkerton work. Many RAM media, specifically shielded hydrophobic phases, semipermeable surfaces, alkyl-diol silica, and superficially hydrophilic reversed phases, have been used. Recent packings comprise polymeric outer coatings such as polyoxyethylene and oligoglycerol on silica gel and C_8, C_{18}, and phenyl inner coatings. Polymeric materials with both hydrophobic and hydrophilic groups seem to work in the same way as restricted-access media. The outer phase forms a semipermeable surface that prevents large molecules from reaching the inner phase, while small molecules can penetrate and interact with the inner phase. The depth of penetration depends on the hydrophobicity of the solute. The solute can also interact with the more hydrophobic outer layer.

These procedures have been used for therapeutic and biological monitoring of compounds of medical interest in body fluids and tissues. The restricted-access media generally have small sample capacities that can limit their applicability for trace analysis of drugs.

In on-line approach techniques for the determination of drugs and their metabolites in biological fluids, unretained proteins are diverted to waste, and the desired drugs and metabolites are directed to the analytical column. In this way, an isolation procedure prior to the chromatographic step is avoided, and the whole procedure is less time-consuming and error-prone [110].

The Bio trap 500 C_{18} column, a small cartridge (20 × 40 mm, 13 × 4 mm, or 20 × 2 mm, distributed by BAS, West Lafayette, IN, USA), is a biocompatible sample extraction column containing porous silica particles that allows direct injection of biological samples into an LC system. The cartridge is installed in a 6-port valve between the autosampler and the analytical column. Two pumps are used, one for the extraction of the mobile phase (pump A) and the other for the analytical mobile phase (pump B) [116,117].

The separation of analyte from protein is based on a combination of size exclusion and reverse phase partitioning. Small analyte molecules enter the pores and are retained by the C_{18} and C_8 groups on the inner surface. Albumin with a molecular mass of about 66,000 with the dimensions 150 × 38 Å is the drug-binding protein that is present in plasma at the highest concentration, and it is involved in the binding of a very broad range of drugs. The pores of the particles in the BioTrap 500 C_{18} column are such as to exclude the plasma proteins from the internal hydrophobic surface of the particles by washing them with the continuously flowing extraction mobile phase. After allowing sufficient time to wash the protein off, the valve is switched so that the elution mobile phase can now go through the BioTrap column. After washing with the extraction mobile phase, the valve is switched to the elution position, and the extraction column is backflushed with the analytical mobile phase. The retained analytes are carried to the analytical column for chromatographic separation and detection. Now the valve can be switched back to the extraction position for reequilibration with the extraction mobile phase. The composition of the mobile phase, its flow rate, and the extraction time must be optimized to obtain the maximum recovery of the analyte. The biocompatibility has been obtained by attachment of the plasma protein α1-acid glycoprotein (AGP) on the external surface of the particles. The pore sizes of the particles have been chosen so that plasma proteins and other large molecules will be excluded from the pores.

Since no manual manipulations are involved, such bioanalytical methods have high accuracy and precision, they save time, they provide good economy, and they enable safer handling of hazardous or infectious samples. However, carryover between injections may be a problem [66,116].

5 SOLID SAMPLE PREPARATION

Solid samples such as biological tissues may be prepared for extraction by a stepwise process that begins with disruption of the gross architecture of the sample. This modifies the physical state of the sample and provides the extracting medium with a greater surface area per unit mass. There are many methods of reducing sample particle size, as listed in Table 15. Additionally samples may be frozen in liquid nitrogen or by exposure to dry ice, or they may be freeze-dried to produce a material that can be mechanically pulverized. These processes produce a finely divided powder that may then be extracted [11,118].

Dissolution is the predominant process that takes place before solid samples are injected into gas or liquid chromatographs in order to be converted into a liquid state. Components of interest are transferred into solution either by dissolving the entire sample matrix or by leaching the analytes from the solid matrix using a suitable solvent.

For many years the traditional sample preparation methods reported in Table 15 were applied as isolation techniques. Most of these methods, for example, Soxhlet extraction, have been used for more than 100 years, and they mostly require large amounts of organic solvents. These methods were tested over time, and analysts were familiar with the processes and protocols required. The trend in sample preparation in recent years, however, is toward automation, short extraction times, and reduced organic solvent consumption. This trend is reflected in the modern advances in sample

TABLE 15 Solid Samples' Preparation Processes

	Isolation techniques	
Preliminary size reduction	Traditional	Modern
Milling	Dissolution	Microwave-assisted extraction (MASE)
Mincing	Solid–liquid extraction	Pressurized liquid extraction (PLE, or ASE)
Blending	Homogenization	Matrix solid-phase dispersion (MSPD)
Macerating	Sonication	Automated Soxhlet extraction
Pulverizing	Forced flow leaching	Supercritical fluid extraction (SFE)
Chopping	Soxhlet extraction	Gas-phase extraction
Grinding		
Sieving		
Freeze-dried		
Crushing		
Cutting		
Homogenizing		

Sources: Refs. 3, 9, 10.

preparation that yielded various techniques, such as microwave-assisted solvent extraction (MASE), pressurized liquid extraction (PLE) or accelerated solvent extraction (ASE), matrix solid-phase dispersion (MSPD), automated Soxhlet extraction, supercritical fluid extraction (SFE), and gas-phase extraction [9].

5.1 Solid–Liquid Extraction

Solid–liquid extraction (*SLE*) refers to classic extraction technology, which is achieved by using the appropriate solvent that dissolves selectively only the analyte of interest. SLE takes on many forms. The most common is the shake-filter method that involves the addition of an organic solvent for organic compounds or dilute acid or base for inorganic compounds, to the sample, and agitation to allow the analytes to dissolve into the surrounding liquid, until they have completely migrated. Heating or refluxing the sample in hot solvent may speed up the extraction process. The shake-filter method can be performed in batches, which helps the overall sample throughput. After the analytes are removed, as is usually determined by analyte measurements over time, the insoluble substances are removed by filtration or centrifugation. Temperature, pressure, and time conditions have to be optimized. High temperature or sonication at ambient temperature can be performed. Extraction may be achieved in an open vessel at the boiling point of the solvent or in a closed vessel at 100°C. In the former case, the extraction needs 1 h to be completed, while in the latter, 12 min are enough. Faster and more complete extraction is achieved by *sonication*, where the sample is immersed into a solvent within a vessel that is placed in an ultrasonication bath. The ultrasonic agitation allows more intimate solid–liquid contact, and the gentle heating generated during sonication can aid the extraction process. Sonication is also recommended for the pretreatment of solid samples for extracting nonvolatile and semivolatile organic compounds. Different extraction solvents and sonication conditions may be used, based on the type of analytes and their concentration in the solid matrix. Homogenization in the presence of solvent is also an effective process to maximize extraction yield [9,119].

If only the volatile constituents are to be determined, headspace sampling, purge and trap sampling, gas extraction in two modes, equilibrium headspace (static) or dynamic extraction (thermal desorption and purge-and-trap techniques), have to be considered. Headspace is suitable for dirty samples. If the whole sample should be analyzed, then it has to be dissolved in the appropriate solvent so that the analytes are not affected. This can be achieved by digestion or by vortexing with an organic solvent or an aqueous solvent for salts. The extract can either be directly injected or treated further as described in Table 3.

5.2 Soxhlet Extraction

This technique is the most widely used for solid-sample pretreatment. The solid sample is placed in a "thimble," that is a disposable, porous container made of stiffened filter paper or Pyrex glass. The thimble is placed in the Soxhlet apparatus, into which refluxing extraction solvent condenses and the soluble components are leached out. The Soxhlet apparatus is designed to siphon the solvent with the extracted components after the inner chamber, holding the thimble, is filled to a specific volume with solution. The siphoned solution containing the dissolved analytes is returned to the boiling flask, and the process is repeated until the analyte is successfully removed from the solid sample. Soxhlet extractions are usually slow, often requiring 24 h or more. However, the process requires little operator involvement after the sample is loaded and refluxing begins until the termination of the extraction. Samples can only be extracted one at a time for each apparatus, but rows of Soxhlet extractors can be used, usually under fume hoods, when the technique is integrated in routine analysis. Compared with modern extraction techniques, this is a low-cost method. The Soxhlet extraction glassware itself is rather inexpensive, though the cost is elevated by the large solvent volumes required most often a few hundred milliliters. Small-volume Soxhlet extractors and thimbles can accommodate milligram-scale sample amounts.

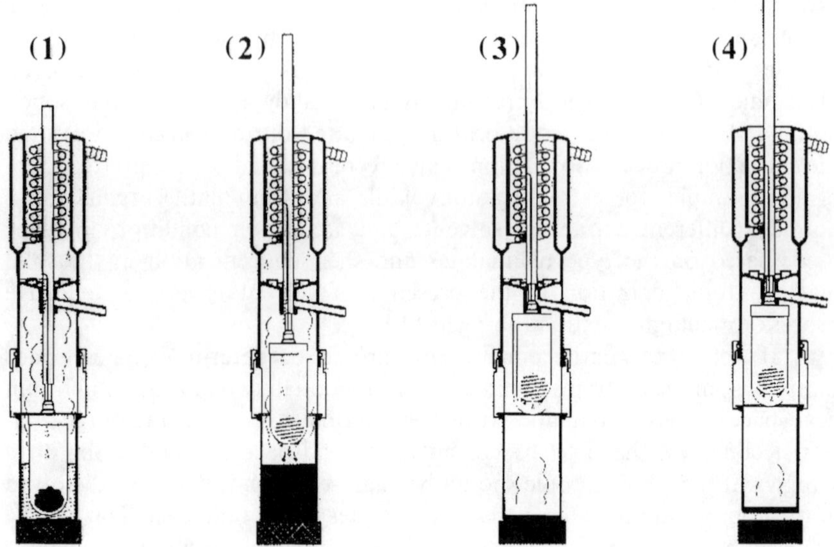

FIGURE 22 Automated Soxhlet extraction in four steps. (1) Boiling. (2) Rinsing. (3) Recovery. (4) Predrying. (Reprinted with permission of Foss Tecator, Sweden.)

Method development involves the choice of the suitable solvent or solvent mixture (azeotrope) of high volatility, so that it is easily removed and has a high affinity for the analyte but a low affinity toward the solid sample matrix [120].

Fully automated Soxhlet extraction has been patented as Soxtec by Foss Tecator (Sweden), based on a four-step technique as shown in Fig. 22. The sample is rapidly dissolved during the first step in boiling solvent. Remaining soluble matter is efficiently removed at the second step, while distilled solvent is collected at the third step. In the fourth step, the sample cup lifts off the hot plate, utilizing residual heat to predry while eliminating boil-dry risk.

5.3 Forced-Flow Leaching

Forced-flow leaching is an extraction technique that can provide nearly quantitative recovery of many organic compounds. In this technique, the sample of interest is packed into a 20 cm × 4 mm stainless steel column. Extraction solvent is pumped under a gas pressure of 2.5 kg/cm^2 through the column, which is heated close to the solvent's boiling point. The results are comparable to Soxhlet extraction, but the extraction time is reduced from 24 to 0.5 h using the forced-flow technique. An advantage of this method is that the sample is subjected continuously to fresh, hot solvent, and the effluent from the column can be collected easily for further treatment [9].

5.4 Supercritical Fluid Extraction (SFE)

Supercritical fluid extraction (SFE) is a modern sample preparation technique of great interest and utility for complex matrices, primarily considered as an alternative for Soxhlet and sonication extraction for solid and semisolid matrices [9,120].

A *supercritical fluid* is defined as a substance above its critical temperature and pressure, which means that it does not condense or evaporate to form a liquid or a gas; it is a fluid with intermediate properties. These properties change from gaslike to liquidlike as the pressure is increased.

The essential tools for supercritical fluid extractors include a supercritical fluid (most often CO_2 or CO_2 with an organic modifier) source, a means of pressurizing the fluid, a pumping system (for the liquid CO_2), an extraction thimble, a device to depressurize the supercritical fluid (flow restrictor), an analyte collection device, temperature-control systems for several zones, and an overall system controller. Figure 23 depicts a modern generic supercritical fluid extractor [120].

Pumps are derived from HPLC pumps, mostly reciprocating, and they often are used with little, e.g., cooling the active pump head, or no adaptation. The substantial vapor pressure of the CO_2 at ambient temperature helps to displace the liquid into the pump.

The CO_2 remains a liquid throughout the pumping or compression zones and passes through small-diameter metal tubing as it approaches the extraction thimble itself. The fluid then passes through the extraction thimble

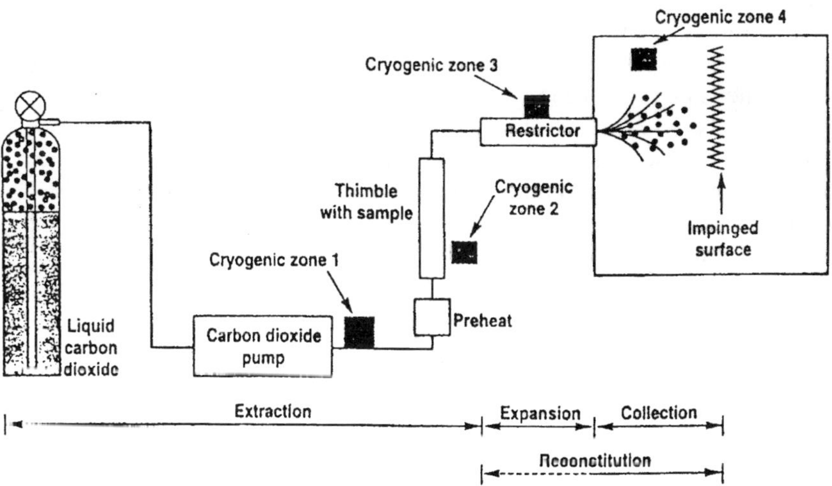

FIGURE 23 Schematic of a typical SFE system. (From Ref. 120. Reprinted with permission of LCGC Europe.)

at a flow rate and a time period predetermined at the method development stage. After passing through the extraction thimble, the supercritical fluid containing extracted analytes flows through additional capillary tubing until it reaches the restrictor zone. After passing the extraction thimble zone, the fluid could return to the liquid state. At this point, maintaining supercritical conditions is unnecessary for analyte solubility.

Temperature is an important but complex parameter for controlling the extraction. A lower temperature (for example, from 32°C down to ambient) causes a significant increase in density. In almost all experimental cases this increase in fluid density causes an increase in the saturation solubility. At the restrictor zone, the temperature is usually raised significantly. Between the pump and the outlet of the restrictor, the pressure (and hence the density) is also increased to a very high level. As the fluid passes out of the restrictor and the conditions are no longer supercritical, this elevated pressure is now lowered to 1 bar in a short period of time and a short linear distance. Joule–Thomson cooling occurs during this rapid expansion of the fluid, so that heating of the restrictor zone is necessary to avoid plugging and precipitation of analytes [120].

Beyond the outlet of the restrictor, within a millimeter or less, there is a region with a small packed trap, or alternatively the analytes can be precipitated and thus collected in an empty thimble or a thimble with an appropriate liquid. The packed trap contains inert packing materials, such as metal or glass beads or packing that has significantly more surface area and internal porosity.

Analytes: SFE can be used for samples such as blood, serum, and plasma, as the nonpolar nature of supercritical fluid CO_2 allows easy fractionation of nonpolar drugs. Applications include the extraction of salts, proteins, carbohydrates, peptides, amino acids, and other interfering polar compounds in a biological matrix.

Matrices: SFE shows its best advantage with extracting analytes from solids and semisolids, rather than from liquids, fluids, and gases, owing mainly to the extraction thimble's design. The extraction thimbles and pieces are made of porous materials such as nickel, chlorofluorocarbon compounds, and stainless-steel materials that are very similar to the frits used in HPLC columns. To extract a liquid sample by SFE successfully, it has to be mixed first with a solid material such as diatomaceous earth or alumina, so that the sample is no longer in a liquid state. An SPE filter bed medium may also be used so that a liquid sample is passed through this first. Then the SPE material can be transferred to the SFE extraction thimble and one proceeds with the extraction directly. Care has to be taken to avoid restrictor plugging by solid carbon dioxide or extract. Using a back-pressure regulator might cause some problems, as the extract could be deposited in the regulator.

An advantage of the technique is that the extract can pass directly to the separation column in hyphenated systems such as SFE-Supercritical Fluid Chromatography (SFC), SFE-GC, and combinations with MS, thus leading to a reduction in sample handling and compatibility between the two solvents. Attempts to automate SFE brought additional problems because the extraction system had to be capable of maintaining the pressure and temperatures needed [121].

Drawbacks. There is a clear need for SFE to be carried out on reference materials of known composition determined by an alternative technique. This approach is not perfect; in many cases SFE gave higher yields than the standard Soxhlet method. SFE like any extraction process is a solubility and diffusion controlled process. Given enough time, even a poorly soluble analyte can be extracted, or an apparently impermeable matrix can be penetrated, yielding to a decreased selectivity, as some undesired material is usually also obtained. Since carbon dioxide is miscible with water and will dissolve in it to a small extent, the extraction of wet or liquid samples, such as blood, urine, and saliva, is precluded; so applications are limited in drug metabolism studies and toxicology. Instead these matrices are more readily examined using SPE or SPME methods [121–122].

The ideal matrix for SFE is a finely powered solid with good permeability, allowing a large surface area for interaction. Typical examples are particulates and powdered dried materials. Some wet samples can be dried or blended with a drying agent, but in routine use this requires additional handling steps, while traditional methods of extraction, such as maceration and LLE methods, are much easier.

The advantages of SFE in comparison to LLE are

1. It uses less organic solvent in a reduced time.
2. Carbon dioxide as an extraction solvent has the advantage of low critical temperature, and it is cheap and nontoxic. It is additionally nonexplosive. It is a relatively fast technique with extraction times of less than 0.5 h. It is classified as a nonpolar solvent that can be modified to make it more polar by the addition of organic solvents (modifiers) such as lower alcohols, e.g., methanol.
3. On-line and off-line analytical scale SFE can be applied. In the former, the coupling step, the transfer and collection of extracted analytes from the SFE to the chromatographic system, is of great importance.

The instrumentation to perform off-line SFE is shown in Fig. 23, while on-line SFE can be achieved by three main approaches:

1. By a sample loop and a switching valve that introduces the

SFE extract to the GC via a heated transfer line by means of GC carrier gas.

2. By a cryogenic or sorbent trap external to the chromatograph that collects the SFE extracted analytes from the depressurized fluid; analytes are transferred by heating the trap and sweeping it with carrier gas or extraction fluid.
3. By the direct approach, where the analytes are directly introduced to the GC, without the use of valve or trap [123].

Extraction modes include dynamic and static. In the first, the sample is continually supplied with fresh supercritical fluid and the extracted analytes are constantly swept into the collection device. In the second, the outlet of the extraction cell is shut off and the cell is pressurized under static (nonflowing) conditions. Following an appropriate extraction time, the analytes are recovered from the static extraction generally by opening a valve at the outlet of the cell and performing a short dynamic extraction [9].

Optimization of extraction conditions for improved selectivity includes the suitable selection of pressure and thus density, temperature, and organic modifier introduction to alter their functionality. Supercritical fluids with low critical temperatures enable extraction under milder thermal conditions so that thermally labile components do not decompose [123].

5.5 Pressurized Solvent Extraction (PSE)—Accelerated Solvent Extraction (ASE)

As an alternative to SFE with carbon dioxide or other supercritical fluids, it was proposed that heating organic solvents under pressure above their boiling points but below their supercritical points would enhance the speed of reaction and solvent strength. This technique is known as *pressurized solvent extraction (PSE)* and provides an easy method for extraction, reducing the amount of solvent required and speeding up the process. The system is marketed as *accelerated solvent extraction (ASE)*, by Dionex Corporation (Sunnyvale, CA, USA). Because PSE represents an extension of existing methods, it attracted attention and is often adopted by official organizations [121,124–125].

5.6 Microwave-Assisted Solvent Extraction (MASE)

Microwave digestion methods have been developed for different sample types, such as environmental, biological, geological, and metallic matrices, as well as for fly ashes and coal. Over the years, procedures based on microwave ovens have replaced some of the conventional hot plate and other thermal digestion techniques that have been used for decades in chemical laboratories. Applications of microwave-assisted techniques in other fields of analytical chemistry, such as sample drying, moisture measurements, chromogenic reactions, speciation, and nebulization of sample solutions, can be found in the literature [126–129].

MASE is a process of using microwave energy rapidly to heat solvents in contact with a sample, in order to partition analytes from the matrix into the solvent. By using closed vessels, the extraction can be performed at elevated temperatures that accelerate the mass transfer of target compounds from the sample matrix. A typical extraction procedure takes 15 to 30 min and uses small solvent volumes in the range of 10 to 30 mL [16].

One of the main advantages of using MAE is the reduction of the extraction time when applying microwaves. This can mainly be attributed to the difference in heating performance employed by the microwave technique and conventional heating. In the latter, a finite period of time is needed to heat the vessel before the heat is transferred to the solution, while microwaves heat the solution directly, thus accelerating the speed of heating. Additionally, MASE allows for a significant reduction in organic solvent consumption, since volumes about 10 times smaller are possible, and one can run multiple samples. Consequently, MASE is an attractive alternative to conventional techniques.

Microwave energy is introduced for both digestion and solvent extraction. In digestion, acidic solutions are combined with organic, inorganic, and biological samples in a closed, chemically resistant, non-microwave-absorbing container. The container is then subjected to microwave radiation for a controlled period of time. Microwave radiation heats the solution to a high temperature, and the hot acid digests the sample matrix, e.g., an organic polymer. This process is much faster and safer than the hot-plate technique [59].

In extraction there are two approaches when using MASE:

1. The use of a microwave-absorption solvent, with a high dielectric constant
2. The use of a non-microwave-absorption extraction solvent, with a low dielectric constant

Chemical compounds absorb microwave energy roughly in proportion to their dielectric constants: the higher the value of the dielectric constant, the higher the level of microwave energy absorption.

In the *microwave absorbing solvent approach*, the sample and the solvent are placed in a closed vessel similar to those used for microwave digestions. Microwave radiation heats the solvent to a temperature higher than its boiling point, and the hot solvent provides a rapid extraction of analyte under moderate pressure, usually a few hundred psi. For these high-pressure extractions, the containers are made of polytetrafluoroethylene (PTFE), quartz, or other materials that combine optimum chemical and temperature resistance and good mechanical properties.

In the *non-microwave-absorbing solvent approach*, the sample and the solvent can be placed in an open or closed vessel. The temperature of the solvent is unchanged, as it absorbs very little microwave radiation. On the other hand, the sample, usually containing water or other compounds possessing a high dielectric constant, absorbs the microwave radiation, and the heated analytes are transferred into the surrounding cool liquid, which is selected according to its solubility characteristics. The second method approach is gentler, as it is performed under atmospheric or low- pressure conditions and can be used with heat-sensitive or thermally labile analytes.

MASE uses less solvent than conventional Soxhlet or LLE. Selecting extraction solvents based on their microwave absorption abilities can control heat exchange between the sample and the solvent. The extraction-solvent power for the analytes of choice can also be an important selection criterion. Optimization in MASE includes parameters such as heating time, pulsed heating vs. continuous heating, stirring vs. nonstirring, closed containers vs. open containers, and outside cooling of vessel vs. noncooling. Multiple samples can be extracted simultaneously, resulting in increased throughput.

The MASE techniques are safe because analysts are not exposed to the extraction solvents in most cases; however, users should be careful when dealing with microwave radiation and highly pressurized closed containers; special care must be taken for safety.

Microwave heating can be employed for gas-phase extraction, since most gases absorb less microwave radiation than liquids and solids. Thus, in the presence of a solid sample, the microwave energy is absorbed directly by the sample that probably processes a larger dielectric constant than the surrounding gaseous medium. Volatile analytes can, therefore, be vaporized. If water is present in the sample, its thermal energy can be transferred to other volatile analytes and enhance their volatilization. This process is analogous to static headspace sampling [126–129].

5.6.1 Dynamic Microwave-Assisted Extraction (DMAE)

An improvement to the techniques listed above is the monitoring of the extraction, which results in a high level of certainty concerning the point at which the extraction process is complete or can be considered quantitative. In the opposite case, there is considerable difficulty in establishing fixed times for extracting standard reference materials. The practical solutions to this problem are usually to use excessively large extraction times to ensure complete extraction, or to compromise with possible losses of compounds due to matrix. Techniques like *dynamic microwave-assisted extraction (DMAE)*, enable the extraction process to be monitored with the aid of a high-performance liquid chromatography (HPLC) detector. Fresh solvent is pumped through the extraction stage, in a process that resembles chromatography, with a rather undefined solid phase. Using a dynamic procedure, the partition equilibrium does not demand that the solutes be forced into the solvent with almost 100% efficiency. The drawback with different types of microwave-assisted extractions is that completely nonpolar solvents cannot generally be used, although the sample matrix can sometimes be used as an energy absorber. The dynamic microwave extractor consists of a solvent delivery system, a microwave oven, an extraction cell, a temperature set point controller with a thermocouple, an HPLC fluorescence detector, and a 300 × 0.25 mm ID fused-silica restrictor (Fig. 24) [130].

5.6.2 Focused Microwave-Assisted Extraction

Extraction by reflux with an organic solvent may last several hours and require several hundreds of milliliters of the organic solvent. As already

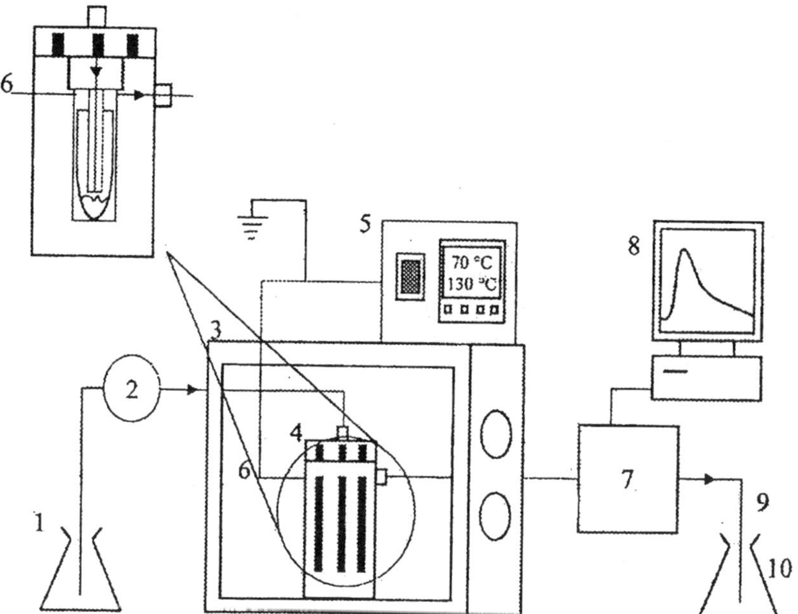

FIGURE 24 System of dynamic microwave-assisted extraction (DMAE). (1) Solvent. (2) Pump. (3) Microwave oven. (4) Extraction cell. (5) Temperature set point controller. (6) Thermocouple. (7) Fluorescence detector. (8) Registering device. (9) Restrictor. (10) Extract. (From Ref. 130. Reprinted with permission of Elsevier Science.)

mentioned, MASE under pressure can be an alternative for conventional extraction techniques such as Soxhlet extraction. *Focused microwave-assisted extraction (FAME)* at atmospheric pressure can be applied for the extraction of contaminants in biological tissues [131].

The principle involves the heating of both the solvent and the matrix by wave/matter interactions. The microwave energy is converted into heat by two mechanisms: dipole rotation and ionic conductance. Heating is therefore selective with only polar or moderately polar compounds. Dichloromethane is selected for its polarity; it is able to absorb and transmit the energy of the microwave beam. Moreover it is a good solvent for aromatic compounds.

The use of the focused microwaves in apparatus such as Soxwave by Prolabo, France (Fig. 25) allows homogenous and reproducible treatment of the sample owing to the focusing of the microwave on the medium. Heating to the boiling temperature is very rapid; it takes place in less than 1 min at 30 W. This method is secure because the extraction is performed at atmospheric pressure and there is no risk of explosion. The temperature of extraction is the boiling point of the solvent. There is no problem with compound degradation owing to high temperatures, and no cooling is required after extraction. A few minutes are sufficient (~10 min) and only 30 mL is necessary for the extraction of 1 g of contaminated matrix [131].

The recoveries and reproducibility are good and comparable with those obtained by conventional techniques.

5.7 Thermal Desorption

Volatile analytes can be collected from solid materials by varying temperatures and carrier-gas flows instead of solvent extraction. The use of thermal desorption in conjunction with GC assumes that some analytes of interest adsorbed or absorbed by some material can be deliberated by heating and that once desorbed they are amenable to analysis by GC.

A wide range of complex sample types is included as application examples, among them predosage clinical samples, for example unwanted volatile compounds such as residual solvents in pharmaceuticals. Thermally reversible desorption is the essential principle in purge-and-trap and dynamic headspace techniques, which rely on selective adsorption to concentrate volatile organic compounds before analysis.

It is of great importance that analytes be desorbed intact from the matrix, which means that the heat applied to the sample must be enough to volatilize the organic compounds without degrading them and without producing unwanted artifacts from the matrix itself. For this reason the control of sample temperature, heating rate, and sampling time is essential [132,133].

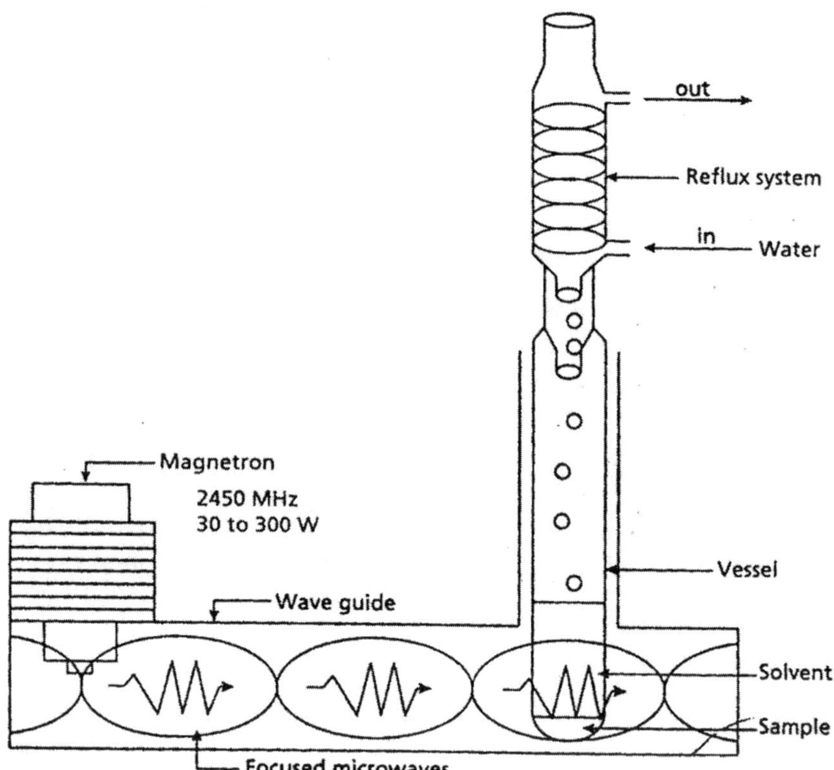

FIGURE 25 Focused microwave apparatus. (From Ref. 131. Reprinted with permission of LCGC Europe.)

The main advantage of this technique is that it permits analysis without the use of solvents and that the sample preparation can be automated. It permits also 100% of the sample to be analyzed, instead of an aliquot, which increases sensitivity. Also the environmental concerns of evaporating extraction solvents and waste disposal are eliminated.

A sample as a pharmaceutical powder is placed into a temperature-controlled chamber and heated; the resulting desorbed organic compounds are transferred either directly or with some intermediate enhancement to a GC. The simplest arrangement is that the thermal desorption is accomplished by heating the sample with a flow of carrier gas directly to the column inlet. Cryogenic traps may also be used to trap more volatile compounds cryogenically onto a GC column before programming when purging or using low temperatures. This can be accomplished by using a GC oven's cryogenic function or by installing a cryogenic focuser, which uses either liquid nitrogen or carbon dioxide as a cooling agent at the head of the column. The latter approach adds the expense of additional equipment and tanks of cryogen. Adsorption tubes or cartridges filled with a sorbent may be used to preconcentrate volatile compounds onto a collection device. Analytes are subsequently desorbed either directly to the GC or to an intermediate trap at first and then to the GC [134].

5.8 Matrix Solid-Phase Dispersion (MSPD)

Matrix solid-phase dispersion (MSPD) is a relatively new technique that appeared in 1989 and remedied many of the complications of dealing with solid samples and their subsequent extraction using solid-phase material. This process can handle a viscous solid, a solid, or a semisolid sample directly by blending, with a solid phase support such as silica or bonded silica, which are similar to those used in SPE columns. This is performed by placing a small quantity, ~ 0.5 g, of the sample in a glass mortar and blending it with a glass or agate pestle as shown in Fig. 26. The bonded phase acts as a lipophilic bind solvent that assists in sample factorization. The best ratio of sample to solid support-bonded phase is 1:4. But this depends on the application. The isolation of polar analytes from biological samples is assisted by the use of polar solid support phases and less-polar analytes by less-polar phase. The presence of the bound organic phase provides the following process: the sample components dissolve and disperse into the bound organic phase on the surface of the particle, leading to the complete disrupting of the sample and its dispersion over the surface. Sample components are distributed over the surface based on their relative polarities [118,121].

Nonpolar compounds disperse into the nonpolar organic phase based on their distribution coefficients with the phase. Highly polar molecules associate with silanols on the surface of the silica particle and inside the pores of the silica solid support, as well as with matrix components capable of hydrogen bonding across the surface of the now biphasic, bonded-phase/dispersed sample lipid structure.

Conditioning of the material to be used for MSPD can greatly enhance analyte recovery. It also speeds the process of blending and dispersal. It is essential to condition the column material with a solvent that breaks the surface tension differences that may exist between the sample and the solid support bonded phase. The ionization or the suppression of ionization of analytes and sample components can greatly affect the nature of interactions of specific analytes with the blend and the eluting solvents.

Once the MSPD blending process is completed, the material is transferred to a column constructed from a syringe barrel or other appropriate device containing a frit that retains the entire sample, which is compressed to

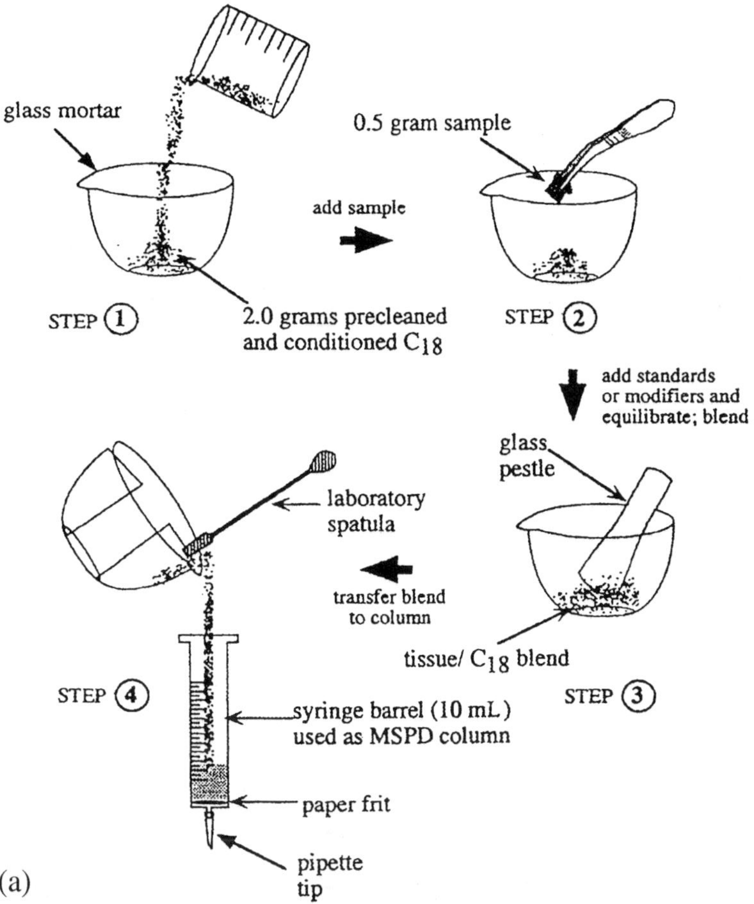

FIGURE 26(a) Schematic representation of the MSPD process. (From Ref. 118. Reprinted with permission of Elsevier Science.)

form a column packing by using a modified syringe plunger. A second frit is placed on top of the material, which is subsequently compressed so that no channels are formed. Addition of eluting solvent to the column may be preceded by the use of some or all of the solvent to backwash the mortar and pestle. Approximately 8–10 mL of solvent is used to perform elution. However most target analytes are eluted in the first 4 mL [68].

Since the entire sample is present in the column, it is possible to perform multiple or sequential elution. For this purpose, MSPD columns prepared with C_{18} supports have been most frequently eluted with a

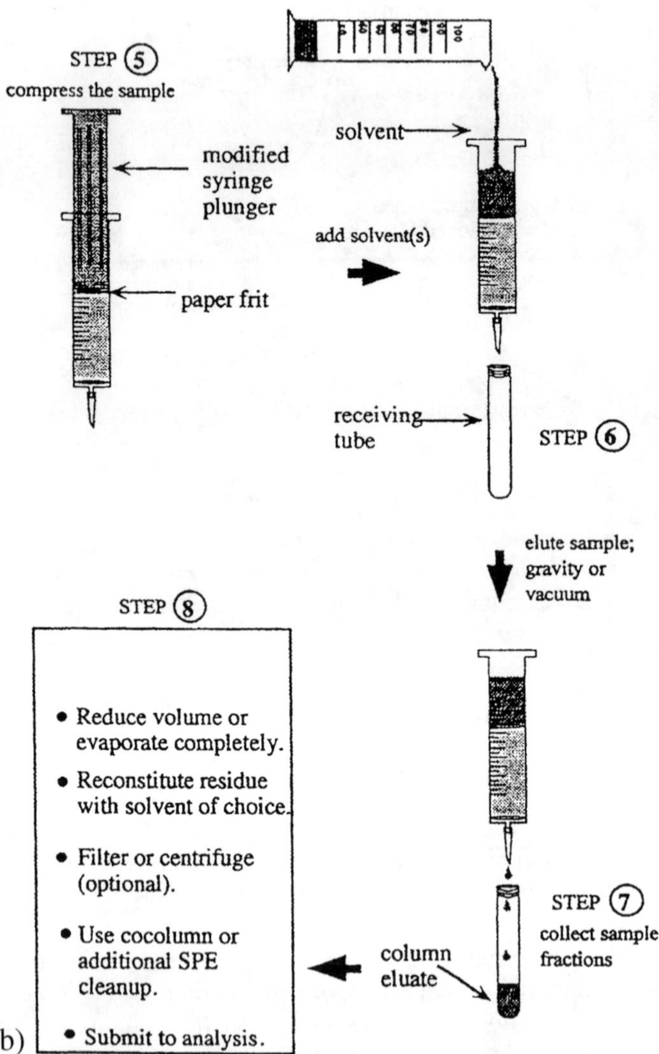

FIGURE 26(b) Schematic representation of the MSPD process. (From Ref. 118. Reprinted with permission of Elsevier Science.)

sequence of solvents, beginning with a least polar (hexane) and increasing the polarity (ethyl acetate, acetonitrile, methanol) up to water. Elution can be conducted by gravity flow, by application of pressure to the head of the column, or by placing the columns on a vacuum manifold and applying suction.

The nature of the MSPD column and the range of interactions permit the isolation of a range of different polarity analytes or an entire class of compounds in a simple solvent or in different solvents passed through the column. In some cases, the eluate from a matrix solid dispersion column is adequately clean for direct injection. However, additional steps are usually necessary to remove the coeluted matrix components that are occasionally present. These can be removed either by using other solid-phase material packed at the bottom of the MSPD column or by eluting analytes from the MSPD column directly onto a second SPE column for sample cleanup and analyte concentration.

In most cases MSPD provides results equivalent to older official methods. However, it generally requires 95% less solvent and 90% less time than classic methods. The use of smaller sample sizes and lower solvent consumption, purchase, and disposal, combine to make MSPD competitive with classic methods on several levels.

MSPD has been applied to the isolation of drugs, herbicides, pesticides, and other pollutants from animal tissue. Animal tissues can simply be covered with the solid support material for a certain time period, leading to dissolution of the sample into the bonded phase. More viscous samples, such as blood, can be blended by placing the sample in a test tube or a syringe barrel that serves as a column.

Similar sorbents of SPE can be used. As with liquid samples to be applied to SPE columns, it is sometimes necessary to alter the ionization state of the sample components to assure that certain interactions occur between the solid support bonded-phase and/or the eluting solvent in MSPD. This can be accomplished by adding acids, bases, salts, chelating or dechelating agents, antioxidants, etc., at the time of sample blending and/or as an additive to the eluting solvent [118].

6 COLUMN SWITCHING TECHNIQUE

Column switching techniques are a group of similar methods whereby complex samples can be selectively determined. Various configurations allow samples of high complexity or low analyte concentration to be analyzed. Such a method is a powerful approach that has been applied in some form to every chromatographic technique. This approach depends on the selection of an appropriately chosen HPLC stationary phase to retain and separate the analytes of interest, while it allows unretained components to be flushed from the column. Having first appeared in 1973, *Column Switching Technique* provides an alternative strategy for sample preseparation to LLE or SPE. The technique usually relies on an initial fractionation or incomplete resolution of the sample on one column. Subsequently, the fraction containing the substances of interest is passed for resolution to a second column. It involves the use of high-pressure switching valves to couple two or more columns that trap either defined volumes or the collected samples, usually in a loop, and direct them in a second column. Valve configuration can be used for conventional HPLC analysis, and it can perform more advanced functions, such as diverting the mobile phase containing the desired solutes from the first column to the second column for defined periods of time, a process called *heart-cutting* or *on-column concentration* (trace enrichment). Additionally it can perform *backflushing* of specific sample components from one column to waste, which leaves only the peak of interest on the second column and front- and end-cutting when needed. Selecting the appropriate valve and actuating it at the correct time cause different fractions of the sample to follow different paths through the column network (Fig. 27) [112].

The in-line design of this type of system can also reduce sample contamination and loss. Automating column-switching systems can usually be accomplished using the timed-event tables included with HPLC instrument-control software.

The procedure differs from serial methods used to improve resolution because only selected fractions rather than the whole sample are passed to the second column. The separation of the sample consequently occurs not just by chromatographic means but also by physical means that are under the control of the analyst and unrelated to the chemical properties of the analyte. These dual modes of separation may result in very efficient separations [23,135].

Column switching techniques have been used to analyze a sample directly without any pretreatment stages, with the advantage of easy automation. Methods using a number of chromatographic separations to clean up the sample can often be relatively easily converted to a column-switching configuration. The resulting increase in safety and the improved analytical speed and accuracy often make up for the initial cost of the additional equipment that is required.

The use of computer based equipment enables a simpler and more accurate control of complex column networks.

As with all methods there are always limitations. The major problem is a possible incompatibility between different solvents. Other limitations of the method relate to its complexity and the attendant need for accurate control, together with the high cost of the switching valves, automated actuators, and all the associated controlling equipment. A computer-based system is a

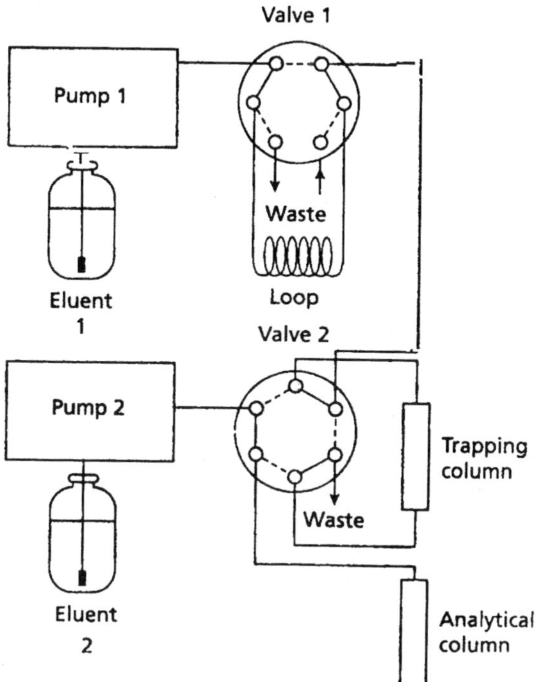

FIGURE 27 Automated column switching system that permits column backflushing, sample heart-cutting, trace enrichment, and front- or end-cutting. (From Ref. 112. Reprinted with permission of LCGC Europe.)

common requirement. In case of treating complex mixtures and attempting to separate a fraction on a column that is identical to the fractionation column can merely lead to a large poorly resolved hump. This occurs because there is no mechanism for separating the very closely eluting interfering substances from the analytes that are present in the chosen fraction. To overcome this, the analytical column must rely on a retention mechanism that produces an elution order very different from that of the fractionation column. Ideally the analyte should elute rapidly from the first column to allow a small fraction to be transferred to the second column. Modes that can be easily connected may include normal phase and normal phase, reversed phase and reversed phase, normal phase and size exclusion, chiral and reversed phases, and ion and reversed phases. A useful method when coupling size exclusion and reversed phase columns involves the on-line dilution of the organic but water miscible phase using a second pump. This method can also be used to achieve the preconcentration of a sample after a number of separation steps. The amount of sample that is passed to the second column must be carefully controlled. If the transfer time is too short, the carryover of the analyte onto the next column will be low, and this will have a severe effect on the limit of detection (LOD). Conversely if the time taken is too long, this will also have an adverse effect on the LOD. This occurs because, even though all the analyte may be transferred, many more interfering substances will be included within the fraction. It is also necessary to consider the distance that the fraction will pass down the second column. If the cut is so large that the head concentration effect is overcome, significant peak dispersion may occur as the zone moves slowly down the column. This effect is more pronounced when different solvents are used to obtain different modes of separation.

One approach for enhanced selectivity and separation of targeted analytes from matrix components involves a backwash of the second column after the targeted compound(s) have been trapped at the inlet of the second column. The benefits of this approach include total automation and quantitative transfer of target components within the column-switching system. Using a "heart-cut" operational mode, a narrow retention time region containing a desired component(s) is "cut" from the chromatogram and transferred onto another HPLC column for additional separation. In this instance, quantitative transfer of the components without adsorptive or degradative losses can be assumed. Another advantage of this approach is the increased selectivity afforded by the judicious choice of two or more HPLC stationary phases. The limitation of column switching includes restricted sample enrichment because of the constrained amount of original sample that may be loaded onto the first column of the HPLC separation. As a result, it may not reach the lower limit of detection required for some analytical problems. Another limitation is that sample throughput using this approach probably will not be as high as for other methods. Finally, it is considered too complicated to be practical, though modern instruments make these procedures relatively straightforward [59,136].

7 DERIVATIZATION

The term derivatization refers to a chemical reaction between analyte and reagent that aims to alter the chemical structure and/or physical properties of the analytes of interest to improve detectability or separation efficiency. This reaction must be rapid and quantitative with minimal byproducts to avoid interference with the analysis. The main challenge in organic analysis originates from polar compounds. These are difficult to extract from biological matrices and difficult to separate on the chromatographic column. Derivatization approaches are frequently used to address this challenge.

Derivatization may change the sample matrix

1. To improve separation, e.g., by adding a nonpolar group to the analyte molecule, as the polar analytes are more easily separated with reversed phase chromatography
2. To improve detectability by adding chromophore or fluorophore groups to the analyte
3. To stabilize an unstable analyte
4. To change the molecular structure, thus altering volatility, polarity, chirality, etc.
5. To change the fragmentation pattern in MS

A derivatization reaction shields several (polar) functional groups that otherwise would disturb an efficient separation process. In addition, derivatization often improves (thermo)stability and volatility, thus improving separation in GC. However, as it enhances the complexity of the method by adding steps that make for errors and are time-consuming, it is considered the last solution when developing a new method. An agent to be used for derivatization should have the following properties:

Selective reaction with specific functional groups
High sensitivity for low concentrations of the derivatives due to high detector response
Good reaction kinetics (rapid reaction)
High reaction yields
Mild reaction conditions
Stable reaction products
Easy removal of excess reagent or no disturbance to the analysis
Nontoxicity [137]

7.1 Derivatization in HPLC

In HPLC thermal instability, peak tailing and ghost peaks in separation of polar compounds are in most cases no problem. However, enhancement of detectability is required in trace analysis, when the analytes do not possess a UV-absorbing, fluorescent or electroactive functionality, therefore derivatization is necessary. Reaction with a fluorotag will produce a highly fluorescing derivative of the compound of interest, so that very low concentrations are detectable. Improvement of UV detectability can also be obtained by a reaction with a chromotag.

Derivatization for HPLC is done either *off-line (precolumn)* before injection into the column or *on-line postcolumn* by mixing the reagent with the column effluent.

Precolumn derivatization offers some advantages over postcolumn as it involves fewer reaction restrictions, simpler equipment, and no time limitation regarding kinetics, provided that all species are stable. It can be performed either manually or automated. However, there are several drawbacks, such as the introduction of contaminants, a possible loss of analyte owing to side reactions, adsorption, degradation, or incomplete reactions.

In postcolumn derivatization the analyte is derivatized after the separation and before detection by using a reaction detector. The simplest way is to add a reagent solution to the column effluent with an extra pump. After the mixing T-piece, a reactor with a suitable holdup volume is inserted to allow reaction to take place. The benefits of this approach are that chromatographic separation is not affected and reaction need not be complete.

The most common *fluorotags (fluorescent reagents)* are

1. Dansyl chloride (Dns-Cl), used in protein sequence analysis and in the determination of small amounts of amines, amino acids, and phenolic compounds with excitation and emission wavelengths of 335–365 nm and about 520 nm, respectively.
2. Dansyl hydrazine, used for aldehydes and ketones that react to form the corresponding hydrazones, useful in the quantitation of ketosteroids with excitation and emission wavelengths of 335–365 nm and about 520 nm, respectively.
3. Coumarin derivatives, e.g., 4-bromomethyl-7-methoxycoumarin (Br-MMC). This under proper conditions reacts with carboxylic acids to form an ester with excitation and emission wavelengths of 400 and 420 nm, respectively.
4. *o*-Phthalaldehyde (OPA), which is widely used to enhance detectability in HPLC for compounds containing primary nitrogen atoms, like histamine, spermidine, nearly all amino acids, and some hydroxy-indole derivatives, with excitation and emission wavelengths depending upon the products formed, that is, 350 nm and 400–460 nm, respectively.

Chromotags are highly purified reagents, specially designed to react with common organic functional groups to form derivatives with enhanced UV adsorption at 254 nm. Chromotags commercially available include

1. *p*-Nitrobenzyloxyamine hydrochloride for aldehydes and ketones.
2. *O*-*p*-Nitrobenzyl-*N*,*N*-disoproplthiourea for carboxylic acids.
3. *p*-Bromophenacyl bromide (PBPB) for derivatization of carboxylic acids (K-salts) with a crown ether catalyst.
4. *p*-Nitrobenzyl-*N*-propylamine hydrochloride for isocyanate monomers.

5. Fluoro-2,4-dinitrobenzene to label free amino groups in amino acids.
6. Ninhydrine for primary amines forming complexes that have their adsorption maxima at about 570 nm.
7. Dabsyl chloride for primary and secondary amines, including amino acids, thiols, imidazoles, phenols and aliphatic alcohols. This has advantages over the most widely used reagent, dansyl chloride, as dabsyl derivatives absorb at 425 nm, thus avoiding interference from other UV-absorbing biological components. Also, the strongly colored dabsyl derivatives are more suitable for thin-layer chromatography (TLC) because they are more stable than the dansyl derivatives [137–138].

7.2 Derivatization in GC

Derivatization in GC is used to increase the volatility and/or reduce the polarity of some analytes and therefore can improve extraction efficiency, selectivity, and subsequent detection.

Silylation, alkylation, and acylation are the most common applied derivatization processes. Silylation is a relatively old technique that has been used with great success in GC analysis, including in MS detection of compounds containing an active hydrogen. Alkylation procedures are used to replace the active hydrogen atoms in molecules by alkyl groups, for example, reaction with diazoalkanes, dialkylacetals, alcohols, alkyl (benzyl) halides, and quaternary ammonium ions. Alkylation is advantageous in the determination of carbohydrates and carboxylic acids. Amines, amides, alcohols, thiols, phenols, enols, glycols, unsaturated compounds, and molecules with aromatic rings may react with anhydrides, acid halides, and reactive acyl derivates as acetylate imidazoles under various circumstances to give stable, highly volatile acylated compounds.

Silylation reagents include bis-trimethylsilylacetamide (BSA), bis-(trimethylsilyl)-trifluoroacetamide (BSTFA), *N*-methyl-trimethylsilyltrifluoroacetamide (MSTFA), *N*-methyl-*N*-*t*-butyldimethylsilyl triofuoroacetamide (MTBSTFA), which react with active hydrogens. Trimethylsilylimidazole (T(M)SIM), hexamethyldisilazane (HMDS), trimethylchlorosilane (TMCS), flophemesyl chloride (pentafluorophenyldimethyl-silylchloride) are also used.

Alkylation reagents include

1. Trimethylanilinium hydroxide (TMAH): trimethyl-(a,a,a,-trifluoro-*m*-tolyl)-ammonium hydroxide (TMTFTH), tetrabutylammonium hydroxide (TBH).
2. Lewis-acid-catalyzed alkylation with alcohols: BF3/methanol for

making methylesters of organic acids, BF3/butanol, similar to BF3/ methanol in reaction rates but giving the *n*-butyl ester.

3. DMF-dialkylacetals: DMF-diethylacetal.
4. Extractive alkylation: pentefluorobenzyl bromide (PFB Br), which is highly reactive. Its reaction products can be detected in low concentrations. Tetrabutylammonium salts (TBAX) are used to extract the acids in an organic solvent, where the alkylation takes place.
5. Diazomethane: diazomethane generator (*N*-methyl-*N'*-nitro-*N*-nitroguanidine).

Acylation reagents include

1. Fluorinated anhydride: trifluoroacetic anhydride (TFAA), pentafluoropropionic anhydride (PFPA). Heptafluorobutyric anhydride (HFBA) gives the highest ECD response of these anhydrides.
2. Acyl imidazoles: trifluoroacetylimidazole, heptafluorobutyrylimidazole.
3. Additional acylation reagents: *N*-Methyl-bistrifluoroacetamide, pentafluorobenzoylchloride [137].

7.3 Chiral Derivatization Is Applied to Improve Separation of Enantiomers

Indirect separation of enantiomers (optical isomers) involves the reaction of the enantiomeric pair with a chiral reagent to form a covalent bond giving the corresponding diastereomeric derivatives of the enantiomeric pair. These diastereoisomers are no longer the mirror images of each other, so each displays different physical and chemical properties in a nonchiral environment. They can be separated on an achiral stationary phase such as silica or C-18 bonded phase [139].

$$\text{Chiral solute } S(\pm) + \text{Reagent}(+) - R \longrightarrow$$
$$\text{Diastereoisomers: } (+) - S - (+) - R, (-) - S - (+) - R \quad (1.4)$$

8 CONCLUSIONS

The analytical process typically consists of several discrete stages such as sampling, preparation, instrumental analysis, quantification, data reporting, and data interpretation; each step is critical in obtaining accurate and reproducible results. A sample preparation step is often necessary to isolate the components of interest from a sample matrix, as well as to purify and concentrate the analytes. The quality of sample preparation is pivotal in the overall quality of the analysis. Despite advances in instrumentation and microcomputer technology, many sample preparation practices are based on nineteenth-century technologies, e.g., the commonly used Soxhlet extraction, which was developed more than 100 years ago.

With more than 60% of total analysis time spent on sample preparation and only 7% on the actual measurement of sample constituents, there is a growing trend toward faster, more cost-effective, more convenient, and safer methods of sample pretreatment. An ideal sample preparation technique should be solvent-free, simple, inexpensive, efficient, selective, amenable to automation, and compatible with a wide range of separation methods and applications. It should also allow for the simultaneous separation and concentration of the components.

There is no universal sample preparation method, as sample pretreatment depends strongly on the analytical demand as well as the size and nature of the sample. What is beyond any doubt is the continuously increasing demand for improved selectivity, sensitivity, reliability, and rapidity in the process of sample preparation. This is especially the case in the field of clinical analysis, where there is a tremendous need for increases in the speed of the whole analysis, and pressure to reduce solvent usage.

Efforts to introduce novel sample preparations concepts are frequently ignored by practitioners, who are used to traditional sample preparation techniques. Problems associated with traditional techniques, such as the use of toxic organic solvents and multistep procedures that often result in loss of analytes during the process, frequently make sample preparation the major source of error in an analysis and prohibit integration with the rest of the analytical process. One reason that keeps practitioners from adopting a new analytical method is expense in both capital cost and training requirements. However, awareness of the pollution and hazards caused by hydrocarbons, including ozone depletion and carcinogenic effects, has resulted in international initiatives to eliminate the production and use of organic solvents upon which many current sample preparation methods depend.

Modern techniques are currently driving out conventional approaches because of their many advantages, including speed, use of less environmentally hazardous solvents, better facilitation for the control of the extraction, and automation and on-line combination of the extraction with other analytical techniques. A major concern when developing techniques for sample preparation is the possibility of automating the entire analytical process, which might lead to increased sample throughput and reduced manual operations with obvious economic benefits. Moreover, automation frequently permits the use of closed analytical systems, leading to better control of contamination in trace analysis and safer handling of contagious or radioactive samples, as well as to higher accuracy and precision. Sample pretreatment is of great significance in clinical chemistry, where complex samples, in particular protein-containing fluids, such as plasma, serum, milk, saliva, and cell and tissue homogenates, have to be processed manually or semiautomatically, prior to analysis, by a number of time-consuming and labor-intensive protocols including extraction, denaturation, dialysis, ultrafiltration, and unspecific adsorption. During the past years, miniaturization has become a dominant trend in analytical chemistry. The development of techniques such as micro LLE (in-vial extraction), ambient static headspace, disc cartridge SPE, SPME, MSPD, MASE, and SBSE, which use smaller sample size, minimize solvent use, and are amenable to automation, is a positive sign for analytical science. The combination of modern sample pretreatment techniques with state-of-the-art analytical instrumentation may result in more cost-effective and faster analysis, higher sample throughput, lower solvent consumption, and less use of manpower while maintaining or even improving sensitivity. Considering selectivity as a prerequisite, the trend of analysis in the new millennium is *speed by all means*.

REFERENCES

1. RE Majors. An overview of sample preparation. LC-GC 9(1):16–20, 1991.
2. RE Majors. Sample preparation perspectives. LC-GC Int 5(2):12–20, 1991.
3. RE Majors. Trends in sample preparation. LC-GC 10(12):912–918, 1992.
4. ESP Bouvier, DM Martin, PC Iraneta, M Capparella, Y-F Cheng, DJ Philips. A novel polymeric reversed-phase sorbent for solid-phase extraction. LC-GC Int 10(9):577–583, 1997.
5. NH Snow. Solid phase extraction of drugs from biological matrices. J Chromatogr A 885:445–455, 2000.
6. M-C Hennion. Solid-phase extraction: method development, sorbent, and coupling with liquid chromatography. J Chromatogr A 856:3–54, 1996.
7. JA Johnsson, L Mathiasson. Membrane based techniques for sample enrichment. J Chromatogr A 902:205–225, 2000.
8. W Huck, GK Bonn. Recent developments in polymer-based sorbents for solid-phase extraction. J Chromatogr A 885:51–72, 2000.
9. RE Majors. The changing role of extraction in preparation of solid samples. LC-GC Int 9(10):638–648, 1996.
10. R A comparative study of European and American trends in sample preparation. LC-GC Int 6 (3):130–140, 1993.
11. R Majors. Particle size reduction in solid sample preparation. LC-GC Int 11(7):434–439, 1998.
12. Phenomenex. SPE Reference Manual and Users Guide. Phenomenex, Torrance, CA, USA.
13. H Lord, J Pawliszyn. Microextraction of drugs. J Chromatogr A 902:17–63, 2000.

14. IN Papadoyannis. In: J Cazes, ed. HPLC in Clinical Chemistry. New York: Marcel Dekker, 1990.
15. GA Mills, V Walker. Headspace solid-phase microextraction procedures for gas chromatographic analysis of biological fluids and materials. J Chromatogr A 902:267–287, 2000.
16. CS Eskilsson, E Bjorklund. Analytical-scale microwave-assisted extraction. J Chromatogr A 902:227–250, 2000.
17. RE Majors. Distillation as a sample preparation and separation technique. LC-GC Int 12(1):19–23,36, 1999.
18. RE Majors. Sample concentration by evaporation. LC-GC Eur 12(10):628–637, 1999.
19. Supelco Chromatography Products Catalog, Supelco, Bellefonte, PA, USA, 2000.
20. J Eriksen. Conducting cooling—safe storage of biological samples. GIT Laboratory Journal 5(2):72–74, 2001.
21. J Blanchard. Evaluation of relative efficacy of various techniques for deproteinizing plasma samples prior to high-performance chromatographic analysis. J Chromatogr 226:455–460, 1981.
22. RE Majors. Liquid extraction techniques for sample preparation. LC-GC Int 10(2):93–101, 1997.
23. J Henion, E Brewer, G Rule. LC-MS sample preparation. Today's Chemist at Work 36–42, February 1999.
24. FQ Yang, TY Zhang, GL Tian, HF Cao, QH Liu, Y Ito. Preparative isolation and purification of hydroxyanthraquinones from rheum-officinale baill by high-speed countercurrent chromatography using pH-modulated stepwise elution. J Chromatogr A 858(1):103–107, 1999.
25. R Majors. Automation of solid-phase extraction. LC/GC Int 6 (6):346–350, 1993.
26. B Kunugi, K Tabei. Kontakte: the extrelut column in organic experiments. Kontakte (Darmstadt) 1:14–21, 1991.
27. Sample Preparation Catalog. Varian, Harbor City, CA, USA, 1996.
28. IN Papadoyannis. ChromBook MERCK, 2nd ed., p. 14, 1997.
29. IN Papadoyannis. Sample preparation prior to HPLC analysis, solid phase extraction techniques. In: M Ochsenkühn-Petropulu, K. Ochsenkühn, eds. International Conference Instrumental Methods of Analysis, Modern Trends and Applications, Chalkidiki, Greece, 1999. Conference Proceedings: 443–451.
30. Macherey-Nagel Chromatography Catalog. Düren, Germany, 2000/2001.
31. ME Leon-Gonzalez, Lv Perez-Arribas. Chemically modified polymeric sorbents for sample preconcentration. J Chromatogr A 902:3–16, 2000.
32. JT Baker. SPE manual. Phillipsburg, NJ: Mallincrodt Baker, 2000.
33. KC van Horn, ed. Handbook of Sorbent Extraction Technology. Harbor City, CA: Analytichem International, 1985.
34. IST Catalogue of SPE products and services. Mid Glamorgam, UK: International Sorbent Technology, 1997.

35. IN Papadoyannis, AC Zotou, VF Samanidou. Simultaneous reversed-phase gradient HPLC analysis of anthranilic derivatives in anti-inflammatory drugs and samples of biological interest. J Liq Chromatogr 15(11):1923–1945, 1992.
36. IN Papadoyannis, VF Samanidou, GD Panopoulou. The use of bamifylline as internal standard in the reversed phase HPLC analysis of mefenamic acid in pharmaceuticals and small volumes of biological fluids. J Liq Chromatogr 15(17):3065–3086, 1992.
37. IN Papadoyannis, VF Samanidou. The use of theobromine as internal standard in the rapid HPLC analysis of theophylline in small blood serum volume. Anal Letters 26(5):851–866, 1993.
38. IN Papadoyannis, AC Zotou, VF Samanidou, G Theodoridis, F Zougrou. Comparative study of different solid-phase extraction cartridges in the simultaneous RP-HPLC analysis of morphine and codeine in biological fluids. J Liq Chromatogr 16(14):3017–3040, 1993.
39. I Papadoyannis, V Samanidou, H Tsoukali-Papadopoulou, F Epivatianou. Comparison of a RP-HPLC method with the therapeutic drug monitoring system TD_X for the determination of theophylline in blood serum. Anal Letters 26(10):2127–2142, 1993.
40. I Papadoyannis, V Samanidou, A Zotou, G Tsioni. Comparative study of solid-phase extraction cartridges in the simultaneous reversed phase HPLC analysis of bamifylline and its major metabolite AC-119 in biological fluids. J Liq Chromatogr 16(17):3827–3845, 1993.
41. I Papadoyannis, A Zotou, V Samanidou, E Georgarakis. Solid-phase extraction and RP-HPLC analysis of atropine sulphate and scopolamine-*N*-butylbromide in pharmaceutical preparations and biological fluids. Instrumentation Science Technology 22(1):83–103, 1994.
42. I Papadoyannis, A Zotou, V Samanidou. Solid-phase extraction study and RP-HPLC analysis of lamotrigine in human biological fluids and in antiepileptic tablet formulations. J Liq Chromatogr 18(13):2593–2609, 1995.
43. I Papadoyannis, V Samanidou, K Georga. Solid-phase extraction study and photodiode array RP-HPLC analysis of xanthine derivatives in human biological fluids. J Liq Chromatogr 19(16):2559–2578, 1996.
44. IN Papadoyannis, GK Tsioni, VF Samanidou. Simultaneous determination of nine water and fat soluble vitamins after SPE separation and RP-HPLC analysis in pharmaceutical preparations and biological fluids. J Liq Chromatogr 20(19):3203–3231, 1997.
45. I Papadoyannis, V Samanidou, K Georga, E Georgarakis. High pressure liquid chromatographic determination of hydrochlorothiazide (hct) in pharmaceutical preparations and human serum after solid phase extraction. J Liq Chromatogr 21(11):1671–1683, 1998.
46. IN Papadoyannis, VF Samanidou, KA Georga. A rapid and simple high pressure liquid chromatographic method for pharmacokinetic study of ciprofloxacin in human serum. Anal Letters 31(10):1717–1729, 1998.
47. IN Papadoyannis, VF Samanidou, ChC Nitsos. Simultaneous determination

of nitrite and nitrate in drinking water and human serum by high performance anion-exchange chromatography and UV detection. J Liq Chrom Rel Technol 22(13):2023–2041, 1999.
48. KA Georga, VF Samanidou, IN Papadoyannis. Simultaneous determination of methyluric acids in biological fluids by RP-HPLC analysis after solid phase extraction. J Liq Chrom Rel Technol 22(19):2975–2990, 1999.
49. VF Samanidou, HG Gika, IN Papadoyannis. Rapid HPLC analysis of thyroid gland hormones triiodothyronine (T_3) and thyroxine (T_4) in human biological fluids after SPE. J Liq Chrom Rel Technol 23(5):681–692, 2000.
50. KA Georga, VF Samanidou, IN Papadoyannis. Improved micro-method for the HPLC analysis of caffeine and its demethylated metabolites in human biological fluids after SPE. J Liq Chrom Rel Technol 23(10):1523–1537, 2000.
51. VF Samanidou, AH Stafylis, IN Papadoyannis. Direct HPLC method for the routine determination of glycine betaine and its metabolite *N*,*N*-dimethylglycine in pharmacokinetic studies during homocystinuria therapy and in renal disorder monitoring. J Liq Chrom Rel Technol 24(1):1–19, 2001.
52. VF Samanidou, IP Imamidou, IN Papadoyannis. Evaluation of solid phase extraction protocols for isolation of analgesic compounds from biological fluids prior to HPLC determination. J Liq Chrom Rel Technol 25(2):185–204, 2002.
53. D Blevins, M Henry, C O'Donell. Advantages of SPEC disk extractions over conventional packed bed SPE. Ansys Inc. Irvine, CA, USA, 1994.
54. DD Blevins, SK Schultheis. Comparison of extraction disk and packed-bed cartridge technology in SPE. LC-GC 12 (1):12–16, 1994.
55. Solid Phase Extraction Disc products and Accessories Catalog. SPEC. Ansys Inc. Irvine, CA, USA, 1997.
56. DD Blevins, DO Hall. Recent advances in disk format solid-phase extraction. LC-GC 13(5):S16–S21, 1998.
57. MP Henry. Reproducibility Monograph. Irvine, CA, USA: Ansys, 1994.
58. D Blevins. Conversion of packed bed SPE procedures to SPEC microcolumns procedures. Irvine, CA, USA: Ansys, 1994.
59. RE Majors. New approaches to sample preparation. LC-GC Int 8(3):128–133, 1995.
60. CF Poole, AD Gunatillelka, R Sethuraman. Contributions of theory to method development in solid-phase extraction. J Chromatogr A 885:17–39, 2000.
61. JS Fritz, JJ Masso. Miniaturized solid-phase extraction with resin disks. J Chromatogr A 909:79–85, 2000.
62. DD Blevins, MP Henry. Pharmaceutical applications of extraction disk technology. American Laboratory, May 1995.
63. JA Bert Ooms, GJM Van Gils, AR Duinkerken, O Halmingh. Development and validation of protocols for solid-phase extraction coupled to LC and LC-MS. Int Lab 11:18–23, 2000.
64. JC Pearce, RD McDowall, A El Sayed, B Pichon. Automated analysis of

drugs in plasma using liquid-solid extraction and HPLC. Inter Lab November/December 34–40, 1990.
65. DT Rossi, N Zhang. Automating solid-phase extraction: current aspects and future prospects. J Chromatogr A 885:97–113, 2000.
66. S Boos, CH Grimm. On-line Integration of solid phase extraction (SPE) into automated bioanalytical systems. BIOforum International 3:125–128, 1998.
67. UATh Brinkman, R Vreuls. SPE for on-line treatment in capillary GC. LC-GC Int 8(12):694–698, 1995.
68. Byoung-Hyoun Kim, Hoon-Joo Kim, Jong Hoa Ok, Seung-Hun Kang. Analysis of a new herbicide (pyribenzoxim) residue in soil using direct-extract-injection HPLC with column switching. J Liq Chrom Rel Technol 24 (5):669–678, 2001.
69. DA Wells. 96-well plate products for solid-phase extraction. LC-GC Europe 12(11):704–715, 1999.
70. Z Zhang, MJ Yang, J Pawliszyn. Solid phase microextraction: solventless sample preparation for monitoring flavor and fragrance compounds by capillary gas chromatography solid-phase microextraction. Anal Chem 66(17):844A–853A, 1994.
71. C Arthur, D Potter, K Buchholz, S Motlagh, J Pawliszyn. Solid phase microextraction for the direct analysis of water: theory and practice. LC-GC 10(9):656–661, 1992.
72. EHM Koster, GJ de Jong. Multiple solid-phase microextraction. J Chromatogr A 878:27–33, 2000.
73. H Lord, J Pawliszyn. Review: Evolution of solid-phase microextraction technology. J Chromatogr A 885:153–193, 2000.
74. P Popp, C Bauer, L Wennrich. Application of stir bar sorptive extraction in combination with column liquid chromatography for the determination of polycyclic aromatic hydrocarbons in water samples. Anal Chim Acta 436:1–9, 2001.
75. Supelco Application Note 98, Supelco, Bellefonte, PA, USA.
76. Supelco Application Note 99, Supelco, Bellefonte, PA, USA.
77. GA Mills, V Walker. Headspace solid-phase microextraction profiling of volatile compounds in urine: application to metabolic investigations. J Chromatogr B 753(2):259–268, 2001.
78. SPME Applications Guide, Supelco, Bellfonte, PA, 1999.
79. T Kumazawa, H Seno, X Lee, A Ishii, O Suzuki, K Sato. Chromatographia 43(7/8):393, 1996.
80. M Krogh, H Grefslie, KE Rasmussen. Solvent modified SPME for the determination of diazepam in human plasma samples by capillary GC. J Chromatogr B 689:357–364, 1997.
81. HG Ugland, M Krogh, K Rasmussen. Aqueous alkylchloroformate derivatisation and solid-phase microextraction: determination of amphetamines in urine by capillary gas chromatography. J Chromatogr B 701:29–38, 1997.

82. C Grote, J Pawliszyn. SPME for the analysis of human breath. Anal Chem 69:587–596, 1997.
83. H Lord, J Pawliszyn. Method optimization for the analysis of amphetamines in urine by solid-phase microextraction. Anal Chem 69(19):3899–3906, 1997.
84. S-W Myung, S Kim, J-H Park, M Kim, J-C Lee, T-J Kim. Solid-phase microextraction for the determination of pethidine and methadone in human urine using gas-chromatography with nitrogen-phosphorus detection. Analyst 124:1283–1286, 1999.
85. F Guan, K Watanabe, A Ishii, H Seno, T Kumazawa, H Hattori, O Suzuki. Headspace solid-phase microextraction and gas-chromatographic determination of dinitroaniline herbicides in human blood, urine and environmental water. J Chromatogr B 714(2):205–213, 1998.
86. T Kumazawa, K Sato, H Seno, A Ishii, O Suzuki. Extraction of local-anesthetics from human blood by direct immersion solid-phase micro extraction. Chromatographia 43(1/2):59–62, 1996.
87. T Kumazawa, X Lee, M Tsai, H Seno, A Ishii, K Sato. Simple extraction of tricyclic antidepressants in human urine using headspace SPME and selected ion monitoring. Jpn J Forensic Toxicol 13:25–30, 1995.
88. T Watanabe, A Namera, M Yashiki, Y Iwasaki, T Kojima. Simple analysis of local anaesthetics in human blood using headspace solid-phase microextraction and GC/MS electron impact ionization selected ion monitoring. J Chromatogr B 709:225–232, 1998.
89. S Ulrich, J Martens. Solid-phase microextraction with capillary gas-liquid chromatography and nitrogen-phosphorus selective detection for the assay of antidepressant drugs in human plasma. J Chromatogr B 696:217–234, 1997.
90. A Namera, T Watanabe, M Yashiki, Y Iwasaki, T Kojima. Simple analysis of tetracyclic antidepressants in blood using headspace-solid-phase microextraction and GC-MS. J Anal Toxicol 22(5):396–400, 1998.
91. S Li, S Weber. Determination of barbiturates by solid-phase microextraction and capillary electrophoresis. Anal Chem 69(6):1217–1222, 1997.
92. B Hall, M Satterfield-Doerr, A Parikh, J Brodbelt. Determination of cannabinoids in water and human saliva by solid-phase microextraction and quadrupole ion-trap gas-chromatography mass-spectrometry. Anal Chem 70(9):1788–1796, 1998.
93. T Kumazawa, K Watanabe, K Sato, H Seno, A Ishii, O Suzuki. Detection of cocaine in human urine by SPME and capillary GC with NPD. Jpn J Forensic Toxicol 13:207–210, 1995.
94. L Junting, C Peng, O Suzuki. Solid phase microextraction (SPME) of drugs and poisons from biological samples. Forensic Sci Int 97:93–100. 1998.
95. J Liao, C Zeng, S Hjerten, J Pawliszyn. Solid-phase micro extraction of biopolymers, exemplified with adsorption of basic proteins onto a fiber coated with polyacrylic acid. J Microcol Sep 8(1):1–4, 1996.
96. P Okeyo, SM Rentz, NH Snow. Analysis of steroids from human serum by SPME with headspace derivatization and GC/MS. J High Resolut Chromatogr 20(33):171–173, 1997.

97. DA Volmer, JPM Hui. Rapid determination of corticosteroids in urine by combined solid-phase microextraction liquid-chromatography mass spectrometry. Rapid Commun Mass Spectrom 11(17):1926–1934, 1997.
98. N Nagasaka, M Yashiki, T Kojima. Rapid analysis of amphetamines in blood using head space–solid phase microextraction and selected ion monitoring. Forensic Sci Int 78(2):95–102, 1996.
99. ACS Lucas, AM Bermejo, MJ Tabernero, P Fernandez, S Strano-Rossi. Use of solid-phase microextraction (SPME) for the determination of methadone and EDDP in human hair by GC-MS. Forensic Sci Int 107(1-3):225–232, 2000.
100. HL Lord, J Pawliszyn. Recent advances in solid-phase microextraction. LC-GC Int 11(12):776–785, 1998.
101. Supelco Guide to Analysis of Pesticides. Bellefonte, PA, USA.
102. S Ulrich. Solid-phase microextraction in biomedical analysis. J Chromatogr A 902:167–194, 2000.
103. P Sandra, E Baltussen, F David, A Hoffmann. Stir bar sorptive extraction (SBSE) applied to environmental aqueous samples. GERSTEL AppNote 2/2000.
104. BM Cordero, JL Perez Pavon, CG Pinto, ME Fernandez Laespada, R Carabias Martinez, E Rodriquez Gonzalo. Analytical applications of membrane extraction in chromatography and electrophoresis. J Chromatogr A 902:195–204, 2000.
105. NC van de Merbel, JJ Hageman, UA Th Brinkman. Membrane based sample preparation for chromatography. J Chromatogr 634:1–29, 1993.
106. NC van de Merbel. Membrane-based preparation coupled on-line to chromatography or electophoresis. J Chromatogr A 856:55–82, 1999.
107. RE Majors. The use of membranes in sample preparation. LC-GC Int 13(5):364–373, 1995.
108. J-F Jen, Y-Y Tsai, TC Yang. Microdialysis of salicyclic acid from viscous emulsion samples prior to high-performance liquid chromatographic determination. J Chromatogr A 912:39–43, 2001.
109. T Buttler, H Jarskog, L Gorton, G Marko-Varga, L Ramnemark. The use of microdialysis for sampling in column liquid chromatography. Inter Lab 23(9):12–18, 1994.
110. Gilson Catalog. The ASTED System. Middleton, WI, USA, 1994.
111. JA Jönsson, L Mathiasson. Membrane-based techniques for sample enrichment. J Chromatogr A 902:205–225, 2000.
112. N Delaunay-Bertoncini, V Pichon, M-C Hennion. Immunoextraction: a highly selective method for sample preparation. LC/GC Europe 14(3):164–172, 2001.
113. K-S Boos, A Rudolphi. The use of restricted-access media in HPLC, Part I. Classification and Review. LC-GC Int 11(2):84–95, 1998.
114. IH Hagestam, TC Pinkerton. Internal surface reversed-phase silica support prepared with chymotrypsin. J Chromatogr 351(2):239–248, 1986.
115. IH Hagestam, TC Pinkerton. Production of internal surface reversed-phase supports—the hydrolysis of selected substrates from silica using chymotrypsin. J Chromatogr 368(1):77–84, 1986.

116. J Hermansson, A Grahn, I Hermanson. Online LC sample preparation with Biotrap 500 C18. Current Separations 16(2):55–59 1997.
117. C Gunaratna. Bio trap sample extraction column: a new approach for on-line sample preparation in drug metabolism studies. Current Separations 17(3): 83–86, 1998.
118. SA Barker. Matrix solid-phase dispersion. J Chromatogr A 885:115–127, 2000.
119. A Clifford. Introduction to supercritical fluid extraction in analytical science. In: SA Westwood, ed. Supercritical Fluid Extraction and Its Use in Chromatographic Sample Preparation. Glasgow, UK: Blackie Academic and Professional, 1993, pp. 1–38.
120. DR Gere, EM Derrico. SFE theory to practice—first principles and method development, Part I. LC-GC Int 7(6)325–330, 1994.
121. RM Smith. Supercritical fluids in separation science—the dreams, the reality and the future. J Chromatogr A 856:83–115, 1999.
122. SB Hawthorne. Methodology for off-line SFE. In: SA Westwood, ed. Supercritical Fluid Extraction and Its Use in Chromatographic Sample Preparation. Blackie Academic and Professional, Glasgow, UK: 1993, pp. 39–64.
123. SB Hawthorne. Coupled (on-line) SFE-GC. In: SA Westwood, ed. Supercritical Fluid Extraction and Its Use in Chromatographic Sample Preparation. Blackie Academic and Professional, Glasgow, UK: 1993, pp. 65–86.
124. BE Richter. The extraction of analytes from solid samples using accelerated solvent extraction. LC-GC 17(6):S22–S28, 1999.
125. BE Richter, BA Jones, JL Ezzell, NL Porter, N Avdalovic, C Pohl. Accelerated solvent extraction—a technique for sample preparation. Anal Chem 68(6):1033–1039, 1996.
126. QH Jin, F Liang, HQ Zhang, LW Zhao, YF Huan, DQ Song. Application of microwave techniques in analytical chemistry. TRAC—Trends in Anal Chem 18(7):479–484, 1999.
127. ZK Guo, QH Jin, GQ Fan, YP Duan, C Qin, MJ Wen. Microwave-assisted extraction of effective constituents from a Chinese herbal medicine radix puerariae. Anal Chim Acta 436(1):41–47, 2001.
128. LG Croteau, MH Akhtar, JMR Belanger, JRJ Pare. High-performance liquid chromatography determination following microwave-assisted extraction of 3-nitro-4-hydroxyphenylarsonic acid from swine liver, kidney, and muscle. J Liq Chromatogr 17(13):2971–2981, 1994.
129. M Letellier, H Budzinski. Microwave-assisted extraction of organic compounds. Analusis 27(3):259–271, 1999.
130. M Ericsson, A Colmsjo. Dynamic microwave-assisted extraction. J Chromatogr A 877:141–151, 2000.
131. M Lettelier, H Budzinski, P Garrigner. Focused microwave assisted extraction of PAH's. LC/GC Int 12(4):222–225, 1999.
132. TP Wampler. Thermal desorption for GC sample preparation. LC GC Int 11(10):653–658, 1998.

133. A Cole, E Woolfenden. Gas extraction techniques fore sample preaparation in GC. LC-GC Int 5(3):8–15, 1992.
134. SA Barker. Matrix solid phase dispersion. LC-GC Int 11(11):719–724, 1998.
135. AJ Packham. Complex sample analysis by column-switching high performance liquid chromatography. LC-GC Int 4(11):26–29, 1991.
136. WS Letter. Automated column selection and switching system for HPLC. LC-GC Int 10(12):799–801, 1997.
137. Chrompack General Catalog. Chrompack International B.V. Middelburg, The Netherlands, 1988.
138. AM Di Pietra, R Gatti, V Andrisano, V Cavrini. Application of high-performance liquid chromatography with diode-array detection and on-line post-column photochemical derivatization to the determination of analgesics. J Chromatogr A 729:355–361, 1996.
139. A Blad. Review: Chiral HPLC stationary phases. In: Advanced Euro HPLC Training Course and Workshop, Patras, Greece, 1992.

2

Separation Science in Therapeutic Drug Monitoring

Zak K. Shihabi
Wake Forest University School of Medicine, Winston-Salem, North Carolina, U.S.A.

1 INTRODUCTION

Interindividual variability in drug response is a major clinical problem. When a drug is administered to a large group of patients, some respond favorably, others do not, while some develop toxic symptoms. This variability is caused by several factors, such as differences in absorption and metabolism. A few decades ago, it was recognized that there was no correlation between the oral dose and the blood level of a drug [1]. More recently, it was recognized that the pharmacogenetic differences in metabolism are often responsible for variable serum concentrations and drug response. The goal of therapeutic drug monitoring (TDM) is to improve patient outcomes by adjusting the drug dose to obtain a certain serum concentration and also to predict those individuals who might develop an adverse drug reaction or toxicity.

TDM evolved within the last four decades as a specialty area in the clinical laboratory. As long as new drugs are introduced, TDM will keep on flourishing. Presently, in our laboratory, we analyze routinely over 30 different drugs for TDM. Furthermore, as we enter the new era of pharmacogenetics, predicting drug response, e.g., patients who are fast metabolizers or the nonrespondents, will become part of TDM in the laboratory. As the patient population ages, medication use increases. The unpredictable interaction of these different drugs becomes important.

The field of TDM encompasses the knowledge of several major areas, the analytical, the pharmacological, and the clinical (interpretive aspects) in order to ensure useful information about drug dosing and utilization. In the analytical aspects, several major advances have occurred in the last decade, e.g., the introduction of capillary electrophoresis (CE), capillary electrochromatography (CEC), and the microchip techniques. Pharmacogenetics is an emerging area in TDM. All these areas will be discussed here in this chapter with emphasis on the role of the separation science in this progress.

2 HISTORICAL

As early as 1927 it was observed that blood level of bromide could be used to differentiate bromide psychosis from other causes. However, TDM can be considered to have started in 1960s. It was limited at that time to a few drugs, such as phenobarbital, phenytoin, and an occasional salicylate performed by spectrophotometry. This method was not specific enough to measure the numerous drugs present at that time. A breakthrough came with advent of gas chromatography (GC). The resolving power of GC avoided the interfering substances and allowed a larger menu of drugs such as theophylline, carbamazepine, and primidone to be analyzed. Although GC could monitor a larger number of drugs, it required tedious sample extraction and skill. Basic drugs were difficult to analyze on the packed column, used at that time, owing to their binding onto the packing material. Derivatization procedures often were required for analysis. The danger of the flame detector and the need for a highly skilled operator were problematic for the routine clinical chemistry laboratory, especially in the face of emergency work requirements.

In the late 1960s and early 1970s, the work of several researchers led to the development of high-performance liquid chromatography (HPLC). This method found wide acceptance in clinical labs because it was able to analyze a much wider variety of compounds than GC. Also during this period, several researchers demonstrated that clinical improvements among patients were better related to the serum level than to the dose of the drug [1]. For example, the number of seizures decreased when the blood level of antiepileptic drug was brought within a therapeutic window [2,3]. Others demonstrated that aminoglycoside antibiotic monitoring decreased both morbidity and length of stay at the hospital. This spurred more interest in TDM. HPLC became the technique of choice for TDM and further increased the menu of analyzed drugs. Several books [4–7] and reviews have been published about TDM and drug analysis by HPLC. The main reason for its popularity was the simplicity of sample preparation (in many cases, protein precipitation with acetonitrile) and the relative ease of instrument operation as compared to GC. At the present time, some drugs like cyclosporine and tacrolimus (FK 506) cannot be prescribed for patient treatment without regular monitoring of their serum level.

3 PRINCIPLES OF TDM: PHARMACOKINETIC AND CLINICAL

The serum level of a drug in a patient is not constant. It changes with many factors, especially with time. Thus an accurate and very precise value for a serum drug level by itself without the clinical information is not meaningful. For proper performance and interpretation of drug level testing, two distinct but important areas are addressed here: the pharmacokinetic/clinical and the analytical aspects.

3.1 Theoretical

Most drugs are administered orally. The extent and rate of drug absorption depends on several parameters of the drug such as its pKa and solubility. Often drug absorption follows first-order kinetics. The fraction of the drug absorbed is referred to as bioavailability and equals the area under the curve (AUC) of concentration vs. the time of the drug orally compared to that by injection. As soon as drug is administered, it is distributed and stored in the body. The apparent "volume of distribution" is a proportionality constant of amount of drug in the body compared to that in the plasma (L/kg). Mathematically,

$$\text{Volume of distribution} = \text{amount of drug in the body/plasma concentration.}$$

After an oral dose, the drug reaches a maximum concentration in blood in several hours (peak level) and then starts to drop and disappear. The peak level with an oral dose is much lower than that by an intravenous injection. Many drugs disappear from the serum following a one-compartment model, in which the concentration drops in a log-linear fashion over time. However, some disappear from the serum following a two-compartment model with an early phase called alpha (distribution), which represents drug distribution in the body, and a later phase called beta (elimination), which represents drug elimination. The elimination is characteristic for each drug and is represented by the half-life ($t_{1/2}$) which can be measured as the time for a particular drug plasma concentration to drop fifty percent. Mathematically,

$$t_{½} = 0.69/K$$

where K is the overall elimination rate constant. Also, K can be stated as

$$K = \text{total (overall) body clearance/volume of distribution}$$

Thus $t_{1/2}$ can be rearranged,

$$t_{1/2} = 0.69 \times \text{volume of distribution/clearance}$$

Most drugs are given not as a single dose but in chronic maintenance dosing. Clearance makes the drug maintenance dose necessary to achieve a therapeutic drug concentration:

$$\text{Dose} = \text{clearance} \times \text{area under the curve for the drug}$$

The therapeutic range for a drug is a range of concentrations within which the probability of desired clinical response is high and the probability of toxicity is low. This indicates that some patients deviate by responding well to the drug outside this range. TDM is useful for drugs with a narrow therapeutic range (upper/lower range = ~2) [8], patients who are not responding well, and individuals with unpredictable pharmacokinetics.

3.2 Practical Aspects

For practical aspects few of the previous parameters are important, especially for the clinical or analytical chemist. The first one is the half-life, which is important for the laboratory to determine when to draw the blood level for TDM and how to interpret the results. TDM relies on measuring the drug concentration at steady state, which usually is achieved after 5–7 half-lives after the drug is started. Also, it takes the same amount of time for plasma concentration to drop to zero when a patient stops drug intake. Most drugs, with the exception of the antiepileptic and tricyclics, have half-lives of a few hours (Table 1). Blood is drawn for TDM at steady state and before the next dose for a trough level measurement. Therapeutic levels in general are based on this level with a few based on the peak concentration [9–11] (Table 1). The half-life is also important in the drug dosing interval.

The second parameter is volume of distribution, which is important for the physician or the pharmacologist in order to determine the loading dose to achieve a certain plasma concentration. It is also useful to calculate the extra amount of drug needed to raise the plasma concentration from one level to another; while the clearance is important for determining the maintenance dose to keep the level at steady state [9–11]. The volume of distribution can be large, indicating that the drug is mainly outside the blood, and it can be small, indicating that it is present mainly in the blood.

TABLE 1 Therapeutic Range and Half-Life (*h*) of Common Drugs

Drug	Therap. range (μg/mL)	$t_{1/2}$ (*h*)
Bromide	800–1500	~300
Carbamazepine	4–12	15 ± 5
Ethosuximide	40–100	45 ± 8
Phenobarbital	10–40	99 + 18
Phenytoin	10–20	6–24
Primidone	5–12	15 ± 4
Valproic acid	30–120	15
Keppra	–*	7
Lamotrigine	1–15*	30
Felbamate	50–100*	21
Topromate	2–20*	22
Gabapentin	2–20*	6
Amiodorane	1–2.5	25 ± 12 days
Digoxin	0.5–2 ng/mL	40 + 13
Disopyramide	2–5	6
Lidocaine	2–5	2
Procainamide	3–10	3
Quinidine	2–5	6
Tocainide	3–15	12
Theophylline	8–20	10
Caffeine	3–20	5 ± 2
Amikacin	10–25	2
Chloramphenicol	10–20	4 ± 2
Gentamicin (peak)	5–10	2.5
(trough)	<2	
Kenamycin (peak)	20–35	2
Sulfanamides	75–125	7
Tobramycin (peak)	5–8	2
Vancomycin (peak)	20–40	6
(trough)	5–10	
Amitriptyline	50–220 ng/mL	21 + 5
Desipramine	50–300 ng/mL	22 ± 5
Doxepin	30–150 ng/mL	17
Fluoxetine	90–300 ng/mL	53 ± 41
Imipramine	100–250 ng/mL	13
Nortryptyline	50–150 ng/mL	31
Trazodone	0.8–1.6*	6
Lithium	0.8–1.2 mmol/L	22
Cyclosporine	100–400 ng/mL	6 ± 2
FK-506	5–20 ng/mL	15 ± 7
Warfarin	1–10	37 ± 15

*Not well established.

The AUC for the drug correlates, in general, better than a single measurement with the clinical effect. However, this requires several measurements, which become costly and impractical. Several shortened alternatives have been proposed [12]. It is important to appreciate that many factors can influence a drug level, an interpretation, and a patient clinical response [8], e.g.,

(a) Food intake and type affects absorption and consequently drug level. For example, grapefruit juice contains compounds that inhibit the metabolism of many drugs in the gastrointestinal tract, leading to higher serum level.

(b) Metabolism and genetics: Most drugs are metabolized by the cytochrome P450 system, which is polymorphic and largely present in the liver.

(c) Protein binding: Some drugs circulate in serum free, such as Li^+, while others such as amitriptyline are almost completely bound to proteins (e.g., albumin or α-1-acid glycoprotein). The free fraction of the drug crosses the cell membranes and binds to the receptor or the active site. Changes in protein binding due to loss or increase in serum protein affects the interpretation of serum concentration of the drug.

(d) Health: Obviously sick patients with renal or liver disease would have altered levels owing to changes in protein binding or clearance.

(e) Age: Pediatric patients have usually increased clearance compared to that in adults, which results in shorter half-life and increased drug dose.

4 ANALYTICAL ASPECTS OF TDM

4.1 General Aspects

TDM, like other analyses, can be performed by a variety of techniques, e.g., immunoassay, spectrophotometry, chromatography, and electrophoresis. From the analytical aspects of TDM, there are two different needs in the laboratory. One need is to perform a rapid and efficient routine analysis including the emergency work; and the other need is to perform research. In routine work, the main considerations are simplicity and reliability of the analysis, while in research the main considerations are collection of new information and the high degree of accuracy. University hospitals, which have need for both aspects face a challenge in choosing a suitable method.

4.2 Immunoassays vs. Separation Methods

Immunoassays are very simple to perform, mainly because of automation and the absence of sample pretreatment steps, but they are subject to cross-reactivity from other compounds. These methods are well suited for routine work. At present, they are very popular in most routine clinical labs. Immunoassays usually are introduced by the commercial companies for those tests with high volume. On the other hand, chromatographic and electrophoretic methods for TDM usually are developed in the laboratories and require skill and time. While the chromatographic methods are difficult to perform, they offer more accurate data and several other advantages, as will be discussed later. Thus chromatographic methods are more suited for research and for university medical centers.

4.3 Sample Preparation

In all chromatographic and electrophoretic methods, proteins present in biological samples such as serum ruin the column or the capillary because of their adsorption. Under special conditions a small amount of protein can be tolerated. In HPLC, either a precolumn with frequent changes or special columns such as the "internal surface reversed-phase" (ISRP) type, which do not retain protein, are used. Acetonitrile deproteinization is a very attractive method for analysis of most drugs by HPLC and also by CE [13]. Acetonitrile treatment does not give very clean chromatograms, but it is a simple procedure to eliminate proteins. We use this method in most of our assays; but care must be taken to ensure that the column and the analytical conditions are properly chosen to separate all the possible interferences, e.g., Figs. 1 and 2. In capillary electrophoresis (CE), acetonitrile induces sample concentration on the capillary "stacking" [14,15]. In micellar electrokinetic chromatography (MEKC), a small amount of serum can be injected directly on the capillary [16] (Fig. 1). As the drug concentration decreases, sample extraction by solvent or solid-phase extraction followed by concentration becomes more important [16,17]. Methods for sample cleanup and concentration are becoming a very specialized subject [16].

4.4 Precision

From a practical point of view, the between-run precision for drug analysis is similar for all the previous methods with CV of 3–10%. Based on our experience, the CV in chromatographic and electrophoretic techniques depends greatly on two factors: the complexity of sample preparation (extraction) steps and the condition of the column/capillary. The fewer the steps used for sample preparation, the better is the precision. Columns (in HPLC) and capillary walls (in CE) tend to adsorb proteins and cationic compounds. Thus the characteristics of the column or the capillary changes with use. A thorough cleaning and preventive maintenance is necessary for good precision. On the other hand, the precision of the immunoassays depends greatly on the degree and type of automation.

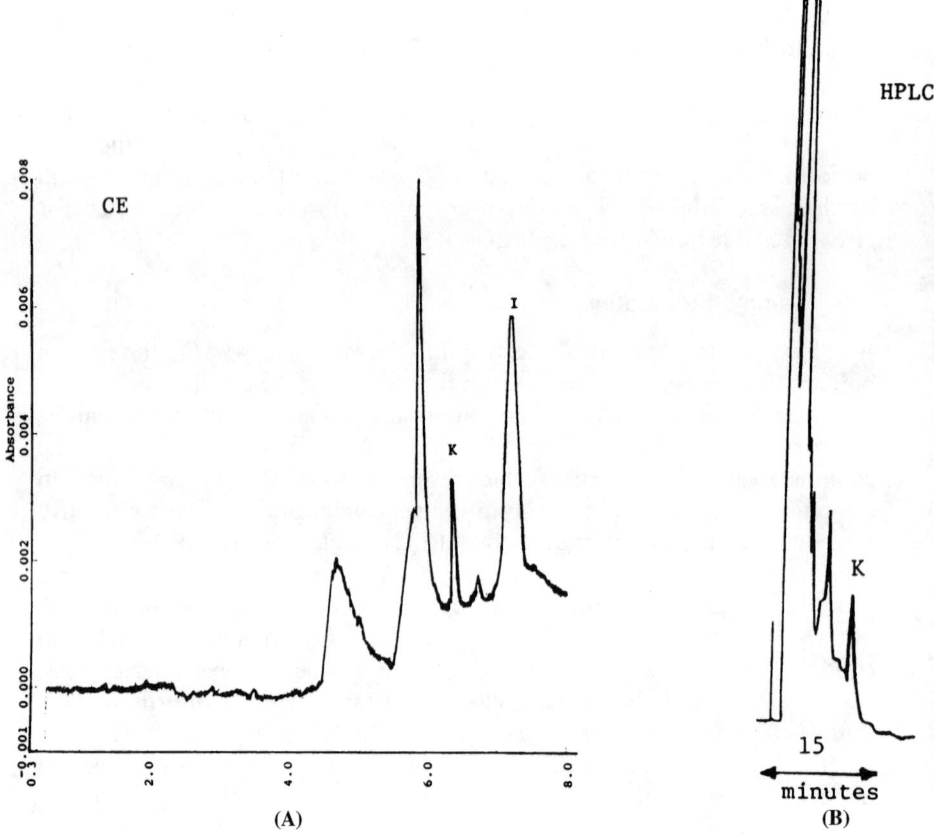

FIGURE 1 Analysis of a serum sample from a patient on levetiracetam (keppra) (25 mg/L, K = keppra, I = iohexol) by (A) CE (MEKC) and (B) HPLC (reversed-phase). CE conditions: serum 100 μL was mixed with 100 μL of the internal standard (200 μL/L of iohexol in water) and injected directly for 15 s on the capillary (50 μm × 30 cm). The sample was electrophoresed for 8 min at 8.8 kV, 214 nm in a buffer composed of 40 mg SDS, and 300 μL methanol/mL of boric acid 140 mmol/L, pH 8.6 (Beckman CE). HPLC conditions: serum 200 μL is mixed with 200 μL of acetonitrile, centrifuged at 14,000 × g for 10 s, and the supernatant is injected on a C_{18} column, Varian Microsorb-MV, 5 μm (250 mm × 4.6 mm). The drug is eluted with 7.5% acetonitrile containing 1 ml/L of phosphoric acid at flow of 1.0 mL/min and detection at 214 nm.

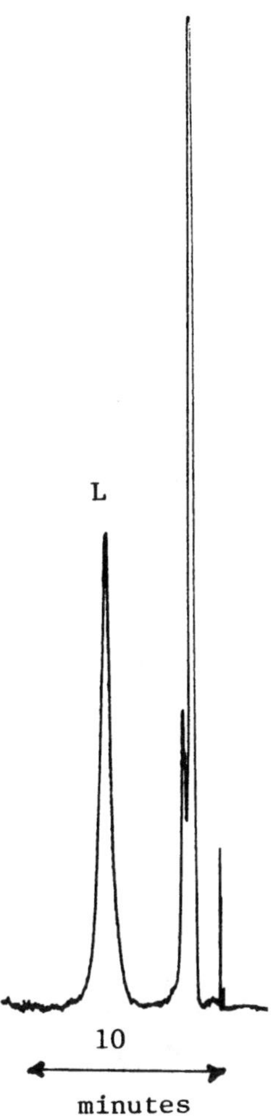

FIGURE 2 Analysis of serum Lamotrigine by HPLC (L, 14.1 mg/L). Conditions of analysis: serum 100 μL is mixed with 200 μL of acetonitrile, centrifuged at 14,000 × g for 10 s, and the supernatant is injected on C_{18} column, Varian Microsorb-MV, 5 μm (150 mm × 4.6 mm). The drug is eluted with 17% acetonitrile in 17 mM potassium phosphate buffer, pH 6.2 at flow rate of 1.0 mL/min and with detection at 313 nm.

4.5 Accuracy

The accuracy of the chromatographic and electrophoretic method in general is better than that for the immunoassays, especially when internal standards are included in the assay. Cross-reactivity, especially for the metabolites, is a problem in immunoassays, which differ depending on the source of the antibody [18]. Chromatographic techniques tend to give better linearity of quantification than the immunoassays.

4.6 Choice of Methods

Immunoassays remain the first choice for routine assays whenever they are available. However, for the new drugs, the separation methods become the main choice. For compounds without a strong chromophore and with good volatility, GC is the preferred instrument. For compounds with a strong chromophore (the majority of drugs), HPLC or CE is preferred. Ionized, highly polar, and large molecular weight compounds are better analyzed by CE, while nonpolar compounds and those present at very low concentration are better analyzed by HPLC. A comparison of the analysis of the same drug by HPLC and CE is presented in Fig. 1 for the new epileptic drug levetiracetam (keppra). In general, CE offers simplicity of analysis, speed, low cost, but the detection limit is not as good as that of HPLC. In practice, some tests tend to be easier to perform by CE, while others are easier by HPLC.

4.7 Routine vs. Research in Chromatographic Analysis

For routine work, it is better to dedicate a simple single instrument for each (or a few) test in order to avoid frequent changes of columns, capillaries, wavelengths, or other parameters of the instruments. In HPLC, isocratic analysis is better suited for quantification and routine work than gradient. In CE, it is better to choose generic methods, i.e., common buffers and capillaries suited for analysis of several drugs to minimize change of instrument settings. Method development in our hands is easier by CE than that by HPLC.

Numerous methods have been described, such as CE [19–21], HPLC [4–7], and GC [22]. GC, at the present time, is limited to few methods such as gabapentin [23,24], topiramate [25], and for TDM and for analysis of drugs of abuse [26–28]. When the GC is combined with mass spectra (MS) it can be considered as the gold standard for confirmation of drugs of abuse. Citing all the described separation methods for TDM will be overwhelming without providing much useful information. However, we will give some examples emphasizing the practical aspects. de Silva [29] has described general strategies for the analysis of drugs by chromatographic techniques.

5 SEPARATION TECHNIQUES IN TDM

Chromatographic and electrophoretic techniques offer several major advantages for TDM, which are not very obvious or well appreciated. These features will be discussed later. Here we describe briefly the types and principles of the chromatographic and electrophoretic techniques most often used in TDM.

5.1 Chromatography

In chromatographic techniques a solute (analyte) is preferentially distributed or partitioned between two phases: a stationary (fixed) and a mobile (moving) phase. Basically there are four different modes of separation in chromatography: adsorption, partition, ion exchange, and size exclusion. However, in many instances more than one mechanism occurs in the same analysis. Some techniques such as thin layer chromatography, paper chromatography, supercritical, and affinity chromatography are seldom used in TDM. Here are the common chromatographic techniques used in TDM:

5.1.1 High-Performance Liquid Chromatography (HPLC)

(a) Reversed phase HPLC. Reversed phase HPLC is the most common type of HPLC used in TDM. Uncoated silica particles (normal phase) yield unsymmetrical peak shapes and poor separation. However, when silica particles are coated with a nonpolar carbon chain, e.g., octadecylsilane group (C_{18}), their adsorption characteristics are altered. The C_{18} gives the particle a nonpolar characteristic, which makes it more suitable for initial elution of the compounds with simple aqueous buffers rather than the expensive organic solvents. A partition system in which the stationary phase is less polar than the mobile phase can be considered as reversed phase. Nonpolar compounds interact more strongly with this stationary phase and thus are retained longer. To elute such compounds, small amounts of organic solvents such as methanol or acetonitrile are added gradually to the aqueous buffers. This type of separation is termed reversed phase. Silica particles coated with cyano groups have an intermediate polarity between the C_{18}-coated and the uncoated silica.

(b) Ion exchange. The column (stationary phase) in this case is packed with particles containing fixed ionic (positive or negative) sites to separate either acidic or basic compounds (ion exchange chromatography). The interaction in the ion exchange mode is based on electrostatic forces, i.e., interaction of opposite charges.

(c) Paired ion chromatography. Ionic compounds are not well retained on the reversed phase column. In order to retain such compounds on the reversed phase column, another ion of an opposite charge is added to the mobile phase so that the two compounds can be eluted as a neutral or ion pair.

(d) Size exclusion (gel permeation). Here the separation is based on molecular size. This type is not used too often in TDM.

5.1.2 Gas Chromatography (GC)

Compounds with low boiling points (volatile, or small in molecular weight) and those that can be rendered volatile after chemical derivatization are suited for analysis by this technique. In the early days, columns were made of glass, 1–2 m long, 2–5 mm i.d. packed with mostly 130–100 μm silica particles coated with high boiling point oils. Drug adsorption on the packed column gives nonlinear data with a low plate number. However, in the last two decades these columns have been replaced by the open tubular or wall-coated (nonpacked) capillary columns, which are much longer (10–50 m) and narrower in diameter (50–300 μm i.d.). The length of these columns allows for much better separation, which is well suited for complex samples.

GC in the 1960s has popularized drug analysis. However, it required sample extraction, concentration, and derivatization steps that were time consuming and not well suited for hospital emergency work. At the present time, GC plays only a very small role in routine TDM.

5.2 Capillary Electrophoresis (CE)

Under the influence of a voltage, ionic compounds move, at different velocities (based on charge/mass), toward the electrode with the opposite charge. As a result of this movement, analytes can be separated from each other. In CE, the separation occurs in capillaries 20–100 μm in diameter and 20–70 cm in length with on-column detection. This technique is relatively new, having emerged in the last decade. CE shares with HPLC many similarities such as separation, on-column detection, and quantification of a wide variety of drugs based not only on charge but also on size, hydrophobicity, and chirality. The boundaries between the two become blurred. CE offers several other advantages for TDM: a high plate number, rapid analysis time, low cost per test, and full automation (Fig. 1). However, CE has several major challenges, such as poor sensitivity of absorbency detection and problems with sample matrix and adsorption to the capillary. A major strength of CE is its ability to modify the selectivity of the separation simply by adding additives to the electrophoresis buffer to alter the velocity of some analytes. Thus different kinds (modes) of separations can be obtained based on these additives.

5.2.1 Modes of CE Separation

(a) Capillary zone electrophoresis (CZE), free solution. The capillary is filled with a buffer to provide separation based on charge/mass ratio differences in velocity without major additives. The sample analytes migrate in the capillary and separate from each other as discrete zones (separated by the buffer). We find this mode is well suited for TDM, which enables an easy stacking, i.e., concentration on the capillary.

(b) Micellar electrokinetic chromatography (MEKC). This method can be considered as a hybrid between chromatography and electrophoresis. A surfactant such as sodium dodecyl sulfate in the form of micelles is added to the buffer in order to partition the different compounds between the buffer and the micelle. The interior region of the micelle is hydrophobic while the outside is hydrophilic (ionic). Neutral and hydrophobic compounds spend more time bound to and migrating with the micelle. Since most of the micelles have a charge, they migrate together with the carried analytes under the influence of the electric field. This mode of CE is suitable for the separation and analysis of neutral drugs [30]. Furthermore, small amounts of serum can be injected directly on the capillary, which makes sample preparation for TDM simple, provided the drug is present in a high concentration [15].

(c) Chiral separation. Here the buffer contains special chiral additives such as cyclodextrins or some macrocyclic antibiotics that can preferentially interact with one isomer and affect its migration. This mode of CE is used often for the separation of optical isomers, especially in the pharmaceutical industry. It is much cheaper and much faster when compared to chiral separations by HPLC.

(d) Size separation. This type of analysis is suitable for separating molecules that differ in size, such as DNA fragments, based on the addition of large molecular weight polymers. It is not used for TDM.

(e) Isoelectric focusing (IEF). This method is not used for TDM but for focusing zwitterionic compounds such as proteins and peptides.

5.2.2 Concentration on the Capillary

Owing to the short light path of the capillaries, the detection limit in CE, based on concentration, is far less than that of HPLC and is not sufficient for many practical applications. Sample concentration on the capillary becomes critical. Fortunately, it can be accomplished easily in CE. Several methods, based on different electrophoretic maneuvers, can concentrate the sample (stack) easily on the capillary before the separation step. These methods incorporate different types of discontinuous buffers as the means for invoking different velocities to the same analyte molecules to produce sharpening of the analyte band [14–17].

6 ADVANCES IN ANALYTICAL TECHNIQUES

The early GC instrumentation in the 1960s and 1970s was difficult and temperamental. This changed over the years. All the separation techniques have witnessed several important changes and advances over the last few decades. A great part of this improvement is due to better electronics and enhanced computerization. The majority of these instruments became fully automated and computer controlled.

In the GC area, capillary columns have completely replaced packed columns. The capillary column offers a very high plate number due to its length. It is suitable for separating complex mixtures. Portable as well as fast GC instruments, which give very rapid separation in a few minutes owing to a more efficient temperature gradient, are commercially available. A wide choice of detectors, e.g., photoionization and thermionic detectors, are available [31]. Mass spectra (MS) such as GC detectors are now widely available with improved stability, wide choices, and enhanced sensitivity.

In the HPLC area, microbore and capillary columns are becoming commonly used. These columns employ reduced flow rates with smaller sample size but give a much better detection signal. They are useful when the sample size is limited, as in the proteomics area, and also for connection to the MS. A wide selection of columns, especially those that are well deactivated (endcapped), the mixed beds, monolithic, small particle, and those that operate at wider pH range, are now available. MS is now very common as a special detector for HPLC.

CE is a new technique that emerged in the last decade for the separation of charged as well as noncharged molecules. The instruments are getting more sophisticated so that they can run about a hundred samples simultaneously (using multicapillaries) with more sensitive and diverse detectors.

All types of detectors for all the chromatographic techniques have improved and have enhanced sensitivity. However, the combination of the chromatograph with MS is getting more popular; MS is becoming much more versatile and sensitive and is able to detect higher molecular weights. In addition to this, a few new hybrid techniques have emerged, as is discussed below.

6.1 Emerging Techniques

6.1.1 Capillary Electrochromatography (CEC)

The same CE instrument is usually used for CEC; however the capillary (~75 μm i.d. × 30 cm) is filled with packing materials (stationary phase) similar to those of the HPLC column. The applied voltage moves both the mobile phase and the sample through the column by electroosmotic flow instead of a mechanical pump [32–35]. The plate number here is about 3–10 times better than that of HPLC owing to the band moving in a flat rather than a parabolic profile. Some practical problems arise from air bubble formation and from clogging of the frits at the column ends. However, the introduction of monolithic columns in this technique avoids many of these problems.

6.1.2 Chip Technology

Smaller capillaries and smaller columns have the advantage of yielding faster separations, in terms of seconds, with a decrease in reagent volume. A new trend, borrowed from the electronic industry, is to perform separations and chemical analysis in small grooves etched on a chip, a piece of glass or plastic where numerous analyses are made simultaneously on a very small surface [36–38]. Two distinct types of chips are used: binding and separation. The binding chip contains on its surface certain molecules that recognize and bind specific analytes such as DNA [39,40] or proteins [41,42]. For example, the DNA binding chip (DNA microarray chip) contains on its surface hundreds of certain oligonucleotides that bind and analyze hundreds of complementary genes. A chip for performing several immunoassay tests simultaneously for drug analysis is being introduced commercially (Randox Laboratories, Antrin, UK).

The separation chip's surface is etched to contain microgrooves (typically about 5–20 μm deep and about 5 cm long) in place of capillaries or columns. Reagents are placed in small wells with connected electrodes, and voltage is applied at selected areas to cause the sample or the reagents to move by electroosmotic flow, and also to mix with other fluid streams. The separation can be accomplished based on electrophoresis or column chromatography. Many channels can be etched so several reactions can be carried out simultaneously in the same chip. Several chemical steps can be performed on the chip such as reagent movement, mixing, separation, and detection, so these systems are sometimes also called Lab-on-a-Chip or micrototal analysis systems (μTAS). Different chips are available now commercially to perform different assays.

6.2 Combination of CE and Immunoassay

Many drugs, such as tacrolimus and digoxin, without extraction and subsequent concentration, remain far below the detection limits of CE and HPLC. A combination of immunoassay, laser induced fluorescence (LIF), and CE has been described for simultaneous determination of several drugs [43,44]. In such a system, prepared fluorescent-labeled drug conjugates were mixed with antibody of known specificity and the unknown samples. After the reaction is complete, the free drug (including the fluorescent-labeled drug) is separated from the bound by CE. Cortisol, an endogenous substance and a drug, has also been quantitated using a similar approach [45,46]. Steinmann and Thormann [47] described an assay using MEKC, LIF, and the commercial fluorescence polarization assay reagents to separate a variety of drugs. The free and bound tracers for the different drugs were separated in phosphate-borate buffer containing SDS. A similar method was also described for the analysis of digoxin using the commercial reagent from an enzyme immunoassay kit [48]. Chiem and Harrison [49] described a microchip capillary electrophoresis method to separate the reaction products of a reaction of an antibody and theophylline within approximately 40 s.

7 CONTRIBUTIONS OF THE SEPARATION METHODS TO TDM

The separation methods offer the TDM several special features, which cannot be achieved easily by immunoassays or colorimetric methods. Some of these features are general. For example, because the separation techniques resolve many closely related compounds, occasionally unexpected information on other compounds in the chromatogram can be discovered. The chromatographic techniques, especially HPLC and TLC, allow for the easy isolation of small amounts of pure compounds for further study, especially the metabolites. Here are more specific features:

7.1 Analysis of Multiple Drugs in the Same Run

Chromatographic and electrophoretic methods can analyze several closely related compounds in the same run. Several procedures have been described for analysis of several closely related drugs in the same run by HPLC, GC, and CE.

7.2 Analysis of Drug Metabolites

The metabolites of a drug can be as pharmacologically active as the parent drug. Many examples of the separation of the parent and the metabolites of drugs have been described using the different separation methods.

7.3 Chiral Separation

Although isomers have very close chemical structures, they can exhibit different biological effects or can be metabolized differently [50]. It is estimated that about 40% of drugs have at least a chiral center. In the past years, most drugs were sold as racemic, but this practice has changed in recent years. HPLC was the main method for chiral separation; recently, however, CE and CEC are becoming more often used for this purpose.

For HPLC, several types of columns based on polysaccharides, cyclodextrins, macrocyclic antibiotics, or proteins are available for the chiral separations [51–54]. These columns can resolve and quantify minor chiral components as low as 0.1%. In CE, because the amount of reagent is very small, the chiral additives, which are expensive, are usually added to the buffer, and the separation is performed by MEKC [55]. In CEC, the chiral separations can be performed by columns similar to those of HPLC or by adding the chiral selector to the mobile phase [56].

Most of the described chiral separations are performed on drugs to check for purity in pharmaceutical preparations. However, a few studies have been performed on separations from biological samples. For example, Srinivasan and Bartlett [57] described a stereoselective method for serum phenobarbital using cyclodextrin and solid phase extraction. Ohara et al. [58] also described a method using CE to determine the enantioselective determination of the basic drug verapamil that was not bound to serum proteins. Nishi [59] reviewed the separation of enantiomers of drugs by electrokinetic chromatography using chiral micelles and proteins. Fanali [60] and Bojarski and Aboul-Enein [61] reviewed the identification of chiral drugs by CE, including those present in biological fluids.

Isomers also can bind to various serum proteins in different ways. Thus several proteins, including those found in human serum, such as transferrin, have been utilized in CE for a chiral separation of a variety of drugs [62]. Giacomini et al. [63] have shown that the coadministration of racemic disopyramide affected the clearance of the d-isomer owing to the more avid binding of its isomer to serum proteins. Also the S form of verapamil has less binding to serum proteins, increasing its plasma clearance to twice that of the R form [64].

The HPLC columns can resolve and quantify a minor component as low as 0.1%. D'Hulst and Verbeke [65] and Altria et al. [66] showed that a limit of detection of $<1\%$ and $\sim 0.1\%$, respectively, can be obtained for the minor enantiomer levels in CE. CE is rapid, excellent, and inexpensive for chiral separations, but the reproducibility falls short of that of HPLC [64,66,67].

7.4 pKa

The pKa of a compound or drug can also be measured using the chromatographic and the electrophoretic methods, taking advantage of the analytical speed and the small volume of sample. In CE, the pKa can be determined by measuring its mobility as a function of pH [68] and by HPLC based on capacity factor of both the dissociated and the undissociated forms of the drug [69]. Several formulas have been described for determining the pKa by HPLC [70–72]. Schmutz and Thormann [73] determined how the physical and chemical properties of 25 drugs would affect their analysis by MEKC. They found that compounds, which did not bind tightly to proteins in addition to those with a low pKa, dissociated easily from the bound proteins and migrated as sharp peaks.

7.5 Predicting Drug–Membrane Interactions

Hydrophobicity is assumed to be the force behind passive drug transport across membranes and binding to the receptor. Thus hydrophobicity affects many of the drug pharmacokinetics such as absorption, volume of distribution, and excretion. Hydrophobicity can be measured by the traditional partition coefficient of octanol-water and also by the capacity factor of the chromatographic data. Several studies have used HPLC and CE to predict interactions of the drugs with biological membranes, to investigate and predict bioavailability distribution, or the blood–brain barrier. Many methods used and compared *n*-octanol/water, immobilized artificial membrane (IAM), or liposomes to characterize these interactions especially by HPLC. Chromatographic surfaces containing phosphocholine or interfacial polar groups such as OH and OCH3 can model drug–membrane interactions better than surfaces lacking these groups (e.g., the ODS surface). Drug partitioning into immobilized octanol correlates with drug partitioning into liposomes [74].

The in vivo tissue distribution of seventeen drugs has been modeled using estimated *n*-octanol/water and membrane/water distribution coefficients. The logarithm of distribution of (*n*-octanol/water–membrane/water), which measures a hypothetical drug equilibrium between *n*-octanol and membrane phase, was found to be a better model of in vivo tissue measure of drugs than *n*-octanol/water distribution coefficients [75]. The retention parameter log *k*(IAM) on the (IAM) column for 30 barbiturates was related to the logarithms of partition coefficients determined in an octanol–water partition system. The parameter log *k*(IAM) was shown to correlate with bioactivity data of barbiturates [76]. A fast method based on gradient elution has been used to measure the hydrophobicity index values of an IAM [77].

The uptake of the compounds across the in vivo brain–blood barrier expressed as a brain uptake index has been correlated with HPLC capacity factors as well as octanol/water partition (and distribution coefficients). log k(IAM) was found to be a better predictor over other parameters when polar and ionizable compounds are included. The predictive value of IAM, combined with the power of HPLC, thus holds great promise for the selection process of drug candidates with high brain penetration [78].

CE, and especially MEKC, has been also used to investigate these interactions similar to HPLC. Using oppositely charged surfactant vesicles as buffer modifiers to estimate hydrophobicity, there was a linear relationship between the log of the capacity factor and the octanol/water partition coefficient for both neutral and basic species at pH 6.0, 7.3, and 10.2. Vesicular electrokinetic chromatography using surfactant vesicles as buffer modifiers is a promising method for the estimation of hydrophobicity [79]. The chiral surfactant dodecoxycarbonylvaline has been used as a pseudostationary phase for the separation of many enantiomeric pharmaceutical compounds. Its hydrophobicity was a good predictor for *n*-octanol-water partition coefficients for 15 beta amino alcohols [80]. Octanol-water log P values for a large number of standards and bioactive molecules have been correlated to the logarithm of the corresponding capacity factors determined by reversed-phase HPLC, using a phase containing phosphatidylcholine. Similarly a correlation was also obtained for capacity factors determined by MEKC involving the use of phosphatidylcholine-bile acid mixed micelles in the separation buffer [81].

7.6 Free and Bound Drugs

As pointed out earlier, drugs bind to various serum proteins at different affinities. However, the free fraction of the drug is thought to be the active form, since it can pass through cellular membranes to exert its action at the active site. In addition to activity, protein binding is important in drug distribution and excretion. Protein binding can be stereoselective and thus affect the different forms of the drug in the blood. Measuring and understanding drug-protein interaction becomes important in disorders such as malnutrition and renal failure where the amount of the free drug changes due to changes in protein concentration. Usually, in routine TDM, the total of a drug (bound + free) is assayed. The free, bound, percentage of binding, and binding constant can be determined by several well-established techniques such as dialysis, filtration, and size exclusion. The problem with many of these methods of measurement is the time, or the membrane leakage, or the low concentration of the free drug, which make the analysis difficult.

Drug binding can be measured quickly by both HPLC and CE based on several principles. In CE, drug binding is based on changes in the electrophoretic mobility [82,83]. Kraak et al. [82] described three different methods for measuring protein–drug binding by CE. The first is based on the Hummel–Dryer method, in which the capillary is filled with a buffer containing the drug, giving a large background signal. The sample, which contains drug, protein, and buffer, is injected. The bound drug migrates differently from the free drug and produces a negative peak, which is a measure of the bound drug. The second method is based on the vacancy method, in which the capillary is filled with a mixture of buffer, drug, and protein. This also causes a large background signal. The sample, which contains only the buffer, is injected. Both the free and the bound drug migrate separately, and each gives a negative peak. The third method depends on the frontal analysis, which can be performed by CE and HPLC. In CE [83], the capillary is filled with buffer. Different concentrations of the drug, in the presence and the absence of a fixed binding protein concentration, are incubated at 25°C and then a large amount of sample (~5–7% of the capillary volume) is injected. The free drug, the complex, and the protein gives each a frontal, plateau-shaped peak. The free drug concentration (D) can be calculated from the height of the frontal peak as follows:

D = concentration/height of the pure standard × drug peak height in presence of the protein [84].

Several approaches in HPLC for determining the binding constants based on special columns are used for this purpose such as "internal surface reversed-phase (ISRP)" and size exclusion columns [85]. Since dissociation takes place on the column for the bound fraction of the protein–drug mixture, direct measurements are not successful. In one method, *zonal analysis*, protein added to the mobile phase and the change in retention time by the addition of protein are measured. In a second method, *frontal analysis*, a very large volume of the protein–drug mixture is injected into the column (preferably an ISRP column). At least two wide plateaux are observed, provided the column has a longer retention time for the drug, the first a large one for the bound drug and the second a smaller one for the free drug [85]. In a third type, the *vacancy method*, the mobile phase contains either drug or the drug–protein mixture. A sample lacking the drug or protein is injected, and a peak and trough is obtained that can be used to estimate free vs. protein-bound drug [85].

The percentage and the binding association constant can be also calculated from a Scatchard plot of this data at the different drug concentrations [83,84,86]. Kraak et al. [82] concluded that the *frontal analysis* appeared to be the preferred method for drug binding. It is more reproducible and gives a smooth Scatchard curve compared to the other techniques.

Protein binding can alter the metabolism or delivery of the drug to the target organ. Naproxen conjugated to albumin is an example of such drug targeting. Albrecht et al. [87] have shown that naproxen, a nonsteroidal anti-inflammatory drug, can be determined in serum using MEKC as free, albumin conjugated, and lysine conjugated. This method has also been extended to measurement of this drug in liver and kidney tissue [88].

8 PHARMACOGENETICS AND TDM

Pharmacogenetics is a new area that is expected in the near future to expand the boundaries of the TDM laboratory. It is known that responses to drugs can vary greatly between individuals and between different ethnic populations. The interindividual variability in drug response is a major clinical problem in therapeutics. Large sums of money are wasted because many drugs are not effective owing to noncompatibility with the patient's genetic makeup. In addition, extra costs are incurred to treat the side effects of these drugs. The goal of pharmacogenetics is to predict a patient's genetic response to a specific drug as a means of designing and delivering the best possible medical treatment. By predicting the drug response of an individual, it will be possible to design better drugs, increase the success of therapies, and reduce the incidence of adverse side effects.

Much of the observed variability in drug response is due to the expression of polymorphic enzymes [89]. Several polymorphic enzymes, such as cytochrome p 450 and UDP glucuronsyltransferase, are involved in drug metabolism. The most important is the cytochrome p 450 family, which shows a well-characterized polymorphism. One of these enzymes, CYP2D6, metabolizes one-quarter of all prescribed drugs including a number of antidepressants, antipsychotics, beta-adrenoreceptor blockers, and anti-arrhythmic drugs. Polymorphism enables division of individuals within a given population into few groups, e.g., poor, average, and extensive metabolizers of certain drugs. About 7% of Caucasians and 1% of Asians are poor metabolizers of CYP2D6 substrates [90], which puts them at risk of developing drug reactions. CYP2C19 enzyme is responsible for the metabolism of propranolol and psychotropic drugs such as hexobarbital, diazepam, imipramine, and amitriptyline. The incidence of poor metabolizers of CYP2C19 substrates is much higher in Asians (15–30%) than in Caucasians (3–6%). Such individuals can be identified or investigated by using DNA-based tests [89], HPLC of the metabolites of selected substrates. Caffeine is used in CYP1A2 phenotyping, but the widely used caffeine urinary metabolic ratios may not be the optimal method of measuring CYP1A2 activity. Mephenytoin has long been considered the standard of CYP2C19 phenotyping probe. Several probes, such as dextromethorphan, debrisoquin, and sparteine, have been used for CYP2D6 with dextromethorphan and may be preferred due to the wide safety margin. Chlorzoxazone has been used as a probe for CYP2E1. Several probes of CYP2C9 activity have been suggested [91]. Furthermore, a commercial reagent for in vitro enzymatic florescence detection of several isoenzymes of cyp450 (PanVera Corporation, Madison, WI) can be useful.

It is expected that the application of pharmacogenetics to improve the proper administration of drugs at the proper dose for the patient thus improves safety and decreases medication costs. It is expected to rescue many old drugs abandoned because a few patients experienced severe adverse effects or because the drug was not effective in some patients. It is expected also to help individuals subjected to environmental toxins.

9 DRUG GROUPS COMMONLY MONITORED

9.1 Antiepileptic Drugs

Antiepileptic drugs include compounds with a wide range of structures. They act by depressing the neural excitability by stabilizing the cell membrane and prevent the spread of the seizure. Many patients do not respond well to the existing drugs, so new ones are always being introduced. Antiepileptic drugs were among the first to illustrate the importance of drug monitoring. The number of seizures decreased as the patient blood levels were brought within the therapeutic ranges [2,3]. Many of the antiepileptic drugs have narrow therapeutic ranges. It is important to monitor such drugs. Table 1 lists many of the antiepileptic drugs among others, with their therapeutic range and half-life. GC [92,93], HPLC [94,95], and CE [96,97] have all been used successfully to analyze these drugs.

Most of the new antiepileptic drugs do not have commercial immunoassays available and thus have to be monitored by separation methods. For example, felbamate has been analyzed by HPLC [98,99] and GC [100]. Zonisamide was determined in serum by (MEKC) with a diode array detector [101]. Lamotrigine has been analyzed by HPLC [102,103] and by CE [104]. Gabapentin [23,24] and topiramate [25] have been analyzed by GC. Here are examples for the analysis of lamotrigine by HPLC (Fig. 2) and the analysis of levetiracetam (keppra) (Fig. 1) by both CE (MEKC) and HPLC (reversed-phase) as performed in our laboratory.

9.2 Cardioactive Drugs

Cardioactive drugs are used to treat arrhythmia and congestive heart failure. Some arrhythmias, such as ventricular arrhythmia and bradyarrhythmias, are characteristic of acute myocardial infarction. Cardioactive drugs contain compounds with different chemical structures. It is important to monitor some drugs such as lidocaine and procainamide. These two drugs are present in high concentrations so they can be measured easily by chromatographic techniques. On the other hand, digoxin, which is monitored frequently because of the danger of toxicity, is present in very low concentrations. Thus from a practical point of view it is more suited for analysis by immunoassays. Some of the cardioactive drugs, like amiodarone, do not have a commercial immunoassay.

Most cardioactive drugs are cationic in nature. Such compounds are more difficult to analyze by the chromatographic and electrophoretic methods because they tend to adsorb to the silica surfaces. A well endcapped or deactivated column is important for the analysis. Several methods have been described for the analysis of the cardioactive and the antihypertensive drugs from tablets and aqueous standards by HPLC [105,106] and CE [107,108]. Figure 3 illustrates the analysis of procainamide and *N*-acetyl procainamide by CE [109]. Figure 4 illustrates the analysis of amiodarone and its metabolite desethylamiodarone by HPLC as performed in our lab.

9.3 Bronchodilators

Methylxanthines such as theophylline, aminophylline, and dyphylline are among the main bronchodilators. They act on the enzyme phosphodiastrase to decrease cAMP breakdown. Caffeine is used for the treatment of apnea in newborns, while theophylline is used for the treatment of asthma. There is a marked interindividual variation in the metabolism of both theophylline and caffeine, which makes therapeutic monitoring of these drugs important. Several methods have been described for the analysis of caffeine and theophylline [110,111].

9.4 Antibiotics

Because of the increased resistance of bacteria to existing antibiotics, new drugs are always being introduced. These drugs have different modes of action. Some, such as the aminoglycosides (amikicin, gentamycin, tobramycin), bind to the 30S ribosomal subunit of the bacterium and inhibit protein synthesis. Other antibiotics, such as the sulfonamides, act as antimetabolites inhibiting p-aminobenzoic acid, which is needed for the synthesis of folic acid. Some antibiotics interfere with the synthesis of the cellular wall, such as vancomycin.

Many of the antibiotics have wide therapeutic ranges or are not highly toxic. Thus there is no need to monitor their level except in special

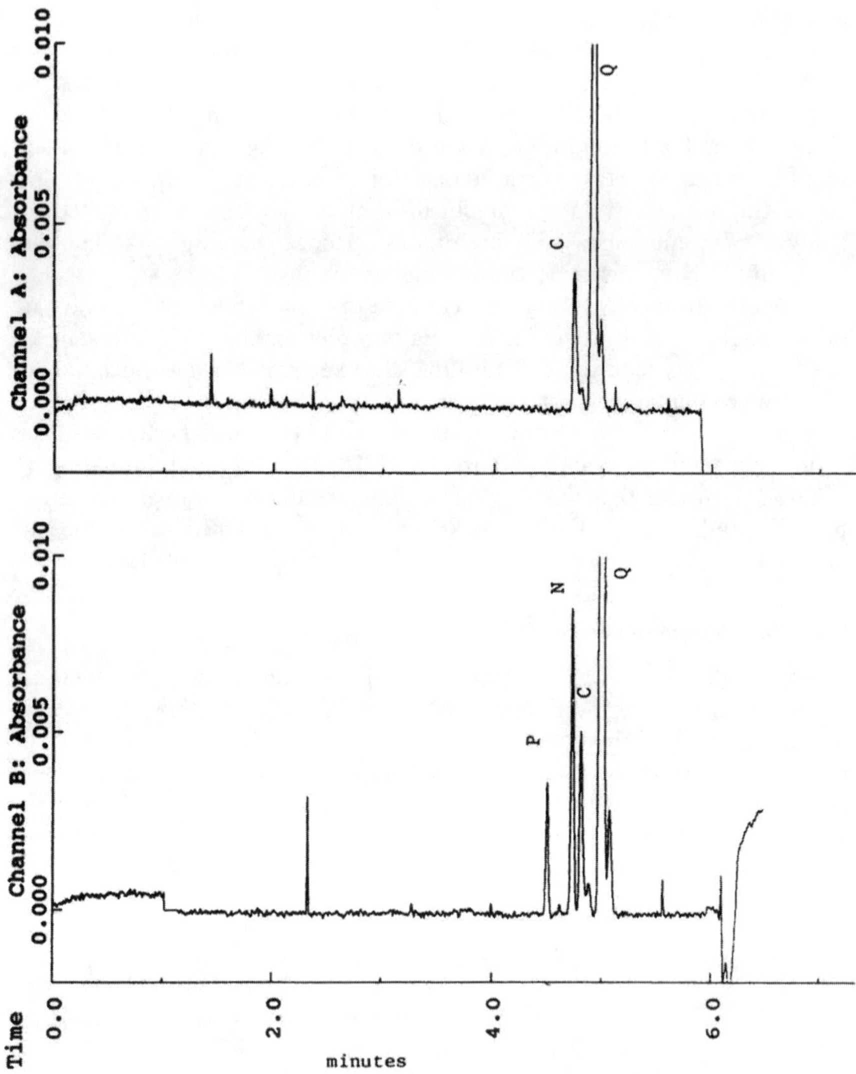

FIGURE 3 Analysis of procainamide and *N*-acetyl procainamide by the CE. (top) Electropherogram of patient free of procainamide. (bottom) A patient receiving procainamide. (P = procainamide 6.3 mg/l, N = *N*-acetylprocainamide, 15.1 mg/L, C = unknown peak, Q = quinine). (With permission from Ref. 109.)

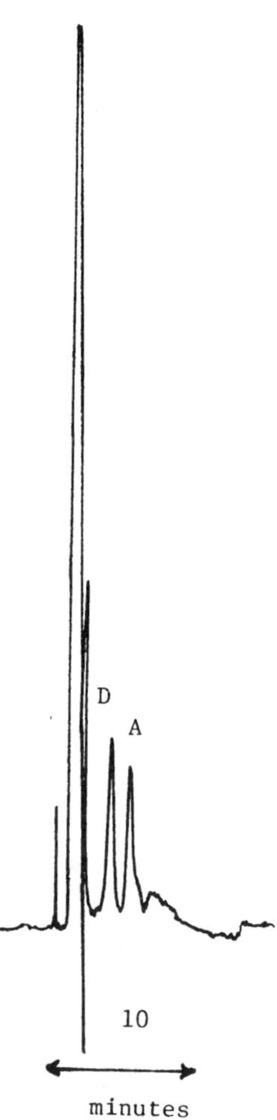

FIGURE 4 Analysis of serum amiodarone (A, 1.2 mg/L) and desethylamiodarone (D, 1.2 mg/L) by HPLC. Conditions of analysis: serum 100 μL is mixed with 200 μL of acetonitrile, centrifuged at 14,000 × g for 10 s, and the supernatant was injected on Waters Novapak CN Cartridge (100 mm × 3.9 mm), 4 μm. The column was eleuted with 37% acetonitrile containing 1.8 mL phosphoric acid and 1.8 mL butylamine/L at flow rate of 1 mL/min, detection at 215 nm.

cases, e.g., in patients with renal function impairment [8]. On the other hand, some, like the aminoglycosides, require monitoring because of their nephrotoxicity [8]. The aminoglycosides are poorly absorbed from the gastrointestinal tract. Thus they are injected intramuscularly or intravenously. They remain in the circulation and have a low volume of distribution. Both peak and trough levels have been advocated for monitoring the aminoglycosides [8]. The peak level (highest concentration) occurs soon after an intravenous infusion and after 45–60 min after intramuscular administration. For gentamicin and tobromycin, collection of samples over 2–3 half-lives after an intravenous injection, and extrapolating the concentration to the time of injection to estimate the peak level, has been advocated. Cloramphenicol is rapidly absorbed from the GI with peak level occurring 1–2 hours after the dose. It is 50% bound to the serum. Because of its toxicity its level is monitored. Several methods have been described for the analysis of antibiotics by HPLC [112–114] and CE [115,116].

9.5 Antidepressants

The introduction of psychotherapeutic drugs three decades ago has changed the thinking in psychiatry. The emphasis for psychiatric disorders shifted from the unconscious to the biochemical aspects. The tricyclics affect the mood (endogenous depression). They are highly protein bound and have a high volume of distribution. The peak level occurs 2–12 hours after an oral dose. The metabolites of these compounds are biologically active. Since the tricyclics are cationic compounds they tend to adsorb to the silica particles of the column and the walls of the capillary. In addition, their low concentration and similarity to many over-the-counter drugs render their analysis by HPLC and CE difficult [117]. Immunoassays are used mainly for the general screening of these compounds as a class. For specific quantification of each compound the HPLC [118–120] and the CE [121] are used.

9.6 Immunosuppressants

Cyclosporine and FK 506 (tacrolimus) are used often in transplantation as immunosuppressants. Cyclosporine is a cyclic peptide, which acts by inhibiting proliferation of the lymphocytes. It interferes with interleukin-2, which is a growth factor for the T-lymphocytes. Cyclosporine is not a water-soluble drug and it is given orally as a suspension in an oil-based medium. Its absorption is quite variable, from 5 to 40%. Thus there is a poor relationship between the dose and blood concentration. Since metabolites of this drug cross-react with antibodies to the parent compounds, immunoassays give falsely elevated values. For example, a polyclonal immunoassay gives a 100%, while the monoclonal immunoassay gives 30% higher values than that for HPLC [18,122]. Many problems are encountered in the analysis of this drug by chromatographic techniques, e.g., low levels in serum, and the absence of a strong absorbency.

FK 506 inhibits interleukin production, thus inhibiting lymphocyte proliferation. It has less nephrotoxicity than cyclosporine. Because the therapeutic level is low it is very difficult to analyze by the separation methods.

9.7 Renal Function and Contrast Agents

Several iodinated compounds, such as iothalamic acid and iohexol, are used to provide a better measurement of renal clearance relative to the common test creatinine clearance. These compounds are used in research as radioisotope-labeled compounds. Iohexol can be assayed rapidly ($<$ 5 minutes) by both CE [123] and HPLC [124], without the administration of radioactive compounds. Isovue, another candidate compound for the measurement of renal function, was also measured by CE after acetonitrile deproteinization [125]. Landers et al. [126], used CE to quantify iothalamic acid in serum and in a timed urine collection to measure the glomerular filtration rate. The test correlated well with an isotopic reference method.

10 CONCLUDING REMARKS

The different areas of separation science had a big impact on TDM, sparked by the advent of GC a few decades ago. Separation science remains in the forefront for TDM and especially for the analysis of the new drugs. The number of drugs measured in the routine lab for TDM has been steadily increasing every year. At the present time we offer in our lab analysis more than 30 drugs. In addition to measuring the drug itself, separation science enables simultaneous analysis of several others including the metabolites. It also provides basic vital information on the drug characteristics, such as binding, pKa, and tissue interactions. The analytical aspects have improved over the years, especially in the areas of computerization, automation, mass spectrometry, and wider selection of columns. New separation techniques have emerged in the last decade, such as capillary electrophoresis, capillary electrochromatography, and chip technology. These methods have several advantages, such as the speed of analysis and the high resolution. New areas for TDM are on the horizon, such as predicting patient response to a drug based on DNA or cytochrome p 450 polymorphism.

REFERENCES

1. MH Jacobs, RM Senior, G Kessler. Clinical experience with theophylline. Relationships between dosage, serum concentration, and toxicity. JAMA 235:1983–1986, 1976.
2. H Kutt, JK Penry. Usefulness of blood levels of antiepileptic drugs. Arch Neurol 31:283–288, 1974.
3. L Lund. Anticonvulsant effect of diphenylhydantoin relative to plasma levels. A prospective three-year study in ambulant patients with generalized epileptic seizures. Arch Neurol 31:289–294, 1974.
4. MK Ghosh. HPLC Methods on Drug Analysis. New York: Springer-Verlag, 1992.
5. RH Liu, DE. Gadzala. Handbook of Drug Analysis: Applications in Forensic and Clinical Laboratories. Washington, DC: Am Chem Soc, 1997.
6. PM Kabra, LJ Marton. Liquid Chromatography in Clincal Analysis. Clifton, NJ: Humana Press, 1981.
7. HY Wong. Therapeutic drug monitoring and toxicology by liquid chromatography. Chromatogr Sci Series 32, 1985.
8. M Bialer, RH Levy, E Perucca. Does carbamazepine have a narrow therapeutic range? Therp Drug Monit 20:56–59, 1998.
9. WJ Taylor, MH Caviness. A Textbook for the Clinical Application of Therapeutic Drug Monitoring. Abbot Laboratories, 1986.
10. M Rowland, TN Tozer. Clinical Pharmacokinetics, Concepts and Applications. Baltimore: Williams and Willkins, 1995.
11. TM Ludden. In: WE Evans, JJ Schentag, WJ Jusko, eds. Applied Pharmacokinetics. San Francisco: Applied Therapeutics, 1980, pp. 4–26.
12. CL Marsh. Abbreviated pharmacokinetics profiles in area-under-the-curve monitoring of cyclosporine therapy in de novo renal transplant patients treated with Sandimmune or Neoral. Therap Drug Monit 21:27–43, 1999.
13. ZK Shihabi. Review of drug analysis with direct serum injection on the HPLC column. J Liq Chromatogr 11:1579–1593, 1988.
14. ZK Shihabi. Stacking in capillary zone electrophoresis. J Chromatogr A 902:107–117, 2000.
15. ZK Shihabi. Effect of sample matrix on capillary electrophoresis. In: JP Landers, ed. Handbook of Electrophoresis. 2nd ed. Boca Raton, FL: CRC Press, 1997, pp. 457–477.
16. Z Shihabi, Z Deyl. Preconcentration and sample enrichment techniques. J Chromatogr 902, 2000.
17. ZK Shihabi. Stacking and discontinuous buffers in capillary zone electrophoresis. Electrophoresis 21:2872–2878, 2000.
18. LP Hackett, LJ Dusci, KF Ilett. A comparison of high-performance liquid chromatography and fluorescence polarization immunoassay for therapeutic drug monitoring of tricyclic antidepressants. Therap Drug Monit 20:30–34, 1998.
19. ZK Shihabi. Serum drug monitoring by capillary electrophoresis. In: JR

Peterson, AA Mohmmad, eds. Clinical and Forensic Applications of Capillary Electrophoresis. Totowa, NJ: Humana Press, 2001, pp. 335–384.
20. ZK Shihabi. Therapeutic drug monitoring by capillary electrophoresis. In: H Shintani, J Polonsky, eds. Handbook of Capillary Electrophoresis Applications. London: Chapman and Hall, 1997, pp. 386–408.
21. RP Oda, ME Roche, JP Landers, ZK Shihabi. Capillary electrophoresis for the analysis of drugs in biological fluids. In: JP Landers, ed. Handbook of Capillary Electrophoresis. 2nd ed. Boca Raton, FL: CRC Press, 1997, pp. 639–674.
22. CE Pippenger, JK Penry, H. Kutt. Antiepileptic Drugs: Quantitative Analysis and Interpretation. New York: Raven Press, 1978.
23. MM Kushnir, J Crossett, PI Brown, FM Urry. Analysis of gabapentin in serum and plasma by solid-phase extraction and gas chromatography-mass spectrometry for therapeutic drug monitoring. J Anal Toxicol 21:1–6, 1999.
24. F Van Lente, V Gatautis. Cost-efficient use of gas chromatography-mass spectrometry: a "piggyback" method for analysis of gabapentin. Clin Chem 44:2044–2045, 1998.
25. CE Wolf, CR Crooks, A Poklis. Rapid gas chromatographic procedure for the determination of topiramate in serum. J Anal Toxicol 24:661–663, 2000.
26. S Paterson, R Cordero, S McCulloch, P Houldsworth. Analysis of urine for drugs of abuse using mixed-mode solid-phase extraction and gas chromatography-mass spectrometry. Ann Clin Biochem 37:690–700, 2000.
27. W Weinmann, M Renz, S Vogt, S Pollak. Automated solid-phase extraction and two-step derivatisation for simultaneous analysis of basic illicit drugs in serum by GC/MS. Intl J Legal Med. 113:229–235, 2000.
28. T Soriano, C Jurado, M Menendez, M Repetto. Improved solid-phase extraction method for systematic toxicological analysis in biological fluids. J Anal Toxicol 25:137–143, 2001.
29. JA de Silva. Analytical strategies for therapeutic monitoring of drugs in biological fluids. J Chromatogr 340:3–30, 1985.
30. JR Mazzeo. Micellar electrokinetic chromatography. In: JP Landers, ed. Handbook of Capillary Electrophoresis. 2nd ed. Boca Raton, FL: CRC Press, 1977, pp. 49–73.
31. M Dressler. Selecttive chromatographic detectors. J Chromatogr Library 36, 1985.
32. A Dermaux, P Sandra. Applications of capillary electrochromatography. Electrophoresis. 20:3027–3065, 1999.
33. AS Rathore, C Horvath. Capillary electrochromatography: theories on electroosmotic flow in porous media. J Chromatogr A 781:185–195, 1997.
34. KD Altria. Overview of capillary electrophoresis and capillary electrochromatography. J Chromatogr A 856:443–463, 1999.
35. Z Deyl, F Svec. Capillary electrochromatography. J Chromatogr Library 62, 2001.

36. V Dolnik, S Liu, S Jovanovich. Capillary electrophoresis on microchip. Electrophoresis. 21:41–54, 2000.
37. LJ Kricka. Miniaturization of analytical systems. Clin Chem 44:2008–2014, 1998.
38. D Figeys, MDS Ocata, D Pinto. Lab-on-a-chip. Anal Chem 72:330A–335A, 2000.
39. TP Reilly, M Bourdi, JN Brady, CA Pise-Masison, MF Radonovich, JW George, LR Pohl. Expression profiling of acetaminophen liver toxicity in mice using microarray technology. Biochem Biophys Res Commun 282:321–288, 2000.
40. K Kudoh, M Ramanna, R Ravatn, AG Elkahloun, ML Bittner, PS Meltzer, JM Trent, WS Dalton, KV Chin. Monitoring the expression profiles of doxorubicin-induced and doxorubicin-resistant cancer cells by cDNA microarray. Cancer Res. 60:4161–4156, 2000.
41. TO Joos, M Schrenk, P Hopfl, K Kroger, U Chowdhury, D Stoll, D Schorner, M Durr, K Herick, S Rupp, K Sohn, H Hammerle. A microarray enzyme-linked immunosorbent assay for autoimmune diagnostics. Electrophoresis 21:2641–2650, 2000.
42. JW Silzel, B Cercek, C Dodson, T Tsay, RJ Obremski. Mass-sensing, multianalyte microarray immunoassay with imaging detection. Clin Chem 44:2036–2043, 1998.
43. FT Chen, RA Evangelista. Feasibility studies for simultaneous immunochemical multianalyte drug assay by capillary electrophoresis with laser-induced fluorescence. Clin Chem 40:1819–1822, 1994.
44. RA Evangelista, FT Chen. Analysis of structural specificity in antibody-antigen reactions by capillary electrophoresis with laser-induced fluorescence detection. J Chromatogr A 680:587–591, 1994.
45. D Schmalzing, W Nashabeh, X Yao, R Mhatre. Capillary electrophoresis-based immunoassay for cortisol in serum. Anal Chem 67:606–612, 1995.
46. D Schmalzing, W Nashabeh, M Fuchs. Solution-phase immunoassay for determination of cortisol in serum by capillary electrophoresis. Clin Chem 41:1403–1406, 1995.
47. L Steinmann, W Thormann. Characterization of competitive binding, fluorescent drug immunoassays based on micellar electrokinetic capillary chromatography. Electrophoresis 17:1348–1356, 1996.
48. X Liu, Y. Xu, MP Ip. Capillary electrophoretic enzyme immunoassay for digoxin in human serum. Anal Chem 67:3211–3218, 1995.
49. N Chiem, DJ Harrison. Microchip-based capillary electrophoresis for immunoassays: analysis of monoclonal antibodies and theophylline. Anal Chem 69:373–378, 1997.
50. IW Wainer. Toxicology through a looking glass. In: SH Wong, I Sunshine, eds. Handbook of Analytical Therapeutic Drug Monitoring and Toxicology. Boca Raton, FL: CRC Press, 1996, pp 21–34.
51. E Yashima. Polysaccharide-based chiral stationary phases for high-perfor-

mance liquid chromatographic enantio-separation. J Chromatogr A 906:105–125, 2001.
52. TJ Ward, AB Farris. Chiral separations using the macrocyclic antibiotics: a review. J Chromatogr A 906:73–89, 2001.
53. J Haginaka. Enantiomer separation of drugs by capillary electrophoresis using proteins as chiral selectors. J Chromatogr A 875:235–254, 2000.
54. HY Aboul-Enein. High-performance liquid chromatographic enantioseparation of drugs containing multiple chiral centers on polysaccharide-type chiral stationary phases. J Chromatogr A 906:185–193, 2001.
55. MR Hadley, P Camilleri, AJ Hutt. Enantiospecific analysis by capillary electrophoresis: applications in drug metabolism and pharmacokinetics. Electrophoresis 18:2322–2330, 1997.
56. B Chankvetadz, G Blaschke. Enantioseparations in capillary electromigration techniques: recent developments and future trends. J Chromatogr A. 906:309–363, 2001.
57. K Srinivasan, MG Bartlett. Capillary electrophoresis stereoselective determination of R-(+)- and S-(−)-phenobarbital from serum using hydroxy propyl-γ-cyclodextrin, solid phase extraction and ultraviolet detection. J Chromatogr B 703:289–294, 1997.
58. T Ohara, T Nakagawa, A Shibukawa. Capillary electrophoresis/frontal analysis for microanalysis of enantioselective protein binding of a basic drug. Anal Chem 67:3520–3525, 1995.
59. H Nishi. Enantiomer separation of drugs by electrokinetic chromatography. J Chromatogr A 735:57–76, 1996.
60. S Fanali. Identification of chiral drug isomers by capillary electrophoresis. J Chromatogr A 735:77–121, 1996
61. J Bojarski, HY Aboul-Enein. Application of capillary electrophoresis for the analysis of chiral drugs in biological fluids. Electrophoresis 18:965–969, 1997.
62. MG Schmid, G Gubitz, F Kilar. Stereoselective interaction of drug enantiomers with human serum transferrin in capillary zone electrophoresis. Electrophoresis 19:282–287, 1998.
63. KM Giacomini, WL Nelson, RA Pershe, L Valdivieso, K Turner-Tamiyasu, TF Blaschke. In vivo interaction of enantiomers of disopuramide in human subjects. J Pharmacokin Biopharm 14:335–356, 1986.
64. AS Gross, B Heuer, M Eichelbaum. Stereoselective protein binding of verapamil enantiomers. Biochem Pharmacol 37:4623–4627, 1988.
65. A D'Hulst, N Verbeke. Quantitation in chiral capillary electrophoresis: theoretical and practical considerations. Electrophoresis 15:854–863, 1994.
66. KD Altria, DM Goodall, MM Rogan. Quantitative applications and validation of the resolution of enantiomers by capillary electrophoresis. Electrophoresis 15:824–827, 1994.
67. SG Penn, G Liu, FT Begrton, DM Goodall, JS Loran. Systemic approach to treatment of enantiomeric separations in capillary electrophoresis and liquid chromatography. I. Initial evaluations using propranolol and dansylated aminoacids. J Chromatogr A 680:147–155, 1994.

68. JA Cleveland, MH Benko, SJ Gluck, YM Walbroehl. Automated pKa determination at low solute concentrations by capillary electrophoresis. J Chromatogr A 652:301–308, 1993.
69. JE Hardcastle, I Jano. Determination of dissociation constants of polyprotic acids from chromatographic data. J Chromatogr B 717:39–56, 1998.
70. FH Clarke. Ionization constants by curve fitting: application to the determination of partition coefficients. J Pharm Sci 73:226–230, 1984.
71. HY Ando, T Heimbach. pKa determinations by using a HPLC equipped with DAD as a flow injection apparatus. J Pharma Biomed Anal 16:31–37, 1997.
72. M Roses, F Rived, E Bosch. Dissociation constants of phenols in methanol–water mixtures. J Chromatogr A 867:45–56, 2000.
73. A Schmutz, W Thormann. Assessment of impact of physico-chemical drug properties on monitoring levels by micellar electrokinetic capillary chromatography with direct serum injection. Electrophoresis 15:1295–1303, 1994.
74. H Liu, S Ong, L Glunz, C Pidgeon. Predicting drug-membrane interactions by HPLC: structural requirements of chromatographic surfaces. Anal Chem 67:3550–3557, 1995.
75. P Barton, AM Davis, DJ McCarthy, PJ Webborn. Drug-phospholipid interactions. 2. Predicting the sites of drug distribution using *n*-octanol/water and membrane/water distribution coefficients. J Pharm Sci 86:1034–1039, 1997.
76. E Kepczynska, J Bojarski, P Haber, R Kaliszan. Retention of barbituric acid derivatives on immobilized artificial membrane stationary phase and its correlation with biological activity. Biomed Chromatogr 14:256–260, 2000.
77. K Valko, CM Du, CD Bevan, DP Reynolds, MH Abraham. Rapid-gradient HPLC method for measuring drug interactions with immobilized artificial membrane: comparison with other lipophilicity measures. J Pharm Sci 89: 1085–1096, 2000.
78. A Reichel, DJ Begley. Potential of immobilized artificial membranes for predicting drug penetration across the blood-brain barrier. Pharm Res 15:1270–1274, 1998.
79. JL Razak, BJ Cutak, CK Larive, CE Lunte. Correlation of the capacity factor in vesicular electrokinetic chromatography with the octanol:water partition coefficient for charged and neutral analytes. Pharm Res 18:104–111, 2001.
80. RJ Pascoe, AG Peterson, JP Foley. Investigation of the chiral surfactant *N*-dodecoxycarbonylvaline in electrokinetic chromatography: improvements in elution range and pH stability via mixed micelles and vesicles, and the hydrophobicity determination of basic pharmaceutical drugs. Electrophoresis 21:2033–2042, 2000.
81. M Hanna, V de Biasi, B Bond, C Salter, AJ Hutt, P Camilleri. Estimation of the partitioning characteristics of drugs: a comparison of a large and diverse drug series utilizing chromatographic and electrophoretic methodology. Anal Chem 70:2092–2099, 1998.

82. JC Kraak, S Bush, H Poppe. Study of protein-drug binding using capillary electrophorsis. J Chromatogr A 608:257–264, 1992.
83. FA Gomez, JN Mirkovich, VM Dominguez, KW Liu, DM Macias. Multiple-plug binding assays using affinity capillary electrophoresis. J Chromatogr A 727:291–299, 1996.
84. PA McDonnell, GW Caldwell, JA Masucci. Using capillary electrophoresis/frontal analysis to screen drugs interacting with human serum proteins. Electrophoresis 19:448–454, 1998.
85. DS Hage, SA Tweed. Recent advances in chromatographic and electrophoretic methods for the study of drug-protein interactions. J Chromatogr B 699:499–525, 1997.
86. A Shibukawa, T Nakagawa. Theoretical study of high-performance frontal analysis: a chromatographic method for determination of drug-protein binding interaction. Anal Chem 68:447–454, 1996.
87. C Albrecht, J Reichen, J Visser, DKF Meijer, W Thormann. Differentiation between naproxen, naproxen-protein conjugates, and naproxen-lysine in plasma via micellar electrokinetic capillary chromatography—a new approach in the bioanalysis of drug targeting preparations. Clin Chem 43:2083–2090, 1997.
88. C Albrecht, W Thormann. Determination of naproxen in liver and kidney tissues by electrokinetic capillary chromatography with laser-induced fluorescence detection. J Chromatogr A 802:115–120, 1998.
89. CR Wolf, G Smith. Pharmacogenetics. Br Med Bull 55:366–386,1999.
90. N Poolsup, PA Li, TL Knight. Pharmacogenetics and psychopharmacotherapy. J Clin Pharm Ther 25:197–220, 2000.
91. DS Streetman, JS Bertino Jr, AN Nafziger. Phenotyping of drug-metabolizing enzymes in adults: a review of in-vivo cytochrome P450 phenotyping probes. Pharmacogenetics 10:187–216, 2000.
92. ZK Shihabi. Emergency gas-chromatographic assay of phenobarbital and phenytoin and liquid-chromatographic assay of theophylline. Clin Chem 24:1630–1633, 1978.
93. K Chen, H Khayam-Bashi. Comparative analysis of antiepileptic drugs by gas chromatography using capillary or packed columns and by fluorescence polarizationimmunoassay. J Anal Toxicol 15:82–85, 1991.
94. KM Matar, PJ Nicholls, A Tekle, SA Bawazir, MI Al-Hassan. Liquid chromatographic determination of six antiepileptic drugs and two metabolites in microsamples of human plasma. Therap Drug Monit 21:559–566, 1999.
95. W Kuhnz, H Nau. Automated high-pressure liquid chromatographic assay for antiepileptic drugs and their major metabolites by direct injection of serum samples. Therap Drug Monit 6:478–483, 1984.
96. W Thormann, P Meier, C Marcolli, F Binder. Analysis of barbiturates in human serum and urine by high-performance capillary electrophoresis—micellar elektrokinetic capillary chromatography with on-column multi-wavelength detection. J Chromatogr 545:445–460, 1991.

97. KJ Lee, GS Heo, NJ Kim, DC Moon. Analysis of antiepileptic drugs in human plasma using micellar electrokinetic capillary chromatography. J Chromatogr 608:243–250, 1992.
98. ZK Shihabi, KS Oles. Felbamate measured in serum by two methods: HPLC and capillary electrophoresis. Clin Chem 40:1904–1908, 1994.
99. TM Annesley, LT Clayton. Determination of felbamate in human serum by high-performance liquid chromatography. Ther Drug Monit 16:419–424, 1994.
100. P Gur, A Poklis, J Saady, A Costantino. Chromatographic procedures for the determination of felbamate in serum. J Anal Toxicol 19:499–503, 1995.
101. K Makino, R Oishi, Y Kataoka, K Futagami, M Sueyasu, Y Goto. Micellar electrokinetic capillary chromatography of therapeutic drug monitoring of zonisamide. J Chromatogr B 695:417–425, 1997.
102. M Torra, M Rodamilans, S Arroyo, J Corbella. Optimized procedure for lamotrigine analysis in serum by high-performance liquid chromatography without interferences from other frequently coadministered anticonvulsants. Ther Drug Monit 22:621–625, 2000.
103. KM Matar, PJ Nicholls, A Tekle, SA Bawazir, MI Al-Hassan. Liquid chromatographic determination of six antiepileptic drugs and two metabolites in microsamples of human plasma. Ther Drug Monit 21:559–566, 1999.
104. ZK Shihabi, KS Oles. Serum lamotrigine analysis by capillary electrophoresis. J Chromatogr B 683:119–123, 1996.
105. RR Bridges, TA Jennison. An HPLC method for the simultaneous quantitation of quinidine, procainamide, *N*-acetylprocainamide, and disopyramide. J Anal Toxicol 7:65–68, 1983.
106. JF Wesley, FD Lasky. Simultaneous analysis of antiarrhythmic drugs and metabolites by high performance liquid chromatography: interference studies and comparisons with other methods. Clin Biochem 15:284–290, 1982.
107. NT Nguyen, RW Siegler. Capillary electrophoresis of cardiovascular drugs. J Chromatogr A 735:123–150, 1996.
108. HJ Gaus, A Treumann, W Kreis, E Bayer. Separation of cardiac glycosides by micellar electrokinetic capillary electrophoresis. J Chromatogr A 635:319–327, 1993.
109. Shihabi ZK. Serum procainamide analysis based on acetonitrile stacking by capillary electrophoresis. Electrophoresis 19:3008–3011, 1998.
110. WJ Hurst, RA Martin, SM Tarka Jr. Analytical methods for quantitation of methylxanthines. Prog Clin Biol Res 158:17–28, 1984.
111. T Umemura, R Kitaguchi, K Inagaki, H Haraguchi. Direct injection determination of theophylline and caffeine in blood serum by high-performance liquid chromatography using an ODS column coated with a zwitterionic bile acid derivative. Analyst 123:1767–1770, 1998.
112. DA Stead. Current methodologies for the analysis of aminoglycosides. J Chromatogr B 747:69–93, 2000.
113. CM Lock, L Chen, DA Volmer. Rapid analysis of tetracycline antibiotics by

combined solid phase microextraction/high performance liquid chromatography/mass spectrometry. Rapid Comm Mass Spectrometry 13:1744–1754, 1999.
114. L Soltes. Aminoglycoside antibiotics—two decades of their HPLC bioanalysis. Biomed Chromatogr 13:3–10, 1999.
115. J Zhou, Y Chen, R Cassidy. Separation and determination of the macrolide antibiotics (erythromycin, spiramycin and oleandomycin) by capillary electrophoresis coupled with fast reductive voltametric detection. Electrophoresis 21:1349–1353, 2000.
116. CL Flurer. Analysis of antibiotics by capillary electrophoresis. Electrophoresis 20:3269–3279, 1999.
117. ZK Shihabi. Stacking of weakly cationic compounds for capillary electrophoresis. J Chromatogr A 817:25–30, 1998.
118. SR Needham, PR Brown, K Duff, D Bell. Optimized stationary phases for the high-performance liquid chromatography—electrospray ionization mass spectrometric analysis of basic pharmaceuticals. J Chromatogr A 869:159–170, 2000.
119. BA Way, D Stickle, ME Mitchell, JW Koenig, J Turk. Isotope dilution gas-chromatographic-mass spectrometric measurement of tricyclic antidepressant drugs. Utility of the 4-carbethoxyhexafluorobutyryl derivatives of secondary amines. J Anal Toxicol 22:374–382, 1998.
120. G Aymard, P Livi, YT Pham, B Diquet. Sensitive and rapid method for the simultaneous quantification of five antidepressants with their respective metabolites in plasma using high-performance liquid chromatography with diode-array detection. J Chromatogr B 700:183–189, 1997.
121. M Kurzawa. B Dembinski, A Szydlowska-Czerniak, P Sandra. Quantitative determination of selected tricyclic biological active compounds by using capillary electrophoresis. Acta Poloniae Pharm 56:415–417, 1999.
122. JM Tredger, N Roberts, R Sherwood, G Higgins, J Keating. Comparison of five cyclosporin immunoassays with HPLC. Clin Chem Lab Med 38:1205–1207, 2000.
123. ZK Shihabi, EN Thompson, MS Constantinescu. Iohexol determination by direct injection of serum on the HPLC column. J Liq Chromatogr 16:1289–1296, 1993.
124. ZK Shihabi, MS Constantinescu. Iohexol in serum determined by capillary electrophoresis. Clin Chem 38:2117–2120, 1992.
125. ZK Shihabi, MV Rocco, ME Hinsdale. Analysis of the contrast agent iopamidol by capillary electrophoresis. J Liq Cromatogr 18:3825–3832, 1995.
126. JH Bergert, RR Liedtke, RP Oda, JP Landers, DM Wilson. Development of a nonisotopic capillary electrophoresis-based method for measuring glomerular filtration rate. Electrophoresis 18:1827–1835, 1997.

3

The Impact of Chirality in Pharmacokinetics and Therapeutic Drug Monitoring

Karla A. Moore and Barry Levine
Office of the Chief Medical Examiner, Baltimore, Maryland, U.S.A.

> All artificial bodies and all minerals have superimposable images. Opposed to these are nearly all organic substances which play an important role in plant and animal life. These are asymmetric and indeed have the kind of asymmetry in which the image is not superimposable with the object ... *This establishes perhaps the only well marked line of demarcation that can at present be drawn between the chemistry of dead matter and the chemistry of living matter.*
>
> —Louis Pasteur, 1860

1 STEREOCHEMISTRY

Thus began the field of study in organic chemistry known as isomerism and stereochemistry. The importance of what Louis Pasteur recognized well over 100 years ago is only now coming to light with the application of the principles of stereochemistry to the fields of pharmacology, pharmacokinetics, and pharmacodynamics. But before this diverse and expanding field in the pharmaceutical industry can be explored, one must have at least a basic knowledge of the principles and, more importantly, the vocabulary, of stereochemistry.

Isomers are simply different compounds having the same molecular formula. The five currently recognized categories of isomers are structural, positional, geometric, conformational, and configurational/stereoisomers. For purposes of this chapter, the only category of concern is *configurational/stereoisomers*. A carbon atom to which four different groups are attached is a *chiral* (asymmetrical) *carbon*. Most (but not all) molecules containing a chiral carbon are *not* superimposable on its mirror image. A molecule that is superimposable on its mirror image is *achiral*. A molecule that contains just one chiral center is always chiral. A compound with chiral carbons/centers yet is superimposable on its mirror image is called a *meso compound*.

Isomers that have the same atoms bonded to one another but differ only in the way atoms are arranged in space are called *stereoisomers*. Stereoisomers are further classified as *enantiomers* or *diastereomers*. Isomers that are mirror images of each other are called *enantiomers*. Enantiomers have identical chemical and physical properties except for the directions of rotation of polarized light and their interactions with optically active substances.

Optically active substances can rotate a plane of polarized light. If the rotation of the plane is clockwise, the substance is *dextrorotatory* (*d*- or (+)-); if the rotation is counterclockwise, the substance is *levorotatory* (*l*- or (−)-). The *d*- and *l*-designations are no longer in common use; thus the (+)- and/or (−)- nomenclature will be used throughout this chapter. Enantiomers of a given substance rotate the plane of polarized light equally but in opposite directions. A mixture of equal amounts of enantiomers (designated by ±) is called a *racemic mixture* and is optically inactive.

Stereoisomers that are not mirror images of each other are called *diastereomers* (e.g., pseudoephedrine and ephedrine; Fig. 1). Diastereomers contain two or more chiral centers and result from inversion around at most $n-1$ of the chiral centers; a compound having n chiral centers will often have 2^n possible stereoisomers. Diastereomers have different chemical and physical properties.

The arrangement of atoms that characterizes a given stereoisomer is called its *configuration*. The most common, currently accepted system of nomenclature is known as the *Cahn–Ingold–Prelog* system (often referred to as simply the Cahn system). In the Cahn system, the atom with the smallest atomic mass is pushed to the back of the observation plane; the remaining three groups are then ranked from largest to smallest atomic number. If

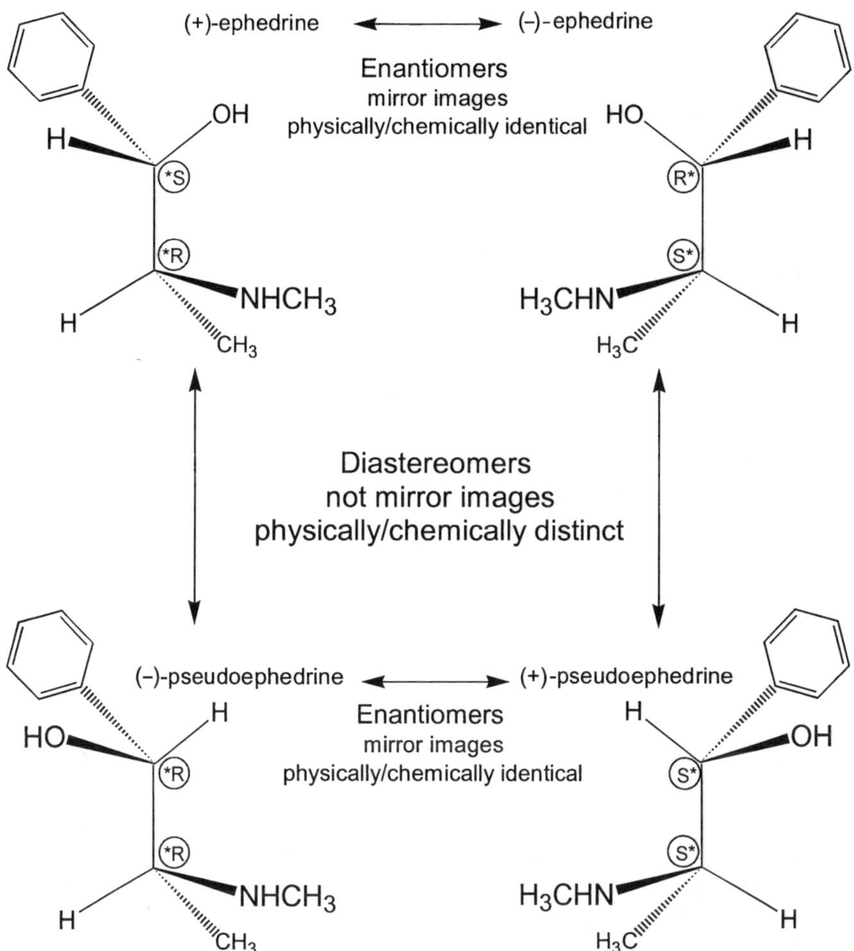

FIGURE 1 Comparison of ephedrine and pseudoephedrine as enantiomers and diasteromers.

the groups (from largest to smallest) go in a clockwise direction around the asymmetrical carbon, the designation of *R* is given to that chiral center. If the remaining three groups (from largest to smallest) go in a counterclockwise direction around the carbon atom, a designation of *S* is given (Fig. 2). This designation has absolutely no relationship to the (+)- or (−)-designation for the rotation of plane-polarized light.

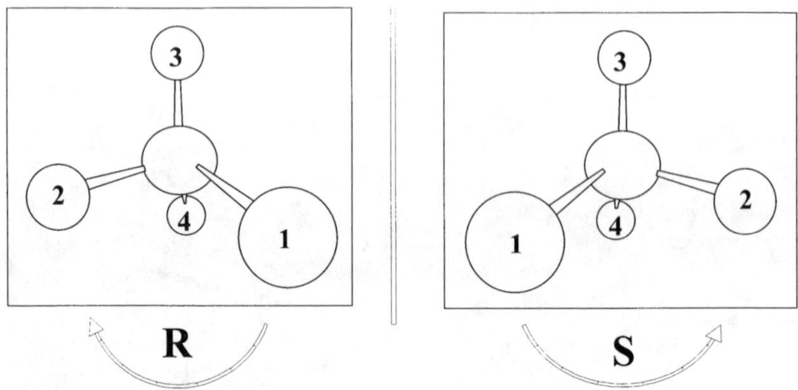

FIGURE 2 Absolute configuration about an asymmetric center as determined by the Cahn–Ingold–Prelog rules. The group with the lowest atomic weight is pushed to the back of the plane (4); the remaining three groups 1, 2, and 3 are weighted from highest atomic number to lowest. The asymmetric carbon is designated R when the order is clockwise and S when the order is counterclockwise.

2 PHARMACOKINETICS

Of 1850 drugs used therapeutically, natural and synthetic, 1045 are chiral [1]. 517 of these chiral drugs are acquired from natural sources (e.g., quinine and quinidine, (−)-morphine, and (+)-digoxin), and 509 of these are available as single isomers [2]. However, many synthetic therapeutic drugs used in clinical practice are racemic mixtures of enantiomers. Of the 528 synthetic, chiral pharmaceuticals listed in the U.S Pharmacopeial Dictionary of Drug Names in 1990, 467 (88%) were sold as racemic mixtures, leaving only 61 that are marketed as a "stereochemically pure" drug. This manufacturing trend continued through the first half of the last decade; if drugs in use through the early 1990s are included, approximately 40% of synthetic compounds are chiral, with 90% of these manufactured as racemates (90% of the β-adrenergic agonists and blockers, antieleptics, and anticoagulants and up to 50% of the antihistamines and antidepressants) [3]. In contrast, owing to new pharmacological and manufacturing capabilities as well as heightened regulatory interest, about 66% of new chiral drugs registered since 1993 are single enantiomers.

Despite having identical physical and chemical properties, they often exhibit in vivo differences, including different affinities at receptor sites, different affinities for tissue and protein binding sites, and different rates of biotransformation. In 1991, Levy and Boddy [4] proposed that stereoselectivity might affect pharmacokinetic parameters at macromolecule, whole organ, and whole body levels. In keeping with Pasteur's original theory of the asymmetry of living matter, at the macromolecular level, an antibody, enzyme, or drug receptor may display stereoselective affinity for interacting with or binding to drug enantiomers. As a result of such "enantioselectivity," it is not difficult to imagine that plasma/tissue protein binding, metabolism, and therapeutic actions of enantiomers may be different. At the organ level, clearance and distribution of enantiomers may be affected. Subsequently, whole body pharmacokinetic parameters such as half-life, total body clearance, and volume of distribution may be affected.

Enantioselectivity has been reported for all pharmacokinetic parameters [5]. Absorption from the gastrointestinal (GI) tract is a passive, mass action process for most drugs. However, receptor mediated active transport can demonstrate stereoselectivity and has been demonstrated with the preferential absorption of L-dopa over D-dopa and D-cephalexin (a β-lactam antibiotic) over the L-isomer [6,7]. Possible stereoselective factors in drug distribution include carrier-mediated transport (e.g., GI, kidney, liver, and brain) and binding to macromolecules such as blood proteins and stereoselective sites on tissues [8]. Acidic drugs seem to be more enantioselective in binding to human serum albumin than basic drugs [9].

L-tryptophan and *S*-(+)-oxazepam bind to serum albumin with 90–100 times the affinity of their isomers [10]. Subsequently, the elimination half-life of the enantiomer with the higher tissue protein binding affinity may be greater than the less tightly bound isomer, For example, *S*-(−)-propranolol is stereoselectively taken up by the heart in the rat [11]. As a result, the *R*-(+)-enantiomer is eliminated more rapidly than *S*-(−)-propranolol.

Finally, stereoselective metabolism and excretion of drugs is a well-recognized phenomenon. Enzymes are themselves asymmetric molecules that often exhibit stereoselectivity in binding to and metabolizing their substrates. Diastereometric complexes are thus formed as a result of a racemic substrate (e.g., drugs made up of two enantiomers) and a chiral enzyme [12]. As diastereomers, the physical and chemical properties of these complexes are different, and the respective metabolic rates may also differ.

Stereoselective biotransformation may be due to the asymmetry of the substrate, the product, or the substrate-product combination [7]. Substrate stereoselectivity refers to the preferential metabolism of chiral substrates where no new asymmetric center is introduced. In product stereoselectivity, a prochiral drug is metabolized into a chiral product with the preferential formation of one particular stereoisomer over that of other stereoisomers [13]. In substrate/product combination, both chiral and prochiral centers are subject to stereoselective metabolism by which additional asymmetric centers are formed. In summary, this gives rise to five stereochemical possibilities for enantiomer metabolism: prochiral to chiral, chiral to chiral, chiral to diastereomer, chiral to nonchiral, and chiral inversion of racemates [13].

The most unusual metabolic pathway, and one that is unique to chiral drugs, is the phenomenon of enantiomer inversion. The mechanism of the inversion reaction is believed to involve the stereospecific formation of a coenzyme-A thioester from one enantiomer [13]. Examples of drugs that undergo rapid and extensive metabolic racemization in vivo include oxazepam and thalidomide, but the phenomenon has been most comprehensively studied in the 2-arylpropionic acid (the "profens") nonsteroidal anti-inflammatory drugs (NSAIDs). These compounds possess a chiral carbon α- to the carboxyl group; the *S* enantiomer possesses the pharmacological activity. The *R* enantiomer forms the coenzyme-A thioester, which then follows one of these pathways: racemization of the chiral center in the profen molecule followed by hydrolysis to yield a mixture of enantiomers; hydrolysis with retention of the *R* configuration; or acyl transfer of the profen moiety to glycerol, resulting in the formation of a hybrid triglyceride. There are clearly a number of enzymatic steps associated with inversions; the formation of any of the intermediaries or the CoA thioesters themselves may have considerable toxicological significance. For example, the formation of the acylglycerols and their subsequent incorporation into membranes can alter membrane structure and can be associated with disordered membrane function.

The extent of the inversion varies considerably with the drug (substrate) and is highly species dependent [7]. In humans, conversion is significant for ibuprofen (Advil®, Motrin®) and fenoprofen (Nalfon®) but negligible for carprofen (Rimadyl®) and flurbiprofen (Ansaid®). This further underscores the general rule when reading scientific literature that great care must be taken when extrapolating animal studies to humans. Ketoprofen (Profenid®, Orudis®) undergoes 80% conversion in the rat, but this process is very limited in humans. This phenomenon must also be considered when evaluating other analytical and pharmacokinetic data. For example, the inversion phenomenon may be underestimated. When researchers administered *R*-ketoprofen alone to prove inversion, the *S* enantiomer was measurable in urine but not in plasma. The undetectable concentrations in the bloodstream do not imply the absence of inversion but may instead be due to the insensitivity of the assay and/or inversion in the excretion system. Conversely, it is also possible to overestimate the extent of the inversion. Reports state that *R*-ibuprofen is 60% inverted in humans. This percentage was calculated by giving equal doses of each enantiomer individually, and the AUC of the *S* enantiomer (as derived from *R*-ibuprofen) is compared to the AUC of its racemate. Linear pharmacokinetics and the absence of an interaction between the two enantiomers was assumed. However, it has been determined that ibuprofen pharmacokinetics are nonlinear owing to saturation of plasma protein binding, and larger AUCs were observed for enantiomers when given alone. Both of these later findings contributed to the overestimation of ibuprofen inversion.

Biological inversion is not unique to the NSAIDs [7]. This phenomenon has also been documented in soil organisms' metabolism of a structurally related herbicide, and *S*-(+)-oxazepam will racemize in aqueous solutions.

Additionally, completely different enzymes can catalyze the metabolism of different enantiomers, in which case stereoselective metabolism will be altered by differential enzyme induction [14]. For example, the CYP2C19 enzyme metabolizes the *S* enantiomer of the antiepileptic mephenytoin; the *R* enantiomer utilizes other pathways. Rifampin, a CYP2C19 inducer, can thus significantly alter the excretion and potentially the therapeutic effects of mephenytoin if given concurrently. UDP-glucuronosyltranferases can also be induced, thus altering stereoselective phase II (conjugation) metabolism as well. Pentobarbital is one example of this enzyme's inducer, and coadministration of this compound with benzodiazepines that are metabolized to oxazepam can increase clearance of oxazepam by 50% while decreasing the urinary (+/−) oxazepam glucuronide ratio.

Competition for the same metabolic pathway can modify drug action and disposition through enantiomer–enantiomer interactions when one enantiomer inhibits metabolism of the other. The best example of this type of interaction is *S*-(−)-and *R*-(+)-propafenone's (Rhythmol®; a class 1C antiarrhythmic) competition for 5-hydroxylation catalyzed by CYP2D6. Both enantiomers are equipotent sodium channel blockers, but the *S* enantiomer has β-blocking properties as well. Inhibition of the metabolism of this enantiomer by its racemate thus could lead to potentially adverse β-blockade after administration of the racemic mixture when compared to the *S* enantiomer alone [14].

The phase II (conjugation) metabolic reactions result in the linkage of a drug or its metabolite to an endogenous conjugating agent (endocons) [13]. Endocons can also be chiral (glucuronic acid, glutathione, glutamine, and glucose) or achiral (methyl groups, acetyl groups, sulphates, glycine, and taurine). Conjugation reactions can take two possible courses: enantiomers can be conjugated at different rates with achiral endocons or the formation of diastereomeric complexes with chiral endocons. Diastereomeric conjugates will have different physicochemical properties, facilitating their relatively easy discrimination on achiral analytical systems. For example, carprofen, a nonsteroidal anti-inflammatory agent, Rimadyl®, undergoes enantioselective glucuronidation that subsequently alters its pharmacokinetics. The renal clearance of the (*R*)-D-glucuronide is 50% less than the (*S*)-D-glucuronide diastereomer. Since (*S*)-carprofen is preferentially conjugated, the result is that plasma concentrations of both (*S*)-carprofen and its conjugate are much lower than (*R*)-carprofen.

Hepatic clearance of drugs with low hepatic extraction is directly influenced by their intrinsic clearance [15]. Intrinsic clearance can be defined as the ability of the liver to remove irreversibly a drug independent of blood flow [16]. Interindividual variations in genetic makeup and liver disease may influence intrinsic clearance. Following oral administration of a racemic drug with low hepatic extraction, differences in intrinsic metabolic clearance of enantiomers may influence plasma concentrations of the drug, thus indirectly determining the oral bioavailability of each enantiomer [15].

First-pass extraction of enantiomers by both the GI tract and the liver may also be stereoselective [17]. Enantioselective first-pass metabolism will influence the plasma enantiomer concentrations of racemic drugs with high hepatic extraction. For example, the oral bioavailability of *S*-(−)-propranolol is greater than the *R*-(+) isomer because the *R* isomer is cleared faster following oral administration.

Both passive diffusion and active transport of drugs and metabolites may play a role in renal and biliary excretion. Owing to similar partition coefficients, passive diffusion of enantiomers is not considered stereoselective. However, protein binding (plasma and tissue) may determine the fraction of enantiomers available for passive or active renal clearance. Active transport does govern the secretion and reabsorption of substances from the renal tubules. Thus, stereoselectivity can directly influence the clearance of isomers based on their preferential binding to the proteins involved with active transport. For example, stereoselective renal clearance for metoprolol, quinine and its isomer quinidine, and ketoprofen glucuronides have been demonstrated [7].

In conclusion, one must realize that, while differences in the "measurable" events of total clearances of stereoisomers may appear to be insignificant, this final event in the pharmacokinetic cascade is the sum of the infinitely complex processes of absorption, distribution, metabolism, and excretion.

3 PHARMACODYNAMICS

The pharmacodynamics of isomers is just as enantioselective as their pharmacokinetics. Drugs elicit their effects by binding to receptors. Like all other proteins, receptors are three-dimensional macromolecules that may impose chirality on both chiral and achiral molecules. Therefore binding affinity and geometry with respect to the target receptor may be different for each individual enantiomer [18]. The enantiomer that has a high degree of affinity for a site of action, and therefore is believed to be the enantiomer primarily responsible for the desired therapeutic effect, is referred to as the *eutomer*, whereas the isomer with poor or no activity is called the *distomer*. The distomer is considered an impurity that may have little or no therapeutic activity, contribute to side effects, or cause paradoxical effects [19]. The ratio of activities of eutomer : distomer is known as the eudismic ratio [20]. This ratio is specific for a given racemate and is only rarely equal to 1.

There are many examples of drug enantiomers displaying different or paradoxical "therapeutic" actions [19]. The (–) isomers of barbiturates have depressant effects, whereas the (+) isomers are convulsants. The *S*-(–) isomer of propranolol decreases cardiac output by 25% in humans, whereas the *R*-(+) isomer has no effect. The sedative effects of thalidomide are demonstrated by both enantiomers, while only the (–) enantiomer is responsible for the teratogenic effects. The antipsychotic thioridazine owes its therapeutic effects to the (+) enantiomer, with (+)-thioridazine demonstrating 2.7 and 4.5 times more affinity than (–)-thioridazine for D_2 and α-1 receptors, respectively [21]. Table 1 lists the pharmacodynamic differences of many of the drugs that have been studied in humans.

The study of the pharmacokinetic and pharmacodynamic differences between drug enantiomers continues to generate interest. The Food and Drug Administration (FDA) now requires that the toxicology, pharmacology, and biological properties of enantiomers of new drugs be studied and documented prior to approval [22].

There are many so-called combination drugs frequently prescribed with little evidence of increased therapeutic benefit yet with well-documented increased toxicological consequences [3]. For example, Darvocet®, the combination of acetaminophen and propoxyphene, is one of the most commonly prescribed analgesics although there has never been any convincing evidence that the analgesic effect is any greater than acetaminophen alone. However, the addition of the narcotic propoxyphene now increases the incidence of adverse effects, ranging from constipation to death from respiratory depression in overdose cases.

Combinations like Darvocet® are clearly labeled as containing more than one active ingredient. However, many drugs prescribed as monotherapies are actually racemates, usually 50 : 50 mixtures of enantiomers. And just as with other combination drugs, the two enantiomers may have entirely different pharmacological effects: one enantiomer can be responsible for both therapeutic and adverse effects while the other is inactive at normal doses (e.g., atenolol); each enantiomer may have activity at the same receptor but differ in potency; enantiomers may interact at different receptors, one responsible for the therapeutic effect, one causing serious adverse reactions (e.g., mianserin); each enantiomer may have multiple effects, both adverse and beneficial (e.g., verapamil).

TABLE 1 Pharmacodynamic Differences of Various Enantiomers

Drug	(+) Isomer	(−) Isomer
Barbiturates	Excitation	Sedation
Dobutamine	β_1 & β_2-adrenoceptor agonist (vasodilation)	α_1-adrenoceptor agonist (+)-inotropic/ vasoconstriction
Epinephrine	Inactive	α- and β-adrenergic
Fenfluramine	*S*-(+): selective serotonin reuptake inhibitor	*R*-(−): norepinephrine/ dopamine reuptake inhibition (adverse effects)
Fluoxetine	Selective serotonin reuptake inhibitor	Minimal effect
Ketamine	*S*-(+): strong anesthetic	*R*-(−): weak anesthetic; adverse effects
Levodopa	Antiparkinsonian	Agranulocytosis
Methadone	*S*-(+): minimal effect	*R*-(−): strong analgesic
Methamphetamine	*S*-(+): CNS stimulant	*R*-(−): peripheral vasodilatation
Morphine	Minimal effect	Strong analgesic
NSAIDs	*S*-(+): anti-inflammatory; analgesic	*R*-(−): minimal effect
Penicillamine	Antirheumatic (Wilson's disease)	Neurotoxic
Pentazocine	Anxiety	Analgesia; respiratory depression
Propafenone	*R*-(+): antiarrhythmic	*S*-(−): antiarrhythmic; β-blocker
Propoxyphene	Analgesia	Antitussive
β-Adrenergic antagonists (e.g., propranolol)	*R*-(+): suppresses ventricular arrhythmias w/o β-blockade	*S*-(−): active β-blocker
Quinine/quinidine	Quinidine: treatment of malaria and cardiac arrhythmias	Quinine: treatment of malaria; minimal cardiac effects
Selegiline	*S*-(+): direct CNS stimulation	*R*-(−): MAO-B inhibitor
Tetramisole	*R*-(+): minimal effect	*S*-(−): anthelmintic (Levamisole®)
Thalidomide	Sedative	Sedative; teratogenic
Thioridazine	D_2 and α-adrenergic receptor antagonist	D_1 receptor antagonist
Thyroxine	Inactive	Natural form; thyrogenic effects
Verapamil	*R*-(+): minimal effect	*S*-(−): (−)-dromotropic; (−)-inotropic (−)-chronotropic
Warfarin	*R*-(+): weak anticoagulant	*S*-(−): anticoagulant
3-Methoxy-*N*-methylmorphinan	"Dextromethorphan" (antitussive)	"Levomethorphan" (narcotic analgesic)

4 CHIRALITY AND THERAPEUTIC DRUG MONITORING

In light of what we know about the impact of chirality on various pharmacokinetic and pharmacodynamic parameters, the therapeutic drug monitoring community must decide whether the current practice of measuring total drug blood/plasma concentrations (the sum of both enantiomers) should be continued or whether there is a need to separate and quantitate individual enantiomers [9]. The ability to discriminate between pharmaceutical enantiomers has numerous implications in both clinical and forensic toxicology [23]. For example, (+)-3-methoxy-*N*-methylmorphinan is the legally available, over-the-counter antitussive "dextromethorphan," whereas (–)-3-methoxy-*N*-methylmorphinan is the illegal substance, levomethorphan. The ability to distinguish between these compounds is of obvious interest in forensic cases but also for the treatment of overdose/intoxication cases in a clinical setting. Furthermore, athletes are allowed the use of dextromethorphan, while the International Olympic Committee has banned the use of levomethorphan and its active metabolite, levorphanol (a powerful narcotic analgesic in its own right). The discrimination between enantiomers is also of both clinical and forensic importance in the detection of (+)-propoxyphene (a controlled narcotic analgesic) and its noncontrolled (–) enantiomer, the antitussive, Novrad®. Moreover, natural cocaine is the (–) enantiomer; thus the detection of the (+) cocaine in a specimen would indicate illicit manufacture and may be helpful in tracking sources.

The "eudismic proportion" is the ratio of the concentrations of the eutomer : distomer in plasma or other body fluids [20]. As described above, after absorption, stereoselective metabolism will steadily change the eudismic proportion until a steady state is achieved. The only way to determine if it is the eutomer or the distomer that is preferentially metabolized is with "chiral assays." The eudismic proportion is often route dependent as well. For example, the eudismic proportion for verapamil is 0.2 after oral consumption and 0.5 after IV administration (the eutomer concentration is much lower after oral than after IV dosing). The implications for therapeutic drug monitoring are obvious: plasma concentrations of a racemate measured by nonchiral means only provide information on the total (eutomer plus distomer) drug concentration. These concentrations will not be consistently related to response, since pharmacokinetic constants (half-life, bioavailability, bioequivalence, etc.) vary with each enantiomer. With the clinical lab only providing information on the total drug concentration, further complications arise with respect to interpretation of side effects, protein binding, metabolic conversion to active metabolites, etc.

From a clinical perspective, there is often large interindividual variation in the plasma ratios of the enantiomers that may have significant therapeutic implications [9]. For example, the *S*-(+)/*R*-(−) plasma ratio of tocainide (Tonocard®, an antiarrhythmic agent) varies from 4:1 to 1.3:1; the intersubject (−)/(+)-verapamil ratio varies from 0.15 to 0.30 (after oral dosing); (−)/(+)-propranolol ratios can range from 2:1 to 1:1. It is easy to see how, when enantiomers have different potencies and effects, therapeutic drug monitoring with conventional analytical methods that do not resolve and quantitate individual enantiomers can be misleading at best, and dangerous at worst.

Perhaps the best example of the consequences of not monitoring enantiomers is amphetamine/methamphetamine. Amphetamine and methamphetamine contain a chiral carbon and therefore exist as enantiomers. *S*-(+)-Methamphetamine and *S*-(+)-amphetamine are used therapeutically for narcolepsy, attention deficit disorder, and weight loss. Because *S*-(+)-methamphetamine is a powerful CNS stimulant, it is a popular drug of abuse.

R-(−)-Methamphetamine has one-tenth the CNS stimulation effect of the (+) isomer but has greater peripheral vasoconstrictive properties and thus has been used in over-the-counter nasal inhalers (Vicks® inhalers). Since methamphetamine and amphetamine abuse is monitored in employee drug testing programs, the ability to test stereospecifically for these compounds is critical.

Further complicating the picture, there are a number of drugs that are metabolized to amphetamine and methamphetamine. This group includes benzphetamine, clobenzorex, deprenyl, famprofazone, fenethylline, and fenproporex. Many of these are marketed as the single enantiomer and thus are metabolized to the single enantiomer of either amphetamine or methamphetamine. There are several ways for a medical review officer to document the use of these drugs as an explanation for a positive urinalysis result. Examination of medical history is a straightforward approach. Alternatively, it may be possible to identify parent drug or a unique metabolite in the urine specimens. Unfortunately, this is not always possible given the short half-lives of many of these drugs. Instead, the identification of a single enantiomer unique to these drugs may provide analytical verification that these drugs were ingested.

For example, enantiomer analysis can be used to document selegiline (Deprenyl®, Eldepryl®) use. Selegiline is a selective monoamine oxidase (MAO) inhibitor that has proven effective in the treatment of Parkinson's disease and shows promise in alleviating the symptoms of dementia and schizophrenia. Prescription selegiline is the *R*-(−)-enantiomer; it is converted in the body exclusively to *R*-(−)-methamphetamine and *R*-(−)-amphetamine. Therefore, any findings of the (+) isomers of these compounds would be inconsistent with Deprenyl® use. Conversely, fenproporex (Antiobes®,

Tegisec®) is an anorectic agent not available in the United States, which is sold as a racemic mixture; thus one would expect to find equal amounts of the (+) and (−) isomers in the urine.

The goals of forensic and clinical laboratories should include preventing such false positive reports and detecting all true positive cases of illicit drug abuse. Many screening/confirmation techniques were failing one or both of these goals. Some commercially available immunoassays (EMIT®, FPIA) and Toxilab® have the required sensitivity but are susceptible to false positives from nasal inhaler use. Radioimmunoassay (RIA) screens are generally less sensitive. Standard gas chromatography/mass spectrometry (GC/MS) procedures are unable to distinguish between enantiomers. It became clear that an additional step was necessary before reporting positive methamphetamine results.

Chromatographic identification of (+) and (−) isomers depends on the enantiomers reacting with optically active substrates to form diastereomers. This can be accomplished in one of two ways: a chiral and optically active stationary phase can be used to form transient diastereomers; the optically active drug causes different partition coefficients, or a chiral derivatizing reagent may be used to achieve separation on an achiral stationary phase. Chiral stationary phases are available for both high-performance liquid chromatography (HPLC) and GC systems. However, chiral derivatizing reagents have gained popularity, since they allow chiral analysis on the same instrument used for other routine analyses. The main disadvantage of using a chiral derivatizing reagent is the difficulty in ensuring the optical purity of the derivatizing reagent; four possible isomers can result from the reaction of an asymmetric sample with an asymmetric reagent rather than the two desired. The four derivative isomers consist of two diasteromeric pairs of chromatographically irresolvable enantiomers. Experiments must also be done to demonstrate both that racemization is not occurring during the derivatizing process and that stereoselective formation of one pair of diastereomers is not occurring.

Amines have been converted to diasteromeric amides with several chiral acid chloride reagents for GC analysis, such as *O*-methylmandelyl chloride, α-phenylbutyryl chloride, or anhydride and *N*-trifluoroacetyl-L-prolyl chloride (L-TPC). L-TPC is the most popular of this type of reagent. First used as a chiral derivatizing reagent for the separation of amino acids, it has since become popular for both the on-column and preinjection derivatization of amphetamine and methamphetamine. There have been some reports of racemization with the use of L-TPC. Amide derivatives can also be prepared with acyclimidazole reagents and with chiral acids in the presence of *N,N′*-carbonyldiimidazole. Diasteromeric sulfonamide derivatives have been created using (+)-camphor-10-sulfonyl chloride and menthyl carbamates, such as menthyl chlorformate. These reagents have been used for sympathomimetic amines and amino acid analysis.

More recently, chiral separations by capillary electromigration (CE) have been studied extensively and have proven to provide excellent resolution relatively inexpensively [23]. The two methods that have shown the most promise are capillary zone electrophoresis (CZE) and micellar electrokinetic capillary chromatography (MECC). For enantiomeric separation under electrokinetic conditions, a chiral selector such as a cyclodextrin and the appropriate buffer conditions (pH, ionic strength, other additives, etc.) are required. CZE uses a plain buffer in which mainly charged solutes can be separated. MECC, as its name implies, is a separation technique that utilizes micelles. By virtue of the micelle's relatively hydrophobic interior, this methodology permits the separation and analysis of neutral solutes. The newest methodology incorporates chiral selectors into packed or gel-filled capillary columns and is known as chiral capillary electrochromatography and, if the selector is bonded to the surface, open tubular electrochromatography (OTEC).

Currently, the most popular detection method is ultraviolet (UV) absorbance [24]. Because CE offers a very short pathlength within the detection cell, the lower limit of detection (LOD) for most analytes (without preconcentration) is the low μg/mL range. Other optical detection techniques include fluorescence and laser induced fluorescence (LIF). These two detectors improve sensitivity about 10 and up to 1,000 times, respectively, over UV absorption detectors.

There are many methods currently under development linking CE to a mass spectrometric (MS) detector [24]. There are few examples of methods utilizing CE-MS that are ready for clinical use, but it has proved to be useful in the research lab for gathering structural information. Methods developed primarily for the clarification of drug metabolism include the urinary determination of *N*-1-hydroxyethylflurazepam, haloperidol, and some of the NSAIDs. Other laboratories have developed a CE-MS method for confirmation of methadone and its metabolite, EDDP. To this date, applications of this methodology have been limited by the development of applicable buffer systems and efficient interfaces between the CE instrument and the MS.

When compared to chiral HPLC or GC, chiral CE's key features of automation, minimal sample preparation, the use of minimal amounts of organic solvents, ease of buffer change, analysis speed, and low cost guarantee its future in the clinical and forensic lab. Additionally, since most CE methodologies require very small sample volumes, this technique has the advantage in postmortem laboratories where sample quantities are often limited. Reliable and automated CE instruments are now commercially available [24]. For example, for drugs of clinical and toxicological interest, the CE assay is claimed to be the only assay that can simultaneously separate the optical isomers of 3-methoxy-*N*-methylmorphinan and 3-hydroxy-*N*-methylmorphinan, thus distinguishing cough syrup consumption from illegal drug use [23]. The chiral CE assay of basic drugs in urine provides a simplified approach for characterization of the enantiomeric "pattern" of sympathomimetic amines, methadone, and EDDP. As described above, the analysis of metabolite isomers of some prescribed drugs is critical to the recognition and discrimination of illegal drug use. Finally, in CE, samples can often be injected directly onto the column with no or minimal pretreatment (dilutions, centrifugation, filtration, protein precipitation, etc.) [24]. This has obvious advantages in the rapid analysis of samples from the emergency room in suspected overdose cases as well as routine therapeutic drug monitoring.

In addition to being a complementary tool to the more commonly applied chromatographic methods of HPLC and GC, chiral CE also has the prospect of bringing chiral separations to the clinical laboratory.

5 SPECIFIC EXAMPLES

5.1 β-Adrenoceptor Antagonists

While most β-blockers are sold as racemates, some are available as single enantiomers with the *S* enantiomer acting as the eutomer [3]. However, the *R* isomer has been shown to have important properties as well. For example, *R*-propranolol has been shown to suppress ventricular arrhythmias independent of β-blockade. *R*-sotalol also has antiarrhythmic properties and prolongs the QT interval. These compounds are examples of single enantiomer therapy that would be beneficial in heart failure patients who require an antiarrhythmic agent but could not tolerate the further depression of cardiac output resulting from β-blockade. Similarly, *R*(+)-timolol is an extremely effective agent for lowering intraocular pressure and could be used by glaucoma patients without the adverse effects of systemic β-blockade.

Labetolol is both an α- and a β-adrenergic antagonist with two asymmetric carbons, thus four isomers. These isomers are present in equal concentrations in the currently marketed product. Recently, it has been demonstrated that the *RR* isomer is responsible for the β-antagonist properties while the *SR* isomer is the α-blocker [3]. The other two isomers appear to be inactive. Despite its α-blocking properties, labetolol does not appear to have a larger effect on blood pressure maintenance than pure β-blockers. The single *RR* isomer, as "dilevalol," had been under evaluation as a nonselective β-blocker with agonist properties at vascular and other β_2-adrenoceptors until the study was halted owing to reports of severe hepatotoxicity. It is unknown why the single racemate was more hepatotoxic than the mixture, but it is possible that one of the other isomers may have inhibited the metabolism of dilevalol to a toxic intermediate. This shows that it cannot be automatically assumed that marketing of new drugs as the single, most active, enantiomer will always be beneficial.

5.2 Barbiturates

It is the (–) isomer of the barbiturates that is responsible for the sedating effects; the (+) enantiomer causes excitation [9]. In one pharmacokinetic study, there was reduced oral clearance of the (–) enantiomer in elderly patients that would tend to produce enhanced sedation [25].

The barbiturates are good examples to show that there is no relationship between optical activity and configuration. While the levorotatory form of the barbiturates acts as the eutomer, the configuration of the various barbiturates varies: *R*(–)-hexobarbital, *S*(–)-secobarbital, and *S*(–)-thiopentone are the anesthetic/anticonvulsant forms [7].

5.3 Fenfluramine

Fenfluramine (Pondimin®), available for over 20 years as an anorectic for the treatment of obesity, was initially marketed as a racemate but was recently successfully marketed as a the single *S*(+) enantiomer, dexfenfluramine (Redux®) [3]. In 1997, both of these compounds were withdrawn from the market by their manufacturer owing to reports that long-term administration may be related to heart valve disease and/or primary pulmonary hypertension. While both enantiomers act mainly by selective inhibition of the reuptake of serotonin into nerve endings, dexfenfluramine has no effect on noradrenergic or dopaminergic systems. It is these effects of racemic fenfluramine on the catecholinergic systems that are believed to be responsible for the adverse effects.

5.4 Ketamine

Ketamine (Ketalar®, Ketaject®) is a short-acting injectable anesthetic primarily used in veterinary medicine although there is still some use in human pediatric anesthesia. Both enantiomers of ketamine have anesthetic action. However, *R*(–)-ketamine is believed responsible for the adverse effects of delirium, agitation, excitement, tachycardia, unpleasant "emergence reactions," and visual disturbances [9]. In studies where equianesthetic doses of the individual enantiomers [*S*(+)-ketamine requiring 60% less than *R*(−)-ketamine] were given to surgical patients, the mean plasma level of *S*(+)-ketamine at emergence was 0.5 μg/mL, while that of *R*(−)-ketamine was 1.7 μg/mL. If one assumes that brain concentrations parallel plasma concentrations, this would confirm that *R*(−)-ketamine has much less anesthetic activity. Drayer [9] cautions, however, that one cannot also unequivocally conclude that this pharmacodynamic difference also explains the adverse reactions attributed primarily to *R*(−)-ketamine. He notes that *R*(−)-ketamine is unable maximally to suppress EEG activity to the same extent as the (+) isomer, and thus inadequate depth of anesthesia rather than intrinsic activity may explain the previously listed side effects. Finally, there may be stereoselective elimination of the (+) isomer, resulting in substantially higher plasma concentrations of *R*(−)-ketamine, and the adverse effects may simply be a concentration effect rather than an isomeric effect.

5.5 Mianserin

Mianserin (Bolvidon®, Norval®) is a tetracyclic antidepressant used clinically since 1975 that apparently does not cause the cardiac toxicity characteristic of the tricyclic antidepressants. However, other adverse reactions such as agranulocytosis, skin rashes, and hepatotoxicity have been reported [3]. Since mianserin is administered as a racemate, it has been suggested that these adverse reactions may be due to one of the enantiomers. Pinder and Van Delft [26] demonstrated that it is most likely *S*-mianserin that is responsible for the antidepressant effect. *R*-mianserin has little effect on alleviating depression but has been shown to be cytotoxic. This "differential cytotoxicity" is believed to be due to the differential metabolism of the enantiomers: *R*-mianserin is primarily metabolized by *N*-demethylation, while *S*-mianserin is metabolized by hydroxylation [3].

5.6 Nonsteroidal Anti-Inflammatory Drugs (NSAIDs)

It is the *S*-enantiomer of these acids that is widely considered to be the active isomer, though the ratio of *S:R* activity varies between compounds of this class [27]. Only naproxen (Aleve®, Naprosyn®) is marketed as the single *S* enantiomer. Studies of the individual NSAID enantiomers are difficult since these drugs serve as the model for "chiral inversion" metabolism of the relatively inactive *R* form to the active *S* enantiomer in vivo. The "reversion" of the *S* to the *R* enantiomer does not seem to occur. Some of the unpredictable problems with toxicity of this group of compounds could be due to the *R* isomer, especially since naproxen is one of the least toxic of these compounds. If this were unequivocally established, a case could certainly be made for the use of the single *S* enantiomers, but more research is required, especially on the importance of chiral inversion, before more development along these lines is possible.

5.7 Opiates/Narcotic Analgesics

The (–) isomer of most of the compounds of this class (morphine, codeine, levomethorphan, etc.) is the analgesic form. Most often the (+) enantiomer acts as an antitussive agent, but while (–)-pentazocine (Talwin®) is the more potent analgesic, (+)-pentazocine produces anxiety in some patients [9].

The only notable exception to this rule is propoxyphene. (+)-propoxyphene (Darvon®) is responsible for the analgesic action of this compound, while the (–) isomer is marketed as an antitussive (Novrad®).

5.8 Quinine/Quinidine

Quinidine and quinine have four asymmetric carbons and are actually diastereomers resulting from inversions at the 8- and 9-carbon. The (+) enantiomer of quinidine is utilized for the treatment of both malaria and cardiac arrhythmias but has the adverse effect of acting as a cardiodepressant [3]. The (–) enantiomer of quinine has minimal effects on the heart and thus can be used to treat malaria with significantly reduced adverse cardiac side effects.

5.9 Verapamil

Verapamil (Calan®, Isoptin®, Verelan®) is a calcium-channel blocker used for the treatment of arrhythmias, angina, hypertension, and, more recently, migraine headaches. This compound is an excellent example of clinically important differences in both pharmacokinetics and pharmacodynamics due to stereochemistry. Both enantiomers are potent vasodilators, but *S*-(–)-verapamil has more potency with regard to negative chronotropic, dromotropic, and inotropic effects than *R*-(+)-verapamil in doses that are equipotent for coronary vasodilatation [28]. One could theorize that it would be safer to use *R*-(+)-verapamil alone for the treatment of angina, hypertension, and migraines, where vasodilator properties are necessary, giving a significantly reduced risk of adverse cardiac events like heart failure or various degrees of heart block.

The isomers of verapamil also differ pharmacokinetically, specifically, with regard to extensive stereoselective first-pass metabolism in the liver. After intravenous (IV) infusion, the (–)/(+) enantiomer ratio is 0.56; following oral dosing, the ratio is 0.23. This property could be taken advantage of therapeutically in the treatment of acute arrhythmias with the relatively greater concentration of *S*-(−)-verapamil after IV administration.

Differential metabolism of verapamil can also have some potentially dangerous consequences. Cimetidine (Tagamet®) is a histamine H_2-receptor antagonist used for the treatment of ulcers and often prescribed to patients who are also taking verapamil. Cimetidine also appears to inhibit preferentially the metabolism of *S*-(−)-verapamil, reducing its clearance and producing potentially dangerous prolongation of the PR interval at otherwise "therapeutic" doses [*S*-(−)-verapamil is 10 to 20 times more potent than *R*-(+)-verapamil in prolonging AV conduction] [14].

5.10 Warfarin

Both warfarin enantiomers are highly beneficial anticoagulants. Despite *S*-warfarin having about three times the potency of the *R* enantiomer, both enantiomers can achieve the same maximum degree of anticoagulation [3]. Adverse effects other than the risk of hemorrhage due to excessive anticoagulation are uncommon and are the same for both enantiomers. Therefore one would not expect the risk of this dose-related effect to be eliminated with the use of a single enantiomer. Interestingly, for two of the coumarin anticoagulants, warfarin and phenprocoumon, the *S*-(−) enantiomers are 2 to 5 times more potent than the *R*-(+) enantiomer, while *R*-(+)-nicoumalone has many times the potency of its *S*-(−) enantiomer [17].

Adverse drug interactions can occur, however, since different enzymes metabolize the enantiomers (*S*-warfarin by CYP2C9; *R*-warfarin by CYP3A4 and CYP1A2) [14]. One would expect that, since the *S* enantiomer possesses most of the pharmacological activity, compounds such as sulfinpyrazone that inhibit the CYP2C9 family of enzymes would decrease the clearance of this enantiomer, thus increasing warfarin's effects.

5.11 Single Enantiomers

Naturally occurring dopa is the (−) isomer and is therefore manufactured as the single isomer, levodopa. Similarly, thyroxine is present in the body as levothyroxine, and this is also the manufactured form. (+)-thyroxine is believed to be inactive. Levamisole, a nematocidal agent, is also marketed exclusively as the (−) enantiomer.

Compounds with several chiral centers are also usually marketed as single enantiomers to reduce the chance of including a potentially toxic enantiomer and to minimize the size and weight of the tablet [3]. For example, perindopril (Aceon®) is an angiotensin converting enzyme (ACE) inhibitor used in the treatment of hypertension. This compound has five chiral centers giving rise to 32 (2^n; n = number of chiral centers) possible enantiomers. The *S,S,S,S,S* enantiomer is used clinically.

6 THE CONTROVERSY: SINGLE ENANTIOMER VS. RACEMIC THERAPY

As you now know, and has been known by the pharmaceutical industry for many years, enantiomers and diastereomers are often readily distinguished in biological systems and frequently exhibit different pharmacokinetic (absorption, distribution, metabolism, and excretion) and pharmacodynamic (pharmacological, therapeutic, and toxicologic) effects. Despite this knowledge, it has been the standard practice to produce chiral drugs as racemates [2].

The opinions in the argument of whether all chiral drugs should be produced as single enantiomers spans the ideological spectrum. Some believe that inactive/toxic enantiomers are simply excess weight and risk and should be considered as "impurities" and eliminated. Others believe that it is more important to consider morbidity and mortality due to a lack of effective treatments when drugs are held up by the slow manufacturing process of producing enantiomerically pure compounds [3].

Among the reasons given for ignoring the presence of distomer in drugs has been the contention that "clean drugs" would be difficult and expensive to manufacture [20]. But over the past decade, selective synthesis/enantiomer separation techniques have vastly improved, and the cost of such production has dropped accordingly. Some have gone so far as to suggest that the labels of racemic drugs announce that the contents are composed of 50% therapeutically inactive compounds that may or may not be harmless. Along these same lines, others feel that the development of mixtures of geometric isomers/diastereomers is not justified unless they represent a reasonable fixed dose combination [2]. In these cases, the question of whether the optimal ratio of the isomers is produced by the synthetic process needs to examined. Other factors favoring the use of pure enantiomers include less complex pharmacokinetics, concentration–effect relationships and drug interactions, a decreased body burden of xenobiotic material, and, in some cases, an improved pharmacological profile [1].

Reasons to continue the production of chiral drugs as racemates include the cases where enantiomers were found to be identical in pharmacological properties or one enantiomer is known to be inert or possess little biological activity [2]. Cayen [29] expressed several reasons why the pharmaceutical industry may want to continue to produce racemic drugs. In addition to the above reasons, he felt that additive or synergistic effects of the enantiomers, high therapeutic indexes, chiral inversions, and indications for life-threatening disease were all factors that should seriously be considered as justifications for the production of racemic compounds. In general, published studies do not unanimously agree that all racemic dugs are more harmful than their single isomer counterparts; many times unforeseen benefits have been realized with the inclusion of the so-called "inactive" isomer [3]. In these cases where both enantiomers exhibit desirable but different properties, development of a fixed combination ratio (not necessarily a "true" racemic mixture of 1:1) should be considered.

The development of chiral drugs has opened up a number of issues to be considered by regulatory policies and guidelines: acceptable manufacturing controls of "impurities"; adequate pharmacological and toxicological assessment by pharmaceutical manufacturers; and proper characterization of metabolism and distribution in the drug development and clinical trial phases [2]. Another question that must be addressed is whether the current compounds manufactured as racemates would be "grandfathered" and not subject to new regulatory guidelines [30]. Regulatory authorities currently require explicit justification for marketing racemic drugs. Special guidelines have been issued for chiral drugs stating that the chemical composition of these drugs must be stated and that the pharmacology/toxicology of each enantiomer be thoroughly explored.

The more pragmatic conclusion is to evaluate each new pharmaceutical on its own merit. Clearly, it is vitally important that the decision as to whether to produce single enantiomer dugs or racemic mixtures must be based on clinical evaluation of both enantiomers [2]. If one enantiomer is shown to produce significant toxicity or has little or no activity and large-scale synthesis is easily achieved, it is obviously more reasonable to produce the compound as the pure, active enantiomer. The first requirement must be that the chirality of the drug be recognized and the eutomer identified [31]. The specifications of the drug substance must include assignment of absolute configuration, an enantioselective analytical method, and a full description of the synthesis. More importantly, the need to justify on chemical, preclinical, and clinical grounds the choice of the form chosen for marketing must be added to the current requirements for drug approval. If the new drug is to be marketed as a single enantiomer, only the reasons for the choice and documentation of synthesis (including purity, interbatch variation, etc.) and confirmation of optical stability (synthetic and in vivo) need be added. If a racemic mixture is being considered, the pharmacokinetics of both enantiomers must be determined in Phase I studies including possible metabolic interconversions. Many authors believe that it is unlikely that the use of a single enantiomer would significantly improve efficacy; the most likely value is the reduction of adverse side effects. However, these same authors believe the consumer's dollar and other health care resources would be better utilized on more general measures to reduce the burden of adverse effects.

REFERENCES

1. M Eichelbaum. Side effects and toxic reactions of chiral drugs: a clinical perspective. Arch Toxicol Suppl 17:514–521, 1995.
2. IK Reddy, TR Kommuru, AA Zaghdoul, M Khan, A Mansoor. Chirality and its implications in transdermal drug development. Crit Rev Ther Drug Carrier Syst 17(4):285–325, 2000.
3. AK Scott. Stereoisomers and drug toxicity. The value of single stereoisomer therapy. Drug Safety 8(2):149–159, 1993.
4. RH Levy, AV Boddy. Stereoselectivity in pharmacokinetics: a general theory. Pharmacol Res 8(5):551–556, 1991.
5. S Jortani. Stereoselective metabolism and disposition of thioridazine enantiomers. PhD dissertation, Virginia Commonwealth University, Richmond, Virginia, 1993.
6. DN Wade, PT Mearrick, JL Morris. Active transport of L-dopa in the intestine. Nature 242:463–465, 1973.
7. F Jamali, R Mehvar, FM Pasutto. Enantioselective aspects of drug action and disposition: therapeutic pitfalls. J Pharm Sci 78(9):695–715, 1989.
8. E Jahnchen, WE Muller. Stereoselectivity in protein binding and drug distribution. In: D Briemer, P Speiser, eds. Topics in Pharmaceutical Sciences. New York: Elsevier Science Publishing, 1983, pp 109–117.
9. DE Drayer. Pharmacodynamic and pharmacokinetic differences between drug enantiomers in humans: an overview. Clin Pharmacol Therap 40(2):125–133, 1986.
10. CAM Van Ginneken, JF Rodrigues de Miranda, AJ Beld. Stereochemistry and drug distribution. In: EJ Ariens, W Soudijn, PBMWM Timmermans, eds. Stereochemistry and Biological Activity of Drugs. Oxford: Blackwell, 1982, pp 55–80.
11. K Kawashima, A Levy, S Spector. Stereospecific radioimmunoassay for propranolol isomers. J Pharmacol Exp Ther 196:517–523, 1976.
12. NPE Vermeulen, DD Breimer. Stereoselectivity in drug and xenobiotic metabolism. In: EJ Ariens, W Soudijn, and PBMWM Timmermans, eds. Stereochemistry and Biological Activity of Drugs. Oxford: Blackwell, 1982, pp 55–80.
13. J Caldwell. Stereochemical determinants of the nature and consequences of drug metabolism. J Chromatogr A 694(1):39–48, 1995.
14. HK Kroemer, MF Fromm, M Eichelbaum. Stereoselectivity in drug metabolism and action: effects of enzyme inhibition and induction. Therap Drug Monitoring 18:388–392, 1996.
15. T Walle, K Walle. Pharmacokinetic parameters obtained with racemates. Trends Pharm Sci 7:155–159, 1986.
16. GR Wilkinson, D Shand. A physiological approach to hepatic drug clearance. Clin Pharmacol Therap 18:377–390, 1975.
17. K Williams, E Lee. Importance of drug enantiomers in clinical pharmacology. Drugs 30:333–354, 1985.

18. ASV Burgen. The stereochemistry of binding to receptors. In: EJ Ariens, W Soudijn, PBMWM Timmermans, eds. Stereochemistry and Biological Activity of Drugs. Oxford: Blackwell, 1982, pp 81–87.
19. EJ Ariens. Stereochemistry, a basis for sophisticated nonsense in pharmacokinetics and clinical pharmacology. Eur J Clin Pharmacol 26:663–668, 1984.
20. EJ Ariens, EW Wuis. Chiral cognisance: a road to safer and more effective medicinal products. J Royal Coll Physicians London 28(5):395–398, 1994.
21. CN Svendson, M Froimowitz, C Hrbek, A Campbell, N Kula, RJ Baldessarini, BM Cohen, S Babb, MH Teicher, ED Bird. Receptor affinity, neurochemistry, and behavioral characteristics of the enantiomers of thioridazine: evidence for different stereoselectivities at D1 and D2 receptors in rat brain. Neuropharmacol 27:1117–1124, 1988.
22. CS Kumkumian. The use of stereochemically pure pharmaceuticals. In: IW Wainer, DE Drayer, eds. Drug Stereochemistry: Analytical Methods and Pharmacology. New York: Marcel Dekker, 1988, pp 119–168.
23. S Zaugg, W Thormann. Enantioselective determination of drugs in body fluids by capillary electrophoresis. J Chromatogr A 875:27–41, 2000.
24. W Thormann, Y Aebi, M Lanz, J Caslavska. Capillary electrophoresis in clinical toxicology. Forensic Sci Int 92:157–183, 1998.
25. MHH Chandler, SR Scott, RA Blouin. Age-associated stereo-selective alterations in hexobarbital metabolism. Clin Pharmacol Therap 43:436–441, 1988.
26. RM Pinder, AML Van Delft. The potential therapeutic role of the enantiomers and metabolites of mianserin. Brit J Clin Pharmacol 15(suppl 2):269S–276S, 1983.
27. WF Kean, CJL Lock, HE Howard-Lock. Chirality in antirheumatic drugs. Lancet 338:1565–1568, 1991.
28. K Satoh, T Yanagisawa, N Taira. Coronary vasodilator and cardiac effects of optical isomers of verapamil in the dog. J Cardiovasc Pharmacol 2:309–318, 1980.
29. MN Cayen. Racemic mixtures and single stereoisomers: industrial concerns and issues in drug development. Chirality 3:94–98, 1991.
30. TJ Maher, DA Johnson. Review of chirality and its importance in pharmacology. Drug Develop Res 24:149–156, 1991.
31. J Caldwell. Importance of stereospecific bioanalytical monitoring in drug development. J Chromatogr A 719(1):3–13, 1996.

4

HPLC in Bioavailability Examination

Zdzisław Chilmonczyk
National Institute of Public Health, Warsaw, and
University of Białystok, Białystok, Poland

Kamila Kobylińska
Pharmaceutical Research Institute, Warsaw, Poland

Hassan Y. Aboul-Enein
King Faisal Specialist Hospital and Research Centre,
Riyadh, Saudi Arabia

1 INTRODUCTION

The term pharmacokinetics refers to the movements of drugs within biological systems as affected by the processes of drug absorption, distribution, metabolism, and elimination (Fig. 1). These processes determine how rapidly, in what concentration, and for how long the drug will appear at the target organ. It is particularly important from a clinical point of view. Pharmacokinetic data are essential to support the safety and efficacy of new compounds. The logic of assessing the metabolism and disposition of a drug and understanding its dose–concentration–response relationships is inescapable. Measurement of the absorption, distribution, and elimination of drugs and their metabolites in the body provides valuable information about their behavior and can lead to rational dosage regimens.

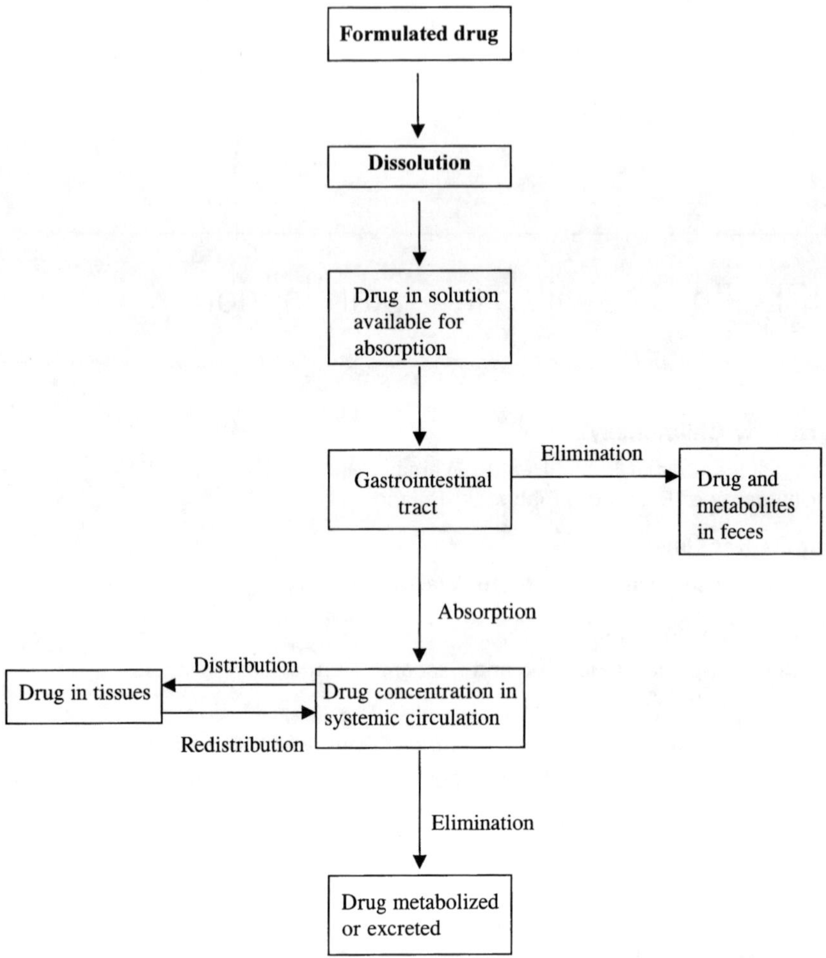

FIGURE 1 Schematic representation of the process of drug absorption, distribution, metabolism, and elimination.

Bioavailability is the pharmacokinetic parameter that describes the completeness of absorption of the intact drug and is defined as the percentage of the fraction of the administered dose that reaches the systemic circulation. The concept of bioavailability was introduced in the 1960s and was based on the observation that two drugs that contained the same active component at the same dose, but formulated in different drug products, often differed therapeutically [1]. Now drug regulatory authorities in the United States (The U.S. Food and Drug Administration), the European Community, and other countries require the characterization of biological availability for all new drugs intended for oral use. It is also necessary to demonstrate a bioequivalence between the innovator product and the proposed generic equivalent.

There are several definitions of bioavailability and bioequivalence. For the U.S. Food and Drug Administration (FDA) [2], bioavailability means the rate and extent to which the active ingredient or active moiety is absorbed from a drug product and becomes available at the site of drug action. For drug products that are not intended to be absorbed into the bloodstream, bioavailability may be assessed by measurements intended to reflect the rate and extent to which the active ingredient or active moiety becomes available at the site of action.

For the European Union [3], bioavailability means the rate and extent to which the active substance or active moiety is absorbed from a pharmaceutical form and becomes available at the site of action. In the majority of cases, substances are intended to exhibit a systemic therapeutic effect, and a more practical definition can then be given, taking into consideration that the substance in the general circulation is in exchange with the substance at the site of action: bioavailability is understood to be the extent to which and the rate at which a substance or its active moiety is delivered from a pharmaceutical form and becomes available in the general circulation. It may be useful to distinguish between the "absolute bioavailability," of a given dosage form, as compared with that (100%) following intravenous administration (e.g., oral solution vs. iv), and the "relative bioavailability," as compared with another form administered by the same or another nonintravenous route (e.g., tablets vs. oral solution).

The European Union's definition of bioequivalence [3] is that two medicinal products are bioequivalent if they are pharmaceutically equivalent or pharmaceutical alternatives and if their bioavailabilities after administration in the same molar dose are similar to such a degree that their effects, with respect to both efficacy and safety, will be essentially the same.

1.1 Absolute Bioavailability

Absolute bioavailability is calculated for all new drugs, and pharmacokinetics data are estimated during clinical study before drug registration. Following intravenous administration, it is assumed that the drug is completely available for systemic circulation—that is, that the bioavailability is equal to 100%. However, following oral administration, this may not always be the same. For some drugs only a percentage reaches the circulation. This may be due to several reasons:

1. The drug may be incompletely absorbed.
2. It may be metabolized in the gut, the gut wall, or the liver prior to entry into the systemic circulation.

3. It may undergo enterohepatic cycling with incomplete reabsorption following elimination into the bile.

The differences in the drug concentration versus time curves obtained from intravenous and oral routes of administration are the results of these processes.

1.1.1 First-Pass Effect

Intestinal and hepatic metabolic systems represent a first-pass effect: a drug is metabolized significantly before reaching the general circulation following its absorption. Many drugs are nonpolar and dissolve readily in lipid, so that they are reabsorbed by the kidney and poorly excreted. The principal function of drug-metabolizing enzymes is to convert these chemicals into highly water-soluble products that can be eliminated. Drug metabolism has classically been divided into Phase I and Phase II reactions [4]. Phase I reactions include oxidations, reductions, and hydrolyses. Phase II reactions involve the conjugation of chemicals with hydrophilic moieties such as glutathione, glucuronic acid, sulfate, or amino acids. Phase I enzymes include cytochromes P450 (CYPs). The CYPs represent a superfamily of heme-containing proteins that catalyze a variety of reactions including C-, N-, and S-hydroxylations and dehalogenations, as well as dealkylations, deaminations, and reductions [5]. Three CYP gene families (i.e., CYP1, CYP2, and CYP3) are involved in metabolism of pharmaceuticals in humans [6]. The CYP1 family contains enzymes CYP1A1 and CYP1A2. CYP1A1 catalyzes the metabolism of relatively few drug classes, i.e., anesthetics (lignocaine), analgesics (i.e., paracetamol, aminopyrine, antipyrine, naproxen, phenacetin), and antiarrhythmics (mexiletine). CYP1A2 is involved in the metabolism of a number of drugs such as analgesics (i.e., paracetamol, aminopyrine, naproxen, phenacetin), antiarrhythmics (verapamil), anticoagulant (R-warfarin), antidepressants (i.e., imipramine, mianserin), antineoplastics (flutamide), antiestrogens (tamoxifen), and anxiolytics (zolpidem) [7].

The CYP2 family includes at least seven gene subfamilies. The CYP2C subfamily is most frequently involved in the metabolism of clinically important drugs. The most abundant enzyme of this family is the CYP2C9 enzyme. This enzyme catalyzes the metabolism of a number of drugs such as analgesics (paracetamol, antipyrine, diclofenac, ibuprofen, naproxen, piroxicam), antiasthmatic (salbutamol), antibacterials (trimethoprim), anticoagulants (*S*-warfarin, dicoumarol), antidepressants (moclobemide), antidiabetic (tolbutamide), antiepileptics (phenobarbital, phenytoin), and antiviral (zidovudine) [7]. CYP2C enzymes show variability due to genetic polymorphism, and the polymorphism is particularly characteristic for CYP2C9 and CYP2C19. Genetic polymorphism has important influences on variability in human pharmacokinetics, including intraindividual differences in drug toxicity and drug interaction. Clinically significant metabolic polymorphism associated with the CYP2C19 enzyme was reported for β-blockers (propranolol), antidepressants (imipramine), anxiolytics (diazepam), proton pump inhibitor (omeprazole), and certain barbiturates [7]. Genetic polymorphism was first demonstrated with CYP2D6 and has become known as sparteine/debrisoquine polymorphism [8]. Debrisoquine is a β-adrenergic blocking agent prescribed in the treatment of hypertension. Analysis of the ability of humans to hydroxylate debrisoquine uncovered different distributions of metabolizer phenotypes. About 5–10% of Caucasians and 1–3% of Asians and Africans excrete the drug virtually unchanged and were termed "poor metabolizers." About 80 to 90% are extensive (excreted large amounts of hydroxylated debrisoquine metabolites) and up to 5% of Caucasians are ultrarapid metabolizers. The polymorphism in CYP2D6 is of great clinical importance, since this enzyme metabolizes many different therapeutic agents, i.e., analgesic and anti-inflammatory drugs, opioid analgesics, antiarrhythmics, and antidepressants. Differing CYP2D6 metabolic phenotypes may account for some of the large interindividual variations seen in the drug clearance and plasma concentrations of patients receiving these agents.

CYP3A is the primary CYP subfamily in humans, responsible for CYP-mediated Phase I metabolism of more than 50% of administered drugs. CYP3A is primarily located in the smooth endoplasmic reticulum of the liver and small intestine. Among the CYP3A subfamily, CYP3A4 is the most abundant and most important enzyme in human drug metabolism. CYP3A4 catalyzes the metabolism of a number of clinically important drugs, for example, the antiarrhythmics (verapamil), antihistamines (i.e., astemizole, mizolastine), antifungals (ketoconazole, miconazole), antidepressants (imipramine), anticonvulsants (carbamazepine), benzodiazepines (alprazolam), calcium channel antagonists, cancer chemotherapeutic agents (etoposide), immunosuppressants (cyclosporin), and HIV protease inhibitors (indinavir, ritonavir) [9]. Although liver and intestinal CYP3A4 activity plays an important role in the first-pass extraction of many CYP3A4 substrates, interindividual variability in the activity of the intestinal transporter P-glycoprotein may be an equally significant determinant of oral bioavailability, for example for cyclosporin [10].

As a consequence, the first-pass effect leads to variable drug excretion rates and intersubject differences in the final serum drug concentrations. For this reason, therapeutic response and side effects vary widely between patients treated with the same dose of drug.

1.2 Relative Bioavailability

Bioequivalence is a kind of relative bioavailability and involves a comparison between a test and the reference drug product. Although bioavailability and bioequivalence are closely related, bioequivalence comparisons require the calculation of a confidence interval for the criterion. Bioequivalence studies are advisable with new dosage forms or new routes of administration, and are necessary in the case of generic product.

The term generic pharmaceutical product refers to a product that is interchangeable with the innovator product and that is usually manufactured without a license from the innovator company and then marketed after the expiry of a patent or other exclusive rights. The innovator product is the original patented drug, with documentation based on chemical, pharmaceutical, and clinical data. Bioequivalence of a generic drug should be compared with the innovator drug.

2 MODELS IN PHARMACOKINETICS

Pharmacokinetics is strictly the science of mathematical assessment of changes in concentrations of drugs. Most commonly, however, the term is used in a general sense to cover the wide area of drug absorption, metabolism, distribution, and excretion. In order to interpret pharmacokinetic data it is necessary to set up certain models so that mathematical equations describing the movement of the drugs can be formulated.

The most important pharmacokinetic parameters illustrating the rate of bioavailability are the blood maximum concentration (C_{max}) and the time to reach the maximum concentration (t_{max}). The area under the concentration–time curve (AUC) is a common measure of the extent of bioavailability (Fig. 2). Pharmacokinetic parameters can be obtained according to standard compartmental models or with noncompartmental analysis. The compartment model is an approximation for a biological system, because variations in physiological distribution, inhomogeneity of the media, and diffusion process are all interrelated with chemical changes. A compartment model is really an average rather than an exact state. The theoretical model involves certain parameters whose numerical values can be varied in order to bring it into the closest possible agreement with experimental results.

2.1 The One-Compartment Model

The body is represented by a single compartment with a volume V (Fig. 3). There is no distribution phase or it is very short. It is assumed that the drug is injected directly into this compartment (e.g., intravenous injection). The

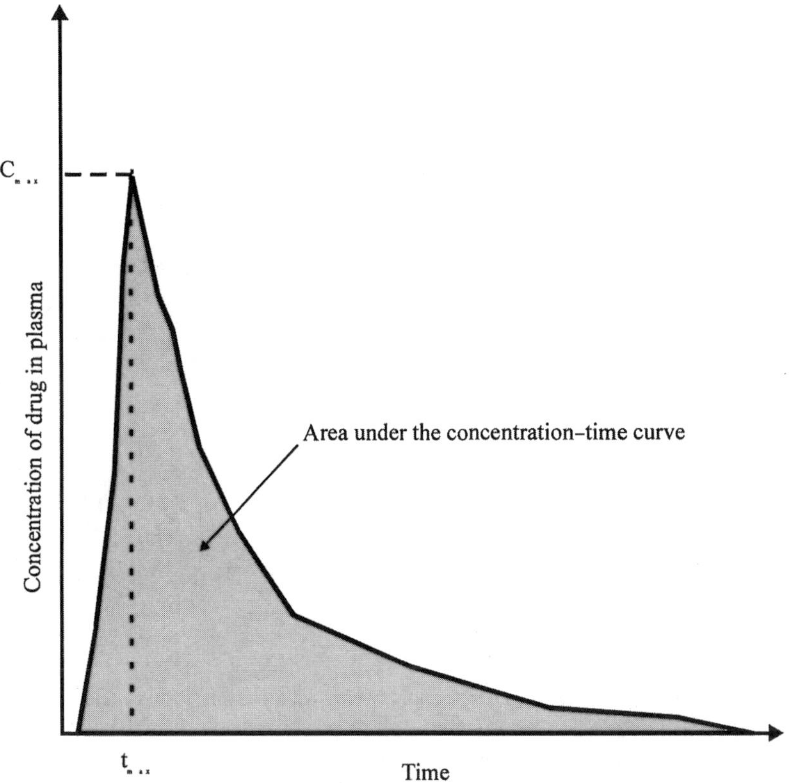

FIGURE 2 Plasma drug time–concentration curve. C_{max} = maximal plasma drug concentration; t_{max} = time corresponding to C_{max}.

concentration of drug at time 0 (C_0) can be calculated according to the equation [11]

$$C_0 = \frac{D}{V}$$

where V = volume of distribution and D = dose of drug.

Following oral administration or by intramuscular injection, the graph of plasma drug concentration against time shows a rising phase and a falling phase. For a one compartment system, the equation is

$$C_t = C_0[\exp(-k_{el}t) - \exp(-k_a t)]$$

where C_t = concentration of a drug at time t, k_a = absorption rate constant, and k_{el} = elimination rate constant.

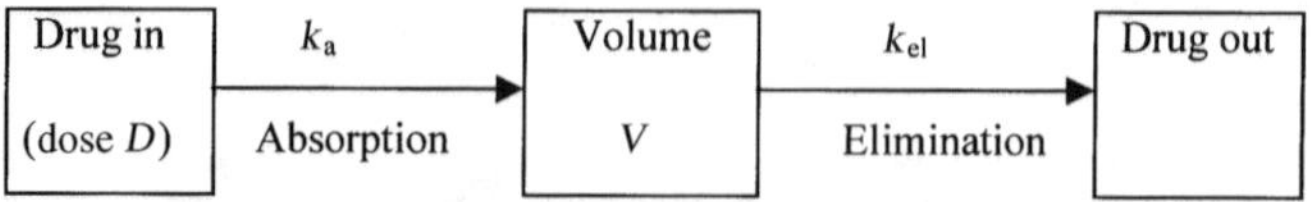

FIGURE 3 The one-compartment model.

The elimination rate constant determines the clearance of the drug from the compartment. For most drugs this clearance is directly related to the concentration of a drug, and the time taken for a decrease in drug concentration to half the original value will always be the same. The plot of log drug concentration versus time yields a straight line.

2.2 The Two-Compartment Model

It is generally considered that double exponential decay of a plot of log drug concentration versus time indicates transfer of the drug through a two-compartmental model (Fig. 4). In this system, there is a rapidly perfused compartment and a slowly perfused compartment. The former comprises the blood, liver, lung, kidney, and sometimes brain. The latter comprises voluntary muscle, fat, and (again) sometimes brain. The equivocal position of brain results from rapid perfusion but relatively slow transfer across the blood–brain barrier for some drugs. It is important to remember that tissue uptake has features of rate of attainment (rapid/slow) and extent (shown by high or low tissue-to-plasma concentration ratios). Rate of attainment is a function of blood perfusion and transfer across tissue membranes, while extent is a function of the ability of tissues to bind drugs. Rate of attainment and extent are not dependent on each other.

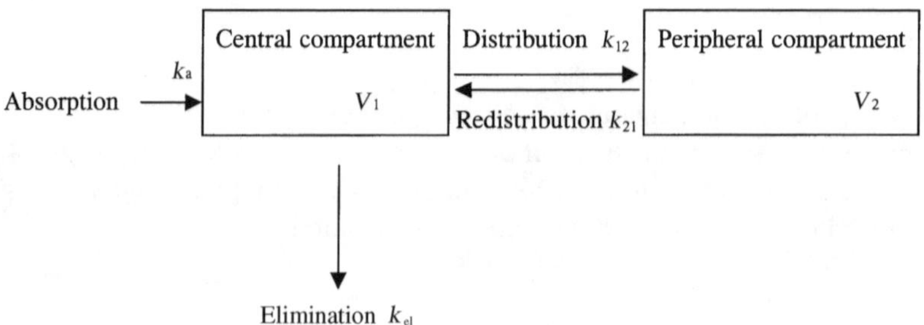

FIGURE 4 The two-compartment model.

Mathematical analysis of a two-compartment model is more complex than that of a one-compartment model. The concentration–time dependance is described according to the equation

$$C_t = A\exp(-\alpha t) + B\exp(-\beta t) - (A + B)\exp(-k_a t)$$

where C_t is drug concentration at time t, A and B are points of the curve used to obtain apparent volumes of two compartments, α and β are the rate constants, and k_a is the absorption constant.

In this system the drug first enters the absorption compartment. Next, the drug is distributed to the central compartment and to the peripheral compartment and excreted. It can be seen that absorption could still be progressing as distribution and elimination start.

2.3 The Multicompartment Model

The pharmacokinetics of drug may involve more than one peripheral compartment (e.g., a shallow compartment and a deep compartment). In this model, the subdivision of the equilibration phase would indicate more than two compartments.

3 SUBJECTS

Bioavailability/bioequivalence studies are usually performed with healthy volunteers. Subjects can be of either gender and should be between 18 and 55 years old with a weight within the normal range according to accepted life tables. The volunteers should be judged to be healthy on the basis of standard laboratory tests, their medical histories, physical examinations, and vital signs. Exclusion criteria for bioavailability/bioequivalence studies include smoking, alcohol intake, consumption of grapefruit juice and caffeinated beverages, concurrent medication, significant medical conditions, positive HIV status, hepatitis B and C, and the presence of illegal drugs in the urine [2,3,12]. Before admission to the study, each subject should be informed of the nature and the risks of the study, and written informed consent should be obtained from all participants in the study. The study should also be approved by the local Ethics Committee. During the study, any adverse effects reported by the subjects should be recorded. The number of subjects required for bioequivalence study should not be smaller than 12 and should be determined using appropriate statistical methods [12,13].

The standard bioequivalence trial is conducted according to a randomized crossover design, in which each subject receives the test and the reference drug with an appropriate washout period between dosing. The washout period is chosen on the basis of the half-life of the drug, and it is usually more than five half-lives [2,3].

In practice, the bioavailability/bioequivalence study involves the measurement of plasma, serum, or whole blood concentrations of drug or metabolite(s) at different times, or alternatively, the measurement of the total amount of parent drug or metabolite(s) excreted in the urine. The number of the blood samples must be appropriate to describe the absorption, distribution, and elimination phases of the drug. For most drugs, 12 to 18 samples, including a predose sample, should be collected per subject per dose. This sampling should be continued for a period not less than three terminal half-lives of the drug [2,3].

4 ASSAY OF DRUGS IN BIOLOGICAL FLUIDS

4.1 HPLC Analysis

Analytical methods for the quantitative determination of drugs and their metabolites in biological samples include

1. Methods based on immunoassays procedure, for example,
   Radioimmunoassay (RIA)
   Enzyme-multiplied immunoassay technique (EMIT)
   Enzyme-linked immunosorbent assay (ELISA)
2. Microbiological methods
3. Capillary electrophoresis and chromatographic methods, for example,
   Thin-layer chromatography (TLC)
   Gas chromatography (GC)
   High-performance liquid chromatography (HPLC)

Thanks to ultraviolet, electrochemical, fluorometric, and mass spectrometry detectors, high-performance liquid chromatography methods are commonly used in the determination of drugs in physiological fluids at levels applicable to pharmacokinetic studies. There are many different types of stationary phase for HPLC analysis, and the selection of stationary phase, mobile phase, and type of detection depends on the specificity of the drug and the goal of the separation [14]. The chosen analytical method should be fully validated with respect to adequate sensitivity, specificity, linearity, recovery, accuracy, and precision [15,16]. Additionally, the stability of the sample under frozen conditions, at room temperature, during freeze-thaw cycles, and during the analytical process should be checked. Chromatograms of the analysis should show that the separation of drug from matrix components is sufficient for reliable quantitation and that no endogenous compounds caused interference with the drug peak.

4.2 Sample Preparation Techniques

Sample preparation is an important step in HPLC analysis for several reasons:

1. It separates the analyte from matrix constituents.
2. It concentrates the analytes to improve sensitivity and the detection limit of the intended method.
3. It removes endogenous interfering compounds to improve the specificity of the method. The final aim of sample preparation is to obtain a reproducible and homogenous solution compatible with subsequent HPLC separation.

The most frequently used procedures are

Protein precipitation
Liquid–liquid extraction (LLE)
Solid-phase extraction (SPE)
Direct injection techniques (i.e., column switching methods)

4.2.1 Precipitation Technique

Protein precipitation is a simple and rather fast method for sample cleanup. In this procedure, precipitating agents (organic modifiers and acid agents such as trichloroacetic, perchloric, tungstic, and methaphosphoric acids) are added to the sample, and the precipitate is removed by filtration or centrifugation [17]. The supernatant can often be injected into the HPLC column without any further treatment. Such methods are highly suitable for use in the clinical therapeutic drug monitoring laboratory. In some cases when the drug is present in high concentration these methods could be used in pharmacokinetic studies. Examples of chromatograms of tamoxifen (antiestrogenic drug), zolpidem (an hypnotic agent), and itraconazole (antifungal drug) obtained by direct injection of plasma samples after precipitation with acetonitrile are shown in Fig. 5, Fig. 6, and Fig. 7, respectively.

4.2.2 Liquid–Liquid Extraction

This is the most common technique used to extract analytes from liquid matrices. The compounds of interest are selectively transferred, on the basis of their different partition coefficients, from one liquid phase into a second immiscible liquid phase (organic solvent) [18]. The organic solvent is then

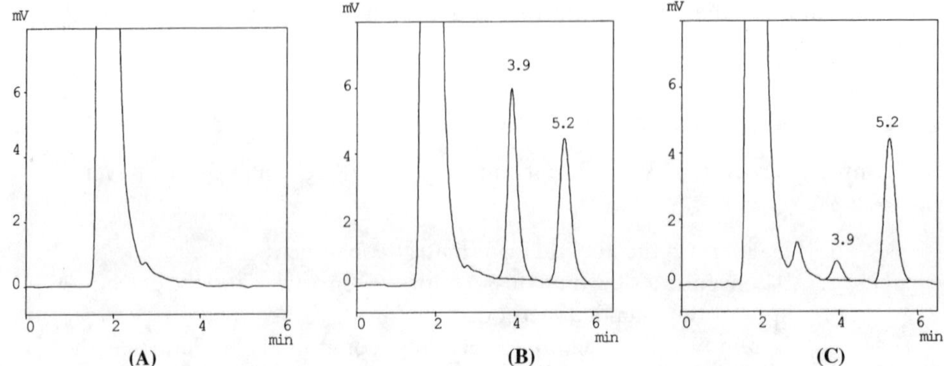

FIGURE 5 Chromatograms obtained by direct injection of 20 μL plasma samples after precipitation with acetonitrile. (A) Blank human plasma, (B) spiked human plasma with 50 ng of tamoxifen and internal standard, and (C) a plasma sample collected 96 h after oral administration of 20 mg of Nolvadex D™ tablets per day to the volunteer. The retention times for tamoxifen and internal standard are 3.9 and 5.2 min, respectively. Chromatographic conditions: Column: 150 × 3.9 mm I.D., 10 μm, μPorasil (Waters); mobile phase: methanol-water-triethylamine-acetic acid (98:2:0.03:0.3 v/v); flow rate: 1.4 mL/min; detection: fluorescence λex = 256 nm and λem = 380 nm after post-column photochemical reaction. (Unpublished data from the Pharmacology Department, Pharmaceutical Research Institute, Warsaw, Poland.)

evaporated to dryness and the residue is reconstituted in a small volume of solvent that is compatible with the HPLC separation. Analytes extracted into the organic phase could be also reextracted into the aqueous phase (i.e., acid) and then directly injected into the reversed-phase column. Figures 8 and 9 illustrate examples of chromatograms obtained after liquid–liquid extraction of carvedilol (an antihypertensive agent) and ondansetron (an antiemetic drug) from human plasma.

4.2.3 Solid-Phase Extraction

Solid-phase extraction (SPE) isolates compounds from a sample by using the same mechanisms as in liquid–solid chromatography: molecules are separated according to their interaction with solid and liquid phase. The small, disposable extraction columns filled with a variety of sorbents are used [19]. The SPE involves several different steps, which are exemplified for the isolation of domperidone (a dopamine antagonist) from human plasma [20] (Fig. 10). The column is first conditioned with an appropriate solvent (i.e., methanol) to solvate the functional groups of the sorbent.

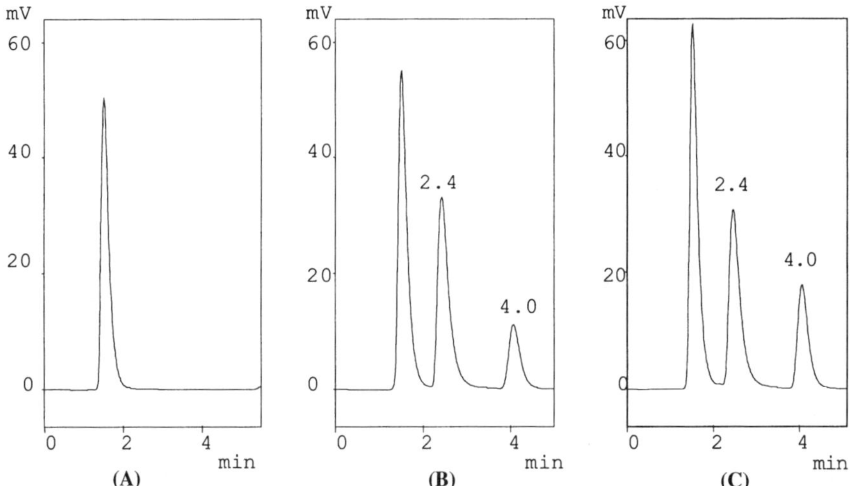

FIGURE 6 Chromatograms obtained by direct injection of 20 μL plasma samples after precipitation with acetonitrile. (A) Blank human plasma, (B) spiked human plasma with 50 ng of zolpidem and internal standard (pefloxacine), and (C) a plasma sample collected 80 min after oral administration of 10 mg of Stilnox™ tablets per day to the volunteer. The retention times for zolpidem and internal standard are 4 and 2.4 min, respectively. Chromatographic conditions: Column: 100 × 4.6 mm I.D., 5 μm, Chromolith Performance Rp-18 (Merck); mobile phase: 0.05 phosphate buffer with 0.1% triethylamine, pH 6.4–acetonitrile (65:35 v/v); flow rate: 1.0 mL/min; fluorescence λex = 244 nm and λem = 388 nm. (Unpublished data from the Pharmacology Department, Pharmaceutical Research Institute, Warsaw, Poland.)

Then the sample is applied to the conditioned column and passed slowly through the column by aspiration or positive pressure. The column containing retained analyte is subsequently washed with an appropriate solvent that selectively elutes the impurities but leaves the analyte on the column. The final step of this process is the quantitative elution of the analyte from the column and collection of the pure concentrated analyte. An example of the chromatogram obtained after SPE extraction of domperidone is illustrated in Fig. 11.

4.3 Detectors

4.3.1 Spectrophotometric Detection

The sensitivity of HPLC methods depends on the type of detection. In most cases, the HPLC method is carried out with ultraviolet (UV) detection using

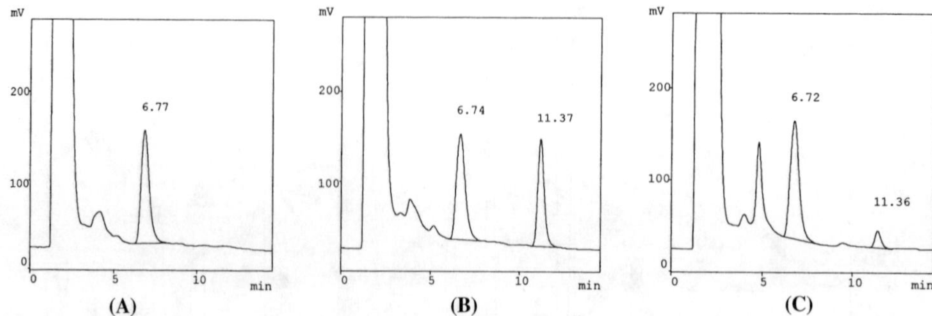

FIGURE 7 Chromatograms obtained by direct injection of 50 μL plasma samples after precipitation with acetonitrile. (A) Spiked human plasma with internal standard, (B) spiked human plasma with 300 ng of itraconazole and internal standard, and (C) a plasma sample collected 12 h after oral administration of 200 mg of Orungal™ capsules per day to the volunteer. The retention times for itraconazole and internal standard are 11.4 and 6.7 min, respectively. Chromatographic conditions: Column: 150 × 4.0 mm, I.D., 5 μm, Novapak C18 (Waters); mobile phase: water–acetonitrile–diethylamine pH 2.2 (50:50:0.01 v/v). The pH was adjusted with 85% phosphoric acid; flow rate: 0.7 mL/min; detection: fluorescence λex = 260 nm and λem = 360 nm. (Unpublished data from the Pharmacology Department, Pharmaceutical Research Institute, Warsaw, Poland.)

either a variable-wavelength or diode- or photodiode-array detector [21]. The instrument must consist of an energy source, a system for selecting the required wavelength, a sample flow cell, and detector electronics converting the signal from diode and reference diode into absorbance, which is subsequently transmitted into the data system. A deuterium lamp as light source is suitable for wavelengths from 190 to 400 nm. When it is necessary to measure absorbance at visible wavelengths, a higher energy tungsten lamp is used. Some detectors have incorporated both deuterium and tungsten halogen lamps. Such a system provides the measure of absorbance at UV and visible light simultaneously. One of these two light sources is selected according to the mirror angle. At the same time as the mirror is switched, the lamp selector photo microsensor detects the change and lights the appropriate lamp, while turning the other lamp off. Analyte concentration C in the flow cell is related to absorbance A, analyte molar absorptivity ε, and flow-cell length L_{fc} by Beer's law:

$$A = C\varepsilon L_{fc}$$

Good analytical results can only be obtained by careful selection of the wavelength used for detection. The wavelength is estimated from spectra of

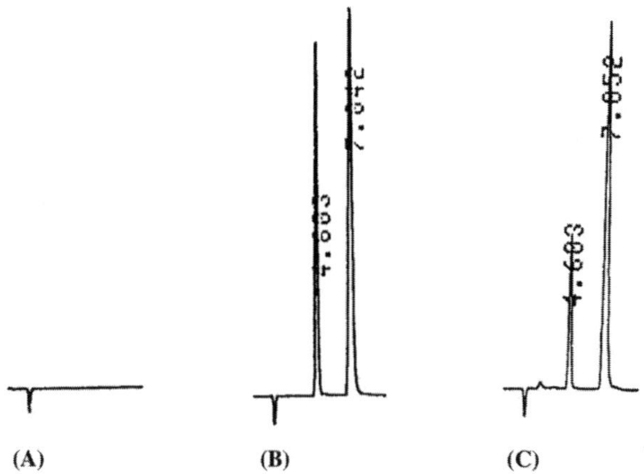

Figure 8 Chromatograms of (A) blank human plasma, (B) spiked human plasma with 20 ng of carvedilol and internal standard, and (C) a plasma sample from a volunteer at 3 h after single administration of Dilatrend 25 mg tablets. The retention times for carverdilol and internal standard are 4.6 and 7 min, respectively. Chromatographic conditions: Column: 150 × 4.6 mm I.D., 5 μm, Supelcosil LC-8-DB (Supelco); mobile phase: 1 M dibutylamine phosphate (pH 2.5): acetonitrile: water (0.5 : 36 : 73.5 v/v); flow rate: 1.5 mL/min; detection: fluorescence λex = 240 nm and λem = 340 nm. (Unpublished data from the Pharmacology Department, Pharmaceutical Research Institute, Warsaw, Poland.)

individual sample components or, alternatively, with a diode array detector, which permits the aquisition of absorbance spectra for all sample components during method development. The wavelength chosen for spectrophotometric detection must provide acceptable absorbance by the analyte in the sample with minimal absorption by sample interferences, combined with acceptable light transmittance by the mobile phase.

The detector signal A is proportional to the molar absorptivity ε of the compound. The trace analysis of compounds with ε below 100 is usually not possible with UV detection. Saturated hydrocarbons and their amino or nitrile derivatives are the only organic compounds for which UV detection is impossible unless they are derivatized. Aromatic compounds usually have ε values above 1000 at wavelengths above 210 nm. Most drugs are aromatic compounds, and they have absorptivity higher than 1000.

The mobile phase absorbance should be less than 0.5 at the wavelength used for detection. If mobile phase absorbance is higher than 1.0, the detection of absorbance is impossible. Some solvents used in normal-phase

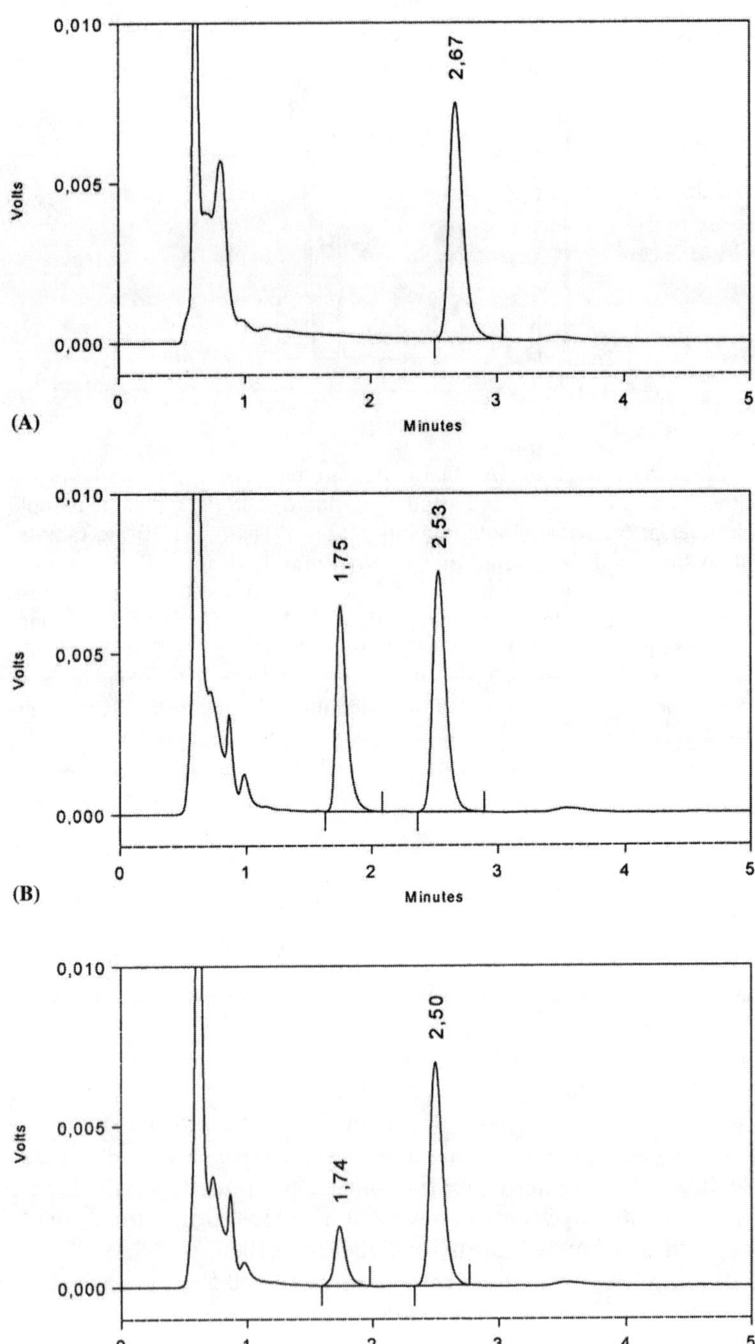
0,010
0,005
0,000
Volts
2,67
0
1
2
3
4
5
(A)
Minutes
0,010
0,005
0,000
Volts
1,75
2,53
0
1
2
3
4
5
(B)
Minutes
0,010
0,005
0,000
Volts
1,74
2,50
0
1
2
3
4
5
(C)
Minutes

chromatography have absorbances higher than 1.0 for lower wavelength values. For instance, ethyl acetate cannot be used for wavelengths below 260 nm.

4.3.2 Fluorescence Detection

Fluorescence detection is more sensitive than UV, so it is often used in assays of drugs whose plasma or serum concentrations are rather small (ng/mL, pg/mL) [21]. Fluorescence is characterized by specific excitation spectra and fluorescence (emission) spectra. The excitation spectrum is, in theory, identical with the absorption spectrum. The fluorescence spectrum is displaced toward higher wavelengths. As an approximation, the fluorescence spectrum has been considered as the mirror image of the absorption spectrum [22]. The intensity of fluorescence is proportional to the concentration of fluorophore only if the solution is diluted and of low absorbance ($A < 0.05$). If the solution is so concentrated that it absorbs all the exciting radiation, the fluorescent intensity must inevitably be maximal. Also, samples may absorb at the wavelength of the emitted light (internal absorbance) in addition to that of the excitation energy, reducing the fluorescence.

The individual features of different fluorescence detectors may lead to major differences in performance, which can cause problems in transferring an HPLC method from one laboratory to another. A further complication in the use of fluorimetric detection is that the fluorescence signal and optimum wavelength for excitation and emission can be dependent on separation conditions: temperature, solvent polarity and viscosity, and pH.

4.3.3 Electrochemical Detection

Electrochemical detectors are more sensitive than fluorimetric detectors, and compounds that exhibit electrochemical activity are more common than compounds that fluoresce. So when greater sensitivity is required, the electrochemical detector is the detector of choice.

FIGURE 9 Chromatograms of (A) spiked human plasma with internal standard, (B) spiked human plasma with 30 ng of ondansetron and internal standard, and (C) a plasma sample collected 6 h after oral administration of 8 mg Zofran tablets per day to the volunteer. The retention times for ondansetron and internal standard are 1.7 and 2.5 min, respectively. Chromatographic conditions: Column: 33 × 4.6 mm I.D., 3 μm, Supelcosil LC-CN (Supelco); mobile phase: acetonitrile-0.02 phosphate buffer pH 3.0 (23:77 v/v); flow rate: 1.4 mL/min; detection: UV, λ = 305 nm. (Unpublished data from the Pharmacology Department, Pharmaceutical Research Institute, Warsaw, Poland.)

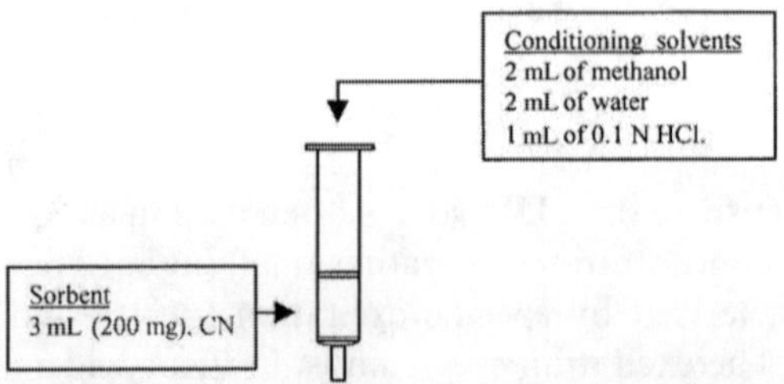

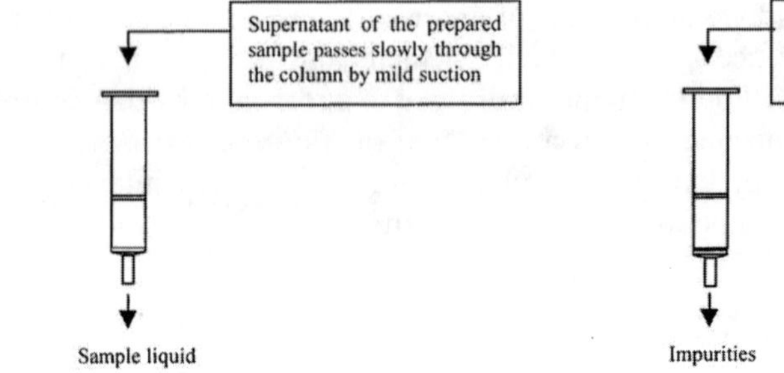

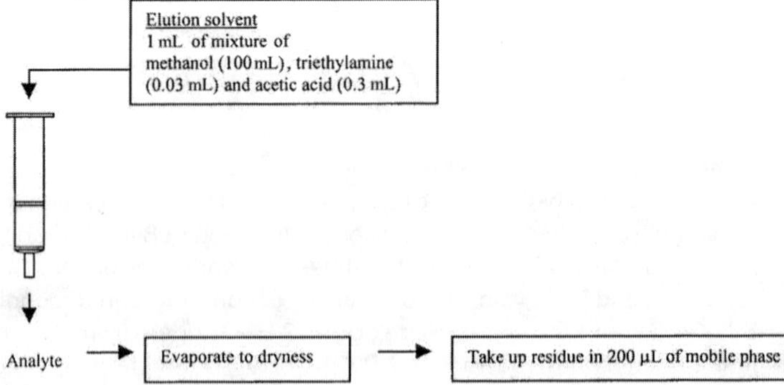

FIGURE 10 Extraction procedure of domperidone from human plasma.

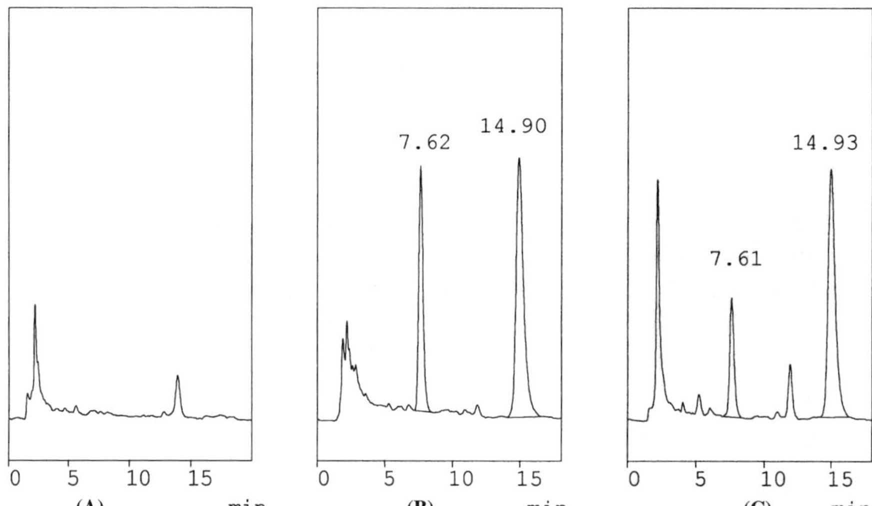

FIGURE 11 Chromatograms of (A) blank human plasma, (B) spiked human plasma with 20 ng of domperidone and internal standard, and (C) a plasma sample from the volunteer after oral administration of 20 mg Motilium tablets. The retention times for domperidone and internal standard are 7.6 and 14.9 min, respectively. Chromatographic conditions: Column: 150 × 4.6 mm I.D., 5 μm, Supelcosil LC-8 (Supelco); mobile phase: methanol–water–triethylamine–acetic acid (60:40:0.02:0.3 v/v); flow rate: 1.4 mL/min; oven temperature: 35°C; detection: fluorescence λex = 282 nm and λem = 326 nm after postcolumn photochemical reaction. (Reprinted from J Chromatog B 744:207–212, 2000, with the permission of Elsevier Science.)

In all electrochemical detectors, current generated at a working electrode is actually being measured. Usually the working electrode is held at a fixed potential. When this electrode is placed in mobile phase, it will generate a background current due to any oxidation or reduction of the mobile phase or contaminants. If an analyte passes the working electrode, it is oxidized (or reduced) by the working electrode, and an increased background current may be observed. Electrochemical detection can be performed in either oxidative or reductive mode, depending on the analyte to be assayed. The most common is oxidative detection. Detector response for a given analyte is dependent on its molecular structure, concentration, and applied potential. It is affected by temperature, pump pulsations, and pH of mobile phase.

4.3.4 Mass Spectrometer Detection

The combined liquid chromatography–mass spectrometry process is commonly abbreviated as LC-MS or, if multiple MS stages are involved such as with triple or quadrupole mass spectrometers, as LC-MS-MS. The use of LC-MS is becoming more popular, despite the high cost of detectors. The LC-MS systems are simple, so more chromatographers are using this technique in routine work. Mass spectrometers are "ion optical" devices that produce a beam of gaseous ions from a sample, separate the resulting mixture of ions according to their mass-to-charge ratios, and provide output signals from which the nominal mass and abundance of each detected ionic species can be determined. Mass spectrometers have three distinct features: ion source, mass analyzer, and detector. The ion source has two functions; the first is to produce a beam of ions representative of the sample and the second is to accelerate the ion beam and cause fragmentation of the weaker bonds within the molecule. Mass analysers resolve ions of the same mass-to-charge ratio (m/z) value from all other ions in a mixture and can focus specific ion beams onto the detector. The number of ions at each m/z value is then detected by converting the ion current into a signal that can be amplified [21].

5 PHARMACOKINETIC ANALYSIS

As previously noted, bioavailability concerns the rate and extent to which a drug is absorbed into the systemic circulation. For evaluating the bioequivalence of two drug products, their bioavailabilities are compared. Both the rate and the extent of absorption of the new formulation and the reference formulation should be the same within a statistical tolerance. The rate of absorption can be characterized by the maximum plasma concentration (C_{max}) and the time of its occurrence (t_{max}). The extent of absorption is usually quantified by the area under the plasma-concentration–time curve (AUC) that is proportional to the fraction of the administered dose reaching the systemic circulation.

According to the EU guideline [3], the following pharmacokinetic parameters should be used for bioequivalence single-dose studies:

C_{max}, t_{max}, AUC_t (AUC from administration to the time of the last measured concentration), AUC_{∞} (AUC extrapolated to infinity)
Ae (cumulative urinary excretion from administration until time t)
Ae_{∞} (cumulative urinary excretion extrapolated to infinite time)
dAe/dt (urinary excretion rate)

$t_{1/2}$ (plasma concentration half-life)
MRT (mean residence time)

For studies in steady state, the following parameters should be determined:*

C_{min} (the minimum concentration after drug administration)
C_{max} (the maximum concentration)
AUC_{τ} (AUC during a dosage interval in steady state where τ is the dosing interval)
Fluctuation (C_{max}- C_{min}/C_{av}, where C_{av} is average plasma concentration)

Pharmacokinetic data may be evaluated by conventional compartmental models [12] or noncompartmental models [23]. Compartmental analysis can produce C_{max}, t_{max}, and AUC results differing markedly from those obtained with the noncompartmental approach [24]. For this reason FDA and EU guidelines recommended the noncompartmental analysis [2,3]. In this case, C_{max} and t_{max} are obtained directly from the plasma-concentration–time data without interpolation. The half-life of the elimination process is evaluated from the equation

$$t_{1/2} = \frac{\ln 2}{k_{el}}$$

where k_{el} is the apparent elimination rate constant calculated by log-linear regression of the terminal phase of the plasma concentration versus time curve.

The area under the concentration/time curve (AUC) is usually obtained by summing up the AUC_t with the extrapolated area AUC_{rest} according to

$$AUC_{rest} = \frac{C_{last}}{k_{el}} \qquad (C_{last}\text{—the last plasma concentration})$$

$$AUC_{\infty} = AUC_t + AUC_{rest}$$

AUC_t is usually calculated by the linear or logarithmic trapezoidal rule up to the last plasma concentration $\geq$ limit of quantification [25,26]. The ratio $100 \cdot AUC_t/AUC_{\infty}$ should be at least 80%. This means that the extrapolated fraction should be $\leq$20% of the total AUC [26].

* The so-called steady state is a result of metabolic equilibrium after multiple drug administration.

The absolute bioavailability (F) is estimated using the equation [24]

$$F = \frac{\text{dose}_{\text{i.v.}} \times \text{AUC}_{\text{oral}}}{\text{dose}_{\text{oral}} \times \text{AUC}_{\text{i.v.}}}$$

The relative bioavailability is calculated as

$$\frac{\text{AUC}_{\infty}\,(\text{test formulation})}{\text{AUC}_{\infty}\,(\text{reference formulation})}$$

6 STATISTICAL ANALYSIS TO ASSESS BIOEQUIVALENCE STUDIES

Assuming a multiplicative model in the statistical analysis of two-period crossover bioequivalence studies, the following hypotheses should be tested [27]:

$$H_0: \quad \mu_T/\mu_R \leq \theta_1 \quad \text{or} \quad \mu_T/\mu_R \geq \theta_2 \quad \text{(bioinequivalence)}$$
$$H_1: \quad \theta_1 < \mu_T/\mu_R < \theta_2 \quad \text{(bioequivalence)}$$

where $0 < \theta_1 < 1 < \theta_2$ and μ_T and μ_R are the geometric means of C_{max} and AUC values after administration of the test formulation and reference formulation, respectively.

The ratio of μ_T over μ_R is called the point estimate, and θ_1 and θ_2 denote the lower and upper limits of the bioequivalence range, which is 0.80–1.25 [27]. Tighter acceptance criteria for C_{max}, such as an acceptance range of 0.9–1.1, have been proposed for narrow-therapeutic-index drugs [28]. In certain cases, a wider range of acceptance (0.75–1.33) could be defined [3]. The assessment of a 90% confidence interval around the point estimate is in a two-way crossover design equal to the application of the Schuirmann's two one-sided test procedure [29]. Thus H_0 is rejected in favor of bioequivalence if the 90% confidence interval around the point estimate is included in the bioequivalence acceptance range ($\alpha = 5\%$).

For a standard bioequivalence trial, AUC and C_{max} values, both log transformed, should be subjected to analysis of variance (ANOVA) techniques, including terms for sequence, subject within sequence, period, and formulation. Using the mean square error obtained from the ANOVA, a 90% confidence interval for the ratio of the test to the reference formulation could be constructed. Bioequivalence can be assumed if the 90% confidence interval of the point-estimate (test over reference formulation) falls inside 0.80–1.25 for AUC and for C_{max} range and when the left and right sides of Schuirmann's t test are both statistically significant ($p < 0.05$) [12].

In the case of t_{max}, the assumption of an additive model seems appropriate, and the test problem is formulated as [27]

$$H_0: \quad \mu_T - \mu_R \leq \theta_1 \quad \text{or} \quad \mu_T - \mu_R \geq \theta_2$$
$$H_1: \quad \theta_1 < \mu_T - \mu_R < \theta_2$$

where $\theta_1 < \theta_2$; the specification of θ_1 and θ_2 should be based on clinical relevance.

In other words, in the case of t_{max} the statistical analysis should be performed on the untransformed data, and the limits of the bioequivalence should be expressed in absolute differences instead of proportionality. A nonparametric test must demonstrate that there are no statistically significant differences between the test and the reference formulation. According to the EU guideline [3], the evaluation of t_{max} for characterization of the rate of absorption is only useful if

> there is a clinically relevant claim for rapid release or action or signs related to adverse effects. The nonparametric 90% confidence interval for this measure of relative bioavailability should lie within a clinically determined range.

In the case of other pharmacokinetic parameters (e.g., C_{min}, fluctuation, $t_{1/2}$, etc.) the EU guideline [3] states,

> considerations analogous to those for AUC, C_{max} or t_{max} apply, taking into consideration the use of log-transformed or untransformed data, respectively.

7 THE BIOEQUIVALENCE OF DIFFERENT GLIPIZIDE FORMULATIONS (AN EXAMPLE OF BIOEQUIVALENCE STUDY)

7.1 Introduction

Glipizide is a second-generation sulfonylurea oral hypoglycaemic drug used in the treatment of diabetes mellitus of the maturity-onset type [30]. The drug is completely absorbed after oral administration with maximum plasma concentration at 1–3 h. The elimination half-life ranges from 2 to 5 h [31,32]. The purpose of this section is to describe how bioavailabilities of different pharmaceutical formulations are compared. As an example, two glipizide formulations, Glipizyd 5 mg tablets, T. Z. F. POLFA S.A., as a test formulation, and Glibenese 5 mg tablets, Pfizer, as a reference formulation, were chosen [33].

7.2 Subjects, Materials, and Methods

7.2.1 Subjects

Twenty-four healthy male volunteers with mean age (±SD) of 26.8 (±4.8) years, mean body weight of 81 (±12.9) kg, and mean height of 179.7 (±6.6) cm were selected for the study. The participants were confirmed to be in good health by physical examination (vital signs, ECG) and laboratory testing (hematology, serum biochemistry, urinalysis) performed within one week of study entry and repeated after completion of the study. All subjects gave written informed consent, and the study protocol was approved by Ethical Committee of the Institute of Tuberculosis and Lung Disease (Warsaw). The subjects were asked not to take any medications for 7 days before the study or during the study.

7.2.2 Procedure

A randomized two-way crossover design with a 1 week washout period was used. After a 12 hour overnight fast, each subject received a single dose of 5 mg tablet of glipizide of either formulation. Blood samples were taken from a forearm vein into heparinized tubes immediately prior to the dose (time zero) and at 0.25, 0.5, 1, 1.5, 2, 2.5, 3, 3.5, 4, 6, 8, 10, 12, and 24 h post dose. Plasma was separated by centrifugation and stored at −20°C until drug analysis.

7.2.3 Analysis of Plasma Samples

Plasma samples were analyzed for glipizide by solid-phase extraction, adopting by a specific and validated HPLC method with UV detection at 225 nm [33]. The analytical column μBondapak (C_{18}, 250 × 3.9 mm. I.D., 10 μm.; Waters) was preceded by a C_{18} guard column (4 × 3.0 mm, Phenomenex). The column was heated at 40°C. A 20 μL aliquot of the plasma extract was injected into the HPLC system by utilizing a mobile phase of 0.01 M phosphate buffer (pH 3.5)-methanol (45:55 v/v), with a flow rate 1 mL/min. The chromatographic conditions and the extraction procedure gave a clean chromatogram for the compound.

7.2.4 Pharmacokinetic and Statistical Analysis

The pharmacokinetic parameters for glipizide were calculated from the plasma concentration time profiles by a noncompartmental method using the Pharm/PCS program [34]. The maximum plasma concentration (C_{max}) and the time to peak (t_{max}) were read directly from the concentration versus-time curves for each volunteer. The area under the curve (AUC_t) was calculated using the trapezoidal rule. The apparent terminal rate constant k_{el} was computed by log-linear regression over the last data points of the concentration–time curve. The value of k_{el} was then used to extrapolate the AUC_t values until infinity. Terminal half-life ($t_{1/2}$) was calculated using the equation $t_{1/2} = \ln 2/k_{el}$.

Bioequivalence between formulations was assessed by calculating individual AUC_t, AUC_∞, C_{max} ratios (test/reference) together with their means and 90% confidence intervals (90% CI). The inclusion of the 90% CI for the ratio into the bioequivalence range was analyzed by using ANOVA for log-transformed data.

7.3 Results and Discussion

The individual plasma concentration/time profiles of glipizide following oral administration of 5 mg tablets of the test and reference formulation are shown in Fig. 12.

The mean plasma glipizide concentration time curves following the administration of the test and reference formulations are presented in Fig. 13. Table 1 summarizes some of the mean pharmacokinetic parameters. The mean plasma concentration–time profiles and the pharmacokinetic characteristics for the two glipizide formulations were similar; no statistically significant differences were observed. The mean area under the curves (AUC_t) were 1828 ± 336 ng · h/mL and 1934 ± 363 ng · h/mL for the test and reference formulations, respectively. The corresponding values for the extrapolated areas (AUC_∞) were 2011 ± 403 ng · h/mL and 2126 ± 381 ng · h/mL for

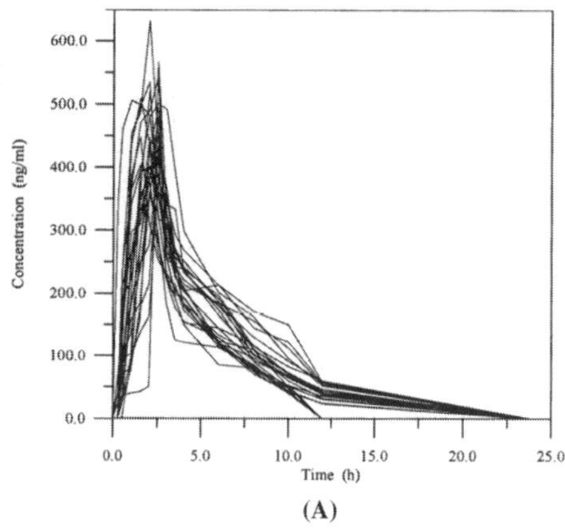

(A)

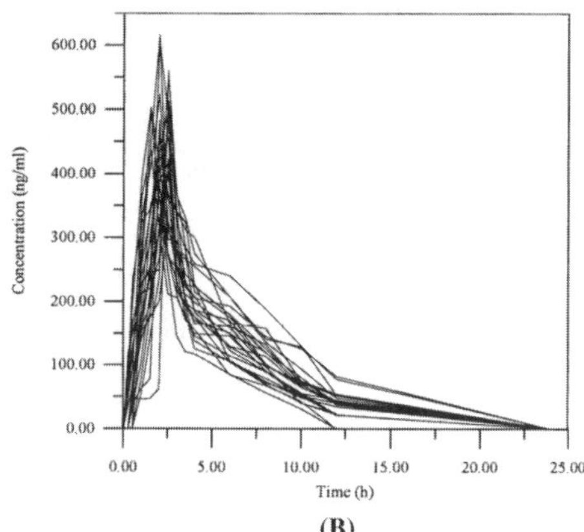

(B)

FIGURE 12 Individual plasma glipizyde concentration/time curves following a single dose of 5 mg tablets of glipizide as test (A) or reference (B) formulation to 24 healthy volunteers.

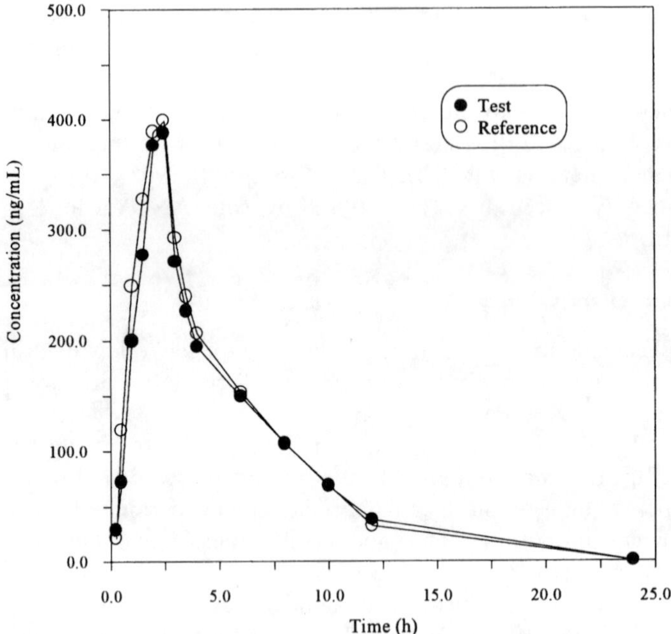

FIGURE 13 Mean plasma glipizide concentrations versus time curves. Data were obtained from 24 healthy volunteers after single oral administration (5 mg) of each formulation.

the test and reference formulations, respectively. Using these values, the relative bioavailability of the test formulation was 94.8 ± 9.5%.

Table 2 shows the geometric mean values and the range corresponding to mean ± standard deviation after logarithmic transformation for AUC_t, AUC_∞, C_{max} as well as 90% confidence intervals and the point estimates (test to reference ratio of geometric means). The 90% confidence intervals

TABLE 1 Pharmacokinetic Parameters (Mean ± S.D.) Obtained in 24 Healthy Volunteers for Two 5 mg Glipizide Formulations

Parameter	Test	Reference
AUC_t (ng · h/mL)	1828.35 ± 336.76	1934.86 ± 363.15
AUC_∞ (ng · h/mL)	2011.48 ± 403.47	2126.13 ± 381.37
C_{max} (ng/mL)	471.07 ± 72.69	463.01 ± 66.89
t_{max} (h)	2.18 ± 0.35	2.15 ± 0.42
$t_{1/2}$ (h)	2.96 ± 0.75	2.99 ± 0.51
k_{el} (h^{-1})	0.25 ± 0.05	0.24 ± 0.04

TABLE 2 Bioequivalence Analysis of Individual AUC_t, AUC_∞, C_{max} Ratios Between the Two Formulations

Parameter	Test	Reference	Pt. estimate	90% conf. interv.
AUC_t geom. mean	1799.5	1902.6	0.95	0.86–1.03
AUC_∞ geom. mean	1978.9	2093.7	0.94	0.86–1.03
C_{max} geom. mean	465.5	458.5	1.02	0.94–1.09

for the test/reference mean ratio of the pharmacokinetic variable AUC_t, AUC_∞, and C_{max} fall within the conventional bioequivalence range of 0.8–1.25.

7.4 Conclusion

In the present comparative bioequivalence study of two tablet formulations of glipizide, there were no significant differences in the values of pharmacokinetic parameters. The 90% confidence intervals for both AUC and C_{max} mean ratios were lying within the bioequivalence acceptance range of 0.8–1.25. It can be concluded that Glipizyd 5 mg tablets (Tarchomińskie Zaklady Farmaceutyczne POLFA S.A.) is bioequivalent to Glibenese 5 mg tablets (Pfizer) for both the extent and the rate of absorption after single oral administration.

REFERENCES

1. JG Wagner. Biopharmaceutics and Relevant Pharmacokinetics. Hamilton, IL: Drug Intelligence Publications, 1971, pp 166–179.
2. Guidance for Industry. BA and BE Studies for Orally Administered Drug Products—General Considerations, U.S. Department of Health and Human Services, Food and Drug Administration, Center for Drug Evaluation and Research (CDER), August 1999.
3. CPMP. Note for guidance on the investigation of bioavailability and bioequivalence. London: CPMP/EWP/QWP/1401/98, 26 July 2001.
4. GG Gibson, P Skett. Pathways of drug metabolism. In: Introduction to Drug Metabolism. Glasgow: Chapman and Hall, 1994, pp 1–34.
5. DW Nebert, DR Nelson, MJ Coon, RW Estabrook, R Feyereisen, Y Fujii-Kuriyama, FJ Gonzalez, FP Guengerich, IC Gunsalus, EF Johnson, JC Loper, R Sato, MR Waterman. The P450 superfamily: update on new sequences, gene mapping, and recommended nomenclature. DNA Cell Biol 10:1–14, 1991.
6. SA Wrighton, M VandenBranden, BJ Ring. The human drug metabolizing cytochromes P450. J Pharmacokinet Biopharm 24:461–73, 1996.
7. S Rendic, FJ Di Carlo. Human cytochrome P450 enzymes: a status report summarizing their reactions, substrates, inducers, and inhibitors. Drug Metab Rev 29:413–580, 1997.

8. FJ Gonzalez, UA Meyer. Molecular genetics of the debrisoquine-sparteine polymorphism. Clin Pharmacol Ther 50:233–238, 1991.
9. SN de Wildt, GL Kearns, JS Leeder, JN van den Anker. Cytochrome P4503A. Ontogeny and drug disposition. Clin Pharmacokinet 37:485–505, 1999.
10. Y Zhang, LZ Benet. The gut as a barrier to drug absorpion. Combined role of cytochrome P4503A and P-glycoprotein. Clin Pharmacokinet 40:159–168, 2001.
11. GG Gibson, P Skett. Pharmacokinetics and the clinical relevance of drug metabolism. In: Introduction to Drug Metabolism. Glasgow: Chapman and Hall, 1994, pp 182–184.
12. A Marzo, P Balant. Bioequivalence. An updated reappraisal addressed to applications of interchangeable multi-source pharmaceutical products. Arzneim-Forsch/Drug Res 45:109–115, 1995.
13. E Diletti, D Hauschke, VW Steinijans. Sample size determination for bioequivalence assessment by means of confidence intervals. Int J Clin Pharmacol Ther Toxicol 29:1–8, 1991.
14. DJ Andersen. High-performance liquid chromatography in clinical analysis. Anal Chem 71:314R–327R, 1999.
15. S Braggio, RJ Barnaby, P Grossi, M Cugola. A strategy for validation of bioanalytical methods. J Pharm Biomed Anal 14:375–388, 1996.
16. F Bressolle, M Bromet-Petit, M Audran. Validation of liquid chromatographic and gas chromatographic methods. Applications to pharmacokinetics. J Chromatogr B 686:3–10, 1996.
17. J Blanchard. Evaluation of the relative efficacy of various techniques for deproteinizing plasma samples prior to high-performance liquid chromatographic analysis. J Chromatogr 226:455–460, 1981.
18. RE Majors. Liquid extraction techniques for sample preparation. LC GC Int 10:93–101, 1997.
19. M-C Hennion. Solid-phase extraction: method development, sorbents, and coupling with liquid chromatography. J Chromatogr A 856:3–54, 1999.
20. M Kobylińska, K Kobylińska. High-performance liquid chromatographic analysis for the determination of domperidone in human plasma. J Chromatogr B 744:207–212, 2000.
21. WI Krull, M Szulc. Detection sensitivity and selectivity. In: LR Snyder, JJ Kirkland, JL Glajh, eds. Practical HPLC Method Development. New York: John Wiley, 1997, pp 59–96.
22. S Udenfriend. Fluorescence Assay in Biology and Medicine. New York: Academic Press, 1966, pp 5–36.
23. WR Gillespie. Noncompartmental versus compartmental modelling in clinical pharmacokinetics. Clin Pharmacokinet 20:253–262, 1991.
24. A Marzo, NC Monti. Acceptable and unacceptable procedures in bioavailability and bioequivalence trials. Pharmacol Res 38:401–404, 1998.
25. H-U Schulz, VW Steinijans. Striving for standards in bioequivalence assessment: a review. Int J Clin Pharmacol Ther Toxicol 29:293–298, 1991.

26. R Sauter, VW Steinijans, E Diletti, A Böhm, H-U Schulz. Presentation of results from bioequivalence studies. Int J Clin Pharmacol Ther Toxicol 30: 233–256, 1992.
27. D Hauschke, VW Steinijans, E Diletti. A distribution-free procedure for the statistical analysis of bioequivalence studies. Int J Clin Pharmacol Ther Toxicol 28: 72–78, 1990.
28. LZ Benet, JE Goyan. Bioequivalence and narrow therapeutic index drugs. Pharmacotherapy 15:433–440, 1995.
29. DJ Schuirmann. A comparison of the two one-sided test procedures and the power approach for assessing the equivalence of average bioavailability. J Pharmacokinet Biopharm 15:657–680, 1987.
30. RN Brogden, RC Heel, GE Pakes, TM Speight, GS Avery. Glipizide: a review of its pharmacological properties and therapeutic uses. Drugs 18:329–353, 1979.
31. BD Prendergast. Glyburide and glipizide, second-generation oral sulfonylurea hypoglycemic agents. Clin Pharmacokinet 3:473–485, 1984.
32. WA Kradjan, KA Kobayashi, LA Bauer, JR Horn, KE Opheim, FJ Wood Jr. Glipizide pharmacokinetics: effects of age, diabetes, and multiple dosing. J Clin Pharmacol 29:1121–1127, 1989.
33. M Kobylińska, M Bukowska-Kiliszek, M Barlińska, B Sobik, K Kobylińska. A bioequivalence study of two brands of glipizide tablets. Acta Pol Pharm Drug Res 57:101–104, 2000.
34. RJ Tallarida, B Murray. Manual of Pharmacological with Computer Programs. New York: Springer Verlag, 1987, p 75.

5

Validation of HPLC Analyses in Therapeutic Drug Monitoring

Zdzisław Chilmonczyk
National Institute of Public Health, Warsaw, and
University of Białystok, Białystok, Poland

Iwona Cendrowska
Pharmaceutical Research Institute, Warsaw, Poland

Hassan Y. Aboul-Enein
King Faisal Specialist Hospital and Research Centre,
Riyadh, Saudi Arabia

1 INTRODUCTION

Any analytical method used for the quantification of drugs and their metabolites in biological samples plays a significant role in the evaluation and interpretation of drug biological fluid levels, bioavailability, bioequivalence, or pharmacokinetic data. To yield reliable results that can be satisfactorily interpreted, it is essential to use well-characterized and fully validated analytical methods. The basic principle of validation is not difficult to understand or apply to individual laboratory practice. But with the increased complexity of the analytical instrumentation marketplace—computerization, multiple applications, regulatory compliance, and a worldwide emphasis on improved product quality—widespread interpretation and implementation of that principle becomes much more complex. Simply stated, the basic principle of validation is "The process of providing documented evidence that something does what it is intended to do" [1].

Thus validation may be considered as

Assurance that the system is operating properly and that methods are in compliance with regulatory requirements
Documented evidence that the system has done, is doing, and/or will do, reliably, what it has been designed to do

Who has to go through a validation procedure and why? Validation has to be carried out by those who must control their manufacturing process, that is, those working in regulated environments such as pharmaceuticals and biotechnology. They are supposed to do so because

Validation ensures quality.
Validation is part of the overall quality process.
Validation can save money and time.
Validation makes good business sense.

2 HPLC AND VALIDATION

High-performance liquid chromatography (HPLC) has emerged as a generally accepted analytical tool with a very broad range of separation modes. It is especially popular in the pharmaceutical, chemical, and food processing industries and more recently in environmental testing. With the aid of HPLC, complex mixtures of different origin (plant extracts, food additives and ingredients, components of pharmaceutical formulations, reaction mixtures) can be separated within several minutes. Moreover, components can also be evaluated in a quantitative manner. Therefore HPLC is a very efficient analytical method, since it can yield excellent separations and quantifications in a short time.

The modern HPLC apparatus is a multicomponent unit. It usually contains (Fig. 1)

A solvent reservoir
High pressure pump(s) equipped with a pulse dumper (to reduce pressure and flow pulsations) and flow and pressure meters
A precolumn—used to remove impurities able to damage a column
A column—where the separation of an analyzed mixture takes place
An injector enabling application of analyzed compounds to a column

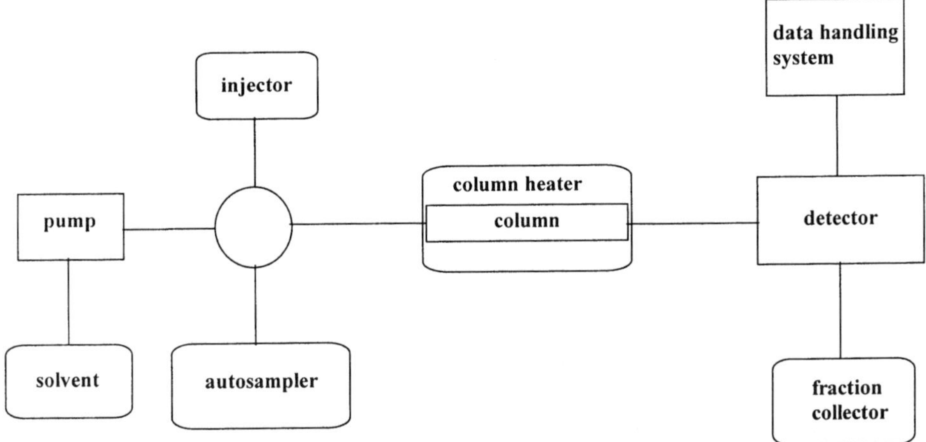

FIGURE 1 Basic modules of an HPLC system.

A detector where an analyte signal is measured
A data acquisition system, which could be a recorder, an integrator, or a computer with a proper program

Recently, more and more HPLC systems are equipped with

A column oven to provide constant column temperature (in cooling and/or heating)
An autosampler (usually thermostated) used to apply to a column several samples in an automated manner
A multicolumn valve able to accommodate several columns and to switch between them
A fraction collector
An HPLC system controller coordinating the work of all the system units (pump, detector, autosampler, thermostat, etc.).

All of these components may exhibit inaccuracies in operation that give rise to misleading results. Therefore the validation of an HPLC system can be considered on several levels:

1. Functional validation relating to how the HPLC equipment is used at the user's site
2. Modular validation, used to measure and evaluate the performance of particular components of the system
3. Holistic validation, where a series of tests to measure and evaluate the performance of the entire computerized LC system under the conditions of its intended use is carried out

4. Method validation, used to confirm that the analytical procedure employed for a specific test is suitable for its intended use
5. System suitability testing, used to verify that the resolution and reproducibility of the chromatographic system are adequate for the analysis to be done

The present chapter will mostly be devoted to the method validation and system suitability testing, although the other points as well as some regulatory problems will also be discussed.

3 FUNCTIONAL VALIDATION

This kind of validation relates to how the product is used at the user's site. The user is responsible for the validation, and vendors can provide assistance. The validation consists of installation, operation, and performance qualification (IQ, OQ, PQ) and test suites. The user has to provide documentation concerning installation and on-site verification tests (IQ-OQ-PQ certification documents). The documents have to concern tests of the on-site separation method performances.

According to current good manufacturing practices (GMP) and good laboratory practices (GLP), equipment used in the laboratory should be of appropriate design, of adequate capacity, and properly maintained. The following validation sketch summarizes the necessary steps according to Maxwell and Sweeny [2]:

3.1 Step 1: Vendor Qualification

The vendor qualification step documents all information that pertains to the design, development, and manufacture of a product used in a regulated environment [3]. All documentation relevant to proper installation, operation, qualification, and maintenance of an HPLC system should be available for review before purchase and installation. The vendor qualification process should also provide an opportunity for management personnel to become involved in approving a system for which they will be responsible.

3.2 Step 2: Installation Qualification

The installation qualification process is specific to instrument components of a modular-based HPLC system or to completely integrated system configurations. The installation qualification protocol should contain all information required properly to install, operate, and maintain the specific product.

Frequently, this step is broken down into two phases: preinstallation and physical installation.

Preinstallation. Before actual installation, all installation, operation, qualification, and maintenance documentation that was shipped with the system must be reviewed and approved. For many regulated laboratories the availability of such validation support is a criterion for choosing instrumentation and software. As part of this review, users should verify evidence of the vendor's quality system and the receipt of validation protocols for operational, performance, maintenance, and calibration qualification.

Physical installation. During the physical installation phase, users must document exactly how the system was installed, including instrument identification, description, and specifications.

3.3 Step 3: Operational Qualification

The operational qualification protocols (module level) and the performance qualification protocol (system level) contain a series of diagnostic and functional tests to ensure that both the specific modules and the complete HPLC system are operating according to defined specifications for accuracy, linearity, and precision.

Even in fully integrated systems, functional testing must take place at the modular level (pump, detector, injector, data processor-controller) to ensure performance at the complete system level. Merely performing a system test, such as a system-suitability test [4], is not sufficient to qualify all functioning components. After the complete HPLC system has been qualified, however, system suitability testing is used to verify system calibration before analyzing unknowns.

3.4 Step 4: Performance Qualification

The system performance qualification protocol and the module qualification tests conducted under the operational qualification protocols complete the testing necessary for documenting the accuracy, linearity, and precision of the complete system under typical operating conditions. The system performance qualification test should use a well-characterized analyte mixture, column, and mobile phase to avoid spurious results caused by column chemistry, mobile phase, and sample interferences. Because most laboratories use both integrated and module-based systems, the performance qualification test should be flexible and rugged enough to test various system configurations as well as to compare results obtained by analysts in several laboratories using different system configurations

After successful completion of a system-level performance qualification test using a test mixture, users can verify system operation for their specific application and operating conditions as part of their method validation requirements. Finally, users should complete instrument-system logs after finishing the performance qualification and, if company standard operating procedures require it, label the instruments to identify the performer and the qualification date.

In practice, the module operation qualification and system performance qualification tests frequently blend together, particularly for linearity and precision (reproducibility) tests that can be conducted more easily at the system level using modified system suitability tests. Also, users should conduct testing under actual running conditions across the anticipated working range.

4 MODULAR VALIDATION

For each module, specific tests confirming its performance may be conducted. Below one can find some examples of modular validation for chosen HPLC modules. Specific tests will not, however, be discussed, because they may differ for different models. The following examples are just to give an impression of the approach to that kind of validation.

4.1 Pump

Verification of the accuracy of the flow rate and the compositional mixing for a gradient system can be performed. Other parameters that may be checked are flow pulsations, functioning of pressure limiter, and the built-in time program.

4.2 UV Detector

For a variable-wavelength UV detector, wavelength accuracy and linearity, lamp light energy, and drift and noise levels can be tested at the modular level. Many modern variable-wavelength UV detectors perform an automated wavelength diagnostic at power-up [5]. Although this is useful as a daily operational calibration check, the wavelength accuracy should be verified at designated intervals using a chemical compound with a known UV spectrum. Linearity and precision testing should be performed at the system level.

4.3 Column Oven

Column oven may be verified for a proper operation of actual temperature in heating and cooling mode using a calibrated digital thermometer.

4.4 System Controller

For a system controller, such parameters as proper functions of disk drives, memory, LED, and keyboard may be checked. Additionally it is also useful to check the version number of a software, that the system controller does control the system, that all screens are displayed as expected, or that memory is backed up when the power is off.

In a similar fashion, users can test injector or autosampler accuracy at the modular level.

5 HOLISTIC VALIDATION

As was already mentioned, during holistic validation, a series of tests to measure and evaluate the performance of the entire computerized LC system under the conditions of its intended use is carried out. The tests in principle should be based on the following criteria:

Single standardized method should be applied to the system.
Rapid results, low cost.
Minimum disturbance of operating HPLC system.
It should provide data on vital component performance.

In this approach a fast, simple gradient HPLC separation utilizing columns, solvents, and standards of high stability and low costs should be used. The determination of the following analytical parameters will make it possible to obtain information on the functioning of vital system modules:

Peak area precision gives information on auto injector, detector, and integrator performance.
Retention time precision gives information about low rate stability and solvent composition accuracy.
Concentration linearity illustrates detector and injector problems.
Signal-to-noise ratio corresponds to detector sensitivity.
Absorbance maxima (detector wavelength calibration against standard).
Sample carryover (injector problems).

Testing specifications should include the following modules and certain specific parameters characterizing the module:

LC pumps
  Safety verification and electrical functional check
  Pressure, flow, and composition calibration
  Leak test, passivation

Detectors
- Lamp alignment, wavelength calibration
- Beam energy, low angle scatter and spectrum check
- Noise and drift test

Auto injector
- Precision (relative standard deviation $<0.5\%$ at 10 μL) and carryover test ($<0.02\%$)
- Linearity test

6 METHOD VALIDATION

Method validation is the process used to confirm that the analytical procedure employed for a specific test is suitable for its intended use. Specific validation criteria are needed for methods intended for analysis of each drug and/or metabolite. Although a validation of each method is usually independent of other methods, there may be situations in which comparison of different methods will be necessary (e.g., when more than one method has been used in a long-term study). When sample analysis is conducted at more than one site, it is necessary to validate the analytical method(s) at each site and provide appropriate validation information for different sites to establish interlaboratory reliability [6–12]. Usually the following parameters are determined during the method validation procedure:

1. Selectivity (specificity)
2. Precision
3. Accuracy
4. Linearity
5. Range
6. Limit of detection
7. Limit of quantification
8. Stability
9. Ruggedness

6.1 Specificity/Selectivity

Definition: The specificity is the ability of a method to determine exclusively the analyte. The selectivity is the ability to separate the analyte from other compounds present in a matrix.

The terms selectivity and specificity are often used interchangeably. Specificity is the ability to assess unequivocally the analyte in the presence of the coexisting compounds in a specific matrix (i.e., biological, etc.), where only the measured analyte is responsible for an observed detector response. Selectivity includes the ability to separate the analyte from any coexisting compounds such as degradation products, metabolites, or co-administered drugs. Real problems arise when analyte metabolites or known degradation products are not available. In this case, the most difficult task is to identify whether the peaks are pure within a sample chromatogram. Different ways should be used to check the selectivity of the method [2]:

The use of diode arrays or multiple-wavelength detectors to ensure peak purity.
The use of a more specific detector, such as a mass spectrometer.
Running a multidimensional chromatography, i.e., the use of an analytical column with a different selectivity could be an alternative.
The use of biological samples from dosed subjects.

These samples should be analyzed under different chromatographic conditions to resolve as many potentially merged peaks as possible. Moreover, examination of chromatograms from subjects' samples that were collected at various times following the drug intake can reveal peaks due to substances that are absent in the predose sample.

The use of standard additions of known quantities of analyte to real sample (from clinical trial for example) that may contain metabolites. The linear relationship between added analyte and response should be verified.

6.2 Precision (Repeatability)

Definition: The precision is a measure of the result-repeatability of the test procedure under normal operating circumstances (Fig. 2).

The characteristic parameter of a precision is its standard deviation expressed as repeatability standard deviation (S_{repeat}). It can be measured either as the standard deviation of multiple determinations of one sample or as the standard deviation of multiple determinations of several samples. Multiple determinations of individual samples are always performed under purely repeatable conditions, i.e., measurement results are obtained immediately one after the other in the same laboratory with the same measuring procedure, the same sample, by the same analyst, and on the same apparatus. From this the repeatability (S_{repeat}) is determined.

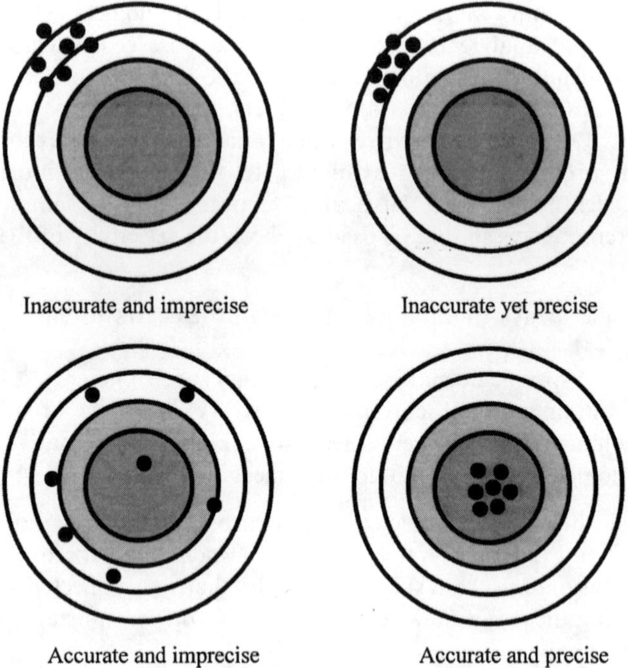

FIGURE 2 Accuracy and precision.

In the general case of multiple determinations of several samples, the absolute standard deviation S_{repeat} can be calculated according to the formula

$$S_{\text{repeat}} = \sqrt{\frac{\sum_{j=1}^{n_1}(X_{1j} - XQ_1)^2 + \sum_{j=1}^{n_2}(X_{2j} - XQ_2)^2 + \cdots \sum_{j=1}^{n_m}(X_{mj} - XQ_m)^2}{n - m}}$$

$$S_{\text{repeat}} = \sqrt{\frac{\sum_{i=1}^{m}\sum_{j=1}^{n_i}(X_{ij} - XQ_i)^2}{n - m}}$$

where

n = total number of determinations
n_i = number of determinations of sample i
m = number of samples

X_{ij} = experimental value X_j of sample i, determination j
XQ_i = mean value of sample i
S_{repeat} = absolute standard deviation

In the case of double determinations of m samples, i.e., $n_i = 2$, the above-mentioned general formula can be reduced and transformed to

$$S_{\text{repeat}} = \sqrt{\frac{\sum_{i=1}^{m}(X_{ij} - X_{i2})^2}{2m}}$$

In the case of multiple determinations of one sample, i.e., $m = 1$, the general formula can be reduced to

$$S_{\text{repeat}} = \sqrt{\frac{\sum_{j=1}^{n}(X_j - XQ)^2}{n - 1}}$$

The relative standard deviation $S_{\text{rel,repeat}}$ can be calculated from the absolute standard deviation S_{repeat} by

$$S_{\text{rel,repeat}}[\%] = \frac{100 \cdot S_{\text{repeat}} \cdot m}{\sum_i XQ_i}$$

Confidence interval of a mean value XQ_i can be calculated from the formula

$$\Delta XQ_i(\text{mean}) = \frac{t(P,f) \cdot S_{\text{repeat},i}}{\sqrt{n_i}}$$

The value of $t(P, f)$ can be read from statistics tables:

P = confidence level (usually 95%)
f = number of degrees of freedom
n_i = number of determinations of sample i

If the estimates s of the standard deviation and the number of degrees of freedom of its determination are known, the confidence interval of a single determination can be calculated for a confidence level P. The confidence interval of a single value is

$$\epsilon x(\text{single}) = t(P, f) A S_{\text{repeat}}$$

Note: Using relative standard deviation instead of absolute standard deviation, the confidence interval ϵx is obtained in relative percent.

Precision is usually assessed on both a within-batch and a between-batch basis (with this terminology being more appropriate than "within-day" and "between-day" for a batch analysis). Between-batch assessment is not always carried out with a single batch per day, and some batches may be of sufficient size that more than one day is required for analysis. Within-batch assessment should be considered as a measure of the precision of a method under optimal conditions. The between-day batch precision is considered to be a better representation of the precision one might observe during routine conduct of a method, because these data are generally subjected to a greater source of variability.

6.3 Accuracy

Definition: The accuracy indicates the deviation between the value found and the true value. This deviation may be caused by constant and/or proportional systematic errors. Accuracy is a measure of the exactness of the result of a determination (Fig. 2).

There are two main methods suitable for the accuracy determination. In the first method (Method 1) of the content determination, the obtained content is compared with that obtained by a second procedure, the accuracy of which is stated and/or defined, and gives the conventional true value as a 100% value. The comparison procedure can be a pharmacopeia one. In another method (Method 2), weighed amounts of analyte (for starting materials) or spiked placebo samples (for finished products) are analyzed according to the procedure given in "Method of Analysis," and the recovery rate is calculated by comparing the weighed amounts of analyte with the measured values (Method 2a). Alternatively (Method 2b), especially for measurements used in a broad concentration range, weighed amounts of analyte or spiked placebo samples covering the necessary concentration range are analyzed according to the procedure given in "Method of Analysis." The results of the analysis are plotted taking

For x : weighed amount of analyte

For y : measured amount of analyte

If there is no systematic error, the measured values lie on a straight line that goes through the origin and has a slope of 1. The function of the regression line and its correlation coefficient are determined according to the method of least squares of weighed amounts of active substance versus test results. The standard deviation S and the confidence intervals of the intercept A of the y axis (ΔA) and of the slope B (ΔB) can be calculated by

$$S = \sqrt{\frac{m\sum_{i=1}(Y_i - Y(\text{calc})_i)^2}{m^{-2}}}$$

$$S_B = \frac{S}{\sqrt{\sum_{i-1}^{m} X_i^2 - \frac{1}{m}\left(\sum_{i=1}^{m} X_i\right)^2}} \qquad \Delta B = t(P,f) \cdot S_B$$

$$S_A = S_B \cdot \sqrt{\frac{\sum_{i=1}^{m} X_i^2}{m}} \qquad \Delta A = t(P,f) \cdot S_A$$

S = residual standard deviation
Y_i = measured amount of analyte in determination i
$Y(\text{calc})_i$ = calculated value of the regression line
m = number of tested concentrations
S_B = standard deviation of the slope of the regression line
X_i = weighed amount of analyte for determination i
S_A = standard deviation of the y-intercept
ΔA = confidence interval of y-intercept A
ΔB = confidence interval of slope B
P = confidence level
f = number of degrees of freedom

The evaluation of the obtained in the both methods results should be based on the following rules:

Method 1: The values of both methods of analysis should not differ significantly.
Method 2a: The average recovery rate should be within 98 and 102%.
Method 2b: The confidence interval ($P=95\%$) of the y-intercept A should contain the origin (no constant systematic error). The confidence interval ($P=95\%$) of the slope B should contain 1 (no proportional systematic error).

However, the precision of the measurement has to be taken into account for a final judgement of this statistical evaluation. The accuracy and the precision should be determined with a minimum of five determinations per quality controlled sample (excluding blank matrix) from an equivalent biological matrix. The precision around the mean value should not exceed 15% of the coefficient of variation, and the mean value should be within a 15% deviation of the nominal value for accuracy. It is desirable that these tolerances be provided both for within-batch and between-batch experiments. At the concentration corresponding to the limit of quantification, 20% is acceptable for both precision and accuracy.

6.4 Linearity

> Definition: The linearity of a test procedure is its ability—within a given range—to obtain test results proportional to the concentration (amount) of analyte in the sample.

The linearity of an analytical method is determined by mathematical treatment (regression line) of test results obtained by analysis of different analyte concentrations or spiked placebo preparations across the claimed range of the method according to the procedure given in "Method of Analysis." Thereby the results are plotted graphically taking

For x : weighed amount of analyte

For y : measured signal response

The standard deviation and the confidence interval of the y-intercept A and the slope B of the regression line are calculated using the same formulas as in the section "Accuracy." The correlation coefficient of the regression line should exceed 0.99. By a visual examination the measured values should lie on the regression line or be uniformly distributed on both sides of it.

The origin should lie within the confidence limits of the y-intercept (after subtraction of a potential blank value).*

Linearity of the method should be demonstrated by showing that the slope of the linear calibration curve is statistically different from 0, that the intercept is not statistically different from 0, and that the regression coefficient is not statistically different from 1. If a significant nonzero intercept is obtained, it should be demonstrated that there is no effect on the accuracy of the method.

In order to generate an accurate analytical calibration curve independent of the possible time effect, prepared fresh quality control (QC) samples are stored frozen at the same temperature as is intended for storage of study samples in order to account for the time effect. A simple approach is to prepare a series of working calibration standards (in purified water, for example) at concentrations that are 10 to 20 times higher than those intended for biological standards. Those working calibration standards may be stored

---

* This point may not be fulfilled with methods that are linear only in a small range.

(at 4°C or −20°C) provided that their stability has been demonstrated previously over the maximum period over which they will be stored. Then, on a daily basis, blank biological matrix is spiked with the working calibration standards at a ratio of e.g. 1:20 working standard biological blank. To compensate for the dilution of biological matrix with working standards, an equal volume of working solution, free of analyte, is added to the study samples. However, in some analytical laboratories, the calibration curves are prepared and frozen for storage with QC samples.

6.5 Range

Definition: The range of a test procedure is the interval between the upper and lower levels of analyte (including these levels) that have been demonstrated to be determined with sufficient precision, accuracy, and linearity using the method as written (Fig. 3).

6.6 Limit of Detection

Definition: The limit of detection (LOD) is the lowest concentration at which a measured value is larger than the uncertainty associated with it.

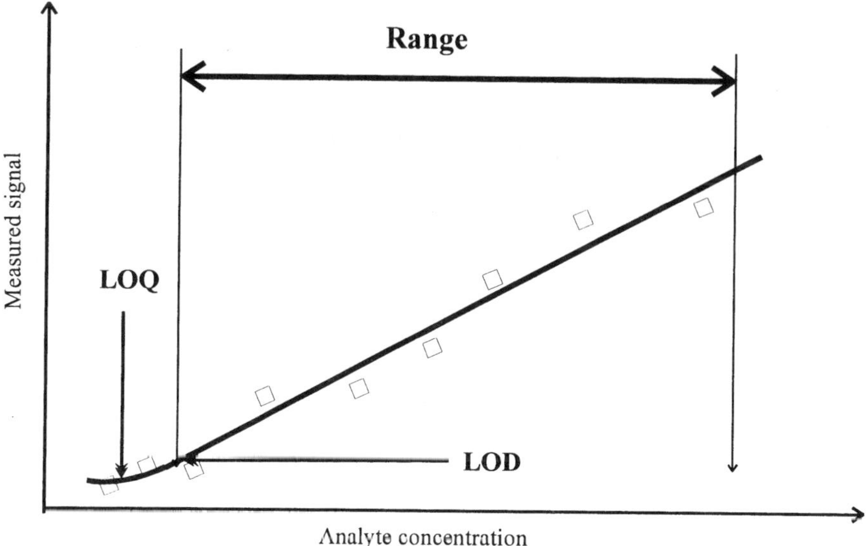

FIGURE 3 Range, limit of detection (LOD), and limit of quantification (LOQ).

The limit of detection is the lowest concentration of analyte in a sample that can be detected, but not necessarily quantified (Fig. 3). In chromatography, the detection limit is the injected amount that results in a peak with a height at least twice that of the baseline noise.

6.7 Limit of Quantification

> Definition: The limit of quantification (LOQ) is the lowest injected amount that results in a reproducible measurement of peak areas (equivalent to amounts).

In order to obtain reasonable quantitative reproducibility, the peak heights are typically required to be about 10 to 20 times larger than the baseline noise. Within-batch and between-batch precision and accuracy of the LOQ can be determined by using at least five QC samples, from a single pool of matrix, independent of standards. The mean values should be within predefined boundaries, normally within ±20% of the nominal concentration, with a coefficient of variation ≤20%. The LOQ should serve as the lowest concentration on the calibration curve.

6.8 Stability

> Definition: Many solutes readily decompose prior to chromatographic investigations, for example during the preparation of the sample solutions, during extraction, and during storage of prepared vials (in an automatic sampler). Under these circumstances, method development should investigate the stability of the analytes.

Stability data are based on duplicate or triplicate determinations of QC samples at two or three concentration levels (low, medium, and high) at multiple time points after the start of storage to allow trends to be detected. However, the issue is not whether there is a trend in degradation, but whether the study samples are adequately preserved at the time of analysis. Each substance should have specified analytical requirements, and the study sample has to fall within those requirements.

6.9 Ruggedness

> Definition: Ruggedness is the combined effect of operational and environmental conditions on the analysis results.

The ruggedness of a method is

> the degree of reproducibility of the test results obtained by the analysis of the same sample under a variety of normal test conditions, such as different laboratories, different analysts, different instru-

ments, different lots of reagents, different elapsed assay times, different assay temperatures, different days, etc. (USP XXIII).

In other words, will other labs be able to reproduce your results? Typically, a valid method must have an RSD of less than 2% between labs.

Thus ruggedness tests examine the effect that the operational and environmental conditions have on the analysis results. The following factors may affect ruggedness:

Different room temperature and humidity in individual laboratories
Analysts with different experience
Instruments from various vendors
Reagents from different suppliers
Columns from different batches

The ruggedness may be measured as the degree of variance in test results obtained by the analysis of the same samples under a variety of different test conditions. A rugged method is one that is not very sensitive to typical abuses, that is, against differences in care, technique, equipment, and conditions. The ruggedness of an analytical method is determined by the analysis of aliquots from homogeneous lots in different laboratories, by different analysts, using operational and environmental conditions that may differ but are still within the specified parameters of the method (inter laboratory tests).

For the determination of methods ruggedness within a laboratory, a number of chromatographic parameters, for example flow rate, column temperature, detection wavelength, or mobile phase composition, are varied within a realistic range, and the quantitative influence of the variables is determined. If the influence of the parameter is within a previously specified tolerance range, the parameter is said to be within the method's ruggedness range.

7 SYSTEM SUITABILITY

7.1 General Description

In the analytical laboratory, validation usually involves the combination of analytical methods, instrumentation, and data systems. This validation, usually referred to as system suitability testing, tests the system for documented performance specifications for the particular analysis method. System suitability tests [13] are an integral part of the liquid chromatographic method and are based on the concept that the equipment, electronics, analytical operations, and samples to be analyzed constitute an integral system that can be evaluated as such. They are used to verify that the resolution and reproducibility of the chromatographic system are adequate for the analysis to be done. They also should be designed to be able to prove whether a particular analytical method achieves the same or better accuracy and precision on the current system as shown during former validation. System suitability testing consists of testing of sample injection, chromatographic separation, and associated data handling. The tests usually include

1. Column efficiency
2. Repeatability of peak areas (system precision)
3. Resolution between two compounds
4. Tailing factor

Acceptance criteria for the individual test points are defined individually within the framework of validation for each method. Many pharmacopeias such as that in United States give detailed information about what is required for system suitability tests for compound analytes. The chromatographic system should demonstrate acceptable resolution of the test solution and system precision. System precision is determined by injecting a standard solution a number of times. The relative standard deviation of the peak responses is measured as either the peak heights or peak areas. When using an internal standard method, the response ratio is calculated. For the USP monographs, five replicate chromatograms are used if the stated relative standard deviation is 2% or less.

7.2 Column Efficiency

Column efficiency is usually described by the number of theoretical plates, N. For Gaussian peaks, it is calculated by the equation

$$N = 16\left(\frac{t}{W}\right)^2 \quad \text{or} \quad N = 5.54\left(\frac{t}{W_{1/2}}\right)^2$$

where

t = the retention time of the substance

W = the width of the peak at its base, obtained by extrapolating the relatively straight sides of the peak to the baseline

$W_{1/2}$ = the peak width at half-height, obtained directly by electronic integrators

Column efficiency may be also specified as a system suitability requirement, especially if there is only one peak of interest in the chromatogram; however, it is a less reliable means to ensure resolution than direct measurement. Column efficiency influences peak sharpness, which is important for the detection of trace components.

7.3 System Precision

System precision is determined by injecting a standard solution a number of times. The relative standard deviation of the peak responses is measured as either the peak heights or peak areas. When using an internal standard method, the response ratio is calculated. For the USP monographs, unless otherwise stated, five replicate chromatograms are used if the stated relative standard deviation is 2% or less. Six replicate chromatograms are used if the relative standard deviation is more than 2.0%. For bioanalytical samples, RSD should not exceed 15% except at the limit of detection, where it should be less than 20%.

7.4 Resolution Between Two Compounds

The resolution, R, is a function of column efficiency, N, and is specified to ensure that closely eluting compounds are resolved from each other, to establish the general resolving power of the system, and to ensure that internal standards are resolved from the drug.

The separation of two components in a mixture reflects their resolution, R, defined by the equation

$$R_s = \left(\frac{2(t_2 - t_1)}{W_2 + W_1}\right)$$

where

t_2 and t_1 = retention times of the two components

W_2 and W_1 = corresponding widths at the bases of the peaks obtained by extrapolating the relatively straight sides of the peaks to the baseline (Fig. 4)

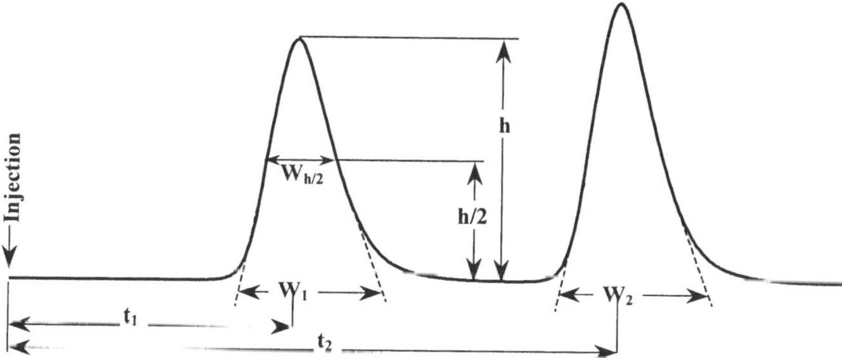

FIGURE 4 Chromatographic separation of two substances.

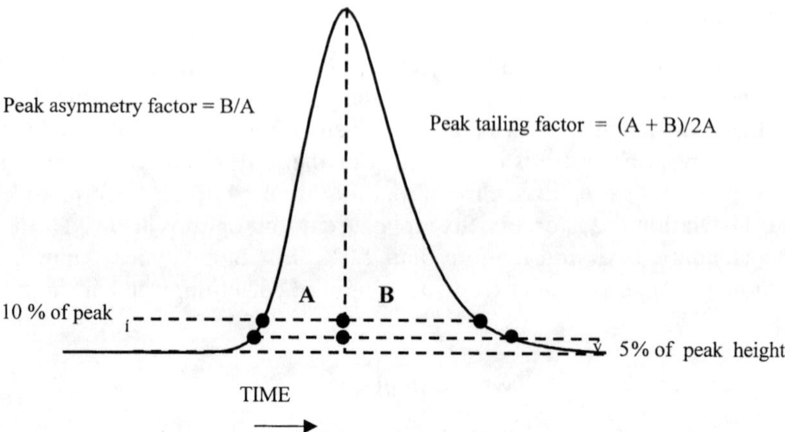

FIGURE 5 Asymmetrical chromatographic peak.

7.5 Tailing Factor

The asymmetry of a peak reflects its deviation from a Gaussian shape. The asymmetry is unity for perfectly symmetrical peaks, and its value increases as the peak tailing becomes more pronounced. It may be defined as the ratio of the distance from the leading edge to the peak maximum and the distance from the tailing edge to the peak maximum at 10% peak height (Fig. 5). For pharmacopeial purposes, the tailing factor is defined as the ratio of the distance from the leading edge to the tailing edge of the peak at 5% peak height, divided by twice the distance from the peak maximum to the leading edge [14]. As peak asymmetry increases, integration, and hence precision, becomes less reliable. Usually the peak asymmetry should lie within the range 0.9–1.5.

8 WHEN TO VALIDATE?

There are different requirements for each level of validation. Instruments or modules should be validated

Upon installation (IQ/OQ)
Before routine use
After major repair
At regular intervals—IPV (on-site Instrument Performance Verification service), e.g., 6 months to 1 year

Data systems and system controllers should be validated

At the end of product development and after software updates; the validation to be performed by vendors
Before routine use; the validation performed by user

A method should be validated

Prior to routine use (includes sample preparation, chemicals, and reagents)
After changing separation parameters (e.g., column)

Analytical system (instrument + method) can be validated in system suitability tests. The validation should be done

Daily; fewer individual determinations are necessary in the daily system suitability test than in method validation; acceptance criteria for the individual test points are defined individually within the framework of validation for each method
Before each single sample analysis
If a series of samples are analyzed within a sequence, before each sequence

9 VALIDATION OF THE ANALYTICAL METHODS IN THE NEW DRUG DEVELOPMENT PROCESS

Validation of an analytical method employed during the development of a new drug candidate may be considered as a two-step process. In the first step of the process, analytical methods are developed to support pharmacological research for the pharmacokinetic and metabolic screening of drug candidates. In this case one can use the so-called preliminary method validation. When the project passes into phase I volunteer clinical trials, a clinical method validation has to be developed.

9.1 Preliminary Method

As only a relatively small number of samples are assayed at this stage, a preliminary validation is performed by testing only the method specificity, recovery, calibration linearity, and stability of the analytes during sample handling. After a compound has been selected for early development, the preliminary method is modified, or a new method is developed for the toxicokinetic monitoring of dose-range-finding and chronic toxicity studies and absorption-distribution-metabolism-excretion (ADME) experiments conducted in animals. The validation of the preclinical method involves the evaluation of intra-assay precision, accuracy, and linearity across an extended range of concentrations to allow the assay of high concentrations of analyte in samples from high-dosage groups without the necessity of error-introducing dilution of samples. Quality control samples are stored and assayed together with unknown samples at this stage to check the daily method performance and also to give an indication of the interassay precision and accuracy.

9.2 Clinical Method

At this stage an attempt is made to design a definitive method that would be used for the entire clinical development of the compound and therefore will already incorporate a high degree of automation. The validation of the method includes tests for interassay precision and accuracy and long-term stability at lower concentrations. When large numbers of samples are expected from the planned phase II and III clinical trials, the method may well be further modified to include or increase automation of the method and so increase batch size. After the validation of a method, a document called a method sheet is produced and made available to all the analysts in the department. The method sheet gives exact details on method procedure and is in a recipe format. The results of each method validation are described in a specific report for submission to the regulatory authorities. It is preferable not to make any changes to a validated method because of the necessity to recheck the validation parameters. However, it is common to make minor changes such as an extension of the calibration range. In this case, it is our practice to perform a within-batch evaluation of precision and accuracy, recovery, and limit of quantification. If, in the case of clinical methods, the validation parameters from the revised method are not significantly different from those of the original method, and the validation parameters are well within the acceptable ranges, then a between-batch evaluation is not performed and the revised method is deemed validated. Otherwise, if more drastic changes in the method are required, such as in the redefinition of extraction or chromatographic parameters owing to an interfering metabolite, then a full validation procedure must be repeated [8].

Before use in clinical studies, a comprehensive validation procedure is performed to monitor both within-batch and between-batch accuracy and precision, and also the results of quality control samples assayed over a number of days. The following samples are assayed in the same batch for intra-assay validation:

- An appropriate set of calibration samples to build the calibration curve
- At least sixfold replicate spiked samples at the same concentration as each of the calibration samples
- A set of 12 quality controls at three concentration levels

The following samples are prepared on at least four sample batches and assayed on different days for the inter-assay validation:

An appropriate set of calibration samples to build the calibration curve
At least triplicate spiked samples at the same concentrations as the calibration standards
A set of six quality controls at three concentration levels
Drug-free matrix samples taken from at least six different subjects

In order to calculate the inter-day precision, the coefficient of variation of the means is determined from the four or more mean concentrations determined on individual days at each level so as not to confuse inter-day and intra-day variability. Alternatively, the analysis of variance can be used to evaluate the inter- and intra-day variations [8,15].

10 HIGH-PERFORMANCE LIQUID CHROMATOGRAPHIC ANALYSIS FOR THE DETERMINATION OF DOMPERIDONE IN HUMAN PLASMA

In this chapter the validation program will be demonstrated by means of a recently developed method for the determination of total domperidone concentrations in human plasma [16]. Domperidone is a dopamine antagonist with antiemetic properties similar to metoclopramide and certain neuroleptic drugs. However, unlike those drugs, domperidone does not readily cross the blood–brain barrier and seldom causes extrapyramidal side effects.

10.1 General Conditions

Separation was performed by reversed-phase high-performance liquid chromatography using a Supelcosil LC-8 analytical column (150 × 4.6 mm I.D., 5 μm; Supelco, Bellefonte, PA, USA), preceded by a 20 × 4.6 mm I.D. LC-8 Supelguard column. The column was heated at 35°C. The mobile phase consisted of methanol-water-triethylamine-acetic acid (60:40:0.02:0.3 v/v) and was delivered at a flow rate of 1.4 mL/min. Quantitative data were obtained with the aid of the fluorescence detection after a postcolumn photoderivatization. Venous blood samples (8 mL) were withdrawn into heparinized tubes. Blood samples were centrifuged at 1000 g for 10 min at 4°C and the plasma obtained was stored at −70°C until analysis. Stock solutions of domperidone maleate (1 mg/mL calculated as free base) and internal standard cisapride (1 mg/mL) were prepared in methanol and stored at −20°C. Calibration was performed by adding known amounts of domperidone to blank human plasma to yield concentrations over the range 1–20 ng/mL. The absolute recovery of domperidone from human plasma was calculated by comparing the peak height obtained from extracts of spiked plasma samples and the peak height obtained from direct injection of known amounts of standard solutions of domperidone. In the range of calibration standards, the absolute recovery of domperidone from human plasma using this method was 83.70 ± 2.17%. The mean recovery for internal standard at a concentration of 0.2 μg/mL was 97.55±4.38% (n=60).

The raw data were calculated using a personal computer with Class-LC10 Chromatography Data System software (Shimadzu).

10.2 Selectivity (Specificity)

No other impurities having similar retention times as domperidone were detected using procedure as described in Sec. 10.1 (Fig. 6).

10.3 Precision

The precision of analysis was assessed by analyzing six 1 mL aliquots of domperidone spiked plasma samples containing 2, 10, and 20 ng/mL of domperidone in duplicates on three separate days, and then the within-day variations were calculated. Between-day variations were also calculated for

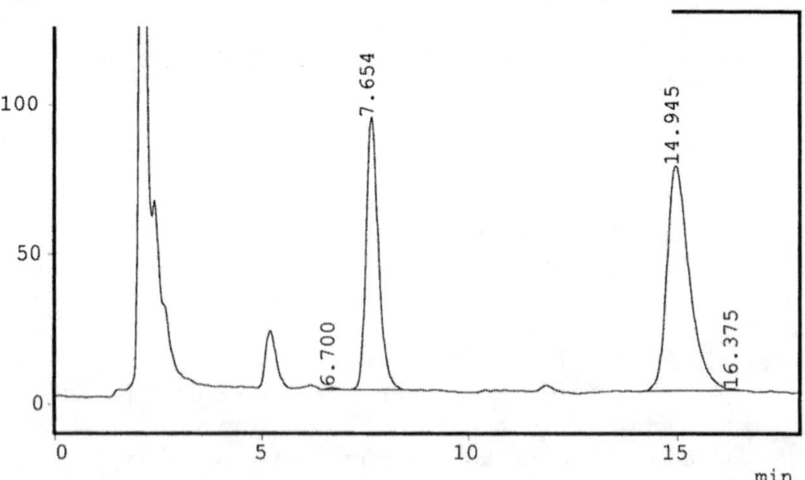

FIGURE 6 Chromatogram of human plasma extracts spiked with domperidone and cisapride (internal standard). The retention times were 7.6 and 14.9 min, respectively.

the above concentrations. In both cases, the RSD was less than 10% at all the concentrations studied (Table 1). Quality control (QC) samples prepared at the same concentrations were placed at random among volunteer samples in each analytical batch. Revalidation was assessed from the standard curves made on days when volunteers' samples were analyzed. For quality control samples, precision ranged from 3% at 29 ng/mL to 7.4% at 2 ng/mL.

10.4 Accuracy

Accuracy was within the range of 94–106% for all concentrations investigated. For quality control samples, accuracy was within the range of 99–101% for all concentration studies (Table 1).

10.5 Linearity and Range

The standard curve for domperidone was linear over the range 1–20 ng/mL. The standard curve was calculated by the linear regression method: $y = ax + b$, where y is the peak height ratio of drug to internal standard, a and b are constants, and x is the domperidone concentration (ng/mL). Typical values for the regression parameters a (slope), b (y-intercept), and correlation coefficient were calculated to be 0.0509364, 0.0209281, and 0.9998, respectively ($n = 6$).

TABLE 1 Precision and Accuracy Data for Domperidone Plasma Extracts

Nominal concentration added (ng/mL)	Concentration found (ng/mL) ± SD	Precision RSD (%)	Accuracy (%)
Intra-day assay ($n = 6$)			
2	1.9 ± 0.1	6.6	95.6
10	10.6 ± 0.4	3.7	105.6
20	20.6 ± 1.5	7.2	103.2
Within-day assay ($n = 6$)			
2	1.9 ± 0.2	9.7	94.7
10	10.4 ± 0.6	5.5	103.8
20	19.5 ± 0.6	3.1	97.6

Source: Reprinted from J Chromatogr B 744:207–212, 2000, with the permission of Elsevier Science.

10.6 Limit of Detection (LOD)

The limit of detection was defined as the sample concentration of domperidone resulting in a peak height of three times the noise level. The minimum detectable concentration of domperidone (LOD) was determined to be 0.2 ng/mL.

10.7 Limit of Quantification (LOQ)

The limit of quantitation was the lowest point on the calibration curve that can be detected with variation below 15%. The quantitative limit was 1 ng/mL, and the relative standard deviation (RSD) of replicate determinations was 13.05% ($n=6$).

10.8 Stability

Stability was determined in samples comprising 10 ng/mL of domperidone in human blank plasma stored at –70°C for 2 months. The obtained data have shown that domperidone was stable in human plasma stored at –70°C at least for 2 months. The RSD was less than 10% at all the concentrations studied.

10.9 Ruggedness

Ruggedness was not determined.

10.10 System Suitability Testing

Of the tests describing system suitability, four, i.e., column efficiency, repeatability of peak areas (system precision), resolution between two compounds, and tailing factor system precision, have already been described in Sec. 10.3. The calculated number of theoretical plates for domperidone was equal to 1500, and the resolution factor between the peak of domperidone and internal standard was 6.2. The tailing factor for domperidone was found to be 1.1.

11 A FEW WORDS ABOUT GLP

GLP (good laboratory practice) deals with the organization, process, and conditions under which laboratory studies are planned, performed, monitored, recorded, and reported. GLP practices are intended to promote the quality and validity of test data. The most important points for GLP are

Quality of test facility
Carefulness with which the study is done
Layout of the report
Archive of samples and reports
Quality assurance program for the test facility

In connection with chromatography, the GLP should provide documented evidence of the system suitability for the study. The documentation should contain

- Structural validation (e.g., certificate of a manufacturer)
- Validation of an instrument (consisting of hardware and software validation)—modular vs. system validation
- Functional validation
  - Validation of installation (IQ)
    - Definition and registration of the instrument
    - Documentation of the installation
  - Operation qualification (OQ)—check of the instrument (specifications)
  - Validation of the system (PQ)
- Validation of a method
- System suitability tests

REFERENCES

1. KG Chapman. A history of validation in the United States: Part I. Pharm Tech, October 1991.
2. W Maxwell, J Sweeney. LC • GC 12(9):678–682, 1994.
3. Validation, compliance and quality for HPLC in the pharmaceutical laboratory. Waters Corp. Milford, Massachusetts, 1993.
4. USP XXII. US Pharmacopeial Convention, Philadelphia, PA, 625–Chromatography, pp 1566–1567, 1990.
5. D Parriott. LC • GC 12(2):132–140, 1994.
6. VP Shah, KK Midha, S Dighe, IJ McGilveray, JP Skelly, A Yacobi, T Layloff, CT Viswanathan, CE Cook, RD McDowall, KA Pittman, S Spector. J Pharm Sciences 81:309–312, 1992.
7. F Bressolle, M Bromet-Petit, M Audran. J Chromatogr B 686:3–10, 1996.
8. S Braggio, RJ Barnaby, P Grossi, M Cugola. J Pharm Biomed Anal 14: 375 338, 1996.
9. D Dadgar, PE Burnett. J Pharm Biomed Anal 14:23–31, 1995.

10. L Huber. Validation of analytical methods: review and strategy. LC • GC International, February 1998:148–156.
11. AF Hirsch. Good Laboratory Practice Regulations. New York: Marcel Dekker, 1989.
12. J Wieling, G Hendriks, WJ Tamminga, J Hempenius, CK Mensink, B Oosterhuis, JHG Jonkman. J Chromatogr A 730:381–394, 1996.
13. VR Meyer. Practical High-Performance Liquid Chromatography. 2d ed. Chichester: John Wiley, 1994.
14. USP XXIII. US Pharmacopeial Convention Inc., Philadelphia, PA, 621-Chromatography, 1995, p 1776.
15. JR Lang, S Bolton. J Pharm Biomed Anal 9:435–442, 1991.
16. M Kobylińska, K Kobylińska. J Chromatogr B 744:207–212, 2000.

6

Analysis of Illicit Drugs with Chromatographic Methods

M. J. Bogusz
King Faisal Specialist Hospital and Research Centre,
Riyadh, Saudi Arabia

1 INTRODUCTION

Substances of abuse form a large group consisting of various compounds, widely differing in their origin, availability, and chemical nature. From the legal point of view, these substances may be divided into legal and illicit drugs. It must be stressed that tobacco products, which are legal in the whole world, are responsible for about 6% of all deaths worldwide. Alcohol, which is legal in most countries, contributes to 1.5%, whereas illicit drugs cause about 0.2% deaths yearly. This review is focused on the last group only.

The analysis of illicit drugs belongs to most important challenges of forensic and clinical toxicology. Like any other branch of forensic sciences, forensic toxicological analysis must apply and keep particularly high standards of quality. This is caused by one reason: an analytical result may have a direct and permanent impact on the fate of the person involved. In consequence, forensic toxicological examinations are subjected to very tight scrutiny and are often performed in several laboratories in the same case.

This requirement should concern not only primarily forensic cases but also all kinds of clinical analysis for illicit drugs. It must be kept in mind that each result of drug analysis may have forensic relevance, irrespective of the primary purpose of the examination.

It is often said that the interpretation of results is the most important and most difficult part of forensic toxicology. Notwithstanding the value of right interpretation, it must be also said that the correct result enables any further action. This is particularly true when the analytical result may serve as an evidence of illegal action. The introduction of "per se" laws in traffic legislation may serve as an example. Such legislation forbids driving a vehicle if a given substance in a given concentration is present in the body fluid of a driver. Per se laws have existed for a long time in regard to ethyl alcohol. In recent years, such laws were introduced, in Germany and in several US states, for other drugs of abuse like amphetamines, cannabis, cocaine, and opiates [1–3]. Belgium and France are preparing similar legal acts. The enforcement of a per se law is possible only when the legal limit of the substance in question is exactly defined and reliably measured. Since in the case of drugs of abuse the law forbids any use of a drug by drivers, the legal limits are defined by the quality of analytical methods and practically reflect the limits of detection. In consequence, only the methods of highest possible selectivity and sensitivity may be applied here.

The analysis of drugs of abuse is important not only in enforcement of road traffic safety. It enables the differentiation between the chronic and the occasional drug user or makes possible the identification of the source of a particular batch of illegal drug. Such an analysis would be not possible without the application of chromatography in various forms. This review will show these applications for the analysis of the most important drugs: opiates, cocaine, amphetamines, and cannabinoids. Very important aspects of quality assurance and quality control, essential in forensic toxicology, will not be covered. The interested reader is referred to the chapter by Aderjan on this topic [4].

2 OPIATE AGONISTS

2.1 Opiates in Plant Material and Illicit Preparations

2.1.1 Composition of Opium and Heroin

All natural and semisynthetic opiates—among them heroin—originate from opium, a dried brown juice obtained from green scratched poppy heads of *Papaver somniferum*. The composition of alkaloids in the plant is variable and depends on multiple factors, like climatic conditions, harvesting time, soil composition, and plant breeding [5–7]. In consequence, the composition of opium and subsequently of heroin reflects the primary variability of plant material. Table 1 shows the content of morphine and codeine in poppy seeds as observed by various authors.

There are two main methods of clandestine heroin production: the lime method used in Southeast Asia and the ammonia method used in Southwest Asia. Both methods give similar yields of morphine, codeine, and thebaine, but the content of noscapine and papaverine is much higher in the ammonia method [14]. Heroin from Southeast Asia ("golden triangle"), known as China white, predominates on the drug market in United States. Southwest-Asian-type heroin, originating from Turkey, Lebanon, Afghanistan, or Iran ("golden crescent") is mainly present on the European market [15–18]. The latter heroin contains over 10% of noscapine and over 2% of papaverine. According to Klemenc, noscapine may be used as an adulterant of street heroin. Recently, heroin samples were seized in Slovenia with noscapine contents over 60%. Also, samples containing only noscapine and papaverine—presumably used as an adulterant—were identified [19]. Ravikumar et al. [20] analyzed in New York City twenty-one heroin samples, which were presumably tainted with some deadly poisons. In one sample obtained from a fatal intoxication, unusually large amounts of noscapine and papaverine were found. In 1995, heroin adulterated with scopolamine appeared on the drug scene in New York City; 370 poisoning cases were then registered [21].

TABLE 1 Morphine and Codeine Content in Poppy Seeds (mg/kg)

Morphine	Codeine	Ref.
39–167	2–44	[5]
6300–20200	900–7700	[6]
2–251	0.3–0.7	[7]
4–200		[8]
17–294	3–14	[9]
17–18	2–3	[10]
964	79	[11]
58–62	28–54	[12]
0.6–11.9	0.3–0.7	[13]

2.1.2 Use of the Chromatographic Method for Analysis of Illicit Heroin

Thin-layer chromatography (TLC) may be used for screening purposes in forensic toxicology and for preliminary examination of illicit heroin samples [14,22,23]. Generally, the sensitivity and selectivity of TLC is not high enough to separate all relevant compounds or to differentiate between several batches of heroin. This may be done using high-performance thin-layer chromatography [24] or various column chromatographic methods, particularly capillary gas chromatography (Fig. 1).

Neumann [25,26] developed a capillary GC method enabling separation and quantitation of 14 of the most important constituents of illicit heroin. The method was applied for identification of particular batches of heroin. Kaa [18] applied capillary GC for long-term observation of adulterant profiles in heroin in Denmark. Beside opiates and adulterants, also some volatile substances occluded in illicit heroin may facilitate a positive identification

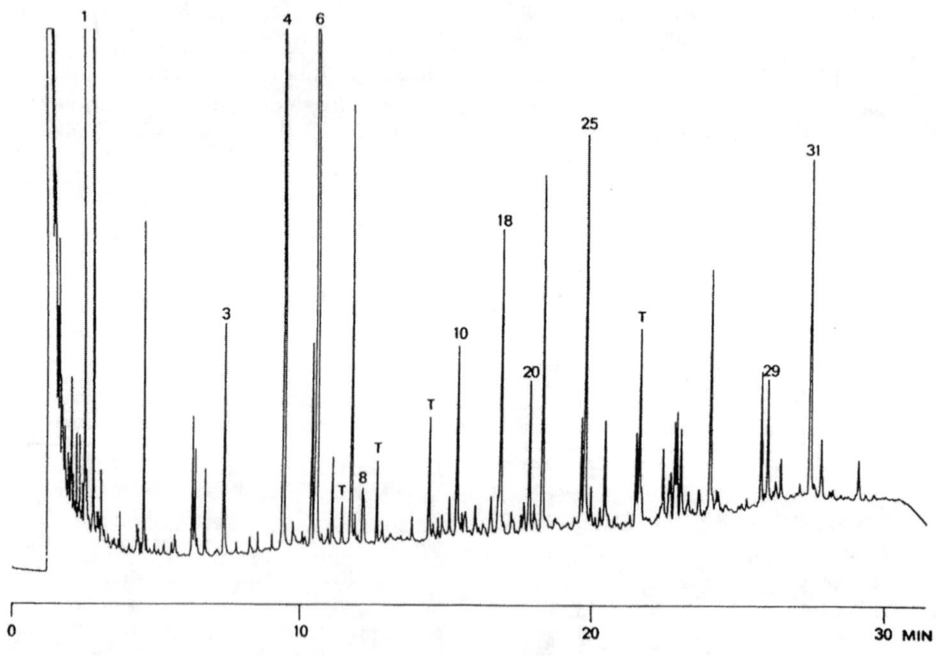

FIGURE 1 Capillary gas chromatographic profile, obtained for Southwest Asian heroin sample. Selected peaks: 4 = thebaol, 6 = acetylthebaol, 10 = papaverine, 25 = noscapine. (From Ref. 14 with permission of the author.)

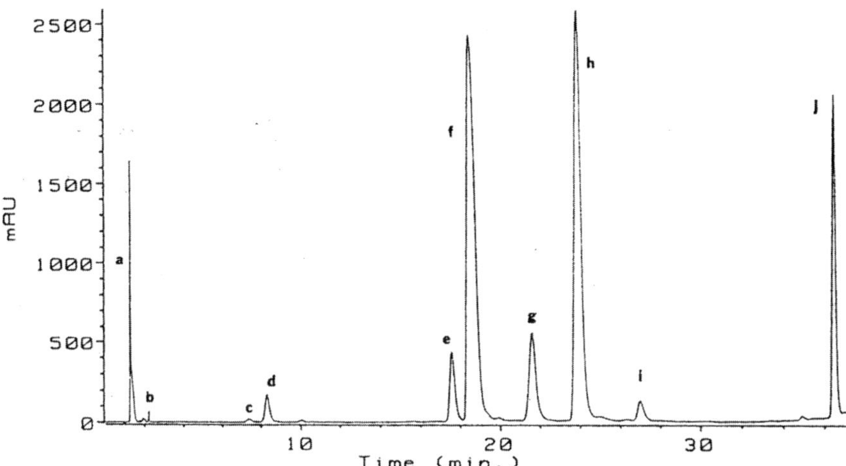

FIGURE 2 HPLC of heroin, its impurities, degradation products, and adulterants. Peaks: a = acetic acid, b = morphine, c = 3-monoacetylmorphine, d = 6-monoacetylmorphine, e = acetylcodeine, f = heroin, g = phenobarbital, h = noscapine, i = papaverine, j = methaqualone. (Reprinted with permission from Ref. 29. Copyright © 1991 American Chemical Society.)

of particular batch of drug [27]. Also, high-pressure liquid chromatography (HPLC) has been successfully applied for heroin profiling. Hays and Lurie [28] used reversed-phase HPLC with multiwavelength detection for quantitation of heroin adulterants. Beside opiates, the following compounds were analyzed: caffeine, methaqualone, nicotinamide, phenobarbital, phenolphthaleine, and *N*-phenyl-2-naphthylamine. Several authors applied capillary electrophoresis (CE) for analysis of illicit heroin seizures [29,30]. This method combines two most important features of other column chromatographic methods: it assures very high resolution, like capillary GC, and is applicable to polar and thermally unstable compounds like HPLC (Fig. 2).

Therefore CE is particularly amenable for the analysis of drug mixtures like illicit heroin samples. In recent years, several authors [31–33] reviewed the use of CE for the forensic profiling of heroin and other illicit drugs. Among various CE techniques, micellar electrokinetic chromatography (MEKC) appeared as most useful for the separation of neutral and electrically loaded compounds and allowed separating several opiates and adulterants in heroin.

Since the composition of illicit heroin is sometimes very complicated and a single method may hardly detect all relevant constituents, a multimethod approach is usually preferred. Huizer [14] in his study of heroin samples used TLC, HPLC (straight and reversed phase), and GC. Hernandez et al. [34] used a combination of HPLC/DAD and GC/NPD for identification of illicit heroin and cocaine constituents. Chiarotti et al. [35] applied headspace GC for detection of volatiles in heroin, then the opiates and adulterants were analyzed with TLC and GC/MS and the sugar diluents were detected with HPLC. Finally, illicit drugs were analyzed on Fe and Zn content with atomic absorption spectroscopy. Another multimethod approach was applied by Besacier et al. [36] who used at first GC/FID for identification of opiates and adulterants in a heroin sample. In the second step, the impurities were analyzed with GC/FID, and in the third step, the isotope ratios $^{13}C/^{12}C$ were calculated using the GC-isotope ratio mass spectrometer.

Table 2 presents an overview of chromatographic methods used for examination of opium and illicit heroin samples.

2.2 Opiates in Biological Samples of Human Origin (Blood, Urine, Organs)

2.2.1 Fate of Heroin in the Human Body

Heroin can be administered intravenously, intranasally ("snorting"), or by smoking. The two latter routes of administration show growing popularity for several reasons. Intravenous application of drug leaves telltale traces and is much more dangerous owing to possible infections and ease of overdose. By smoking, the fractionation of the dose is feasible. However, smoking or intranasal application is less economical, because only part of the administered drug reaches the bloodstream. Therefore smoking or snorting is usually used when the price of street heroin drops and the users can afford such wasteful routes.

TABLE 2 Chromatographic Methods Used for Opium and Illicit Heroin

Method	Conditions	Detection	Analytes	Ref.
TLC	15 TLC systems	Color reactions	8 opiates, 5 adulterants	[23]
GC	Temp. program	Silylation, FID	8 opiates, 5 adulterants	[26]
GC	Temp. program	FID	8 opiates, 6 adulterants	[37]
GC	Temp. program	FID	5 opiates, 5 adulterants	[38]
HPLC	Gradient elution	UV 280 nm	5 opiates	[39]
HPLC	Isocratic elution	UV 4 wavelengths	8 opiates, 3 adulterants	[40]
HPLC	Isocratic elution	Diode array	7 opiates, 3 adulterants	[41]
CE	MEKC	UV, diode array	9 opiates, 4 adulterants	[42]

After administration by any route, heroin is rapidly deacetylated to 6-monoacetylmorphine (6-MAM) and then to morphine. Morphine is consecutively coupled with glucuronic acid to 3-and 6-glucuronides (M3G and M6G). Since the first deacetylation step is very rapid (the half-life of heroin is estimated at 3–8 min), usually only metabolites of heroin are found in blood [43–46]. In urine, a very small fraction of unchanged heroin is excreted, as well as 6-MAM, morphine, and both its glucuronides [47].

It must be stressed that heroin after deacetylation follows metabolic routes that are common with morphine and—to some extent—also with codeine. Therefore only 6-MAM can be regarded as a heroin-specific metabolite and as a marker of heroin use. This analyte is usually detectable in blood and urine samples of heroin users [48,49]. Figure 3 shows the biotransformation of heroin, morphine, and codeine in the human body.

2.2.2 Heroin Metabolites and Associated Drugs in Body Fluids of Heroin Abusers

Thin-Layer Chromatography. TLC can be used for detection of heroin metabolites in urine extracts, using typical TLC plates [50,51] or the commercial Toxi-Lab system [52]. TLC densitometry has been applied for quantitation of heroin metabolites in urine [53]. However, this technique is not sensitive enough for examination of blood or serum.

Gas Chromatography. GC, usually with mass spectrometric detection [GC-MS], is the most frequently used technique for detection and quantitation of 6-MAM, morphine, and other associated opiates in biological material. Fehn and Megges introduced the GC-MS determination of 6-MAM in urine as a marker of heroin use [54]. Since then, several authors have published GC-MS methods for the determination of 6-MAM in urine as a specific heroin metabolite. The concentration of 6-MAM in urine varied in a very broad range of 2 to over 10,000 μg/L [55–57]. This compound can be also find in blood samples taken a short time after heroin abuse [58–61].

Strictly speaking, the presence of 6-MAM in urine, blood, or other body fluids evidences the intake of diacetylmorphine [DAM]. The differentiation between the intake of pure DAM and illicit heroin became relevant since introducing heroin prescription programs in some countries, like Switzerland, Great Britain, Germany, and the Netherlands. One of the basic requirements of these programs is that the participants must not use any illicit drugs, particularly illicit heroin. In illicit heroin not only DAM is present but also several other opiates, like 6-MAM, acetylcodeine, codeine, papaverine, noscapine, and various adulterants [14–19]. Therefore the substances that

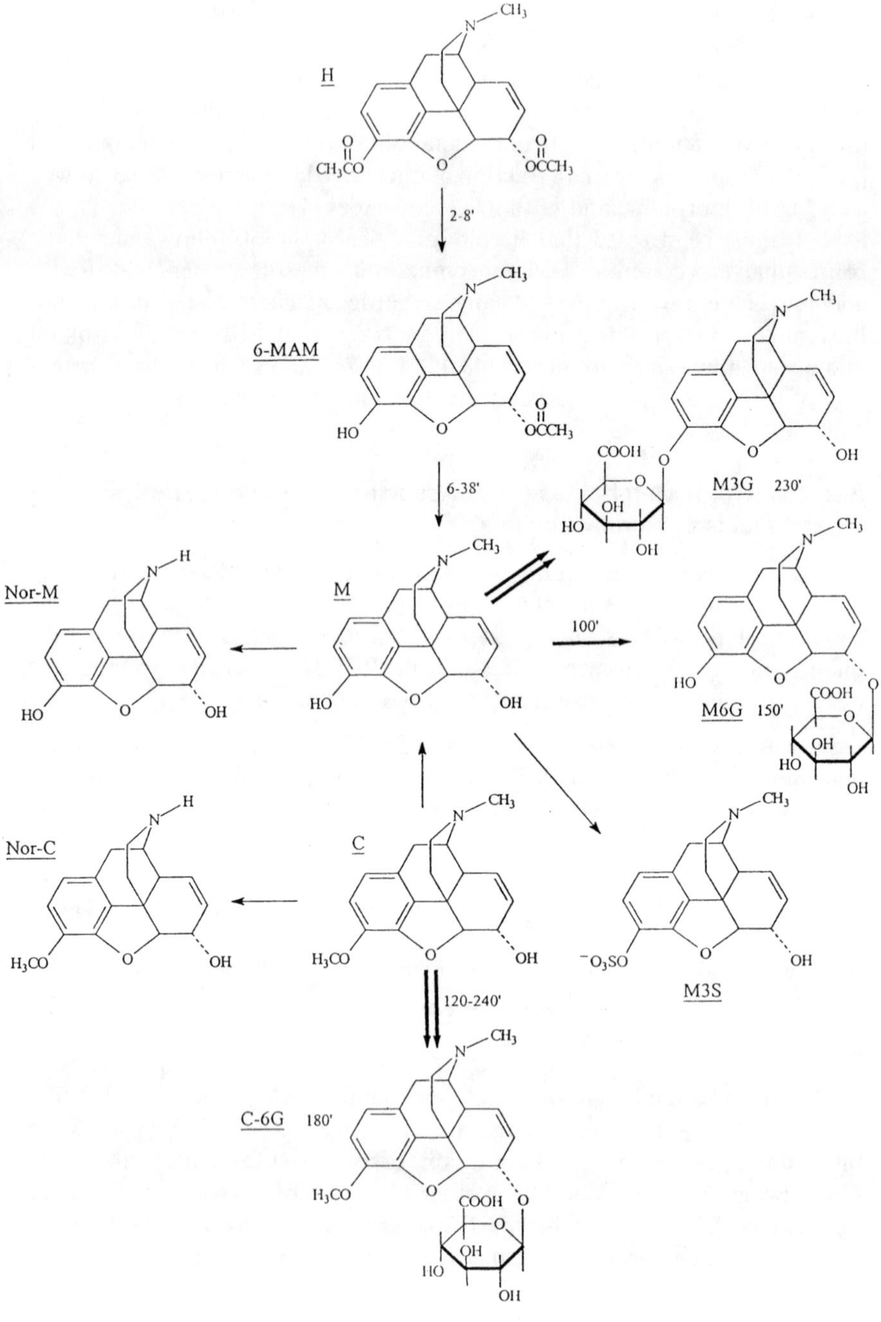

H

2-8'

6-MAM

6-38'

M3G 230'

Nor-M

M

100'

M6G 150'

Nor-C

C

M3S

120-240'

C-6G 180'

are excreted with urine and may serve as prospective markers of heroin use may be divided into two groups:

Markers of DAM use (DAM itself and its metabolite 6-MAM)
Markers of illicit heroin use (acetylcodeine, noscapine, and papaverine).

It must be stressed that only acetylcodeine (AC) can be regarded as a specific marker of illicit heroin. AC is produced from codeine during the acetylation of opium. Its content in illicit heroin ranges from 2 to 7% [14,15]. O'Neal and Poklis [62,63] developed a GC-MS method for the detection of AC in urine and found this drug in over 30% of morphine-positive specimens in concentrations ranged from 1 to 4600 μg/L. 6-MAM was found in over 70% of these samples. Codeine—a possible metabolite of AC—was found in all urine samples. Staub et al. [64] detected AC in over 85% and 6-MAM in over 94% of 71 urine samples obtained from illegal heroin consumers. Bogusz et al. [65] determined heroin markers in 25 morphine-positive urine samples with atmospheric pressure chemical ionization liquid chromatography-mass spectrometry. Codeine-6-glucuronide was found in all samples, codeine in 24, noscapine in 22, 6-MAM in 16, papaverine in 14, DAM in 12, and AC in 4 samples.

Heroin metabolites and associated substances in blood samples have been usually determined simultaneously with GC-MS. Schuberth and Schuberth [58] published a GC-MS method for the determination of 6-MAM, morphine, and codeine in blood. The blood samples were precipitated with methanol, and the drugs were extracted with SPE C_{18} cartridges and derivatized with PFPA. The method was applied for forensic samples, and in six cases of fatal heroin overdose 6-MAM concentrations of 1.6–6.1 μg/L blood were found. Cone and Darwin [66] reviewed in 1992 the GC-MS methods for opiates, cocaine, and metabolites, including also 6-MAM. Wasels and Belleville [59] presented in 1994 an overview of GC-MS procedures for the identification of 6-MAM, morphine, and codeine. Several derivatization methods were reviewed: acetylation, propionylation, acylation (TFA, PFPA, HFBA) and silylation. Wang et al. [60] published a method for the simultaneous determination of heroin and its metabolites 6-MAM, morphine, and normorphine as well as cocaine and its metabolites in hair, plasma, saliva, and urine. The drugs were derivatized with BSTFA/TMCS before GC-MS-SIM analysis. The detection limits for heroin and 6-MAM were 1 μg/L of saliva or urine. Heroin, 6-MAM, and morphine levels were monitored in saliva after experimental admin-

FIGURE 3 Biotransformation of heroin, morphine, and codeine. (From Ref. 48 with permission of Elsevier Science Publishers.)

istration of intranasal heroin. Goldberger et al. [67] developed a GC-MS method for the determination of heroin, 6-MAM, and morphine in body fluids and organs of 21 victims of heroin overdose. The samples were extracted with SPE cartridges and partially derivatized with MBTFA (for 6-MAM and morphine). Heroin was determined without derivatization. 6-MAM was detected in all 21 urine samples and in 14 blood samples. Heroin was not detected in blood but was present in 17 urine samples. The authors used the concentration ratios of drugs for the evaluation of the rapidity of death. Moeller et al. [61] determined 6-MAM in serum, urine, and hair of heroin users by GC-MS. Twenty five urine samples were examined, which showed positive immunochemical reaction on opiates. In 19 urine samples, 6-MAM was detected, the concentration ranging from 1 to 9950 μg/L. In five serum samples 6-MAM levels of 2–9 μg/L were observed. Guillot et al. [68] developed a GC-MS method for the determination of heroin, 6-MAM, and morphine in postmortem blood, urine, and vitreous humor. The drugs were isolated with alkaline solvent extraction and subjected to propionylation. The quantitation limits were 2 μg/L for morphine and 6-MAM and 5 μg/L for heroin.

Liquid Chromatography. HPLC has one important advantage over GC in regard to opiate analysis: it makes possible the determination of polar metabolites, mainly glucuronides, simultaneously with the primary drugs. The hydrolysis step, which has to be applied for glucuronide cleavage before GC analysis, can be omitted, and important differentiation between the active M6G and the inactive M3G is possible. HPLC with DAD detection was applied by Swiss authors [69], who developed an HPLC-DAD method for heroin, 6-MAM, 3-MAM, morphine, M3G, M6G, normorphine, codeine, and codeine-6-glucuronide determination in human serum. SPE with C_{18} cartridges was applied. The LOQ was 25 μg/L for each compound. The method was applied for the determination of heroin and its metabolites in serum samples taken from addicts who participated in the Swiss Heroin Maintenance Program and who were given pure heroin intravenously. The advent of LC-MS brought very important progress in the determination of heroin and its metabolites in biological fluids. Zuccaro et al. [70] developed an electrospray (ESI) LC-MS method for simultaneous determination of heroin, 6-MAM, morphine, M3G, and M6G in serum. The drugs were extracted with SPE C_2 cartridges and separated on a straight phase silica column in a methanol-ACN-formic acid mobile phase. The authors used a silica column in order to separate all substances in one run under isocratic conditions. The LOD for heroin was 0.5 μg/L, for 6-MAM 4 μg/L. The method was applied for pharmacokinetic study on heroin-treated mice. Bogusz et al. [71] used atmospheric pressure chemical ionization (APCI) LC-

MS for the determination of heroin metabolites (6-MAM, morphine, M3G, and M6G) in blood, cerebrospinal fluid, vitreous humor, and urine of heroin victims. The drugs were extracted with C_{18} cartridges; the LOD for 6-MAM was 0.5 μg/L. Low molar ratios of M3G/morphine and M6G/morphine in blood indicated short survival times after drug intake. In the next study, Bogusz et al. [72] extended the LC-APCI-MS method for the determination of 6-MAM, M3G, M6G, morphine, codeine, and C6G, using deuterated internal standards for each compound. The detection limits ranged from 0.5 to 2.5 μg/L. This method was applied for routine forensic examination of blood samples taken from suspected heroin abusers [49].

Capillary Electrophoresis. Capillary zone electrophoresis and micellar electrokinetic chromatography was applied for the determination of M3G in urine, using direct injection or SPE extraction [73]. Tagliaro used capillary electrophoresis with diode array detection for the determination of morphine in hair samples of drug abusers [74,75]. Wu [76] used the same method for the detection of morphine and M3G in the urine of heroin addicts. More recently, the ion trap mass spectrometer was used a detector for morphine separated with capillary zone electrophoresis. The sensitivity was lower than of LC-MS methods [77].

Tables 3 and 4 show selected chromatographic methods for the determination of heroin metabolites and related substances in body fluids.

2.2.3 Other Opioids of Forensic Significance in Body Fluids

Methadone. Methadone is a synthetic opioid of very long elimination half-life (15–55 h) which is used mainly as a heroin substitute in the treatment of heroin addicts. Methadone is metabolized to inactive 2-ethylidene-1,5-dimethyl-3,3-diphenylpyrrolidine (EDDP) and to a lesser extent to 2-ethyl-5-methyl-3,3-diphenyl-1-pyrroline (EMDP). All these compounds contain a chiral center, and the enantiomers could be separated. This is of pharmacological relevance since (R)-(−)-methadone (levomethadone) is 25–50 times more potent than the (S)-(+)-methadone. Commercial methadone preparations contain racemic form or levomethadone.

Several GC-MS procedures have been described for the determination of methadone, EDDP, and EMDP in plasma, urine, and hair. Usually, solvent or solid-phase extraction is applied [78–80]. The detection limits for all substances were 5 μg/L. Sporkert and Pragst [82] applied headspace solid phase microextraction coupled with EI GC-MS for the determination of methadone, EDDP, and EMDP in human hair. Absolute extraction yields were in the range of 10.5% to 17.4% for all examined compounds; the detection limits were 0.03 μg/g for methadone and 0.05 μg/g for metabolites.

TABLE 3 Gas Chromatographic Methods for Heroin, 6-MAM, and Other Opiates

Drug	Sample	Isolation	Derivatization	Column, conditions	Detection	LOD (μg/L)	Ref.
6-MAM	Urine	SPE C_{18}	PFPA	OV-1, 230°	EI-MS (SIM)	2	54
6-MAM	Urine	SPE or L/L	Propionylation	DB 5, 130°–250°	EI-MS (SIM)	0.8	55
6-MAM	Urine	L/L alkaline	Propionylation	RSL 200, 146°–246°	EI-MS (SIM)	n.s.	56
6-MAM	Urine	SPE Certify	TFA	HP-1, 150°–300°	EI-MS (SIM)	n.s.	57
6-MAM	Serum, urine, hair	SPE C_{18}	PFPA	n.s.	EI-MS (SIM)	n.s.	61
AC, 6-MAM, M, C, NC	Urine	SPE	Propionylation	HP-1, 170°–280°	EI-MS (SIM)	0.5	62
6-MAM, M, C	Blood	SPE C_{18}	PFPA	DB 5, 150°–256°	EI-MS (SIM)	0.5	58
Heroin, 6-MAM	Serum, saliva, urine, hair	SPE	BSTFA/TMCS	HP-1, 70°–250°	EI-MS (SIM)	1	66
Heroin, 6-MAM, M	Body fluids, organs	SPE	MBTFA	Rtx-5, 150°–290°	EI-MS (SIM)	1	67
M, C	Urine	L/L pH 9	Acetylation	DB-5, 240°C	EI-MS (SIM)	2 ng on-column	96
M	Blood	L/L pH 9	PFPA	DB-5, 100–300°C	EI-MS-MS	1	97
M, C	Blood	SPE C_{18}	PFPA	CP-Sil5, 200–300°C	NCI-MS (SIM)	2-5	98
M, C, NM	Plasma	L/L pH 9.5	HBFA	HP-1, 100–257°C	NCI-MS (SIM)	1 pg on-column	99
M	Plasma	L/L pH 9	PFPA	HP-5, 150–250°C	EI-MS (SIM)	0.2	100

M = morphine, NM = normorphine, C = codeine, AC = acetylcodeine, NC = norcodeine, n.s. = not stated.

TABLE 4 Liquid Chromatographic Methods for Heroin, 6-MAM, Morphine, and Other Opiates

Drug	Sample	Isolation	Column, elution conditions	Detection	LOD (μg/mL)	Ref.
Heroin, 6-MAM, M, C, pholcodine	Urine	SPE BondElut	Silica, CH_2Cl_2-pentane-MeOH	UV	4–20	101
6-MAM, M	Plasma	solvent	Silica, hexane-2PropOH-NH_3	FI	10–25	102
Heroin, 6-MAM, M, M3G, M6G, C, C6G	Plasma	SPE C_{18}	ODS, ACN-H_2O-H_3PO_4	DAD	25	69
Heroin, 6-MAM, M, M3G, M6G, C	Serum	SPE C_2	Silica, ACN-MeOH-H_2O-HCOOH	ESI-MS	0.5–4	70
6-MAM, M, M3G, M6G	Body fluids	SPE C_{18}	ODS, ACN-$HCOONH_4$	APCI-MS	0.1–1	71
6-MAM, M, M3G, M6G, C, C6G	Body fluids	SPE C_{18}	ODS, ACN-$HCOONH_4$	APCI-MS	0.5–100	72
M, M3G, M6G, NM	Plasma	SPE C_{18}	ODS, ACN-phosphate buffer pH 2.1	FI + EC	1–5	103
M, M3G, M6G	Plasma	SPE C_2	Phenyl, MeOH-phosphate buffer pH 4.0	FI + EC	1–10	104
M, M3G, M6G	Plasma	SPE C_{18}	ODS, ACN-phosphate buffer pH 2.1	UV	4–50	105
M, M3G, M6G, NM	Plasma	SPE C_8	ODS, ACN-phosphate buffer pH 2.1	FI	10–40	106
M, M3G, M6G	Plasma	SPE C_8	ODS, ACN-H_3PO_4	FI	5–10	107
M, M3G, M6G	Plasma	SPE C_2	RP, MeOH-H_2O	ESI-MS	10–100	108
M, M3G, M6G	Plasma	SPE C_{18}	ODS, ACN-HCOOH	ESI-MS	0.8–5	109
M, M3G, M6G	Plasma	SPE C_2	Phenyl, MeOH-HCOOH	ESI-MS-MS	3.8–12	110
M, M3G, M6G, NM	Plasma, urine	SPE C_2	ODS, H_2O-ACN-THF-HCOOH	ESI-MS-MS	0.5–2.5	111
MAM, M, M3G, M6G, C, C6G	Blood	SPE C_8	ODS, ACN-$HCOONH_4$	ESI-MS	0.5–5	112
M, M3G, M6G	Serum	SPE C_{18}	ODS, ACN-$HCOONH_4$	ESI-MS	1–5	113
M, M3G, M6G	Plasma	SPE C_{18}	ODS, H_2O-ACN-HCOOH	ESI-MS-MS	0.25–0.5	114
M, M3G, M6G	Plasma	SPE C_{18}	Silica, H_2O-ACN-HCOOH	ESI-MS-MS	0.5–1	115

M = morphine, M3G = morphine-3-glucuronide, M6G = morphine-6-glucuronide, C = codeine, C6G = codeine-6-glucuronide, NM = normorphine, FI = fluorescence detection, EC = electrochemical detection, ACN = acetonitrile.

The results in forensic investigations were in agreement with the results obtained with GC-MS after solid phase extraction.

Several authors used the possibility of chiral separation of methadone enantiomers. HPLC with UV detection was used for separation of methadone and EDDP enantiomers in urine samples obtained from methadone maintenance patients and from patients with chronic pain [82]. The drugs were separated on a RP8 column coupled serially with a chiral AGP column. This combination improved the separation and prolonged the lifetime of the chiral column. The LOD was 9 ng/mL. Kintz et al. [83] published the first LC-MS method for enantioselective separation of methadone and EDDP in hair, using deuterated analogues of all compounds for quantitation. Ortelli et al. [84] applied LC-MS for enantioselective determination of methadone in saliva and serum. The method was applied for analysis of samples taken from heroin addicts participating in methadone maintenance program. The correlation between the levels of enantiomers in serum and saliva was much better than for total methadone values.

Buprenorphine. Buprenorphine (BP) is a synthetic opiate agonist/antagonist, which was initially used as a potent analgesic and later has been also used in the treatment of heroin addicts [85]. This drug possesses primary addiction potential and is frequently abused. BP as well as its active metabolite norbuprenorphine (NBP) may be determined in serum or urine with GC-MS, usually after derivatization [86–88]. BP and NBP may be determined without derivatization with HPLC with electrochemical detection [89,90]. However, the limits of detection reported for this method (5–35 μg/L) are about 10 times higher than those for GC-MS. Much better results were obtained with LC-MS or LC-MS-MS. These methods appeared more sensitive than GC-MS and could be applied for pharmacokinetic studies [91–94].

Dihydrocodeine. Dihydrocodeine (DHC) is a semisynthetic opiate, which has used been at first as an analgesic and antitussive drug. In the late 1980s, DHC was extensively used in Germany in the treatment of heroin addicts, and in consequence a number of fatal poisonings were observed [127]. Like all opiates, DHC possesses a primary addiction potential and can be abused [128]. DHC in the human body undergoes *N*-demethylation to nor-DHC and *O*-demethylation to very toxic dihydromorphine (DHM). All these drugs are being conjugated to appropriate glucuronides [129]. The excretion of DHC metabolites in urine was studied by Balikova et al., who applied GC/MS after solid-phase extraction and cleavage of conjugates [130]. All metabolites of DHC, nor-DHC, DHM, DHC-6-G, DHM-3-G, DHM-6-G, and nor-DHM-3-G, can be detected directly in urine extract using LC/MS, without hydrolysis [49].

Tables 5 and 6 review chromatographic methods for buprenorphine and methadone determination.

3 COCAINE

Cocaine belongs to the most important hard drugs of modern society. According to the National Institute of Drug Abuse, in 1997 about 1.5 million Americans were identified as current cocaine users [131]. The indicators of cocaine use in USA are stable [132]. The primary source of cocaine is the shrub *Erythroxylum coca*, domestic in the Andes Mountains in Peru and Bolivia, and later cultivated in various South American states. The word coca is derived from the Aymara khoka, meaning tree. The documented history of cocaine use began about 600 A.D., when coca leaves were found in the graves of pre-Incan tribes in Peru. At that time, coca leaves were mixed with lime and ashes and chewed. Cocaine consumed by this route enhanced physical performance and decreased hunger and the need of rest. The custom of cocaine chewing persisted among South American Indians, known as coqueros [133].

3.1 Cocaine and Related Compounds in Preparations

3.1.1 Production Methods of Cocaine

Although the synthesis of cocaine is possible, essentially all cocaine in the illicit traffic is produced by extraction from the leaves of *Erythroxylum coca*. The coca leaves are mixed with water and lime, crushed, and extracted with kerosene or gasoline for several hours. The leaf mulch is then filtered out and discarded. Kerosene extract is mixed with dilute sulfuric acid, and the aqueous layer is collected and made basic with ammonia or lime. The precipitate, known as coca paste, is isolated and dried. Coca paste contains 40 to 70% cocaine as a mixture of base and salt, as well as kerosene, sulfuric acid, and other impurities. Coca paste, a very dangerous intermediate product, is smoked in South America [134]. To form pure cocaine base, the coca paste is dissolved in dilute sulfuric acid, and potassium permanganate is added to oxidize the cinnamoyl cocaines. After filtration, ammonia is added to the solution. Precipitated cocaine base is collected by filtration, washed with water, and dried in the sun. In the last step, dried cocaine base is dissolved in ether or acetone and converted to cocaine hydrochloride by adding of hydrochloric acid. The resulting white, microcrystalline precipitate ("snow") is filtered,

Table 5 Gas Chromatographic Methods for Buprenorphine and Methadone

Drug	Sample	Isolation	Derivatization	Column, conditions	Detection	LOD (μg/L)	Ref.
BP	Blood	Extrelut + SCX	Silylation	CPSil-5, 180°–300°	PCI-MS (SIM)	1 pg on column	87
BP	Plasma	L/L pH 10.5	PFPA	DB-1,160°–310°	PCI-MS (SIM)	0.5 LOQ	116
BP	Plasma	L/L pH 9.1	HFBA	HP 1, 150°–325°	ECD	0.1 BP	93
BP, NBP	Plasma	SPE	HFBA	DB-5, 125°–300°	NCI-MS-MS	0.15 BP 0.016 NBP	117
BP, NBP	Urine	L/L alkaline	Methylation	HP 2, 247°–310°	EI-MS (SIM)	0.2 both	86
Meth	Plasma, urine, CSF	L/L pH	—	SE-52	NPD	0.5 LOQ	118
Meth, EDDP	Urine	L/L alkaline	—	DB-5, 190°	EI-MS (SIM)	50	119
Meth, EDDP, EMDP	Hair	L/L alkaline	—	DB-5, 80°–280°	PCI-MS-ITD	0.5 ng/mg	78
Meth, EDDP, EMDP	Plasma, urine, liver	SPE	—	HP-1, 80°–280°	PCI-MS (SIM)	10	79

BP = buprenorphine, NBP = norbuprenorphine, Meth = methadone, EDDP = 2-ethylidene-1,5,dimethyl-3,3-diphenylpyrrolidine, EMMP = 2-ethyl-5-methyl-3,3-diphenyl-pyrroline.

TABLE 6 Liquid Chromatographic Methods for Buprenorphine and Methadone

Drug	Sample	Isolation	Column, elution conditions	Detection	LOD (μg/mL)	Ref.
BP, NBP	Hair	L/L pH 8.5	CN, ACN-phosphate buffer	EC	0.02 ng/mg BP, 0.01 NBP	89
BP, NBP	Plasma	SPE	C18, ACN-phosphate buffer	EC	25 BP, 5 NBP	90
BP, NBP	Blood, urine, hair	L/L pH 8.4	C18, ACN-NH_4COOH	ESI-MS (SIM)	0.1 BP, 0.05 NBP	91
BP, NBP	Blood	Extrelut pH 9	C18, ACN-NH_4COOH	ESI-MS (SIM)	0.1 LOQ BP, NBP	92
BP	Plasma	L/L pH 10.5	C8, H_20-MeOH-ACN-HCOOH	ESI-MS-MS	0.1 LOQ	93
BP, NBP, BPG	Plasma	SPE	C18, ACN-NH_4COOH	ESI-MS-MS	0.1 LOQ	94
BP	Blood	SPE	C18, ACN-NH_4COOH	APCI-MS (SIM)	0.5	120
BP	Hair	L/L pH 8.5	CN, ACN-phosphate buffer	EC	0.02 ng/mg BP, 0.01 NBP	121
Meth, EDDP	Urine, meconium	L/L pH 9	C18, ACN-phosphate buffer + TEA	DAD 204 nm	76 M, 127 EDDP	122
R/S-Meth, R/S-EDDP	Hair	SPE C18	Chiral-AGP, PropOH-NH_4COOH	ESI-MS (SIM)	0.2 M, 0.1 EDDP	83
R/S-Meth	Serum	SPE mixed	Chiral-AGP	UV 205 nm		123
Meth	Blood	SPE Certify	C18, MeOH-NH_4COOH	TSP-MS-MS	50 pg on-column	124
R/S-Meth	Serum	L/L	Chiral-AGP + CN, ACN-phosphate buffer	UV 200 nm	1.5 LOQ	125
R/S-Meth	Plasma	L/L	Chiral-AGP	UV 215 mm	2.5 LOQ	126

BP = buprenorphine, NBP = norbuprenorphine, BPG = buprenorphine glucuronide, Meth = methadone, EDDP = 2-ethylidene-1,5,dimethyl-3,3-diphenylpyrrolidine, EC = electrochemical detection.

dried, and packaged for sale. Street cocaine preparations are usually adulterated with some other local anesthetics and diluted with neutral compounds. Table 7 shows the substances, that may be found in street cocaine samples.

Cocaine hydrochloride is water-soluble and may be applied intranasally (snorting) or intravenously. In order to obtain a smokable form of drug, cocaine salt must be converted again to a volatile base. This procedure, known as freebasing, involves dissolving cocaine hydrochloride in water, alkalization, and extraction with ether and evaporation. Freebasing creates an extremely hazardous situation owing to the risk of fire and explosion and in 1980s was replaced by "crack." Crack is produced by heating cocaine hydrochloride solution with sodium carbonate, cooling, and

TABLE 7 Compounds Occurring in Illicit Cocaine Samples

Cocaine byproducts	Adulterants	Diluents
3′,4′,5′-trimethoxy-*cis*-cinnamoylcocaine	Lidocaine	Mannitol
3′,4′,5′-trimethoxy-*trans*-cinnamoylcocaine	Benzocaine	Lactose
3′,4′,5′-trimethoxycocaine	Procaine	Dextrose
3′,4′,5′-trimethoxytropacocaine		Sucrose
Alpha-truxinic acid		
Anhydroecgonine		
Anhydroecgonine methyl ester		
Benzoylecgonine		
Beta-truxinic acid		
cis-Cinnamoyl ecgonine methyl ester		
cis-Cinnamoylcocaine		
Ecgonine		
Ecgonine methyl ester		
Namoylcocaine		
N-benzoylnorecgonine methyl ester		
N-formylnorcocaine		
N-formylnorecgonine methyl ester		
Norcocaine		
Norecgonine methyl ester		
Nortropacocaine		
Pseudococaine		
trans-Cinnamoyl ecgonine methyl ester		
trans-Cinnamic acid		
trans-Cinnamoylcocaine		
Tropacocaine		
Truxillines		

filtering. The term "crack" refers to the crackling sound when the drug is smoked. Nowadays, crack is the most prevalent and most dangerous form of street cocaine [131].

3.1.2 Cocaine and Associated Compounds in Street Samples and Other Materials

From the data in Table 7 is obvious that illicit cocaine is a very complex mixture of various substances that may exist in different proportions in samples originating from different sources or even from different production charges. This complexity makes it possible to identify particular samples of cocaine through analytical "fingerprinting." Several such methods were developed. Ensing et al. [135] measured the relative abundance of six congeners present in illicit cocaine with GC/PND. Janzen et al. [136] calculated area ratios of four congeners also using GC/PND. Headspace GC-MS combined with microwave irradiation was applied for the analysis of residual solvents in illicit cocaine hydrochloride samples [137].

In recent years, the transport of internally concealed cocaine, so-called body packing, became a standard smuggling procedure, especially in airline transport [138]. Cocaine containers are usually machine-made and consist of several (up to 7) layers of wax, latex, and foil. One drug courier may swallow more than 100 containers with a total gross weight more than 1000 g. [139]. Body packing is an extremely hazardous smuggling method owing to the possibility of container rupture. In such a case, acute, fatal intoxication occurs, with a very high level of nonmetabolized drug [140,141]. Cocaine trafficking, i.e., the exchange of illicit drug for money, is an everyday occurrence. As a consequence, American currency became entirely contaminated with cocaine. Ubiquitous presence of cocaine on $1 and $20 bills was demonstrated by Oyler et al. [142] and Negrusz et al. [143]. The amount of drug on a single bill ranged from some nanograms to over one milligram. Capillary electrophoresis (CE) has been used for the analysis of street cocaine. CE coupled with electrospray time-of-flight MS was applied for the detection of various drugs of abuse including cocaine [144]. Cocaine and adulterants were quantified by free zone CE [145] (Fig. 4).

3.2 Cocaine and Metabolites in Biological Fluids and Organs

3.2.1 Metabolism of Cocaine in Humans

After administration in humans, cocaine is extensively metabolized. The most prevalent metabolites are benzoylecgonine (BZE) and ecgonine methyl

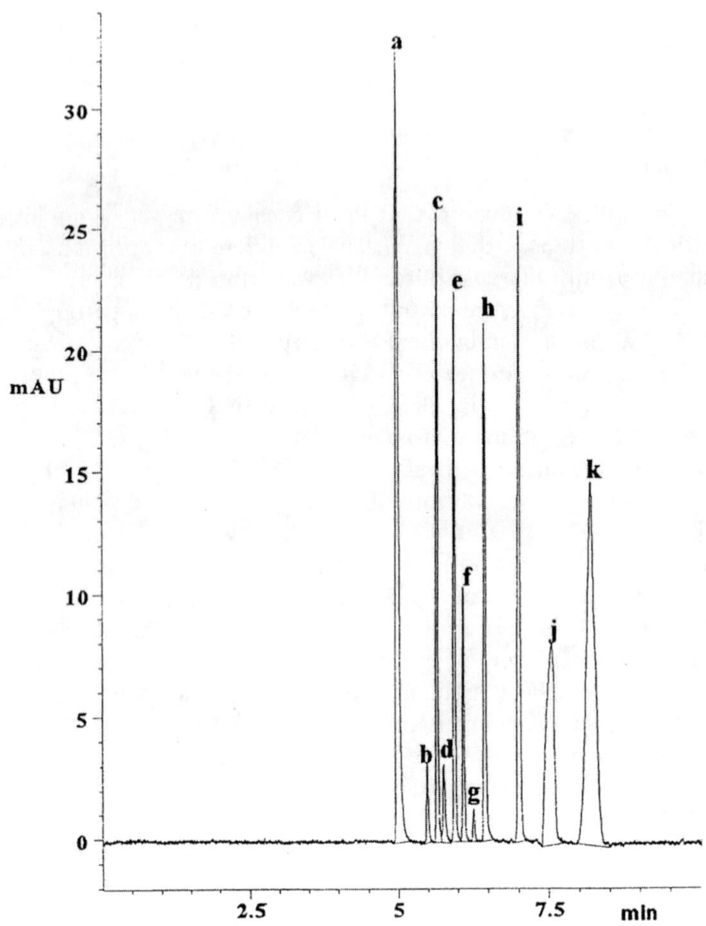

FIGURE 4 Free solution electropherogram of cocaine, adulterants, and impurities. Peaks: a = histamine (internal standard), b = ephedrine, c = procaine, d = phenylpropanolamine, e = cocaine, f = *trans*-cinnamoylcocaine, g = *cis*-cinnamoylcocaine, j = benzocaine, k = acetaminophen. (Reprinted with permission from Ref. 145, Copyright © ASTM.)

ester (EME). Both these compounds are then degraded to ecgonine. The biological half-life of cocaine in plasma ranged from 11 to 87 min and was dose and route dependent [146,147]. Biological half-life for BZE and EME was 7.5 and 3.6 h, respectively [146,148,149]. Therefore the detection window for cocaine metabolites is much broader than for cocaine itself. Some metabolites give specific indications concerning the route or mode of cocaine administration. Anhydroecgonine methyl ester is formed during cocaine pyrolysis and may serve as a marker of crack use [150]. Simultaneous ingestion of cocaine and alcoholic beverages results in formation of cocaethylene, ecgonine ethyl ester, and norcocaethylene [148] (Fig. 5).

Cocaine metabolism occurs not only in corpore, but also in vitro. Enzymatic hydrolysis of cocaine to EME and partially to BZE may be inhibited by the addition of esterase inhibitors like sodium fluoride [151,152]. Also, some drugs that inhibit the activity of serum cholinesterase, like amitriptyline or procainamide, may inhibit the degradation of cocaine or cocaethylene and prolong the half-life of these drugs in serum [153]. The final product of cocaine degradation, ecgonine, may be detected much longer than BZE and EME, even in nonpreserved blood samples [154].

3.2.2 Analytical Methods for Cocaine and Metabolites

Gas Chromatography. Cocaine is often used together with opiates, particularly heroin. Therefore several authors developed methods for simultaneous determination of cocaine, heroin, and metabolites of both compounds in blood or serum [60,66] or in hair [155]. In recent years, all authors have used almost exclusively the mass spectrometer as a detector after GC separation. One of exceptions was the study of Watanabe et al. [156], who applied GC with surface ionization detection (SID) to measurement of cocaine and cocaethylene in blood extracts. The LODs were 1.5 to 3 μg/L. Since cocaine is rapidly metabolized, all modern GC-MS methods are targeted on a broad spectrum of cocaine transformation products. Crouch et al. [157] developed a GC-MS method for determination of cocaine and metabolites in biological fluids and tissues. Derivatization was done with tert-butyldimethylsilane effecting in stable derivatives with mass spectral ions with higher mass than TMS derivatives. Hernandez et al. [158] analyzed cocaine and metabolites in brain using mixed-phase SPE and ion-trap GC-MS. A LOD of 50 ng/g brain tissue was reported. An important problem, the detectability of ecgonine—a polar, final metabolite of cocaine, was addressed by Logan et al., who developed a GC-MS method for the determination of this compound [159]. Extractive propylation, followed by solvent extraction of ecgonine well as cocaine and its major biotransformation products was applied. This method was used for examination of urine samples. Among 11 metabolites, BZE, EME, and ecgonine were excreted in highest concentrations [160]. Ecgonine was also detected as only cocaine metabolite in old, unpreserved blood sample [154].

Methylecgonidine (anhydroecgonine methyl ester) was identified as specific marker of cocaine (crack) smoking [150]. The hydrolytic product of

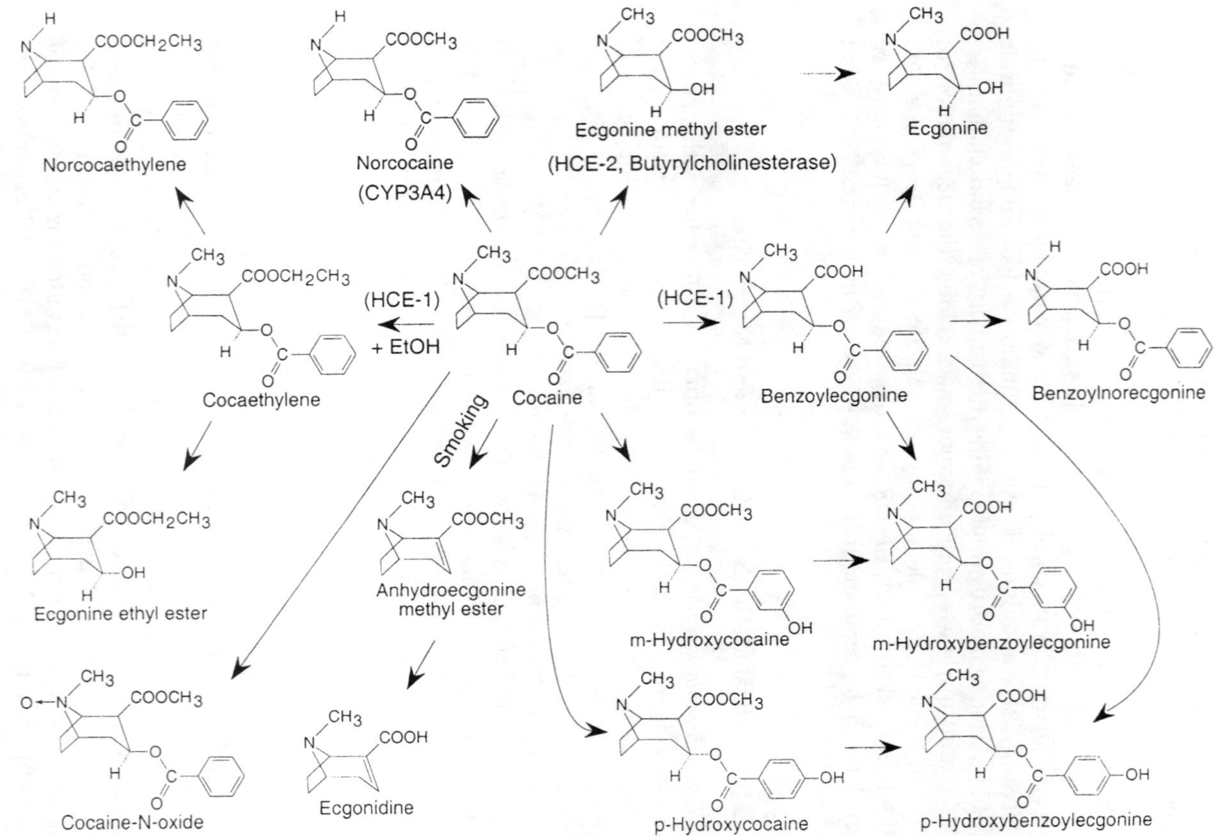

FIGURE 5 Biotransformation and thermal degradation products of cocaine. (From Ref. 148 with permission of Elsevier Science Publishers.)

methylecgonidine—ecgonidine—was extracted with SPE at acidic pH and detected with GC-MS after butyldimethylsilylation. Ecgonidine was detected in 22 out of 23 BZE-positive urine specimens [161]. Methylecgonidine was determined in blood with GC/MS-SIM after derivatization with MBDSTFA. Cocaine, BZ, and EME were analyzed simultaneously. The concentrations of methylecgonidine ranged from 3 to 34 μg/L blood [162].

Cocaine and metabolites were frequently determined in hair or nails. Moeller et al. [163] analyzed cocaine, BZE, and EME by GC-MS in hair samples taken from 20 Bolivian coca chewers. All compounds were found; the cocaine concentration was the highest one. Cocaine and metabolites were also detected frequently in fingernails and toenails of suspected cocaine users subjected to autopsy. The detectability was much higher than in postmortem blood or urine samples [164,165].

The problem of intrauterine exposition to cocaine was addressed from an analytical point of view. Cocaine metabolites, predominantly BZE, were detected in meconium [166,167] in vernix caseosa [168], and in amniotic fluid and umbilical cord tissue [169] (Fig. 6).

Table 8 shows gas chromatographic methods used recently for cocaine analysis.

Liquid Chromatography. Reversed-phase columns were usually used for the separation of cocaine and metabolites with HPLC. The UV detector is still in use even in more recent studies. The LODs ranged from 10 to 20 μg/L [170,171]. Phillips et al. advocated HPLC with UV detection as a viable alternative to GC-MS [172]. The advent of LC-MS brought substantial progress in the sensitivity of detection of cocaine metabolites, which may be detected without any derivatization. Bogusz et al. used LC-APCI-MS for the examination of various drugs of abuse, among them cocaine, BZE, and EME in biological fluids [49,98]. The LOD ranged from 0.2 to 0.5 μg/L. The same technique was applied by Nishikawa et al. [173] for cocaine, norcocaine, BZE, EME, and ecgonine. Tandem LC-ESI-MS was used for simultaneous determination of cocainics and opiates in biological fluids, and an LOQ of 5 μg/L was reported [174]. Sosnoff et al. detected BZE in dried blood spots with tandem LCMS [175]. A LOD of 2 μg/L was achieved from a 12 μL sample. This technique was used for epidemiological screening in a study involving newborns. In a study of Jeanville et al., LC-MS was applied for automated urinalysis [176]. Centrifuged urine samples were injected into an LC/MS/MS system equipped with an on-line extraction unit. The total analysis time was below 4 min. The LODs for EME, BZ, and cocaine were 0.5, 2.0, and 0.5 μg/L, respectively. Skopp et al. carried out the study on the stability of cocaine and metabolites at different temperatures using LC-MS [177]. Only ecgonine

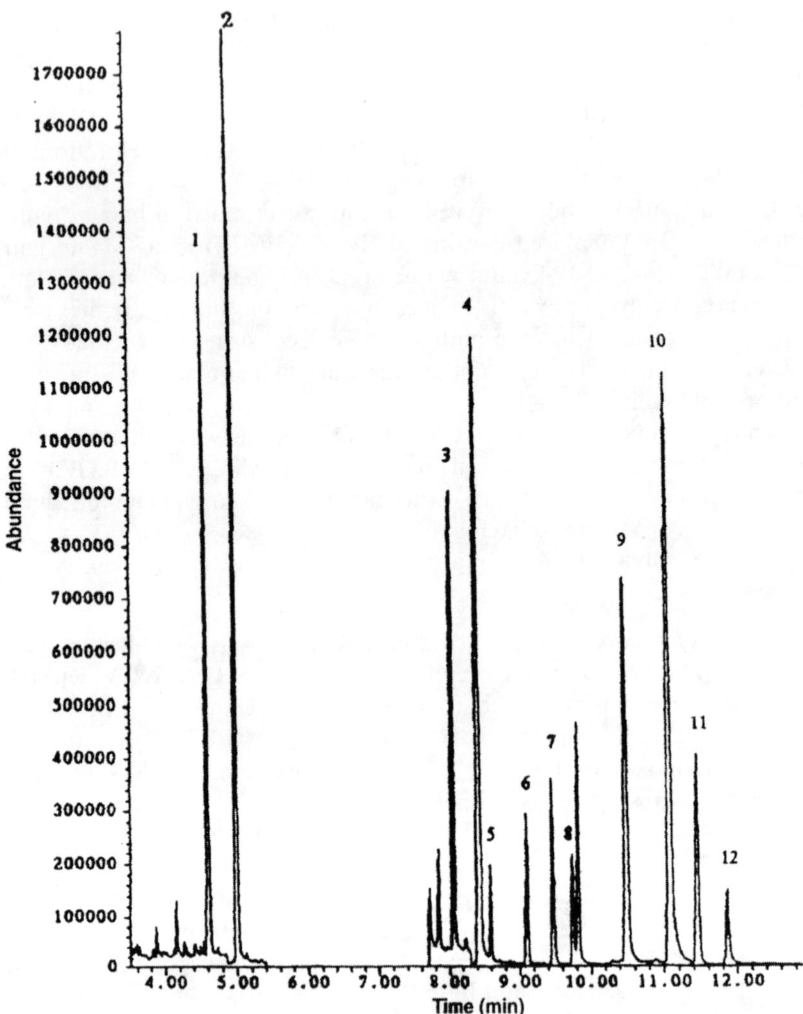

FIGURE 6 GC-MS total ion chromatogram of cocaine and metabolites. Peaks: 1 = ecgonine methyl ester, 2 = ecgonine, 3 = cocaine, 4 = cocaethylene and benzoylecgonine, 5 = norcocaine, 6 = norbenzoylecgonine, 7 = *o*-OH-cocaine, 8 = *o*-OH-benzoylecgonine, 9 = *m*-OH-cocaine, 10 = *m*-OH-benzoylecgonine, 11 = *p*-OH-cocaine, 12 = *p*-OH-benzoylecgonine. (Reproduced from the Journal of Analytical Toxicology, Ref. 167, by permission of Preston Publications, a Division of Preston Industries, Inc.)

TABLE 8 Gas Chromatographic Methods for Cocaine and Metabolites

Drug	Sample	Isolation	Derivatization	Column, conditions	Detection	LOD (μg/L)	Ref.
COC	US currency	SPE	—	HP-5, 130°–280°	EI-MS (f.sc.)	1 ng	143
COC, BZE, EME, ECG, CE, NCOC	Blood	L/L	Alkylation	BP-5, 80°–295°	EI-MS (SIM)	<10	159
COC, BZE, EME, ECG	Urine	SPE	Propylation	BP-5, 100°–295°	EI-MS (SIM)	100	160
COC, CE	Blood	SPE	—	DB-17, 150°–280°	GC-SID	1.5–3	156
COC, BZE, EME, NCOC, CE	Fluids/tissues	SPE	BSTFA	HP-1, 70°–250°	EI-MS (SIM)	1–5	60
COC, BZE, EME, CE	Fluids/tissues	SPE	MBDSTFA	DB-5, 115°–280°	PCI-MS (SIM)	2	157
COC, BZE, EME, ECG, AEME	Serum	SPE	MBDSTFA	HP-5, 55°–310°	EI-MS (SIM)	1	162
COC, BZE, CE	Brain tissue	SPE	PFPA/HFIP	DB-5, 105°–290°	IT-MS (f.sc.)	25	158
COC, BZE, NCOC, CE	Toenails	SPE	MSTFA	DB-5, 150°–260°	EI-MS (SIM)	0.3 ng	165
COC, BZE, EME, CE	Hair	SPE	BSTFA	CPSIL8, 50°–310°	EI-MS (f.sc.)	0.1–0.3 ng/mg	155

COC = cocaine, BZE = benzoylecgonine, EME = ecgonine methyl ester, CE = cocaethylene, NCOC = norcocaine, AEME = anhydroecgonine methyl ester, f.sc. = full-scale mass spectrum.

appeared to be stable at room temperature. The conversion of cocaine to the final metabolite ecgonine was stoichiometric (Fig. 7).

Table 9 presents liquid chromatographic methods published recently for cocaine analysis.

Capillary Electrophoresis. Capillary electrophoresis (CE) has usually been applied as technique for multiple drug detection. Tagliaro et al. [178,179] described a method for the determination of cocaine, morphine, and methylenedioxymethamphetamine (MDMA) in hair by CE with UV detection (single-wavelength or full spectrum 190–400 nm). The LOD for cocaine was 8 ng/mL. The method was applied to the screening of hair

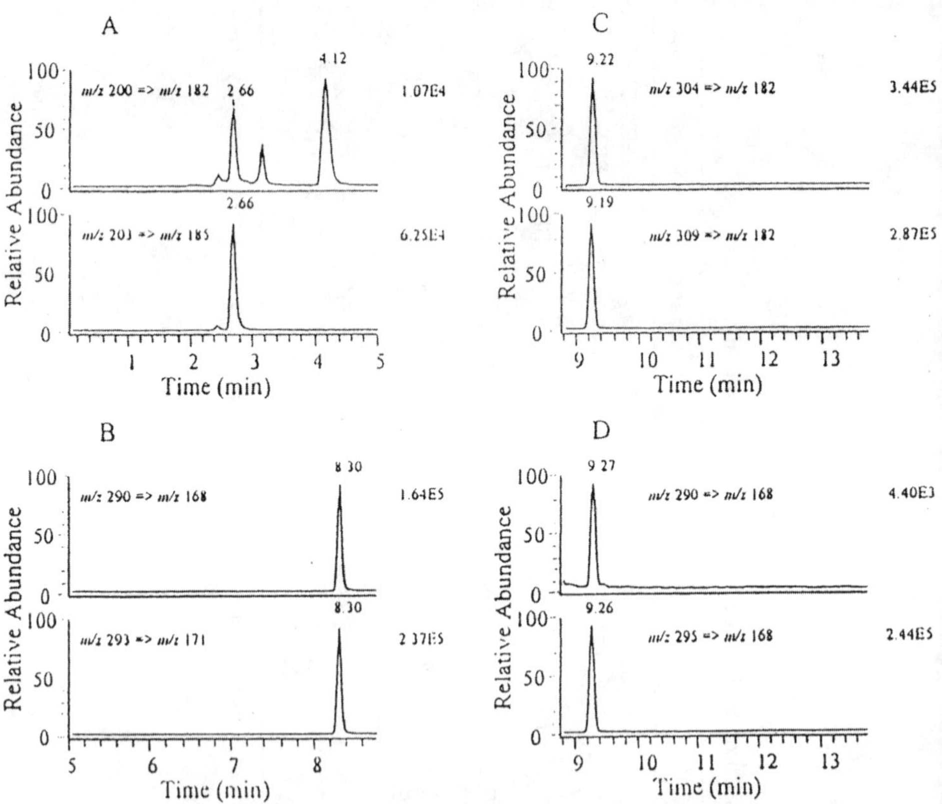

FIGURE 7 LC-APCI-MS-MS chromatograms of cocaine and its metabolites, and corresponding deuterated internals standards in plasma extracts. A = ecgonine methyl ester, B = benzoylecgonine, C = cocaine, D = norcocaine. (Reprinted with permission from Ref. 182. Copyright © 1999 American Chemical Society.)

TABLE 9 Liquid Chromatographic Methods for Cocaine and Metabolites

Drug	Sample	Isolation	Column, elution conditions	Detection	LOD (μg/L)	Ref.
COC, BZE, EME, NCOC, ECG	Urine	SPE	ODS, ammonium acetate, gradient	APCI-MS	1–16 ng	173
COC, BZE, EME, CE, AEME	Blood, urine	L/L	C8, ammonium formate, isocratic	ESI-MS-MS	5	174
COC, BZE, EME	Fluids, tissues	SPE	ODS,ammonium formate, isocratic	APCI-MS	0.2–0.5	120
BZE	Blood spots	Methanol	ODS, ammonium acetate, isocratic	APCI-MS-MS	2	175
COC, EZE	Urine	Filtration	C8, ammonium formate, gradient	ESI-MS-MS	2.5	181
COC, BZE, EME, NCOC	Plasma	Acetonitrile	ODS, ammonium acetate, gradient	APCI-MS-MS	2–5	182
COC, BZE, CE	Hair	SPE	ODS, ammonium acetate, gradient	ESI-MS-MS	0.02 ng/mg	183

COC = cocaine, BZE = benzoylecgonine, EME = ecgonine methyl ester, CE = cocaethylene, NCOC = norcocaine, AEME = anhydroecgonine methyl ester, ECG = ecgonine.

samples taken from applicants for driving licenses. Caslavska et al. developed an electrokinetic capillary-based immunoassay for BZE, opiates, methadone, and amphetamines [180]. 25 μL urine was incubated with polyvalent antibody and subjected to capillary electrophoresis. Unbound fluorescein-labeled drug traces were detected with laser-induced fluorescence.

4 AMPHETAMINE AND OTHER ILLICIT PSYCHOACTIVE PHENETHYLAMINES

4.1 Amphetamine and Ecstasy in Illicit Drugs

4.1.1 Recent Developments on the Illicit Amphetamine Drug Market

Amphetamine and methamphetamine are powerful stimulants used for over 100 years. Some of their methylenedioxy analogues like methylenedioxymethamphetamine (MDMA), methylenedioxyethylamphetamine (MDEA), or methylenedioxyamphetamine (MDA) were synthesized already in 1914 and tested in the treatment of psychiatric disorders. In contrast to amphetamine, MDMA did not become widely abused until the late 1970s. This drug is known under the names Ecstasy, XTC, or Adam. Paradoxically, the widespread abuse of psychoactive phenethylamines was propagated by the book of the eminent pharmacologist Alexander Shulgin, who in his book *Pihkal, A Chemical Love Story* [184] published procedures for the synthesizing of 179 various drugs of this group. Also, information concerning recommended dosage and expected symptoms was given. The book is freely available and certainly helped to proliferate psychoactive phenethylamines in society. Amphetamine derivatives are usually manufactured as tablets and distributed illegally in discotheques. These drugs, besides stimulating action, may alter thermoregulation and have caused a growing number of deaths, mainly due to heat stroke at rave parties [185–188]. Also, chronic abuse of psychoactive phenethylamines causes detectable impairment. High-dependent amphetamine users performed worse than controls on attention/concentration and memory tests [189]. Recreational users of MDMA performed worse than the controls in regard to memory and learning tasks and in tasks reflecting general intelligence [190]. It must be stressed that the recreational use of Ecstasy is very often associated with consumption of other psychotropic drugs or alcohol. Examination of urine samples taken from 64 attendees of rave parties revealed that in the majority of cases MDMA was present in combination with amphetamine, methamphetamine, or other designer amphetamines [191]. This suggests that the majority of ravers are multidrug users. A similar observation was also made by Bogusz [49].

4.1.2 Analytical Methods Used for Psychoactive Phenethylamines in Street Samples

NICI-MS was applied for the study of fragmentation of HFB- and PFB-derivatized amphetamine, methamphetamine, MDMA, MDA, MDEA, MBDB, and deuterated analogues [192]. Solvent-free SMPE combined with GC/PND was applied for impurity profiling of illicit ecstasy and amphetamine samples [193]. HPLC with UV detection at 200 nm was applied for the analysis of illicit amphetamine and ecstasy drug samples. The drugs were separated on a base-deactivated column. The LODs ranged from 0.06 to 0.1 μg/g [194]. Amphetamine, methamphetamine, MDMA, MDEA, MDA, and diethylpropione and ephedrine were determined in street samples by HPLC/DAD and CE/DAD. For the latter method, a chiral column was used for the separation of enantiomers [195]. Backofen et al. [196] applied non-aqueous CE with electrochemical detection for the determination of amphetamines, cocaine, and benzoylecgonine in illicit drugs. Amphetamine, methamphetamine, cathinone, methcatinone, ephedrine, norephedrine, and pseudoephedrine were labeled with 4-fluoro-7-nitrobenzofurazane and subjected to MEKC with laser-induced fluorescence detection. For chiral separations, cyclodextrins with SDS were applied [197].

4.2 Psychoactive Phenethylamines in Biological Fluids and Tissues

4.2.1 Metabolism and Pharmacokinetics of Phenethylamines

All amphetamines undergo extensive metabolism and are excreted with urine. Since all these drugs belong to strong bases, their elimination is dependent on the pH of urine. Enslin et al. [198] studied the metabolism of racemic MDE after oral administration to healthy volunteers. Urinary metabolites were analyzed with GC-MS and HPLC. It was demonstrated that in the first phase *O*-dealkylation to 3,4-dihydroxymetabolites occurs with consecutive methylation at the hydroxyl group at position 3. The second pathway lead through *N*-dealkylation to corresponding primary amines. All first-phase hydroxy metabolites were then partially conjugated. Maurer studied urinary excretion of metabolites of racemic MDMA, MDA, MDEA, *N*-methylbenzoxodialolylbutanamine (MBDB), and benzodioxoazolylbutanamine (BDB). Urine samples were taken from clinical poisoning cases, and the metabolites were identified with GC-MS. All drugs undergo two overlapping metabolic pathways: *O*-dealkylation to dihydroxymetabolites and *N*-dealkylation to amines [199]. Metabolism of designer amphetamines (MDMA MDA, MDEA, MBDB, BDB) was studied by Maurer et al. [200] after oral administration in human volunteers and in rats. Human plasma and urine as well as rat liver microsomes were examined. In the first phase, *N*-dealkylation and demethylenation were observed; in the second phase, conjugation with glucuronic and sulfuric acid (Figs. 8 and 9).

De la Torre et al. [201] investigated MDMA metabolism and pharmacokinetics in 14 human volunteers who were given MDMA in doses of 50–150 mg orally. The subjects were phenotyped for CYPP2D6 activity using dextromethorphan and were classified as extensive metabolizers. It was found that while the dose of MDMA rose threefold, the area of drug under the curve increased up to 10-fold. Also the urinary excretion of 3,4-dihydroxymethamphetamine, a demethylenated MDMA-metabolite, was constant for all doses, but the excretion of MDMA increased in a nonproportional way. This indicated that MDMA metabolism might be saturated even at moderate doses of drug, enhancing its toxicity.

The duration of amphetamine excretion after termination of drug abuse was tested on 22 prison inmates, starting their sentences in Norway [202]. The last positive-screened urine specimen was observed after 9 days of imprisonment. The excretion showed distinct interindividual variability, which was accompanied by variations in pH and creatinine. These data are in agreement with previous observations of Baselt, who wrote that with larger doses of drug, or if urine remains alkaline even with small doses, the amphetamine immunoassay produce positive results up to 7 or 10 days after the administration of amphetamines has ceased [203].

It must be stressed that all amphetamines have a single symmetric center and therefore exist as two enantiomers, each of which has different pharmacological activities. Lim et al. [204] studied the disposition of the enantiomers of MDMA after administration of racemic drug to rats and mice. Quantitative difference in urinary excretion of enantiomers was observed. Fallon et al. [205] determined the enantiomers of MDMA and MDA in plasma and urine after the oral administration of 40 mg of racemic MDMA to eight male volunteers. The plasma concentration of (R)-MDMA was over two times higher than that of the (S)-MDMA. The plasma half-life of the R-enantiomer was significantly longer than that of the S-enantiomer. The results showed that more active (S)-MDMA has a reduced AUC and a shorter half-life.

Besides illicit amphetamine and its analogues, several therapeutic drugs exist that are metabolized to amphetamine. This may create difficulties in the interpretation of results. For example, mefenorex is an anorectic, which is partially metabolized to amphetamine. Kraemer et al. identified thirteen metabolites in urine after oral administration of this drug to volunteers. In the late phase of elimination neither mefenorex nor its specific metabolites were detectable, but amphetamine was still present. Therefore the discrimination between mefenorex and amphetamine intake is not always possible

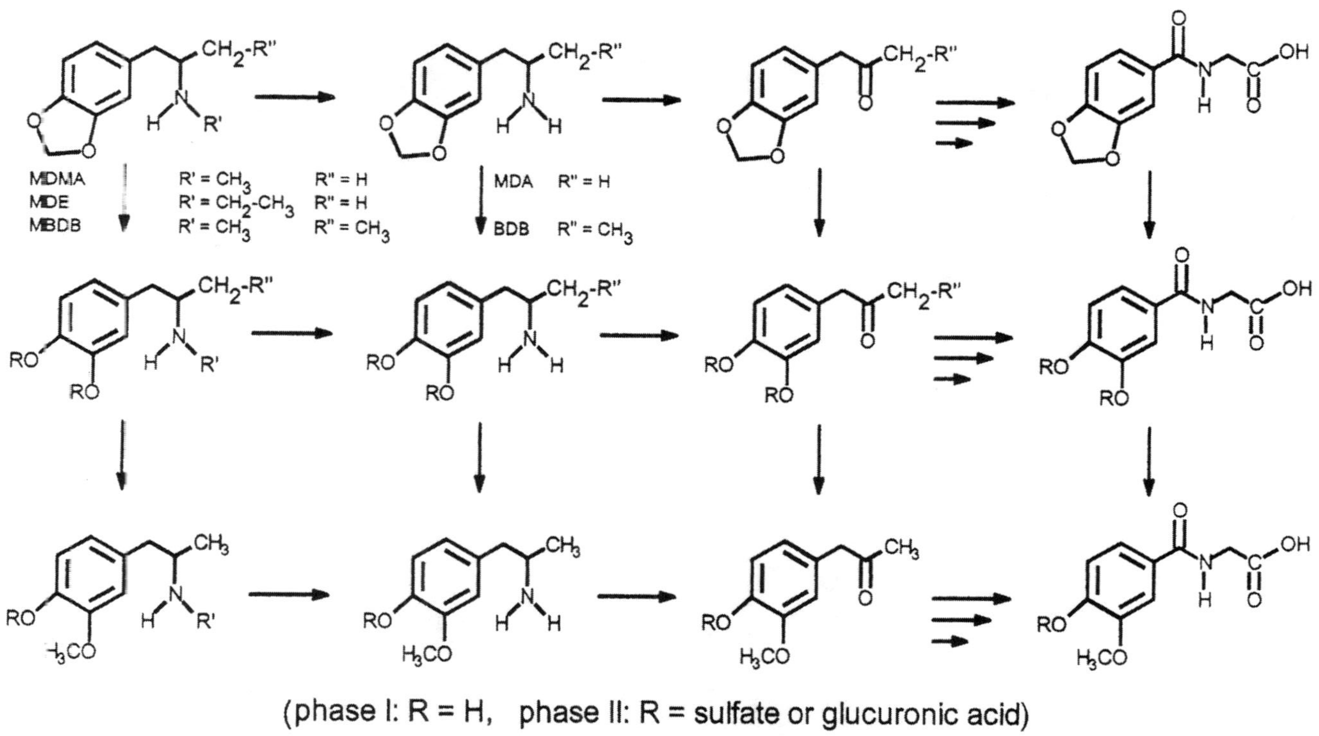
MDMA
MDE
MBDB
R' = CH3 R'' = H
R' = CH2-CH3 R'' = H
R' = CH3 R'' = CH3
MDA R'' = H
BDB R'' = CH3
(phase I: R = H, phase II: R = sulfate or glucuronic acid)

FIGURE 8 Biotransformation of amphetamine derivatives ("Ecstasy" group): MDMA, MDE, MDA, MBDB, and BDB. (Reprinted from Ref. 200 with permission of Elsevier Science Publishers.)

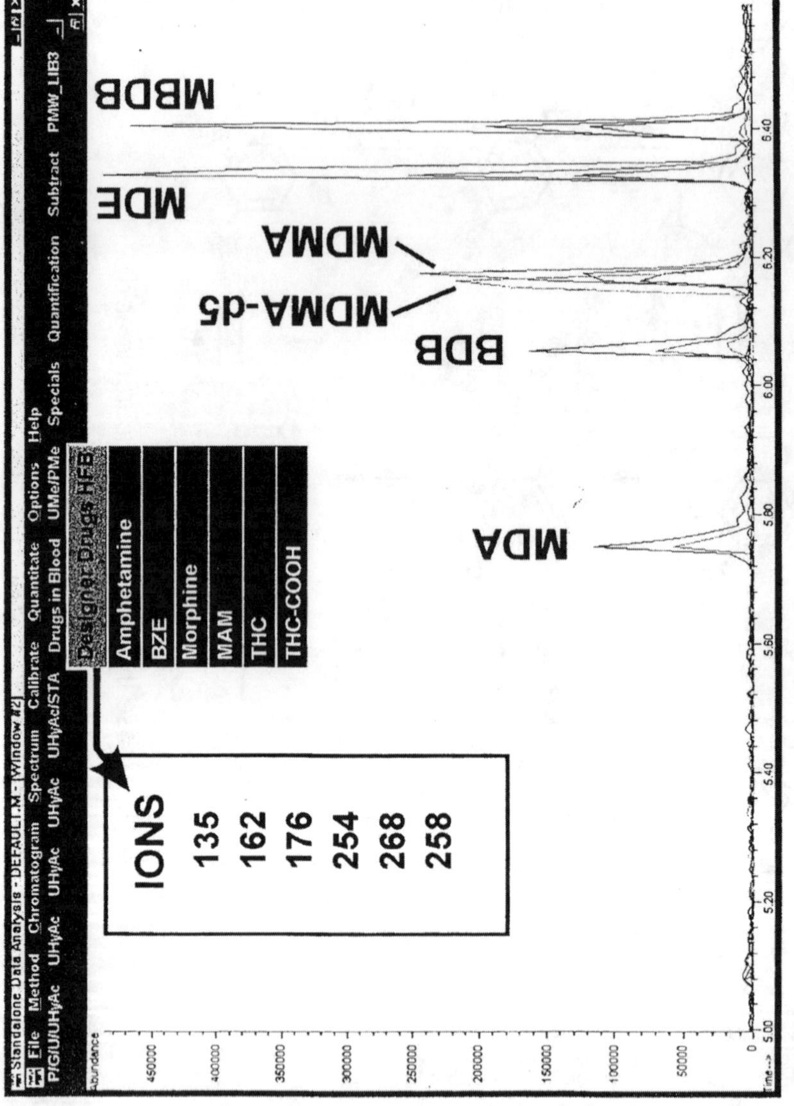

FIGURE 9 GC-MS-SIM chromatogram of MDA, BDB, MDMA, MDE, and MBDB extracted from plasma. (Reprinted from Ref. 200 with permission of Elsevier Science Publishers.)

[206]. Clobenzorex is an another anorectic drug that is metabolized to amphetamine. In order to differentiate between clobenzorex intake and amphetamine abuse, Cody et al. analyzed urine samples for metabolite 4-hydroxyclobenzorex with GC-MS. This compound was always present in urine samples besides amphetamine after administration of clobenzorex, also in cases when the parent drug was not detectable [207].

4.2.2 Analysis of Phenethylamines in Blood, Plasma, and Tissues

Gas Chromatography. Psychoactive phenethylamines present a large group of compounds of very different forensic relevance. Some of them, like MDMA or MDA, are illicit drugs with no therapeutic applications, others, like diethylpropione or methylphenidate, are prescription drugs, and some, like pseudoephedrine or phenylpropanolamine, occur in over-the-counter preparations. Therefore the identification of all drugs should be done preferably with GC-MS. Several authors [208–210] approached the problem of distinguishing sympathomimetic phenethylamines from illicit amphetamine and methamphetamine. An overview of all chromatographic methods for amphetamines and related drugs was published recently by Cody [211]. Valentine et al. [212] studied the identification of phenethylamines for emergency ward needs. Twenty-three illicit and licit phenethylamines were isolated from urine with chloroform/isopropanol, derivatized on-column with HFBA, and subjected to GC-MS examination. The total analysis time was 30 min, making the method suitable for clinical emergency toxicology purposes. Vorce et al. [213] compared the ion ratios, precision, and quantitative accuracy of ion-trap and quadrupole mass detectors for the analysis of different classes of abused drugs extracted from urine. Amphetamine, methamphetamine, THCCOOH, morphine, codeine, and 6-MAM were determined. It was demonstrated that the sensitivity and accuracy of the ion-trap mass analyzer were equivalent to quadrupole SIM measurements for all compounds except for 6-MAM. Kunsman et al. extracted MDMA and MDA from immunoassay-positive urine samples with butyl chloride and derivatized with chlorofluoroacetic anhydride. GC-MS (SIM) analysis was performed on DB-5 column in temperature-programmed run [214]. Helmlin et al. [215] analyzed MDMA and its metabolites MDA, HMMA, and HHA in urine using HPLC-DAD and GC-MS. The drugs were extracted from plasma and urine with SPE SCX columns and derivatized with HFBA before GC-MS examination. Amphetamines belong to relatively volatile bases. The advantage of this feature was taken in application of solid-phase microextraction (SPME) for isolation of these drugs from biological material. Lee et al. [216] applied

SPME for isolation of amphetamine and methamphetamine from serum. The drugs were then subjected to headspace derivatization with HFBA vapor and quantified with GC-MS. In a study of Jurado et al. [217], amphetamine, methamphetamine, MDMA, and MDA were isolated from urine with SPME using the headspace technique. The drugs were derivatized with TFA by exposing the fibers in the headspace of another vial and desorbed in the injection port of GC-MS. The LOQs ranged from 10 to 20 μg/L, using SIM. Okajima et al. [218] isolated amphetamine and methamphetamine from the whole blood using headspace-SPME. The drugs were derivatized with pentafluorobenzyl bromide in headspace vial and subjected to GC-MS (SIM). The LOD was 0.5 μg/L. The method was applied for autopsy cases. The excretion of MDMA enantiomers with urine was studied after administration of MDMA [219]. The drugs were extracted from urine with solvent, derivatized with trifluoroacetyl-l-prolyl chloride and analyzed by GC-MS (Fig. 10).

Liquid Chromatography. Amphetamine and 8 other phenethylamines were analyzed in 250 μL plasma using automatic on-line extraction, column switching, and HPLC with a UV detector set at 210 nm. LODs ranged from 50 to 250 μg/L [220]. Much better sensitivities were achieved using LC-MS methodology. Amphetamine, methamphetamine, MDMA, MDA, MDEA, and eight other phenethylamines were extracted with ether from serum and urine, derivatized with phenylisothiocyanate, and subjected to HPLC with APCI-MS or DAD detection. LC-APCI-MS assured about 10 times higher detection with LODs ranged from 1 to 5 μg/L [221]. In the next study of the same group [222], the derivatization was abandoned. Fourteen amphetamines and related compounds were isolated from biofluids with SPE cartridges and subjected to LC-APCI-MS examination in SIM mode. Again the limit of detection ranged from 1 to 5 μg/L. This method was applied in routine casework [49]. Benzphetamine and its metabolites benzylamphetamine, hydoxybenzphetamine, hydroxybenzylamphetamine, methamphetamine, and amphetamine were isolated from urine with SPE, separated on alkaline mobile phase, and detected with ESI-MS (SIM). The LODs were from 0.3 to 10 μg/L urine [223]. Dimethylamphetamine and its metabolites dimethylamine-*N*-oxide, methamphetamine, and amphetamine were determined in urine by SI-LC-MS (SIM) after isolation with SPE. LODs were 5–50 μg/L urine [224]. Kataoka et al. [225] applied SPME for isolation of amphetamine, methamphetamine, MDMA, MDEA, and MDA from urine. The drugs were desorbed by mobile phase flow and detected with ESI-MS (SIM), with a LOD below 2 μg/L urine. MDMA, MDEA, and MDA were determined in whole blood, urine, and vitreous humor of rabbits using HPLC with fluorometric and tandem mass spectrometric

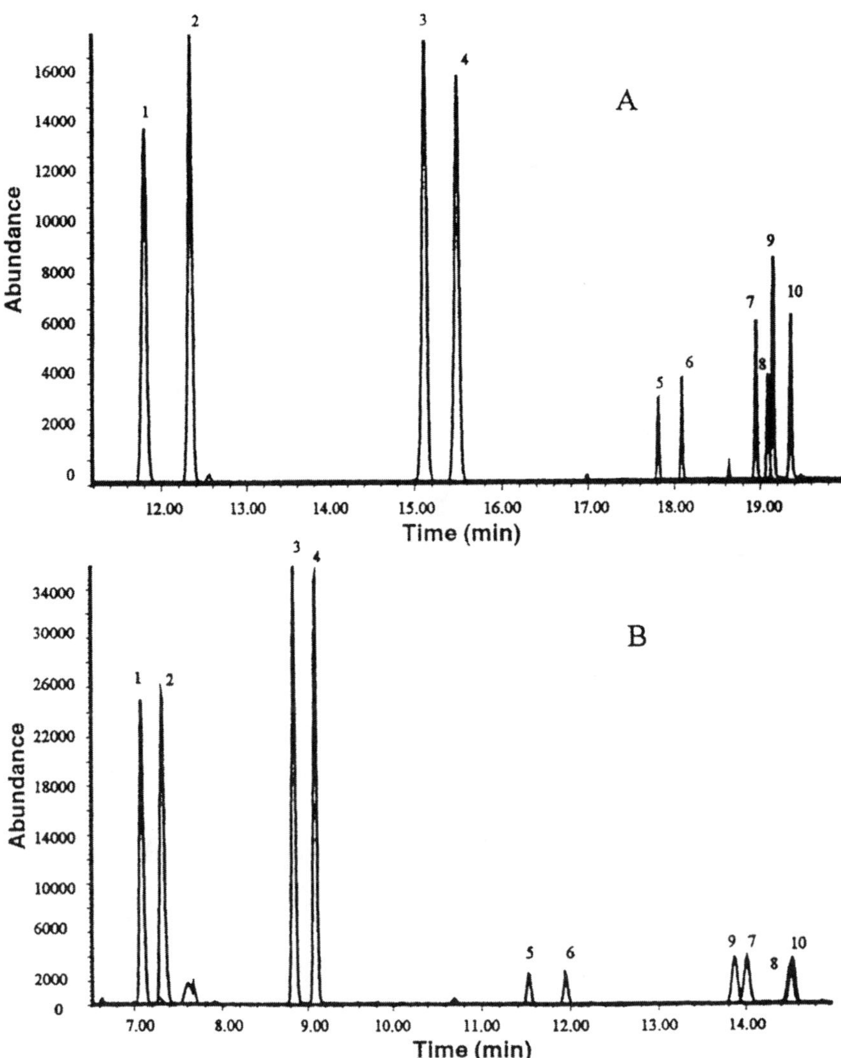

FIGURE 10 GC of enantiomers of amphetamines on an HP-1 column (A) and on a DB-17 column (B). Peaks: 1 = *l*-amphetamine, 2 = *d*-amphetamine, 3 = *l* methamphetamine, 4 = *d*-methamphetamine, 5 = *l*-MDA, 6 = *d*-MDA, 7 = *l*-MMA, 8 = *d*-MDMA, 9 = *l*-MDEA, 10 = *d*-MDEA. (Reproduced from the Journal of Analytical Toxicology, Ref. 219, by permission of Preston Publications, a division of Preston Industries, Inc.)

TABLE 10 Gas Chromatographic Methods for Amphetamines

Drug	Sample	Isolation	Derivatization	Column, conditions	Detection	LOD (μg/L)	Ref.
A, MDMA, impurities	Illicit drugs	L/L or SPME	—	SPB-1, 80°–300°	FID		193
A	Urine	SPE	HFBA	DB-5, 65°–300°	EI-MS (Q,IT)	—	213
MDMA, MDA, MDEA	Urine	L/L	HFBA	HP-1, 80°–280°	EI-MS	—	212
A, MA, MDMA, MDA	Urine	SPME	TFA	HP-1, 60°–290°	EI-MS (SIM)	3–6	217
MDMA, MDA, MDEA enantiomers	Urine	L/L	l-TPC	DB-17, 120°–260°			219

A = amphetamine, MA = methamphetamine, MDMA = methylenedioxymethamphetamine, MDEA = methylenedioxyethylamphetamine, MDA = methylenedioxyamphetamine.

TABLE 11 Liquid Chromatographic Methods for Amphetamines

Drug	Sample	Isolation	Column, elution conditions	Detection	LOD (μg/L)	Ref.
MDA, MDMA, MDEA, MBDB	Illicit drugs	Dissolving in HCl	ODS, ACN-phosphate isocratic	FI	2–3	231
13 amphetamines and rel. drugs	Serum, urine	SPE, derivatization	ODS, ACN-ammonium formate	APCI-MS (SIM)	1–5	221
14 amphetamines and rel. drugs	Serum, urine	SPE	ODS, ACN-ammonium formate	APCI-MS (SIM)	1–5	222
MDMA, MDA, MDEA	Blood, urine	L/L	ODS, CAN-MeOH-ammonium acetate	FI, ESI-MS-MS	0.8–2	226
MDMA, MDA, MDEA	Blood, serum	L/L	ODS, ACN-MeOH-H_2O	ESI-MS-MS (TOF)	1	232
A, MA, MDMA, MDA, MDEA	Urine	SPME	CN, CAN-ammonium acetate	ESI-MS (SIM)	2	225

A = amphetamine, MA = methamphetamine, MDMA = methylenedioxymethamphetamine, MDEA = methylenedioxyethylamphetamine, MDA = methylenedioxyamphetamine, MBDB = *N*-mehylbenzoxodialolylbutanamine.

detection. The LOD for fluorometric detection was 0.8 μg/L in blood and 2 μg/L in urine. Very good correlation between both methods was found [226].

Capillary Electrophoresis. Capillary electrophoresis was used for the determination of MDMA together with other drugs of abuse in the hair of drivers [178,179]. Capillary electrophoretic immunoassay was applied for detection of amphetamines and other drugs of abuse in urine. The results were validated with GC-MS [180]. Geiser et al. applied nonaqueous CE (acetonitrile-methanol 8 : 2, containing ammonium formate/formic acid) coupled to UV and ESI-MS detector for the determination of MDMA and metabolites in urine after solvent extraction. LOD ranged from 20 to 70 μg/L urine in SIM mode [227]. The prescription drug selegiline is metabolized to R(−)A and R(−)-MA but not to S(+)A or S(+)MA. In order to differentiate between selegiline and illicit A/MA use, Heo et al. [228] developed a CE method for quantitation of A and MA enantiomers in urine extracts. Stereoisomers were separated in phosphate buffer containing carboxymethyl-beta-cyclodextrin. The method was applied e.g. for the differentiation of selegiline and illicit amphetamine intake. The enantiomer A, MA, MDMA, MDA, methadone, and its metabolite EDDP were separated using CE with a buffer containing (2-hydroxypropyl)-beta-cyclodextrin [229]. Capillary electrophoresis using the ion-trap mass spectrometric detector was applied for the determination of amphetamine, methamphetamine, MDA, and MDMA in urine extracts [230]. The sensitivity was comparable to that achieved with UV detection, but the selectivity was much higher, enabling detection of amphetamines even in unresolved peaks. Chromatographic methods applied for amphetamines used recently for cocaine analysis are shown in Tables 10 and 11.

5 CANNABINOIDS

Cannabis products—marijuana and hashish—are surpassed only by ethyl alcohol as the most commonly abused drugs. The known history of cannabis abuse began about 3,000 B.C. in India. Later, the cultivation of the *Cannabis sativa* plant and use of cannabis was gradually spread worldwide. Of the three known cannabis preparations in the illicit drug market, herbal cannabis (marijuana), cannabis resin (hashish), and liquid cannabis (cannabis oil), marijuana is the most popular one. It is the most widely used illicit drug in the world. Beside the illicit use of cannabis as a recreational drug, the attempts to use cannabis preparations or its active compound tetrahydrocannabinol (THC) in medicine have an equally long tradition [233].

The discovery of cannabinoid receptors as well as endogenous receptor agonists stimulated the research on the role and possible therapeutic use of cannabis. THC as well as its synthetic analogs (nabilone, levonantradol, and marinol) was tried in treatment of several disorders. It was demonstrated that cannabinoid receptor agonists were effective as antiemetics and appetite stimulants and were applied in HIV patients [234,235]. Among other possible uses, the suppression of muscle spasticity associated with multiple sclerosis, the relief of chronic pain, and the therapy of behavior disorders in Tourette's syndrome were mentioned [236–238]. Martin et al. [239] synthesized several analogies of THC, modifying its alkyl side chain. This manipulation produced high-affinity ligands with antagonist, partial agonist, or full agonist effects on the cannabinoid receptor. It seems that the development of THC analogs, which would be devoid of unwanted psychotropic action, may be of great interest for clinical medicine.

5.1 Cannabinoids in Plant Material (Hashish, Marijuana)

The determination of cannabinoids in plant material has several purposes. The first and most important one is to identify and quantify the main components, like active THC and other cannabinoids—cannabinol (CBN) and cannabidiol (CBD). On the basis of quantitative determination is possible to classify the plant material as resin type (illicit drug) and fiber type, which is used for industrial purposes. The other task of analysis is to identify the production site and sometimes a particular batch of drug. This may be done using a high-resolution chromatography. It must be stressed that *Cannabis sativa* contains several hundred constituents whose proportions may differ in relation to geography, climate, and cultivation mode, etc. [240]. Also, the quantitation of THC in plant material allows monitoring the potency of drug.

Thin-layer chromatography (TLC) was applied frequently for analysis of plant material, and the use of overpressured TLC was reported [241,242]. Debruyne et al. [243] proposed a TLC or HPLC separation of the main cannabis constituents (THC, CBN, and CBD) with subsequent measurement of the UV spectrum as an alternative to GC-MS identification. GC-FID and GC-MS as well as HPLC were applied for differentiation of cannabis samples of different origin. Over 100 compounds could be identified. The composition as well as peak ratios of particular compounds were used for tracking the origin of confiscated cannabis samples [244]. Lehmann et al. [245] used HPLC with DAD for classification of cannabis chemotypes on three types: drug type, intermediate type, and fiber/industrial type (Fig. 11).

Supercritical fluid chromatography coupled with APCI-MS was applied for determination of THC, CBN, and CBD in cannabis products. The LODs (on column) were from 0.55 to 2.1 ng [246]. Lurie et al. [247] presented a capillary electrochromatography of cannabinoids. Seven cannabinoids were separated and detected using DAD.

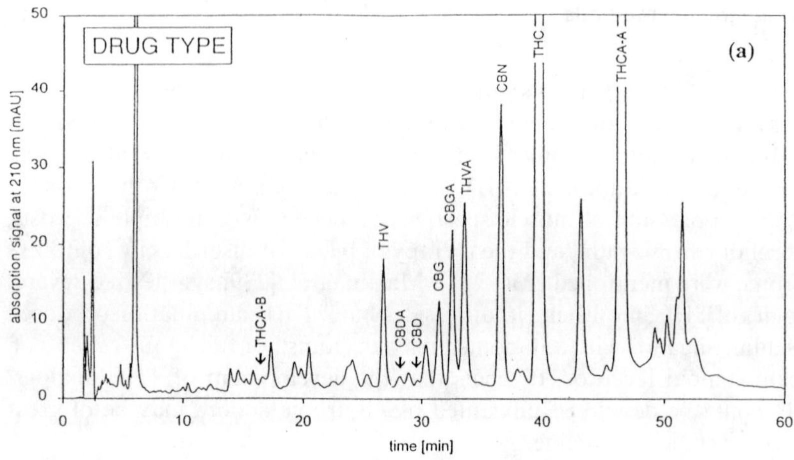
DRUG TYPE
(a)
THCA-B
THV
CBDA
CBD
CBG
CBGA
THVA
CBN
THC
THCA-A
absorption signal at 210 nm [mAU]
time [min]

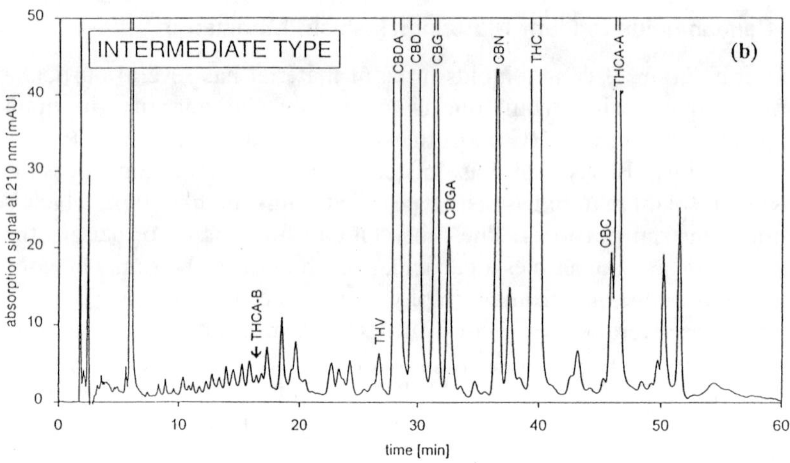
INTERMEDIATE TYPE
(b)
THCA-B
THV
CBDA
CBD
CBG
CBGA
CBN
THC
CBC
THCA-A
absorption signal at 210 nm [mAU]
time [min]

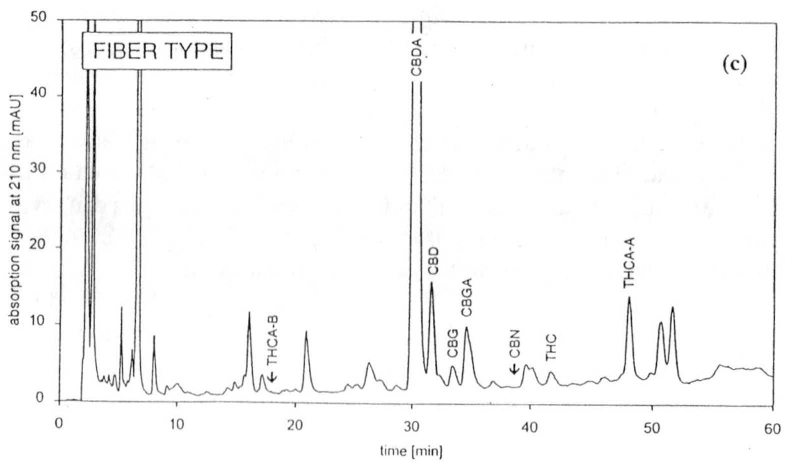
FIBER TYPE
(c)
THCA-B
CBDA
CBD
CBG
CBGA
CBN
THC
THCA-A
absorption signal at 210 nm [mAU]
time [min]

Some cannabis products found their way to the food market. E.g., hemp oil products are being marketed as a healthy source of essential omega fatty acids and their use significantly increased. The study was conducted to check whether the consumption of commercially available hemp oil might be associated with the urinary excretion of relevant amounts of cannabinoids. The content of THC in hemp oils, measured with GC-MS, ranged from 11 to 117 mg/g. After administration of 15 g of oil to volunteers the level of urinary THC-COOH was always below the cutoff value of 50 μg/L [248]. Zoller et al. [249] developed an HPLC method for food control purposes allowing the determination of THC and THC-A (Δ^9-tetrahydrocannabinolic acid) in food containing hemp products. Legal limits of THC in food products in Switzerland range from 0.2 mg/kg in beverages to 50 mg/k in hempseed oil. GC-MS analysis was applied for determination of THC content in drug-and fiber-type cannabis seeds. Drug-type seeds contained much more THC (35–124 μg/g) than fiber-type (0–12 μg/g). Most of THC was located on the surface and could be easily washed with chloroform. It seems that the seeds are contaminated with THC from the plant [250].

5.2 Tetrahydrocannabinol and Metabolites in Biological Material

5.2.1 Metabolism and Pharmacokinetics of Tetrahydrocannabinol

Although the history of cannabis abuse is very long, its major active constituent, Δ^9-tetrahydrocannabinol (THC), was not characterized by Mechoulam and Gaoni until 1967 [252]. The identification and isolation of THC stimulated studies on the pharmacokinetics and pharmacodynamics of cannabis. It was demonstrated that THC is first oxidized to an active 11-OH-THC and then to the THC-carboxylic acid (THC-COOH) which is the main and longest detectable metabolite of THC is blood and urine. The reviews on the metabolism and detection possibilities can be found elsewhere [233,253–257]. It is of forensic importance that the concentrations of active compounds, THC and 11-OH-THC, reach maxima 5 to 15 min after smoking and afterwards undergo biphasic kinetics. In the first phase, THC concentration decreases very rapidly (in about 2–3 hours) from about 100–

FIGURE 11 HPLC profiles of chemotypes of *Cannabis sativa* L. (a) Drug type, (b) intermediate type, (c) fiber/industrial type. (Reprinted from Ref. 245 by courtesy of Marcel Dekker, Inc.)

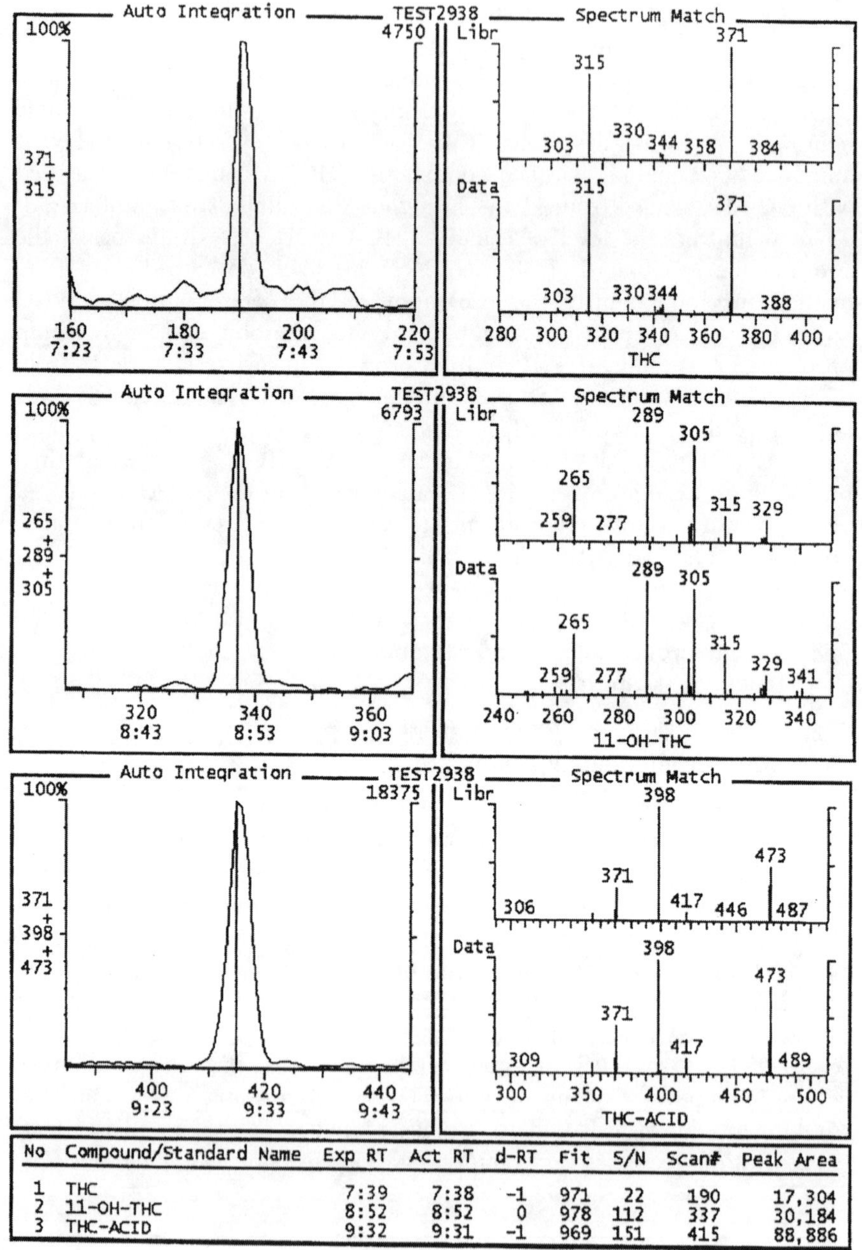

No	Compound/Standard Name	Exp RT	Act RT	d-RT	Fit	S/N	Scan#	Peak Area
1	THC	7:39	7:38	-1	971	22	190	17,304
2	11-OH-THC	8:52	8:52	0	978	112	337	30,184
3	THC-ACID	9:32	9:31	-1	969	151	415	88,886

FIGURE 12 GC-MS-IT mass chromatograms and resulting daughter spectra, retention, and library fit data from serum extract containing THC 0.25 μg/L

150 μg/L to about 5 μg/L. In the further phase, the terminal half-life of THC is about 25 hours [258]. The concentration of THC-COOH reaches maximum about 1 h after smoking and then slowly decreases. This substance may be detected in blood up to two days and in urine up to 5 days after single exposure. For this reason, all immunochemical urine tests on cannabis use are developed for detection of THC-COOH [255]. It must be stated, however, that the detection of THC-COOH gives only the evidence of cannabis use, but not of the acute impairment. This can be done by an analysis of THC and 11-OH-THC.

5.2.2 Tetrahydrocannabinol and Metabolites in Blood and Urine

A review of analytical possibilities for cannabinoids up to 1997 was done by ElSohly and Salem [259]. It this chapter more recent publications will be covered (1998–2001).

Gas Chromatography. Gas chromatography-electron impact mass spectrometry (positive ions in selected ion monitoring mode) is a standard method for the determination of cannabinoids in biological fluids, and most of the previous papers on this topic can be found in the review of ElSohly and Salem [259]. In recent years, some papers appeared, expanding analytical possibilities of GC-MS.

Weller et al. [260] applied ion-trap GC-MS for analysis of THC, 11-OH-THC, and THC-COOH in serum. The drugs were extracted with SPE cartridges and silylated (TMS). The GC-MS-MS (ion-trap) analysis allowed obtaining full-scan daughter spectra at the concentrations 0.25–2.5 μg/L serum. The selectivity of this method was regarded as superior in comparison to that of SIM GC-MS (Fig. 12).

Chiarotti et al. [261] achieved detection of THC-COOH at the ng/L level using GC-MS-MS analysis for biological fluids.

Liquid Chromatography. New applications of liquid chromatography for the analysis of cannabinoids concern the use of LC-MS for this purpose. In the study of Mireault [262], blood or urine samples were extracted with an SPE cartridge and analyzed on a C-8 column in methanol–ammonium acetate. An ion-trap instrument equipped with an APCI source was used in MS/MS mode (positive ions). A detection limit of 1 μg/L was achieved for

---

(upper), 11-OH-THC 0.5 μg/L, and THC-COOH 2.5 μg/L (lower). (Reproduced from the Journal of Analytical Toxicology, Ref. 260, by permission of Preston Publications, a Division of Preston Industries, Inc.)

THC, 11-OH-THC, and THC-COOH. In the second study of Mireault et al. [263], an APCI triple stage quadrupole mass instrument was applied. The limit of quantitation of 0.25 μg/L was reported for all three compounds. In biological extracts, the limit of detection was 10–40 times lower for the quadrupole instrument than for an ion trap (Fig. 13).

Breindahl and Andreasen applied ESI-LC-MS for the determination of THC-COOH in urine [264]. Urine was subjected to basic hydrolysis and solid phase extraction. THC-COOH and its deuterated analog were analyzed with HPLC-ESI-MS, using C8 reversed-phase column, gradient elution in acetonitrile–formic acid and SIM detection (positive ions). In-source collision-induced dissociation was applied, and protonated quasi-molecular ion as well as two fragment ions were monitored. LOD was 15 μg/L. Tai and Welch [265] determined THCCOOH with HPLC-ESI-MS (negative ions). The drug was extracted from urine with SPE and subjected to isocratic separation on ODS column in methanol–ammonium acetate mobile phase. Only deprotonated quasi-molecular ions of the drug and its deuterated analog were monitored. An unusually low detection limit of 5 ng/L urine was reported, but without proper experimental evidence.

Weinmann et al. used LC-MS-MS for simultaneous determination of THC-COOH and its glucuronide in urine [266]. In this method the cleavage of conjugates was omitted. THC-COOH and its glucuronide were extracted from urine with ethyl acetate/ether (1 : 1) and separated on a C8 column in a gradient of ammonium formate buffer and acetonitrile. ESI/MS/MS was used for detection using protonated quasi-molecular ions as precursor ions and two fragment ions for each drug as product ions. The specificity of the method was checked using enzymatic hydrolysis of THC-COOH-glucuronide.

Recent chromatographic methods for cannabinoids are depicted in Tables 12 and 13.

5.2.3 Tetrahydrocannabinol and Metabolites in Alternative Samples

The analysis of classical biological fluids like serum, blood, and urine is useful as evidence of recent exposure to a given drug or even impairment. For many years, however, the applicability of other biological material for drug analysis was examined. For these efforts there are two main reasons. The first is sampling: blood collection belongs to invasive methods, and urine sampling needs cooperation and privacy. The second reason is that both blood and urine results have quite narrow detection windows ranging from hours to days. Among alternative materials used, saliva and sweat can be collected in a noninvasive way and may be used i.e. for roadside testing. Meconium analysis

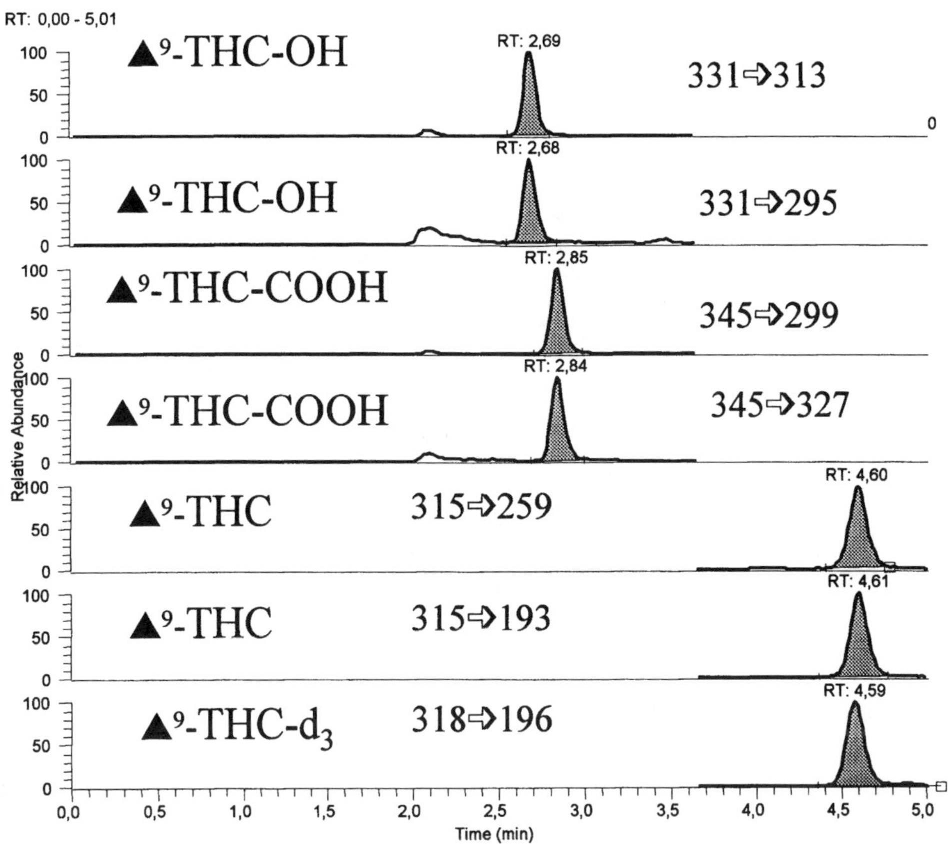

FIGURE 13 LC-APCI-MS-MS of blood extract containing THC, THC-COOH, and 11-OH-THC, 10 μg/L each. (Reprinted from Ref. 263 with permission of the author.)

may give evidence of intrauterine exposition, whereas hair or nail analysis is able to document chronic drug exposure with a detection window of many months. A recent review of analysis of cannabinoids and other drugs of abuse in alternative matrices was given by Kintz and Samyn [267]. This chapter deals with studies published in the last three years.

It is known that after regular abuse of cannabis, THC, CBN, and CBD can be easily detected in hair. The detection of the metabolite THCCOOH is much more difficult and can be done mainly with GC-MS-MS (NCI). Sachs et al. [268] developed a method for the determination of THCCOOH in

TABLE 12 Gas Chromatographic Methods for Cannabinoids

Drug	Sample	Isolation	Derivatization	Column, conditions	Detection	LOD (μg/L)	Ref.
Cannabis constituents (> 100)	Cannabis	L/L	BSTFA + TMCS	DB-1, 70°–250°	FID, EI-MS		244
THC, CBN, CBD	Cannabis	L/L	—	HP-5, 220°	FID	20	251
THC, 11-OH-THC, THCCOOH	Serum	SPE	MSTFA	DB-1, 120°–310°	EI-MS (IT)	0.2–2.5	260
THC	Saliva, sweat	L/L	Methylation	HP-5, 60°–290°	EI-MS (SIM)	1	269
THC, 11-OH-THC, THCCOOH	Plasma, meconium	L/L	BSTFA + TMCS	HP-5, 200°–280°	EI-MS (SIM)	1–2.5	272

THC = Δ^9-tetrahydrocannabinol, CBN = cannabinol, CBD = cannabidiol, THC-COOH = 11-nor-Δ^9-THC-9-carboxylic acid.

TABLE 13 Liquid Chromatographic Methods for Cannabinoids

Drug	Sample	Isolation	Column, elution conditions	Detection	LOD (μg/L)	Ref.
Cannabis constituents (>100)	Cannabis	L/L	ODS, MeOH-acetic acid	UV		244
THC, THC-A, CBD, CBN	Cannabis	L/L	ODS, ACN-phosphoric acid grad.	UV-DAD	1 ng	245
THC, THC-A	Hemp product	L/L	ODS, ACN-phosphoric acid isocr.	UV + FI	0.01 ng	249
THC, 11-OH-THC, THCCOOH	Blood, urine	SPE	C8, MeOH-formic acid isocr.	MS (APCI)	1	262
THCCOOH	Urine	SPE	C8, ACN-formic acid grad.	MS (ESI)	15	264
THCCOOH	Urine	SPE	ODS, MeOH-ammonium acetate isocr.	MS (ESI)	0.005	265

THC = Δ^9-tetrahydrocannabinol, THC-A = tetrahydrocannabinolic acid, CBN = cannabinol, CBD = cannabidiol, THC-COOH = 11-nor-Δ^9-THC-9-carboxylic acid.

hair, using alkaline digestion, followed by solvent extraction and HPLC purification. The fraction after HPLC was collected and extracted with hexane/ethyl acetate, derivatized, and subjected to GC-MS examination. The LOD was 0.3 ng/g. Fingernail clippings taken from 14 cannabis users were analyzed for THC and THCCOOH. The mean value for THC was 1.44 μg/g, and for THCCOOH it was 19.85 μg/g [269]. Blood, urine, saliva, and forehead wipes (sweat) were simultaneously collected from 198 drivers admitted to an emergency ward in Strasbourg, France. Of the 22 subjects positive for THCCOOH in urine, 14 were positive for THC in saliva and 16 for THC in sweat. THCCOOH and 11-OH-THC were not detected in saliva and sweat. As the main limitations of saliva and sweat the small amount of sample and low concentrations of analytes were mentioned [270].

Italian authors collected saliva specimens by the EPITOPE system and centrifuged. The ultrafiltrates were subjected to SPME and analyzed by GC-MS [271]. The analysis of meconium may detect fetal exposure to drugs of abuse. ElSohly et al. developed a GC-MS method for the determination of THCCOOH, benzoylecgonine, morphine, and amphetamines in meconium specimens [272]. An immunoaffinity extraction procedure was developed for isolation of THC and its hydroxymetabolites from urine, plasma, and meconium. After silylation, the drugs were determined with GC-MS. The LODs of 1–2.5 ng/g meconium were achieved [273].

5.2.4 Problems of Sample Adulteration or Contamination

Several urine adulterants are commercially available that are advertised as capable of making cannabinoids undetectable. They contain usually nitrite ion. Addition of these preparations to urine may heavily lower the original THC-COOH concentrations in GC-MS analysis, particularly when the pH of the urine is acidic [274]. Also, some preparations are available that should wash the cannabinoids incorporated in hair. Hair samples collected from persons with a known history of cannabis abuse were analyzed after washing with Ultra Clean shampoo and without treatment. The following drugs were determined with GC-MS: THC, cocaine, amphetamine, MDMA, MDA, heroin, 6-MAM, morphine, codeine, DHC, and methadone. Some decrease in drug concentrations was observed after washing, but a single treatment with Ultra Clean could not remove drugs from hair to an undetectable level [275]. The question arose whether the use of Cannabio shampoo, which is produced on the base of cannabis oil, can result in positive results for cannabis in hair. The Cannabio shampoo contained THC (412 μg/L), CBD (4079 μg/L), and CBN (380 μg/L). Three volunteers washed their hair daily for three weeks, and no cannabinoids were detected in the hair afterwards. Also, an in vitro study with hair incubated with Cannabio gave a negative result [276].

6 CONCLUDING REMARKS

Looking at the perspectives of the analysis of drugs of abuse it is certain that the chromatographic methods will maintain the most important role. Gas chromatography is the technique that is most mature and shows less room for future improvements. This is certainly not the case for high-performance liquid chromatography. HPLC is in the stage of rapid development, and the introduction of new columns and new packing materials with higher efficiency will improve separation possibility in this technique. Also, electrically driven separation methods will probably reach the stage of technical maturity and may be more generally used.

For all separation methods, the use of mass spectrometric detection in drug analysis is a must. In the near future, the use of bench-top tandem MS will be ubiquitous and assure high specificity and sensitivity. Also, the use of mass analyzers of high mass accuracy, like TOF, in combination with GC or HPLC will contribute to better recognition of drugs and metabolites of similar structure.

REFERENCES

1. R Wennig, A Verstraete. Drugs and driving. In: MJ Bogusz, ed. Forensic Science. Handbook of Analytical Separations, Vol. 2. Amsterdam: Elsevier Science, 2000, pp 439–457.
2. R Aderjan, W Bonte, T Daldrup, H Kaferstein, G Kauert, H Joachim, MR Moeller, G Reinhard, G Schewe, J Wilske. Änderung des § 24a des Strassenverkehrsgesetzes und Bericht der Grenzwertkommission. Toxichem Krimtech. 65:70–73, 1998.
3. Legal text, 625 Illinois compiled statutes, § 11501, USA.
4. R Aderjan. Aspects of quality assurance in forensic toxicology. In: MJ Bogusz, ed. Forensic Science. Handbook of Analytical Separations, Vol. 2. Amsterdam: Elsevier Science, 2000, pp 489–530.
5. BD Paul, C Dreka, ES Knight, ML Smith. Gas chromatographic/mass spectrometric detection of narcotine, papaverine and thebaine in seeds of *Papaver Somniferum*. Planta Med 62:544–547, 1996.
6. Z Fater, Z Samu, M Szatmary, S Nyiredy. Combinations of liquid chromatographic methods for the breeding of the high alkaloid content poppy. Acta Pharmaceut Hung 67:211–219, 1997.
7. MG Pelders, JJW Ros. Poppy seeds: differences in morphine and codeine content and variation in inter- and intraindividual excretion. J Forensic Sci 41:209–212, 1996.

8. G Fritschi, WIR Prescott. Morphine levels in urine subsequent to poppy seed consumption. Forensic Sci Int 27:117–117, 1985.
9. LW Hayes, WG Krasselt, PA Mueggler. Concentrations of morphine and codeine in serum and urine after ingestion of poppy seeds. Clin Chem 33: 806–808, 1987.
10. BC Pettitt, SM Dyszel, LV Hood. Opiates in poppy seed: effect on urinalysis results after consumption of poppy seed cake-filling. Clin Chem 33:1251–1152, 1987.
11. RE Struempler. Excretion of codeine and morphine following ingestion of poppy seeds. J Anal Toxicol 11:97–99, 1987.
12. DS Lo, TH Chua. Poppy seeds: implications of consumption. Med Sci Law 32:296–302, 1992.
13. C Meadway, S George, R Braithwaite. Opiate concentrations following the ingestion of poppy seed products—evidence for the "poppy seed defense." Forensic Sci Int 96:29–38, 1998.
14. H Huizer. Analytical studies on illicit heroin. PhD thesis, Technical University Delft, The Netherlands, 1988.
15. TA Gough. The examination of drugs in smuggling offences. In: TA Gough, ed. The Analysis of Drugs of Abuse. Chichester, UK: John Wiley, 1991, pp 511–566.
16. A Johnston, LA King. Heroin profiling: predicting the country of origin of seized heroin. Forensic Sci Int 95:47–55, 1998.
17. K Janhunen, MD Cole. Development of a predictive model for batch membership of street samples of heroin. Forensic Sci Int 102:1–11, 1999.
18. E Kaa. Impurities, adulterants and diluents of illicit heroin. Changes during a 12-year period. Forensic Sci Int 64:171–179, 1994.
19. S Klemenc. Noscapine as an adulterant in illicit heroin samples. Forensic Sci Int 108:45–49, 2000.
20. PR Ravikumar, P Backman, A Ramon. Analysis of street heroin—a case study of fatal overdose. Presented on PittCon 2001 Conference, New Orleans, USA.
21. RJ Hamilton, J Perrone, R Hofman, FM Henretig, EH Karkevandian, RD Shih, B Blok, K Nordenholz. A descriptive study of an epidemic of poisoning caused by heroin adulterated with scopolamine. Clin Tox 38:579–608, 2000.
22. H Kalasz, L Kerecsen, T Csermely, H Gotz, T Friedmann, S Hosztafi. TLC investigation of some morphine derivatives. J Planar Chromatogr 8:17–22, 1995.
23. NK Nair, V Navaratnam, V Rajananda. Analysis of illicit heroin. An effective thin-layer chromatographic system for separating of opiates and five adulterants. J Chromatogr 366:363–372, 1986.
24. E Della Casa, G Martone. A quantitative densitometric determination of heroin and cocaine samples by high-performance thin-layer chromatography. Forensic Sci Int 32:117–120, 1986.
25. H Neumann. Comments on the routine profiling of illicit heroin samples. Forensic Sci Int 44:85–87, 1990.

26. H Neumann. Comparison of heroin by capillary gas chromatography in Germany. Forensic Sci Int 69:7–16, 1994.
27. J Cartier, O Gueniat, MD Cole. Headspace analysis of solvents in cocaine and heroin samples. Science Justice 37:175–182, 1997.
28. PA Hays, IS Lurie. Quantitative analysis of adulterants in illicit heroin samples via reversed phase HPLC. J Liq Chromatogr 14:3513–3517, 1991.
29. R Weinberger, IS Lurie. Micellar electrokinetic capillary chromatography of illicit drug substances. Anal Chem 63:823–827, 1991.
30. IS Lurie, KC Chan, TK Spratley, JF Casale, HJ Isaaq. Separation and detection of acidic and neutral impurities in illicit heroin via capillary electrophoresis. J Chromatogr B 669:3–13, 1995.
31. IS Lurie. Application of micellar electrokinetic capillary chromatography to the analysis of illicit drug seizures. J Chromatogr A, 780:265–284, 1997.
32. F Tagliaro, S Turrina, FP Smith. Capillary electrophoresis; principles and applications in illicit drug analysis. Forensic Sci Int 77:211–229, 1996.
33. F von Heeren, W Thormann. Capillary electrophoresis in clinical and forensic analysis. Electrophoresis 18:2415–2426, 1997.
34. AF Hernandez, A Pla, J Moliz, F Gil, MC Goncalvo, E Villanueva. Application of the combined use of HPLC/diode array detection and capillary GC/nitrogen phosphorous detection for the rapid analysis of illicit heroin and cocaine samples. J Forensic Sci 37:1276–1282, 1992.
35. M Chiarotti, N Fucci, C Furnari. Comparative analysis of illicit heroin samples. Forensic Sci Int 50:47–55, 1991.
36. F Besacier, H Chaudron-Thozet, M Rousseau-Tsangaris, J Girard, A Lamotte. Comparative chemical analyses of drug samples: general approach and application to heroin. Forensic Sci Int 85:113–125, 1997.
37. C Barnfield, S Burns, DL Byrom, AV Kemmenoe. The routine profiling of forensic heroin samples. Forensic Sci Int 39:107–117, 1988.
38. A Sperling. Determination of heroin and some common adulterants by capillary gas chromatography. Chromatogr 538:269–275, 1991.
39. L Krenn, S Glantschnig, U Sorgner. Determination of five major opium alkaloids by reversed phase high-performance liquid chromatography on a base-deactivated stationary phase. Chromatographia 47:21–24, 1998.
40. G Theodoridis, I Papadoyannis, G Vasilikiotis, H Tsoukali-Papadopoulou. Reversed-phase high-performance liquid chromatography—photodiode array analysis of alkaloid drugs of forensic interest. J Chromatogr B 668:253–263, 1995.
41. HA Biliet, R Wolters, L De Galan, H Huizer. Separation and identification of illicit heroin samples by liquid chromatography using an alumina and C18 coupled column system and photodiode array detection. J Chromatogr 368:351–361, 1986.
42. IS Lurie. Application of capillary electrophoresis to the analysis of seized drugs. International Laboratory, March, 21–29, 1996.
43. E Gyr, R Brenneisen, D Bourquin, T Lehmann, D Vonlanthen, I Hug. Pharmacodynamics and pharmacokinetics of intravenously, orally and

rectally administered diacetylmorphine in opioid dependents, a two-patient pilot study within a heroin-assisted treatment program. Int J Clin Pharmacol Ther 38:468–491, 2000.
44. EJ Cone, BA Holicky, TM Grant, WD Darwin, BA Goldberger. Pharmacokinetics and pharmacodynamics of intranasal "snorted" heroin. J Anal Toxicol 17:327–337, 1993.
45. AJ Jenkins, RM Keenan, JE Henningfield, EJ Cone. Pharmacokinetics and pharmacodynamics of smoked heroin. J Anal Toxicol 18:317–330, 1994.
46. G Skopp, B Ganssmann, EJ Cone, R Aderjan. Plasma concentrations of heroin and morphine-related metabolites after intranasal and intramuscular administration. J Anal Toxicol 21:105–111, 1997.
47. EJ Cone, P Welch, JM Mitchell, BD Paul. Forensic drug testing for opiates: detection of 6-acetylmorphine in urine as a indicator of recent heroin exposure; drug and assay considerations and detection times. J Anal Toxicol 15:1–7, 1991.
48. MJ Bogusz. Opiate agonists. In: MJ Bogusz, ed. Forensic Science. Handbook of Analytical Separations, Vol. 2. Amsterdam: Elsevier Science, 2000, pp 3–65.
49. MJ Bogusz. Liquid chromatography–mass spectrometry as a routine method in forensic sciences: a proof of maturity. J Chromatogr B 748:3–19, 2000.
50. K Wolff, MJ Sanderson, AW Hay. A rapid horizontal TLC method for detecting drugs of abuse. Ann Clin Biochem 27:482–488, 1990.
51. J Vecerkova. Proof of diacetylmorphine (heroin) application by thin layer chromatography analysis in urine. Soud Lek 33:32–38, 1997.
52. DJ Dietzen, J Koenig, J Turk. Facilitation of thin-layer chromatographic identification of opiates by derivatization with acetic anhydride or methoxyamine. J Anal Toxicol 19:299–303, 1995.
53. R Jain, R Rav, BM Tripathi, C Singh. Opiate excretion profile among heroin dependent subjects by TLC densitometry. Indian J Pharmacol 28:220–223, 1996.
54. J Fehn, G Megges. Detection of O^6-monoacetylmorphine in urine samples by GC/MS as evidence for heroin use. J Anal Toxicol 9:34–138, 1985.
55. BD Paul, JM Mitchell, LD Mell, J Irving. Gas chromatography/electron impact mass fragmentographic determination of urinary 6-acetylmorphine, a metabolite of heroin. J Anal Toxicol 13:2–7, 1989.
56. RW Romberg, VE Brown. Extraction of 6-monoacetylmorphine from urine. J Anal Toxicol 14:58–59, 1990.
57. DC Fuller, WH Anderson. A simplified procedure for the determination of free codeine, free morphine and 6-acetylmorphine in urine. J Anal Toxicol 16:315–318, 1992.
58. J Schuberth, J Schuberth. Gas chromatographic–mass spectrometric determination of morphine, codeine and 6-monoacetylmorphine in blood extracted by solid phase. J Chromatogr 490:444–449, 1989.
59. R Wasels, F Belleville. Gas chromatographic–mass spectrometric procedures used for the identification of morphine, codeine and 6-monoacetylmorphine. J Chromatogr A 674:225–234, 1994.

60. WL Wang, WD Darwin, EJ Cone. Simultaneous assay of cocaine, heroin and metabolites in hair, plasma, saliva and urine by gas chromatography–mass spectrometry. J Chromatogr B 660:279–290, 1994.
61. MR Moeller, C Muller. The detection of 6-monoacetylmorphine in urine, serum and hair by GC/MS and RIA. Forensic Sci Int 70:125–133, 1995.
62. CL O'Neal, A Poklis. Simultaneous determination of acetylcodeine, monoacetylmorphine and other opiates in urine by GC-MS. J Anal Toxicol 21:427–432, 1997.
63. CL O'Neal, A Poklis. The detection of acetylcodeine and 6-acetylmorphine in opiate positive urines. Forensic Sci Int 95:1–10, 1998.
64. C Staub, M Marset, A Mino, P Mangin. Detection of acetylcodeine in urine as an indicator of illicit heroin use: method validation and results of a pilot study. Clin Chem 47:301–307, 2001.
65. MJ Bogusz, RD Maier, M Erkens, U Kohls. Detection of non-prescription heroin markers in urine with liquid chromatography–atmospheric pressure chemical ionization mass spectrometry. J Anal Toxicol 25:431–438, 2001.
66. EJ Cone, WD Darwin. Rapid assay of cocaine, opiates and metabolites by gas chromatography–mass spectrometry. J Chromatogr 580:43–61, 1992.
67. BA Goldberger, EJ Cone, TM Grant, YH Caplan, BS Levine, JE Smialek. Disposition of heroin and its metabolites in heroin-related deaths. J Anal Toxicol 18:22–28, 1994.
68. JG Guillot, M Lefebvre, JP Weber. Determination of heroin, 6-acetylmorphine, and morphine in biological fluids using their propionyl derivatives with ion trap GC-MS. J Anal Toxicol 21:127–134, 1997.
69. D Bourquin, T Lehmann, R Hammig, M Buhrer, R Brenneisen. High-performance liquid chromatographic monitoring of intravenously administered diacetylmorphine and morphine and their metabolites in human plasma. J. Chromatogr B 694:233–238, 1997.
70. P Zuccaro, R Ricciarello, S Pichini, RI Altieri, M Pellegrini, G D'Aszenzo. Simultaneous determination of heroin, 6-monoacetylmorphine, morphine, and its glucuronides by liquid chromatography-atmospheric pressure ionspray-mass spectrometry. J Anal Toxicol 21:268–277, 1997.
71. MJ Bogusz, RD Maier, S Driessen. Morphine, morphine-3-glucuronide, morphine-6-glucuronide and 6-monoacetylmorphine determined by means of atmospheric pressure chemical ionization-mass spectrometry-liquid chromatography in body fluids of heroin victims. J Anal Toxicol 21:346–355, 1997.
72. MJ Bogusz, RD Maier, M Erkens, S Driessen. Determination of morphine and its 3-and 6-glucuronide, codeine, codeine-6-glucuronide and 6-monoacetylmorphine in body fluids by liquid chromatography-atmospheric pressure chemical ionization mass spectrometry. J Chromatogr B 703: 115–127, 1997.
73. P Wernly, W Thormann, D Bourquin, R Brenneisen. Determination of morphine-3-glucuronide in human urine by capillary zone electrophoresis and micellar electrokinetic capillary chromatography. J Chromatogr 616:305–310, 1993.

74. F Tagliaro, C Poesi, R Aiello, R Dorizzi, S Ghieli, M Marigo. Capillary electrophoresis for the investigation of illicit drugs in hair: determination of cocaine and morphine. J Chromatogr 638:303–309, 1993.
75. G Manetto, F Crivellente, F Tagliaro. Capillary electrophoresis: a new analytical tool for forensic toxicologists. Ther Drug Monit 22:84–88, 2000.
76. WS Wu, JL Tsai. Analysis of morphine and morphine-3-glucuronide in human urine by capillary zone electrophoresis with minimal sample pretreatment. Biomed Chromatogr 13:216–219, 1999.
77. JL Tsai, WS Wu, HH Lee. Qualitative determination of urinary morphine by capillary zone electrophoresis and ion trap mass spectrometry. Electrophoresis 21:1580–1586, 2000.
78. ME Alburges, W Huang, RL Foltz, DE Moody. Determination of methadone and its *N*-demethylated metabolites in biological specimens by GC-PICI-MS. J Anal Toxicol 20:362–368, 1996.
79. DG Wilkins, PR Nasagawa, SP Gygi, RL Foltz, DE Rollins. Quantitative analysis of methadone and two major metabolites in hair by positive chemical ionization ion trap mass spectrometry. J Anal Toxicol 20:355–361, 1996.
80. GA Cooper, JS Oliver. Improved solid-phase extraction of methadone and its two major metabolites from whole blood. J Anal Toxicol 22:389–392, 1998.
81. F Sporkert, F Pragst. Determination of methadone and its metabolites EDDP and EDMP in human hair by headspace solid-phase microextraction in gas chromatography–mass spectrometry. J Chromatogr B 746:255–264, 2000.
82. HR Angelo, N Beck, K Kristensen. Enantioselective high-performance liquid chromatographic method for the determination of methadone and its main metabolite in urine using AGP and C8 column coupled serially. J Chromatogr B 724:35–40, 1999.
83. P Kintz, HP Eser, A Tracqui, M Moeller, V Cirimele, P Mangin. Enantioselective separation of methadone and its main metabolite in human hair by liquid chromatography/ion spray–mass spectrometry. J Forensic Sci 42:291–295, 1997.
84. D Ortelli, S Rudaz, AF Chevalley, JJ Deglon, L Balant, JL Veuthey. Enantioselective analysis of methadone in saliva by liquid chromatography–mass spectrometry. J Chromatogr A 871:163–172, 2000.
85. L Amass, JB Kamien, SK Mikulich. Efficacy of daily and alternate-daily regimes with the combination buprenorphine–naloxone tablet. Drug Alcohol Depend 58:143–152, 2000.
86. AM Lisi, R Kaslauskas, GJ Trout. Gas chromatographic–mass spectrometric quantitation of urinary buprenorphine and norbuprenorphine after derivatization by direct extractive alkylation. J Chromatogr B 692:67–77, 1997.
87. KA Hadidi, J Oliver. Stability of morphine and buprenorphine in whole blood. Int J Legal Med 111:165–167, 1998.
88. JJ Kuhlman, J Magluilo Jr, EJ Cone, B Levine. Simultaneous assay of buprenorphine and norbuprenorphine by negative chemical ionization tandem mass spectrometry. J Anal Toxicol 20:229–235, 1996.

89. P Kintz, V Cirimele, Y Edel, C Jamey, P Mangin. Hair analysis for buprenorphine and its dealkylated metabolite by RIA and confirmation by LC/ECD. J Forensic Sci 39:1497–1503, 1994.
90. A Salem, TL Pierce, W Hope. Analysis of buprenorphine in rat plasma using a solid-phase extraction technique and high-performance liquid chromatography with electrochemical detection. J Pharmacol Toxicol Methods 37:75–81, 1997.
91. A Tracqui, P Kintz, P Mangin. HPLC/MS determination of buprenorphine and norbuprenorphine in biological fluids and hair samples. J Forensic Sci 42:111–114, 1997.
92. H Hoja, P Marquet, B Verneuil, H Lofti, JL Dupuy, G Lachaitre. Determination of buprenorphine and norbuprenorphine in whole blood by liquid chromatography–mass spectrometry. J Anal Toxicol 21:160–165, 1997.
93. DE Moody, JD Laycock, AC Spanbauer, DJ Crouch, RL Foltz, JL Josephs, L Amass, WK Bickel. Determination of buprenorphine in human plasma by gas chromatography–positive chemical ionization mass spectrometry and liquid chromatography–tandem mass spectrometry. J Anal Toxicol 21:406–414, 1997.
94. A Polettini, MA Huestis. Simultaneous determination of buprenorphine, norbuprenorphine, and buprenorphine–glucuronide in plasma by liquid chromatography–tandem mass spectrometry. J Chromatogr B 754:447–459, 2001.
95. F Musshoff, T Daldrup. Evaluation of a method for simultaneous quantification of codeine, dihydrocodeine, morphine and 6-monoacetylmorphine in serum, blood and postmortem blood. J Legal Med 106:107–109, 1993.
96. BD Paul, LD Mell, JM Mitchell, J Irving, AJ Novak. Simultaneous identification and quantitation of codeine and morphine in urine by capilary gas chromatography and mass spectrometry. J Anal Toxicol 9:222–227, 1985.
97. WJ Phillips, K Ota, NA Wade. Tandem mass spectrometry (MS/MS) utilizing electron impact ionization and multiple reaction monitoring for the rapid, sensitive and specific identification and quantitation of morphine in whole blood. J Anal Toxicol 13:268–273, 1989.
98. G Schmitt, M Bogusz, R Aderjan, C Meyer. Zum Nachweis von Morphin un Codein mit GC/MS (NCI und PCI) und zur Unterscheidung einer Codeineinnahme von Heroin-oder Morphinkonsum. Z Rechtsmed 103: 513–521, 1990.
99. DG Watson, Q Su, JM Midley, E Doyle, NS Morton. Analysis of unconjugated morphine, codeine, normrphine and morphine as glucuronides in small volumes of plasma from children. J Pharm Biomed Anal 13:27–32, 1995.
100. B Fryirs, M Dawson, LE Mather. Highly sensitive gas chromatographic mass spectrometric method for morphine determination in plasma that is suitable for pharmacokinetic studies. J Chromatogr B 693:51–57, 1997.

101. AS Low, RB Taylor. Analysis of common opiates and heroin, metabolites in urine by high performance liquid chromatography. J Chromatogr B 663: 225–233, 1995.
102. DA Barrett, PN Shaw, SS Davis. Determination of morphine and 6-acetylmorphine in plasma by high performance liquid chromatography with fluorescence detection. J Chromatogr 566:135–145, 1991.
103. P Joel, RJ Osborne, ML Slevin. An improved method for the simultaneous determination of morphine and its principal glucuronide metabolites. J Chromatogr 430:394–399, 1988.
104. Y Rothsteyn, B Weingarten. A highly sensitive assay for the simultaneous determination of morphine, morphine-3-glucuronide, and morphine-6-glucuronide in human plasma by high-performance liquid chromatography with electrochemical and fluorescence detection. Ther Drug Monit 18:179–188, 1996.
105. RW Milne, RL Nation, GD Reynolds, AA Somogyi, JT Van Crutgen. High-performance liquid chromatographic determination of morphine and its 3- and 6-glucuronide metabolites: improvements to the method and application to stability studies. J Chromatogr 565:457–464, 1991.
106. PA Glare, TD Walsh, CE Pippenger. A simple, rapid method for the simultaneous determination of morphine and its principal metabolites in plasma using high performance liquid chromatography and fluorometric detection. Ther Drug Monit 13:226–232, 1991.
107. R Hartley, M Green, M Quinn, MI Levene. Analysis of morphine and its 3- and 6-glucuronides by high performance liquid chromatography with fluorimetric detection following solid phase extraction from neonatal plasma. Biomed Chromatogr 7:34–37, 1994.
108. R Pacifici, S Pichini, I Altieri, A Caronna, AR Passa, P Zuccaro. High-performance liquid chromatographic–electrospray mass spectrometric determination of morphine and its 3- and 6-glucuronides: application to pharmacokinetic studies. J Chromatogr B 664:329–334, 1995.
109. N Tyrefors, B Hyllbrant, L Ekman, M Johansson, B Langstrom. Determination of morphine, morphine-3-glucuronide and morphine-6-glucuronide in serum by solid-phase extraction and liquid chromatography–mass spectrometry with electrospray ionization. J Chromatogr A 729:279–285, 1996.
110. M Zheng, KM McErlane, MC Ong. High-performance liquid chromatography–mass spectrometry analysis of morphine and morphine metabolites and its application to a pharmacokinetic study in male Sprague–Dawley rats. J Pharmaceut Biomed Anal 16:971–980, 1998.
111. G Schanzle, S Li, G Mikus, U Hofmann. Rapid, highly sensitive method for the determination of morphine and its metabolites in body fluids by liquid chromatography–mass spectrometry. J Chromatogr B 721:55–65, 1999.
112. A Dienes-Nagy, L Rivier, C Giroud, M Augsburger, P Mangin. Method for quantification of morphine and its 3- and 6-glucuronides, codeine, codeine glucuronide and 6-monoacetylmorphine in human blood by liquid chroma-

tography–electrospray mass spectrometry for routine analysis in forensic toxicology. J Chromatogr A 854:109–118, 1999.
113. M Blanchet, G Bru, M Guerret, M Bromet-Petit, N Bromet. Routine determination of morphine-3-β-d-glucuronide in human serum by liquid chromatography coupled to electrospray mass spectrometry. J Chromatogr A 854:93–108, 1999.
114. MS Slawson, DJ Crouch, DM Andrenyak, DE Rollins, JK Lu, PL Bailey. Determination of morphine, morphine-3-glucuronide and morphine-6-glucuronide in plasma after intravenous and intrathecal morphine administration using HPLC with electrospray ionization and tandem mass spectrometry. J Anal Toxicol 23:468–473, 1999.
115. W Naidong, LW Lee, X Jiang, M Wehling, JD Hulse, PP Lin. Simultaneous assay of morphine, morphine-3-glucuronide and morphine-6-glucuronide in human plasma using normal-phase liquid chromatography–tandem mass spectrometry with a silica column and an aqueous organic mobile phase. J Chromatogr B 735:255–269, 1999.
116. ET Everhardt, P Cheung, P Schwonek, K Zabel, EC Tisdale, P Jacob, J Mendelson, RT Jones. Subnanogram concentration measurement of buprenorphine in human plasma by electron capture capillary gas chromatography: application to pharmacokinetics of sublingual buprenorphine. Clin Chem 43:2292–2302, 1997.
117. JJ Kuhlman, S Lalani, J Magluilo, B Levine, WD Darwin, RE Johnson, EJ Cone. Human pharmacokinetics of intravenous, sublingual and buccal buprenorphine. J Anal Toxicol 20:369–378, 1996.
118. N Schmidt, R Sittl, K Brune, G Geisslinger. Rapid determination of methadone in plasma, cerebrospinal fluid and urine by gas chromatography and its application to routine drug monitoring Pharm Res 10:441–454, 1993.
119. LD Baugh, RH Liu, AS Walia. Simultaneous gas chromatography-mass spectrometry assay of methadone and 2-ethyl-1,5-dimethyl-3,3-diphenylpyrrolidine (EDDP) in urine. J Forensic Sci 36:548–555, 1991.
120. MJ Bogusz, RD Maier, KD Kruger, U Kohls. Determination of common drugs of abuse in body fluids using one isolation procedure and liquid chromatography–atmospheric pressure chemical ionization mass spectrometry. J Anal Toxicol 22:549–558, 1998.
121. P Kintz. Determination of buprenorphine and its dealkylated metabolite in human hair. J Anal Toxicol 17:443–444, 1993.
122. LM Stolk, SM Coenradie, BJ Smit, HL van As. Analysis of methadone and its primary metabolite in meconium. J Anal Toxicol 21:154–159, 1997.
123. S Rudaz, JL Veuthey. Stereoselective determination of methadone in serum by HPLC following solid-phase extraction on disk. J Pharm Biomed Anal 14:1271–1279, 1996.
124. AM Verweij, ML Hordijk, PJ Lipman. Quantitative liquid chromatographic–thermospray tandem mass spectrometric analysis of some analgesics and tranquilizers of the methadone, butyrophenone or diphenylbutylpiperidine groups in whole blood. J Anal Toxicol 19:65–68, 1995.

125. K Kristensen, HR Angelo, T Blemmer. Enantioselective high-performance liquid chromatographic method for the determination of methadone in serum using an AGP and a CN column as chiral and analytical column, respectively. J Chromatogr A 666:283–287, 1994.
126. N Schmidt, K Brune, G Geisslinger. Stereoselective determination the enantiomers of methadone in plasma using high-performance liquid chromatography. J Chromatogr 583:195–200, 1992.
127. G Skopp, K Klinder, L Potsch, G Zimmer, R Lutz, R Aderjan, R Mattern. Postmortem distribution of dihydrocodeine and metabolites in a fatal case of dihydrocodeine intoxication. Forensic Sci Int 95:99–107, 1998.
128. M Balikova, V Maresova. Fatal opiates overdose. Toxicological identification of various metabolites in a blood sample by GC-MS after silylation. Forensic Sci Int 94:201–209, 1998.
129. R Aderjan, G Skopp. Formation and clearance of active and inactive metabolites of opiates in humans. Ther Drug Monit 20:561–569, 1998.
130. M Balikova, V Maresova, V Habrdova. Evaluation of urinary dihydrocodeine excretion in human by gas chromatography–mass spectrometry. J Chromatogr B 752:179–186, 2001.
131. NIDA Research Report. Cocaine Abuse and Addiction. NIH Publ No. 99–4342, 1999.
132. Epidemiologic Trends in Drug Abuse, Advance Report, June 2000 National Institutes of Health, National Institute of Drug Abuse (Internet).
133. SB Karch. Cocaine: history, use, abuse. J R Soc Med 92:893–897, 1999.
134. MA ElSohly, R Brenneisen, AB Jones. Coca paste: chemical analysis and smoking experiments. J Forensic Sci 36:93–103, 1991.
135. JG Ensing, C Racamy, RA de Zeeuw. A rapid gas chromatographic method for the fingerprinting of illicit cocaine samples. J Forensic Sci 37:446–459, 1992.
136. KE Janzen, L Walter, AR Fernando. Comparison analysis of illicit cocaine samples. J Forensic Sci 37:436–445, 1992.
137. DR Morello, JF Casale, ML Stevenson, RF Klein. The effects of microwave irradiation on occluded solvents in illicitly produced cocaine. J Forensic Sci 45:1126–1132, 2000.
138. TA Gough, GF Phillips. Scientific support for a custom service. In: TA Gough, ed. The Analysis of Drugs of Abuse. Chichester: John Wiley, 1991, pp 477–510.
139. MJ Bogusz, H Althoff, M Erkens, RD Maier, R Hofmann. Internally concealed cocaine: analytical and diagnostic aspects. J Forensic Sci 40:811–815, 1995.
140. C Wettli, R Mittleman. J Forensic Sci 26:492–500, 1981.
141. Y Balash, L Pollak, J Hiss, MJ Rabey. Eur J Neurol 7:555–558, 2000.
142. J Oyler, WD Darwin, EJ Cone. Cocaine contamination of United States currency. J Anal Toxicol 20:213–216, 1996.
143. A Negrusz, JL Perry, CM Moore. Detection of cocaine in various denominations of United States currency. J Forensic Sci 43:626–629, 1998.

144. IM Lazar, G Naisbitt, ML Le. Capillary electrophoresis—time-of-flight mass spectrometry of drugs of abuse. Analyst 123:1449–1454, 1998.
145. AS Krawczeniak, VA Bravenec. Quantitative determination of cocaine in illicit powders by free zone capillary electrophoresis. J Forensic Sci 43: 738–743, 1998.
146. EJ Cone. Pharmacokinetics and pharmacodynamics of cocaine. J Anal Toxicol 19:459–478, 1995.
147. EJ Cone, J Oyler, D Darwin. Cocaine disposition in saliva following intravenous, intranasal and smoked administration. J Anal Toxicol 21: 465–476, 1997.
148. RA Jufer, WD Darwin, EJ Cone. Current methods for the separation and analysis of cocaine analytes. In: MJ Bogusz, ed. Forensic Science. Handbook of Analytical Separations, Vol. 2. Amsterdam: Elsevier Science, 2000, pp 67–106.
149. RA Jufer, A Wstadik, SL Walsh, BS Levine, EJ Cone. Elimination of cocaine and metabolites in plasma, saliva, and urine following repeated oral administration to human volunteers. J Anal Toxicol 24:467–477, 2000.
150. AJ Jenkins, BA Goldberger. Identification of unique cocaine metabolites and smoking by-products in postmortem blood and urine specimens. J Forensic Sci 42:824–827, 1997.
151. DS Isenschmid, BS Levine, YH Caplan. A comprehensive study of the stability of cocaine and its metabolites. J Anal Toxicol 13:250–256, 1989.
152. WC Brogan, PM Kemp, RO Bost, DB Glamann, RA Lange, LD Hillis. Collection and handling of clinical blood samples to assure the accurate measurement of cocaine concentration. J Anal Toxicol 16:152–154, 1992.
153. DN Bailey. Amitriptyline and procainamide inhibition of cocaine and cocaethylene degradation in human serum in vitro. J Anal Toxicol 23: 99–102, 1999.
154. BK Logan. Ecgonine is an important marker for cocaine use in inadequately preserved specimens. J Anal Toxicol 25: 219–220, 2001.
155. Y Gaillard, G Pepin. Simultaneous solid-phase extraction on C18 cartridges of opiates and cocainics for an improved quantitation in human hair by GC-MS: one year of forensic applications. Forensic Sci Int 86:49–59, 1997.
156. K Watanabe, H Hattori, M Nishikawa, T Kumazawa, H Seno, O Suzuki. Simultaneous determination of cocaethylene and cocaine in blood by gas chromatography with surface ionization detection. Chromatographia 44: 55–58, 1997.
157. DJ Crouch, ME Alburges, AC Spanbauer, DE Rollins, DE Moody. Analysis of cocaine and its metabolites from biological specimens using solid-phase extraction and positive ion chemical ionization mass spectrometry. J Anal Toxicol 19:352–358, 1995.
158. A Hernandez, W Andollo, WL Hearn. Analysis of cocaine and metabolites in brain using solid phase extraction and full-scanning gas chromatography/ion trap mass spectrometry. Forensic Sci Int 13:149–156, 1994.

159. D Smirnow, BK Logan. Analysis of ecgonine and other cocaine biotransformation products in postmortem whole blood by protein precipitation-extractive alkylation and GC-MS. J Anal Toxicol 20:463–467, 1996.
160. KL Paterson, BK Logan, GD Christian. Detection of cocaine and its polar transformation products and metabolites in human urine. Forensic Sci Int 73:183–196, 1995.
161. BD Paul, LK McWhorter, ML Smith. Electron ionization mass fragmentometric detection of urinary ecgonidine, a hydrolytic product of methylecgonidine, as an indicator of smoking cocaine. J Mass Spectrom 34:651–660, 1999.
162. SW Toennes, AS Fandino, G Kauert. Gas chromatographic-mass spectrometric detection of anhydroecgonine methyl ester (methylecgonidine) in human serum as evidence of recent smoking of crack. J Chromatogr. B 735: 127–132, 1999.
163. MR Moeller, P Fey, S Rimbach. Identification and quantitation of cocaine and its metabolites, benzoylecgonine and ecgonine methyl ester, in hair of Bolivian coca chewers by gas chromatography/mass spectrometry. J Anal Toxicol 16:291–296, 1992.
164. D Garside, JD Ropero-Miller, BA Goldberger, WF Hamilton, WR Maples. Identification of cocaine in fingernail and toenail specimens. J Forensic Sci 43:974–979, 1998.
165. DA Engelhart, ES Lavins, CA Sutheimer. Detection of drugs of abuse in nails. J Anal Toxicol 22:314–318, 1998.
166. J Oyler, WD Darwin, KL Preston, P Suess, EJ Cone. Cocaine disposition in meconium from newborns of cocaine-abusing mothers and urine of adult drug users. J Anal Toxicol 20:453–462, 1996.
167. MA ElSohly, W Kopycki, S Feng, TP Murphy. Identification and analysis of the major metabolites of cocaine in meconium. J Anal Toxicol 23:446–451, 1999.
168. C Moore, D Dempsey, D Deitermann, D Lewis, J Leikin. Fetal cocaine exposure: analysis of vernix caseosa. J Anal Toxicol 20:501–511, 1996.
169. RE Winecker, BA Goldberger, I Tebbett, M Behnke, FD Eyler, M Conlon, K Wobie, J Karlix, RL Bertholf. Detection of cocaine and its metabolites in amniotic fluid and umbilical cord tissue. J Anal Toxicol 21:97–104, 1997.
170. P Fernandez, N Lafuente, AM Bermejo, M Lopez-Rivadulla, A Cruz. HPLC determination of cocaine and benzoylecgonine in plasma and urine from drug abusers. J Anal Toxicol 20, 224–228, 1996.
171. P Wei-jian, AM Hedaya. Sensitive and specific high-performance liquid chromatographic assay with ultraviolet detection for the determination of cocaine and its metabolites in rat plasma. J Chromatogr B 703:129–138, 1997.
172. DL Phillips, IR Tebbett, RL Bertholf. Comparison of HPLC and GC-MS for measurement of cocaine and metabolites in human urine. J Anal Toxicol 20, 305–308, 1996.
173. M Nishikawa, K Nakajima, M Tatsuno, F Kasuya, K Igarashi, M Fukuim, H Tsushihashi. The analysis of cocaine and its metabolites by liquid

chromatography/atmospheric pressure chemical ionization–mass spectrometry (LC/APCI-MS). Forensic Sci Int 66:149–158, 1994.

174. A Cailleux, A Le Bouil, B Auger, G Bonsergen, A Turcant, P Allain. Determination of opiates and cocaine and its metabolites in biological fluids by high-performance liquid chromatography with electrospray tandem mass spectrometry. J Anal Toxicol 23:620–624, 1999.
175. CS Sosnoff, Q Ann, JT Bernert Jr, MK Powell, BB Miller, LO Henderson, WH Hannon, P Fernhoff, EJ Sampson. Analysis of benzoylecgonine in dried blood spots by liquid chromatography–atmospheric pressure chemical ionization tandem mass spectrometry. J Anal Toxicol 20:179–184, 1996.
176. PM Jeanville, ES Estape, I Torres-Negron, A Marti. Rapid confirmation/quantitation of ecgonine methyl ester, benzoylecgonine, and cocaine in urine using on-line extraction coupled with fast HPLC and tandem mass spectrometry. J Anal Toxicol 25:69–75, 2001.
177. G Skopp, A Klingmann, L Potsch, R Mattern. In vitro stability of cocaine in whole blood and plasma including ecgonine as a target analyte. Ther Drug Monit 23:174–181, 2001.
178. F Tagliaro, G Manetto, F Crivellente, D Scarcella, M Marigo. Hair analysis for abused drugs by capillary zone electrophoresis with field-amplified stacking. Forensic Sci Int 92:201–211, 1998.
179. F Tagliaro, R Valentini, G Manetto, F Crivellente, G Carli, M Marigo. Hair analysis by using radioimmunoassay, high-performance liquid chromatography and capillary electrophoresis to investigate exposure to heroin, cocaine and/or ecstasy in applicants for driving licences. Forensic Sci Int 107:121–128, 2000.
180. J Caslavska, D Allemann, W Thormann. Analysis of urinary drugs of abuse by a multianalyte capillary electrophoretic immunoassay. J Chromatogr A 838:197–211, 1999.
181. PM Jeanville, ES Estape, SR Needham, MJ Cole. Rapid confirmation/quantitation of cocaine and benzoylecgonine in urine utilizing high performance liquid chromatography and tandem mass spectrometry. J Am Soc Mass Spetrom 11:257–263, 2000.
182. G Singh, V Arora, PT Fenn, B Mets, IA Blair. A validated stable isotope dilution liquid chromatography tandem mass spectrometry assay for the trace analysis of cocaine and its major metabolites in plasma. Anal Chem 71:2021–2027, 1999.
183. KM Clauwaert, JF Van Bocxlaer, WE Lambert, EG Van den Eeckhout, F Lemiere, EL Esmans, AP De Leenheer. Narrow-bore HPLC in combination with fluorescence and electrospray mass spectrometric detection for the analysis of cocaine and metabolite in human hair. Anal Chem 70:2336–2344, 1998.
184. A Shulgin, A Shulgin. Pihkal, A Chemical Love Story. Berkeley: Transform Press, 1995.
185. AS Henry, KJ Jeffrys, S Dawling. Toxicity and daths from 3,4-methylenedioxymethamphetamine ("ecstasy"). Lancet 340:384–387, 1992.

186. Lora-Tamayo, T Tena, A Rodriguez. Amphetamine derivative related deaths. Forensic Sci Int 85:149–157, 1997.
187. CM Milroy, JC Clark, ARW Forrest. The pathology of deaths associated with "Ecstasy" and "Eve" misuse. J Clin Pathol 49:149–153, 1996.
188. N Carter, GN Rutty, CM Milroy, ARW Forrest. Deaths associated with MBDB misuse. Int J Legal Med 113:168–170, 2000.
189. R McKetin, RP Mattick. Attention and memory in illicit amphetamine users: comparison with non-drug-using controls. Drug Alcohol Depend 50:181–184, 1998.
190. E Gouzoulis-Mayfrank, J Daumann, F Tuchtenhagen, S Pelz, S Becker, H-J Kunert, B Fimm, H Sass. Impaired cognitive performance in drug free users of recreational ecstasy (MDMA). Neurol Neurosurg Psychiatry 68:719–725, 2000.
191. H Zhao, R Brenneisen, A Scholer, AJ McNally, MA ElSohly, TP Murphy, SJ Salamone. Profiles of urine samples taken from ecstasy users at rave parties: analysis by immunoassays. J Anal Toxicol 25:258–269, 2001.
192. MS Kaufman, AC Hatzis, JG Stuart. Negative-ion chemical ionization of amphetamine derivatives. J Mass Spectrom 31:913–920, 1996.
193. KE Kongshaug, S Pedersen-Bjergaard, KE Rasmussen, M Krogh. Solid-phase microextraction/capillary gas chromatography for the profiling of confiscated ecstacy and amphetamine. Chromatographia 50:247–252, 2000.
194. F Sadeghipour, C Giroud, L Rivier, J-L Veuthey. Rapid detection of amphetamines by high-performance liquid chromatography with UV detection. J Chromatogr A 761:71–78, 1997.
195. AM Di Pietra, R Gotti, E Del Borello, R Pomponio, V Cavrini. Analysis of amphetamine and congeners in illicit samples by liquid chromatography and capillary electrophoresis. J Anal Toxicol 25, 99–105, 2001.
196. U Backofen, FM Matysik, W Hoffman, CE Lunte. Analysis of illicit drugs by nonaqueous capillary electrophoresis and electrochemical detection. Fresenius J Anal Chem 367:359–363, 2000.
197. SR Wallenborg, IS Lurie, DW Arnold, CG Bailey. On-chip chiral and achiral separation of amphetamine and related compounds labeled with 4-fluoro-7-ntrobenzofurazane. Electrophoresis 21:3257–3263, 2000.
198. HK Ensslin, HH Maurer, E Gouzoulis, L Hermle, KA Kovar. Metabolism of racemic 3,4-methylenedioxyethylamphetamine in humans. Drug Metab Disp 24:813–820, 1996.
199. HH Maurer. On the metabolism and the toxicological analysis of methylenedioxyphenylalkylamine designer drugs by gas chromatography–mass spectrometry. Therap Drug Monit 18:465–470, 1996.
200. HH Maurer, J Bickeboeller-Friedrich, T Kraemer, FT Peters. Toxicokinetics and analytical toxicology of amphetamine-derived designer drugs ("Ecstasy"). Toxicology Letters 112–113:133–142, 2000.
201. R de la Torre, M Farre, J Ortuno, M Mas, R Brenneisen, PN Roset, S Segura, J Cami. Nonlinear pharmacokinetics of MDMA ("ecstasy") in humans. J Clin Pharmacol 49:104–109, 2000.

202. AS Smith-Kielland, B Skuterud, J Morland. Urinary excretion of amphetamine after termination of drug abuse. J Anal Toxicol 21:325–329, 1997.
203. RC Baselt. Urine drug screening by immunoassay: interpretation of results. In: RC Baselt, ed. Advances in Analytical Toxicology, Vol. 1. Foster City, CA: Biomedical Publications, 1984, pp 81–124.
204. HK Lim, Z Su, RL Foltz. Stereoselective disposition: enantioselective quantitation of 3,4-(methylelendioxy)methamphetamine and three of its metabolites by gas chromatography/electron capture ion chemical ionization mass spectrometry. Biol Mass Spectrom 22:403–411, 1993.
205. JK Fallon, AT Kicman, JA Henry, PJ Milligan, DA Cowan, AJ Hutt. Stereospecific analysis and enantiomeric disposition of 3,4-methylenedioxymethamphetamine (Ecstasy) in humans. Clin Chem 45:1058–1069, 1999.
206. T Kraemer, I Vernaleken, HH Maurer. Studies on the metabolism and toxicological detection of the amphetamine-like anorectic mefenorex in human urine by gas chromatography–mass spectrometry and fluorescence detection immunoassay. J Chromatogr B 702:93–102, 1997.
207. Cody JT, Valtier S. Amphetamine, clobenzorex, and 4-hydroxyclobenzorex levels following multidose administration of clobenzorex. J Anal Toxicol 25:158–165, 2001.
208. EM Thurman, MJ Pedersen, RL Stout, T Martin. Distinguishing sympathomimetic amines from amphetamine and methamphetamine in urine by gas chromatography/mass spectrometry. J Anal Toxicol 16:19–27, 1992.
209. BD Paul, MR Past, RM McKinley, JHD Foreman, LK McWhorter, JJ Snyder. Amphetamine as an artifact of methamphetamine during periodate degradation of interfering ephedrine, pseudoephedrine and phenylpropanolamine: an improved procedure for accurate quantitation of amphetamines in urine. J Anal Toxicol 18:331–340, 1994.
210. A Dasgupta, C Gardner. Distinguishing amphetamine and methamphetamine from other interfering sympathomimetic amines after various fluro derivatization and analysis by gas chromatography–chemical ionization mass spectrometry. J Forensic Sci 40:1077–1081, 1995.
211. JT Cody. Amphetamines. In: MJ Bogusz, ed. Forensic Science. Handbook of Analytical Separations, Vol. 2. Amsterdam: Elsevier Science, 2000, pp 107–141.
212. JL Valentine, R Middleton. GC-MS identification of sympathomimetic amine drugs in urine: rapid methodology applicable for emergency clinical toxicology. J Anal Toxicol 24:211–222, 2000.
213. SP Vorce, JH Sklerov, KS Kalasinsky. Assessment of the ion-trap mass spectrometer for routine qualitative and quantitative analysis of drugs of abuse extracted from urine. J Anal Toxicol 24:595–601, 2000.
214. GW Kunsman, B Levine, JJ Kuhlman, RL Jones, RO Hughes, CI Fujiyama, ML Smith. MDA-MDMA concentrations in urine specimens. J Anal Toxicol 20:517–521, 1996.
215. HJ Helmlin, K Bracher, D Bourquin, D Vonlanthen, R Brenneisen, J Styk.

Analysis of 3,4-methylenedioxymethamphetamine (MDMA) and its metabolites in plasma and urine by HPLC-DAD and GC-MS. J Anal Toxicol 20:432–440, 1996.
216. MR Lee, YS Song, BH Hwang, CC Chou. Determination of amphetamine and methamphetamine in serum via headspace derivatization solid-phase microextraction–gas chromatography–mass spectrometry. J Chromatogr A 896:256–273, 2000.
217. C Jurado, MP Gimenez, T Soriano, M Menendez, M Repetto. Rapid analysis of amphetamine, methamphetamine, MDA and MDMA in urine using solid-phase microextraction, direct on-fiber derivatization, and analysis by GC-MS. J Anal Toxicol 24:11–16, 2000.
218. K Okajima, A Namera, M Yashiki, I Tsukue, T Kojima. Highly sensitive analysis of methamphetamine and amphetamine in human whole blood using headspace solid-phase microextraction and gas chromatography–mass spectrometry. Forensic Sci Int 116:15–22, 2001.
219. D Hensley, JT Cody. Simultaneous determination of amphetamine, methamphetamine, methylenedioxyamphetamine (MDA), methylenedioxymethamphetamine (MDMA), and methylenedioxyethylamphetamine (MDEA) enantiomers by GC-MS. J Anal Toxicol 23:518–552, 1999.
220. R Herraez-Hernandez, P Campins-Falco. Automated trace enrichment for screening and/or determination of primary, secondary and tertiary amphetamines in biological samples by liquid chromatography. Analyst 124:239–244, 1999.
221. MJBogusz, M Kala, RD Maier. Determination of phenylisothiocyanate derivatives of amphetamine and its analogues in biological fluids by HPLC-APCI-MD or DAD. J Anal Toxicol 21:59–69, 1997.
222. MJ Bogusz, KD Kruger, RD Maier. Analysis of underivatized amphetamines and related phenethylamines with high-performance liquid chromatography–atmospheric pressure chemical ionization mass spectrometry. J Anal Toxicol 24:77–84, 2000.
223. M Sato, T Mitsui, H Nagase. Analysis of benzphetamine and its metabolites in rat urine by liquid chromatography–electrospray ionization mass spectrometry. J Chromatogr B 751:277–289, 2001.
224. M Katagi, M Tatsuno, A Miki, M Nishikawa, H Tsushihashi. Discrimination of dimethylamphetamine and methamphetamine use: simultaneous determination of dimethylamine-*N*-oxide and other metabolites in urine by high-performance liquid chromatography–electrospray ionization mass spectrometry. J Anal Toxicol 24:354–358, 2000.
225. H Kataoka, HL Lord, J Pawliszyn. Simple and rapid determination of amphetamine, methamphetamine, and their methylenedioxy derivatives in urine by automated in-tub solid-phase microextraction coupled with liquid chromatography–electrospray ionization mass spectrometry. J Anal Toxicol 24:257–265, 2000.
226. KM Clauwert, JF VanBocxlaer, EA De Letter, S Van Calenbergh, WE Lambert, AP De Leenheer. Determination of the designer drugs 3,4-

methylnedioxymethamphetamine, 3,4-methylenedioxyethylmphetamine, and 3,4-methylenedioxyamphetamine with HPLC and fluorescence detection in whole blood, serum, vitreous humor, and urine. Clin Chem 46:1968–1977, 2000.
227. L Geiser, S Cherkaoui, JL Veuthey. Simultaneous analysis of some amphetamine derivatives in urine by nonaqueous capillary electrophoresis coupled to electrospray ionization mass spectrometry. J Chromatogr A 895:111–121, 2000.
228. YJ Heo, YS Whang, MK In, KJ Lee. Determination of enantiomeric amphetamines as metabolites of illicit amphetamines and selegiline in urine by capillary electrophoresis using modified beta-cyclodextrins. J Chromatogr B 741:221–230, 2000.
229. A Ramseier, J Caslavska, W Thormann. Stereoselective screening for and confirmation of urinary enantiomers of amphetamine, methamphetamine, designer drugs, methadone and selected metabolites by capillary electrophoresis. Electrophoresis 20:2726–2738, 1999.
230. A Ramseier, C Siethoff, J Caslavska, W Thormann. Confirmation testing of amphetamines and designer drugs in human urine by capillary electrophoresis–ion trap mass spectrometry. Electrophoresis 21:380–387, 2000.
231. F Sadeghipour, JL Veuthey. Sensitive and selective determination of methylenedioxylated amphetamines by high-performance liquid chromatography with fluorimetric detection. J Chromatogr A 787:137–143, 1997.
232. KM Clauwaert, LF Van Bocxlaer, HJ Major, JA Claereboudt, WE Lambert, E Van den Eeckhout, CH Van Peteghem, AP De Leenheer. Investigation of the quantitative properties of the quadrupole orthogonal acceleration time-of-flight mass spectrometer with electrospray ionization using 3,4-methyleden-dioxymethamphetamine. Rapid Commun Mass Spectrom 15:1540–1545, 1999.
233. LE Hollister. Health aspects of cannabis. Pharmacol Revs 38:2–20, 1986.
234. MR Tramer, D Carroll, FA Campbell, DJM Reynolds, RA Moore, HJ McQuay. Cannabinoids for control of chemotherapy induced nausea and vomiting: quantitative systematic review. BMJ 323:16–21, 2001.
235. NA Darmani. Delta(9)-tetrahydrocannabinol and synthetic cannabinoids prevent emesis produced by the cannabinoid CB(1) receptor antagonist/inverse agonist SR 141716A. Neuropsychopharmacology 24:198–203, 2001.
236. FA Campbell, MR Tramer, D Carroll, DJM Reynolds, RA Moore, HJ McQuay. Are the cannabinoids an effective and safe option in the management of pain? A quantitative systematic review. BMJ 323:13–16, 2001.
237. EM Williamson, FJ Evans. Cannabinoids in clinical practice. Drugs 60: 1303–1314, 2000.
238. KR Muller-Vahl, A Koblenz, M Jobges, H Kolbe, HM Emrich, U Schneider. Influence of treatment of Tourette syndrome with delta9-tetrahydrocannabinol (delta9-THC) on neuropsychological performance. Pharmacopsychiatry 34:19–24, 2001.

239. BR Martin, R Jefferson, R Winckler, JL Wiley, JW Huffmann, PJ Crocker, B Saha, RK Razdan. Manipulation of the tetrahydrocannabinol side chain delineates agonists, partial agonists, and antagonists. J Pharmacol Exp Ther 290:1065–1079, 1999.
240. CE Turner, MA ElSohly, EG Boeren. Constituents of *Cannabis sativa L.* A review of the natural constituents. J Natur Prod 43:169–234, 1980.
241. K Lavanya, TR Baggi. An improved thin-layer chromatographic method for the detection and identification of cannabinoids in cannabis. Forensic Sci Int 47:165–171, 1990.
242. P Oroszlan, G Verzar-Perti, E Mincovics, T Szekely. Separation, quantitation and isolation of cannabinoids from *Cannabis sativa L* by overpressured thin layer chromatography. J Chromatogr 38:217–224, 1987.
243. D Debruyne, F Albessard, MC Bigot, M Moulin. Comparison of three advanced chromatographic techniques for cannabis identification. Bull Narc 46:109–121, 1994.
244. R Brenneisen, MA ElSohly. Chromatographic and spectroscopic profiles of Cannabis of different origin. J Forensic Sci 33:1385–1404, 1988.
245. T Lehmann, R Brenneisen. High performance liquid chromatographic profiling of cannabis products. J Liq Chromatogr 18:689–700, 1995.
246. B Backstrom, MD Cole, MJ Carrott, DC Jones, G Davidson, K Coleman. A preliminary study of the analysis of Cannabis by supercritical fluid chromatography with atmospheric pressure chemical ionization mass spectroscopic detection. Science Justice 37:91–97, 1997.
247. IS Lurie, RP Meyers, TS Conver. Capillary electrochromatography of cannabinoids. Anal Chem 70:3255–3260, 1998.
248. TZ Bosy, KA Cole. Consumption and quantitation of Δ^9-tetrahydrocannabinol in commercially available hemp seed oil products. J Anal Toxicol 24:562–566, 2000.
249. O Zoller, P Rhyn, B Zimmerli. High-performance liquid chromatographic determination of Δ^9-tetrahydrocannabinol and the corresponding acid in hemp containing foods with special regard to the fluorescence properties of Δ^9-tetrahydrocannabinol. J Chromatogr A 872:101–110, 2000.
250. SA Ross, Z Mehmedic, TP Murphy, MA ElSohly. GC-MS analysis of the total delta9-THC content of both drug- and fiber-type cannabis seeds. J Anal Toxicol 24:715–717, 2000.
251. X Peng. Analysis of the major components in cannabis by capillary gas chromatography. Se Pu 16:170–172, 1998.
252. R Mechoulam, RY Gaoni. The absolute configuration of delta-1-tetrahydrocannabinol, the major active constituent of hashish. Tetrahedron Lett 12:1109–1111, 1967.
253. RL Foltz. Analysis of cannabinoids in physiological specimens by gas chromatography/mass spectrometry. In: RC Baselt, ed. Advances in Analytical Toxicology, Vol. 1. Foster City: Biomedical Publications, 1984, pp 125–158.
254. MA Peat. The analysis of Δ^9-tetrahydrocannabinol and its metabolites by

immunoassay. In: RC Baselt, ed. Advances in Analytical Toxicology, Vol. 1. Foster City: Biomedical Publications, 1984, pp 59–80.

255. RC Baselt. Urine drug screening by immunoassay: Interpretation of results. In: RC Baselt, ed. Advances in Analytical Toxicology, Vol. 1. Foster City: Biomedical Publications, 1984, pp 81–124.
256. M Bogusz, G Schmidt. Forensisch-toxikologische Aspekte des Cannabis-Missbrauchs. Zbltt Rechtsmedizin 33:383–398, 1990.
257. PX Iten. Fahren unter Drogen- oder Medikamenteneinfluss. Institut fur Rechtsmedizin, Universität Zurich, 1994, pp 99–121.
258. E Johannsson, S Agurell, LE Hollister, M Halldin. Prolonged apparent half-life of Δ^9-tetrahydrocannabinol in plasma of chronic marijuana users. J Pharm Pharmacol 40:374–375, 1988.
259. MA ElSohly, M Salem. Cannabinoids analysis: analytical methods for different biological specimens. In: MJ Bogusz, ed. Forensic Science. Handbook of Analytical Separations, Vol. 2. Amsterdam: Elsevier Science, 2000, pp 189–193.
260. JP Weller, M Wolf, S Szidat. Enhanced selectivity in the determination of delta9-tetrahydrocannabinol and two major metabolites in serum using ion-trap GC-MS-MS. J Anal Toxicol 24:359–364, 2000.
261. M Chiarotti, L Costamagna. Analysis of 11-nor-9-carboxy-delta(9)tetrahydrocannabinol in biological samples by gas chromatography tandem mass spectrometry (GC/MS-MS). Forensic Sci Int 114:1–6, 2000.
262. P Mireault. Analysis of (Δ^9-tetrahydrocannabinol and its two major metabolites by APCI-LC/MS. Poster at the 46th ASMS conference, Orlando, FL, 1998.
263. P Picotte, P Mireault, G Nolin. A rapid and sensitive LC/APCI/MS/MS method for the determination of Δ^9-tetrahydrocannabinol and its metabolites in human matrices. Poster at the 48th ASMS conference, Long Beach, CA, 2000.
264. T Breindahl, K Andreasen. Determination of 11-nor-delta9-tetrahydrocannabinol-9-carboxylic acid in urine using high-performance liquid chromatography and electrospray ionization mass spectrometry. J Chromatogr B 732:155–164, 1999.
265. SS Tai, MJ Welch. Determination of 11-nor-delta9-tetrahydrocannabinol-9-carboxylic acid in a urine-based standard reference material by isotope-dilution liquid chromatography–mass spectrometry with electrospray ionization. J Anal Toxicol 24:385–389, 2000.
266. W Weinmann, S Vogt, R Goerke, C Muller, A Bromberger. Simultaneous determination of THC-COOH and THC-COOH-glucuronide in urine samples by LC/MS/MS. Forensic Sci Int 113:381–387, 2000.
267. P Kintz, N Samyn. Unconventional samples and alternative matrices. In: MJ Bogusz, ed. Forensic Science. Handbook of Analytical Separations, Vol. 2. Amsterdam: Elsevier Science, 2000, pp 459–488.
268. H Sachs, U Dressler. Detection of THCCOOH in hair by MSD-NCI after HPLC clean-up. Forensic Sci Int 107:239–247, 2000.

269. NP Lemos, RA Anderson, JR Robertson. Nail analysis for drugs of abuse: extraction and determination of cannabis in fingernails by RIA and GC-MS. J Anal Toxicol 23:147–152, 1999.
270. P Kintz, V Cirimele, B Ludes. Detection of cannabis in oral fluid (saliva) and forehead wipes (sweat) from impaired drivers. J Anal Toxicol 24:557–561, 2000.
271. N Fucci, N De Giovanni, M Chiarotti, S Scarlata. SPME-GC analysis of THC in saliva samples collected with "EPITOPE" device. Forensic Sci Int 119:318–321, 2001.
272. MA ElSohly, DF Stanford, TP Murphy, BM Lester, LL Wright, VL Smeriglio, J Verter, CR Bauer, S Shankaran, HS Bada, HC Walls. Immunoassay and GC-MS procedures for the analysis of drugs of abuse in meconium. J Anal Toxicol 23:436–445, 1999.
273. S Feng, MA ElSohly, S Salamone, MY Salem. Simultaneous analysis of delta9-THC and its major metabolites in urine, plasma, and meconium by GC-MS using an immunoaffinity extraction procedure. J Anal Toxicol 24:395–402, 2000.
274. LS Tsai, MS ElSohly, SF Tsai, TP Murphy, B Twarowska, SJ Salamone. Investigation of nitrite adulteration on the immunoassay and GC-MS analysis of cannabinoids in urine specimens. J Anal Toxicol 24:708–714, 2000.
275. J Rohrich, S Zorntlein, L Potsch, G Skopp, J Becker. Effect of the shampoo Ultra Clean on drug concentration in human hair. Int J Legal Med 113:102–106, 2000.
276. V Cirimele, P Kintz, C Jamey, B Ludes. Are cannabinoids detected in hair after washing with Cannabio shampoo? J Anal Toxicol 23:349–351, 1999.

7

Biogenic Amine Neurotransmitters: Their Importance and Measurement in Human Tissue and Body Fluids

Adarsh M. Kumar
University of Miami School of Medicine, Miami, Florida, U.S.A.

1 INTRODUCTION

Biogenic amines are four monoamines, synthesized and secreted within many mammalian tissues, including various regions in the brain, sympathetic nervous system, enterochromaffin cells of the digestive tract, and adrenal medulla. These biogenic amines include catecholamines and an indoleamine, which have been associated with many physiological and behavioral functions in animals and humans. Catecholamines include two primary amines, dopamine (dihydroxyphenylethylamine, DA) and norepinephrine (NE, β-hydroxy dihydroxyphenylethylamine, noradrenaline) and a secondary amine, epinephrine (E, *N*-methyl-β-hydroxy dihydroxyphenylamine, adrenaline). All three catecholamines, NE, E, and DA, are low-molecular-weight substances that contain a catechol nucleus (a benzene ring with two adjacent hydroxyl groups) and an amino acid group, alanine.

Serotonin (5-hydroxytryptamine, 5-HT), an indolemaine, is also a low-molecular-weight monoamine that contains a phenyl ring with an attached indole nucleus and the amino acid alanine, attached at the indole ring. The biogenic amines present in the brain's neuronal and peripheral systems are synthesized intracellularly from their precursor amino acids. The amino acid *L*-tyrosine is the precursor for the synthesis of dopamine, norepinephrine, and epinephrine, and *L*-tryptophan is the precursor for 5-HT synthesis. The two essential amino acids are obtained from dietary sources, and after being absorbed from the digestive tract are transported to various peripheral organs and to the brain via blood circulation. Structures of these biogenic amines are given below in Fig. 1.

Although only a small percentage of total circulating amino acid precursors are transported into the brain, they play a very important role in regulating the synthesis and functions of biogenic amines in the central nervous system (CNS).

The monoamines have unique importance in neurobiological research because of (1) their presence in specific brain regions as well as in the peripheral organ systems that control various body functions; (2) their property of plasticity for neurotransmission in response to environmental stimuli and their ability to regulate simultaneously a number of physiological functions and behaviors in animals and human; (3) their secretion into the body fluids, which can be used for investigating their turnover rate during life and thus providing markers for the diagnosis of physiological dysfunctions and psychiatric disorders and a possible basis for suitable treatment strategies. Biogenic amines are thus an important category of transmitter sub-

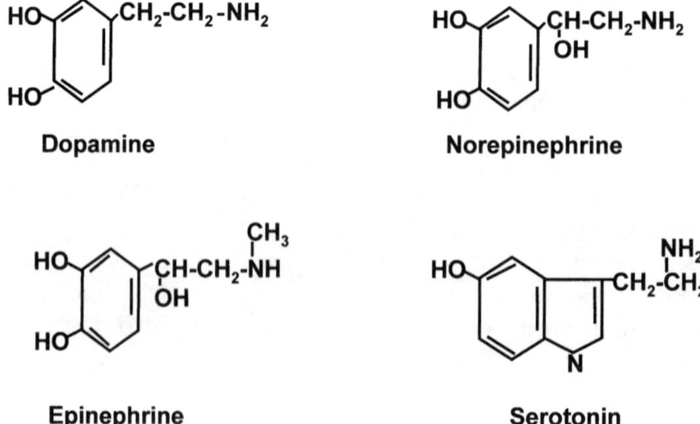

FIGURE 1 Structures of dopamine, norepinephrine, epinephrine, and serotonin.

stances as identified by their property of transmitting signals by their interactions between neurons and between neurons and effector cells, in response to stimuli [1].

A number of other neurotransmitter substances have also been identified, and many of their functions have been delineated. These include acetylcholine (Ach), one of the first substances to be identified with neurotransmitter properties [2]. Acetylcholine is a mediator of impulse transmission at all autonomic ganglia, in the adrenal medulla, and at motor endings in the skeletal muscles. Deficiency of Ach in the brain has been associated with Alzheimer's disease [3]. Gama-aminobutyric acid (GABA) is an inhibitory neurotransmitter in some parts of the CNS, particularly in cerebellum, basal ganglia and spinal cord [4], and deficiency of GABA has been associated with Huntington's disease [5]. Glutamic acid, another important excitatory neurotransmitter is as effective as acetylcholine in activating neurons [4]. Moreover, some peptides, such as substance P, endorphins, and prostaglandins, also possess transmitter properties [6–8]. However, discussion on these peptide neurotransmitters is beyond the scope of this chapter. In this chapter a brief description of biogenic amines and their role in regulation of CNS and peripheral functions is discussed. Also presented in this chapter are some of the technologies with high sensitivity for measurement of ultramicro concentration of biogenic amines in the brain tissues, cells, and different body fluids including blood, plasma, urine, and cerebrospinal fluid (CSF). These technologies include histochemical, electronmicroscopy, fluorometry, gas–liquid chromatography, and high-performance liquid chromatography with electrochemical detection (HPLC-ECD). (In this chapter, the names of biogenic amines, dopamine, norepinephrine, epinephrine, and serotonin are used alternatively with their abbreviations, DA, NE, E, and 5-HT, respectively.)

2 LOCALIZATION OF BIOGENIC AMINES

The biogenic amines present in the CNS and peripheral systems are localized in the aminergic neurons in specific brain regions. Their presence was demonstrated by the earlier investigators using histochemical fluorescent techniques [9], and both catecholamines (DA, NE, and E) and serotonin were identified in the brain by the intense fluorescence of isoquinolines formed during treatment of brain tissue sections with formaldehyde in humid environment. Since then this method has been used for mapping of the neuronal cell bodies, axons, and nerve terminals containing these amines throughout the brain. Subsequently, the same fluorescent technique was used to identify the path of neuronal projections and tracts of the aminergic terminals by causing lesions to the selective brain regions, and following the anterograde loss or retrograde accumulation of amines [10]. In the histochemical fluorescent method, serotonin can be distinguished from catecholamines by the difference in its wavelength of fluorescence, which appears bright yellow for serotonin and bright green for catecholamines [11]. Although fluorescence of both dopamine and norepinephrine appears bright green, they can be differentiated by their response to various drugs, which either deplete or enhance the concentration of one or the other amine [10]. Using this technology, the cell bodies of most of the aminergic neurons have been localized within the brain stem, whereas axons and terminals of the amine-containing neurons have been found to ascend or descend throughout the brain with projections extending to the subcortical and cortical areas, including the basal ganglia, hippocampus, thalamus, hypothalamus, ventral tegmental area, red nucleus, and cerebral cortex and spinal cord. Specific regions of the brain with similar neurons containing predominantly one type of amine have been identified. Thus dopamine is the predominant amine of the substantia nigra, norepinephrine is present mainly in the locus coeruleus, and serotonin is present in the raphe nuclei, composed of dorsal, lateral, and ventral raphe nuclei, and projections extending from raphe to substantia nigra and to many other parts of the brain. In the peripheral system, catecholamines, present in the autonomic system, regulate stress-induced autonomic responses such as heart rate, blood pressure, and glucose levels, whereas serotonin present in the circulating blood and platelets is produced in the enterochromaffin cells of the gut and is involved in regulating blood vessel contractility, blood coagulability, and digestive tract motility as well as pain and appetite. Circulating serotonin binds to its 5-HT_2 receptors present on platelet membrane and is transported by the specific 5-HT membrane transporter for storage in the platelet vesicles during normal resting situations, leaving only a small quantity in plasma. 5-HT is released from platelets in response to stimuli by a similar release mechanism as in presynaptic neuron [12].

In order to delineate various anatomical and functional aspects of these monoamines, studies have been carried out by investigators in various scientific disciplines including histochemistry, electron microscopy, neurophysiology, enzymology, pharmacology, biochemistry, psychology, and clinical psychiatry. Integration of the findings obtained by these investigative approaches have provided a vast body of evidence regarding distribution of these amines in the brain and peripheral systems, their concentration in tissues and body fluids, and their specific regulatory roles played by these monoamines in various physiological and behavioral functions.

3 BIOGENIC AMINES AND NEUROTRANSMISSION

One of the important characteristics of monoamines is their ability of neurotransmission, the key mechanism for maintaining homeostasis of various physiological and behavioral functions. The mechanisms involved in the transmission of signals between neurons and between neurons and effector cells are highly complex processes and are not fully understood. It is, however, well established that in all neurons, whether from leech, lobster, or human, the responses to given stimuli occur through interneuronal communication with the release of specific transmitter substances stored in the granular vesicles in the presynaptic nerve terminal [13]. After synthesis of biogenic amine by multistep processes in the CNS as well as in the peripheral system, the transmitter is stored in the storage granules and are protected against destruction by the enzymes monoamine oxidases (MAO-A, MAO-B). The transmitters are released into the synaptic space, "the synaptic gap," also called the synaptic cleft, a space approximately 20 nm wide between the presynaptic axon terminal and the postsynaptic neuron. The storage granules containing amines are themselves carried by the axoplasmic (intra-axonic fluid) flow to the presynaptic nerve terminal where the granules accumulate and release the transmitter into the synapse in response to a given stimulus. After its release into the synapse, the transmitter molecule generates a number of activities. It initiates an exchange between extracellular and intracellular sodium, potassium, calcium, and chloride ions, and it diffuses across the synaptic gap to the postsynaptic neuron and binds to the specific postsynaptic receptors. These processes cause a change in the permeability of the membrane and generate a signal that is transmitted along the entire length of the nerve terminal. Binding of transmitter to the postsynaptic receptors activates the intracellular second messenger system, either the cyclic-adenosine monophosphate (cAMP system) or the phosphatidyl-inositol system, based on the characteristic of the neurotransmitter and its binding to the specific receptors. Activation of these second messenger systems is an important step in the final expression of response to a given stimulus [14]. Inactivation of the neurotransmitter action occurs by its reuptake through autoreceptors and its transport back into the presynaptic terminal as well as by the intracellular enzymatic degradation into their respective metabolites. The reuptake system of biogenic amine is a universal means of terminating transmitter action, and it is an important step for preventing the amines from enzymatic degradation by the extracellular enzymes present in the synaptic space. Thus, release mechanism of the transmitter in response to a stimulus sets into motion a cascade of events for either an excitatory or inhibitory effect on a specific region of the membrane of presynaptic axonal terminals, dendrites or somatic cells as well as the postsynaptic neuron. In Fig. 2 is a diagrammatic presentation of the pre- and postsynaptic contact of a neuron and the release of a neurotransmitter in the synapse.

Normally, the neuronal function is regulated by maintaining a balance between synthesis, release, binding to the postsynaptic receptors, reuptake,

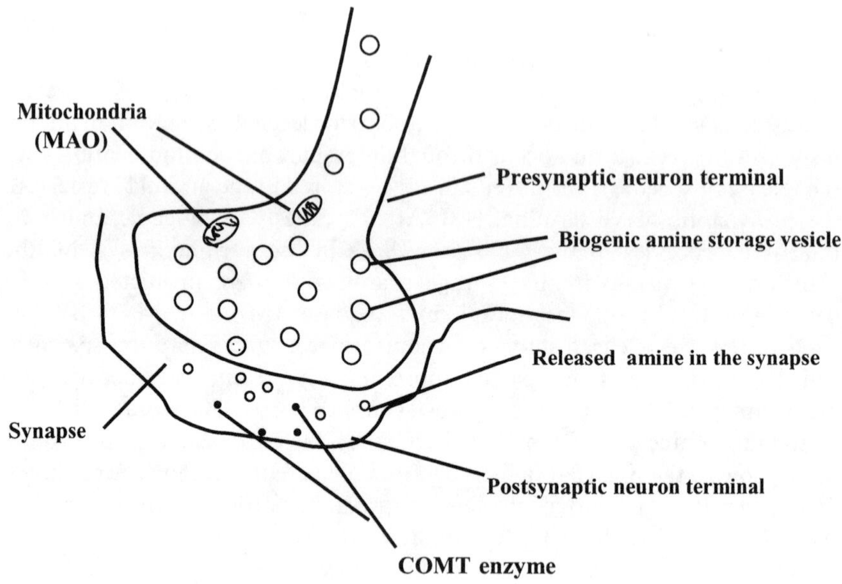

FIGURE 2 Presynaptic and postsynaptic neuron terminals with synaptic space. Biogenic amines containing storage vesicles are shown in the presynaptic terminal.

and enzymatic destruction of the neurotransmitter. However, this balance can be disturbed when the demand by the organism exceeds the normal functional capacity of the neurons. During stressful situations such as severe psychological trauma there may occur excessive stimulation of neurons leading ultimately to depletion of intracellular stores of neurotransmitters.

For example, heightened norepinephrine activity has been reported in posttraumatic stress disorder [15], and both norepinephrine and serotonin dysfunction has been associated with depression [16]. The use of the inhibitory effect of tricyclic antidepressants and serotonin-specific reuptake inihibitors (SSRIs) on the biogenic amine reuptake mechanism has become an important step for the treatment of depression and related psychiatric problems, as well as for the development of newer therapeutic drugs and their efficacy. These drugs allow an increased availability of the amines in the synapse as a part of their regulatory function.

4 BIOGENIC AMINES AND BEHAVIORS

To date a large body of evidence has accumulated that demonstrates an association between disturbance of the functions of DA, NE, and 5-HT and physiological, behavioral, and psychiatric disorders [17,18]. For example, alteration in the turnover rate and changes in receptor densities, increased metabolic activity, alteration in uptake kinetics of NE, DA, and 5-HT have been reported in a number of human conditions including normal aging processes [19], sleep disorders [20], eating disorders [21], obsessive-compulsive disorders [22], aggressive behaviors [23], and suicide [24,25]. A decrease in the availability of neurotransmitters can also occur due to degenerative processes of specific neuronal cells as in age-related diseases such as senile dementia [26], Alzheimer's disease [27–30], and Parkinson's disease [31]. In Parkinson's disease, DA neurons are degenerated in substantia nigra causing motor disorders characterized by rigidity, weakness, and tremors. Other behavioral and psychiatric problems due to genetic defects in aminergic functions include schizophrenia, depression [32–35], and autism [36]. There is an increase in the activity of dopamine in schizophrenia, which is primarily a disorder of thought, and patients with schizophrenia are usually responsive to treatment with neuroleptics, dopamine receptor blocking agents [37]. However, prolonged administration of neuroleptics to schizophrenic patients may induce dopamine deficiency and lead to symptoms similar to that of Parkinson's disease. In posttraumatic stress disorder (PTSD), there is heightened nocturnal activity of catecholamines interfering with sleep pattern [16]. Increased synthesis of catecholamines has been reported in neurogenic tumors [38]. Recent studies show that in human immunodeficiency virus type-1 (HIV-1) infection there are deficits in dopamine as well as in serotonin, which may be responsible for cognitive and motor dysfunctions and depressive disorder [39–42]. Many of the behavioral and psychiatric disorders are linked with deficits in one or more of these neurotransmitters. Thus 5-HT in concert with NE has been associated with depression, which responds to drugs capable of manipulating functions of both 5-HT and NE. Functions of these biogenic amines are also greatly influenced in alcoholism [26] and by the drugs of abuse such as LSD, cocaine, and opioids, leading to craving and addiction and ultimately to cognitive deficits, and locomotor and behavioral dysfunctions [43]. Various lines of investigation also provide convincing evidence of functional interactions among these neurotransmitters [44,45]. A brief description of the individual biogenic amines NE, E, DA, and 5-HT is given below.

4.1 Dopamine

Dopamine is an important neurotransmitter and a precursor for the synthesis of both norepinephrine and epinephrine. In addition to being the immediate precursor of NE and E, it is involved in the regulation of motor functions, feelings of reward and euphoria [46], cognitive functions, spatiovisual and working memory, and executive functions [47]. These functions are regulated by dopamine present in the densely populated zona compacta of substantia nigra in the midbrain region and its transport via their axons and a network of tracts, the nigrostriatal tract from substantia nigra to striatum. The dopaminegic ascending tracts originating from the substantia nigra transport dopamine to various subcortical brain structures including basal ganglia, thalamus, frontocortical regions, and descending tracts to the pons, reticular formation, and spinal cord. In the basal ganglia, most dopamine is localized in corpus striatum, which is composed of caudate nucleus, putamen, and globus pallidus as well as in the other basal gangliar structures. Of the total amount of dopamine found in the brain, approximately 80% is found in the basal ganglia, mainly in the nerve endings in the striatum [48]. In Parkinson's disease, the deficiency of dopamine in substantia nigra results in an insufficient supply to nigrostriatal tracts leading to their progressive degeneration. The characteristic symptoms of the disease are akinesia (difficulty in initiating movement), rigidity, and tremor at rest. Further evidence of a link between dopamine deficiency and Parkinson's disease was provided by the studies showing that many of the symptoms of Parkinson's disease were alleviated by treatment with L-dopa, the amino acid precursor of dopamine and an indirect dopamine agonist, which can cross the blood–brain barrier, in contrast to dopamine, which cannot. As described above, decreases in the brain's dopamine concentration have been reported during normal aging [19], Alzheimer's disease [29], Hungtington's disease [49], and HIV-1 infection [39–40].

Dopamine is synthesized (Fig. 3) both in the periphery and in the large pigmented cells of substantia nigra from the precursor amino acid, L-tyrosine obtained from diet. L-tyrosine is normally present in circulation in the levels of 10–15 mg/L, and in the whole brain at a level of 1.2 mg per 100 g of fresh weight of tissue. It is taken up from blood by an active uptake mechanism and is concentrated within the brain and other tissues. Once inside the neuron or in the chromaffin cells in the peripheral system, L-tyrosine is hydroxylated by a series of enzymatic reactions initiated by tyrosine hydroxylase [50], the rate-limiting enzyme for the synthesis of catecholamines, to form L-dihydroxyphenylalanine (L-dopa), the immediate precursor for dopamine synthesis. Tyrosine hydroxylase is present in neuronal cell bodies as well as in the nerve terminals, indicating thereby that catecholamine synthesis takes place in both these sites. The enzyme requires biopterin, a pteridin compound that enhances its activity [51]. L-dopa is decarboxylated by the enzyme dopa-decarboxylase present in the nerve terminals to form dopamine. Dopamine is metabolized to homovanillic acid (HVA) in two step processes that occur via two pathways. In one pathway, MAO converts dopamine to dihydroxyphenylacetic acid (DOPAC), which is then methoxylated by catechol-*o*-methyltransferase

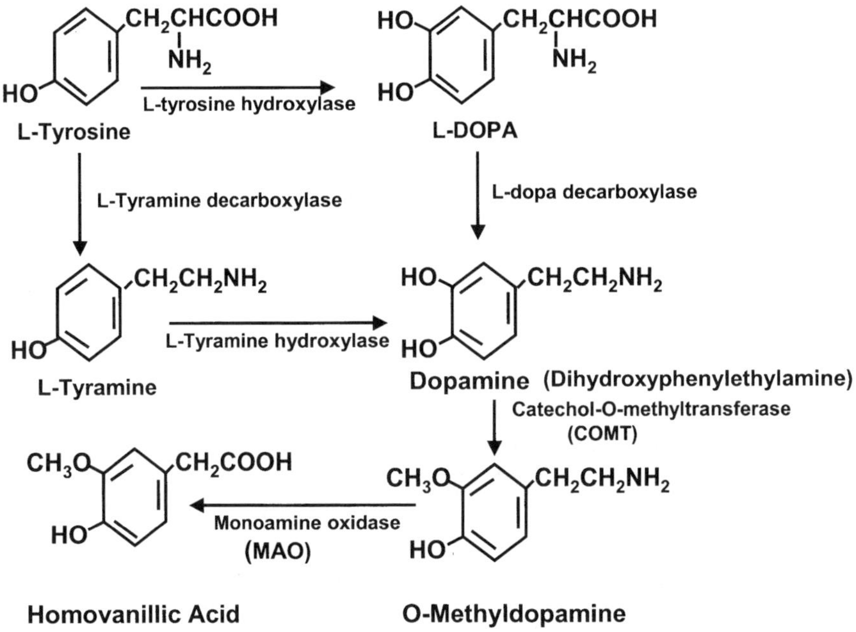

L-Tyrosine
L-tyrosine hydroxylase
L-DOPA
L-Tyramine decarboxylase
L-dopa decarboxylase
L-Tyramine
L-Tyramine hydroxylase
Dopamine (Dihydroxyphenylethylamine)
Catechol-O-methyltransferase (COMT)
Monoamine oxidase (MAO)
Homovanillic Acid
O-Methyldopamine

FIGURE 3 Dopamine synthesis and metabolism.

(COMT) enzyme to HVA, the major metabolite of dopamine. In the other pathway, dopamine is first methoxylated to methoxytyramine, which is then oxidized to HVA by the enzyme, MAO. Measurements of HVA levels in CSF and its excretion in urine are considered as markers of CNS dopaminergic activity. It has been reported that plasma HVA concentration represents only a small percentage of CNS dopaminergic activity.

Dopamine does not cross the blood–brain barrier and therefore cannot be transported from the peripheral system to CNS or vice versa. After its synthesis, dopamine is stored within the vesicles that protect it from oxidation by the mitochondrial monoamine oxidase (MAO). Dopamine is released into the synapse in response to action potential (a stimulus), as described above, and the low steady state concentration in the synapse is regulated primarily by its reuptake mechanism into the presynaptic neuron. Various functions of dopamine are mediated through its transporters as well as receptor systems. Dopamine transporters are localized on the presynaptic terminals, and they serve as one of the markers of dopamine neuronal functions. Dopamine receptors are present on the presynaptic as well as postsynaptic sites. At the postsynaptic site, the receptors maintain cell-to-cell communication, and at the presynaptic site, they modulate the release and synthesis of dopamine [52].

Using newer technologies such as positron emission tomography (PET), recent studies have demonstrated that dopamine has five type of receptors. These receptors have been grouped according to their ability to stimulate or inhibit the intracellular messenger system, the cyclic-adenosine monophosphate (cyclic-AMP) system. Thus receptors D1 and D5 stimulate the cyclic-AMP messenger system, whereas D2, D3, and D4 inhibit the cyclic-AMP system [53]. Using radioautography, Kessler et al. [54] found the highest density of D1 and D2 receptors in striatum. The other family of receptors, D3, D4, and D5, are present in the limbic and cortical areas of the brain, which regulate most of fine cognitive and motor activity [55–57].

4.2 Norepinephrine and Epinephrine

Norepinephrine is a neurotransmitter of the autonomic nervous system. It is present in the central nervous system (CNS) as well as in the peripheral systems. In the peripheral systems, NE is released by the sympathetic nerve terminals, spinal cord, and internal organs, including gut, spleen, and heart [58]. In the CNS, the cell bodies containing norepinephrine are localized in the densely packed neurons in the locus coeruleus in the midbrain area, and in a number of other clusters of neurons in the medulla oblongata, pons, and other midbrain areas where NE-containing neurons are seen mostly scattered throughout the reticular formation. Axons and terminals from NE-containing neurons descend in the sympathetic columns of the spinal cord and terminate at various levels along the cord. The ascending terminals are spread in all areas of the brain, the highest density being present in the hypothalamus and the lowest density in the limbic and cerebral cortex and cerebellar areas.

NE is synthesized in the adrenergic neuronal cell bodies of locus cerouleus as well as nerve terminals in the brain, in the cells of sympathetic ganglia, in the nerve endings of the sympathetic system, and chromaffin cells of the adrenal and other peripheral tissues [59]. In the adrenal gland as well as in the brain, NE is formed from dopamine by its hydroxylation at the beta carbon by the enzyme dopamine β-hydroxylase, which requires copper for its optimal functioning [60]. NE is then converted to E in cytoplasm by the enzyme phenylethanolamine-*N*-methyltransferase (PNMT), which catalyzes the transfer of a methyl group from *S*-adenosylmethionine. The enzyme PNMT, confined mainly to the adrenal glands, is activated by high concentrations of cortisol, which is found only in the portal blood draining the adrenal cortex and supplying the adrenal medulla. Thus the regulatory role of cortisol for epinephrine synthesis in the adrenal medulla shows the adrenocortical and hypothalamic-adrenocortical axis (HPA axis) control of adrenomedullary functions. After synthesis, epinephrine returns to chromaffin granules, where it is stored until its release in response to a stimulus.

Norepinephrine and epinephrine are secreted into circulation from the adrenal medulla, and norepinephrine is also secreted locally as a neurotransmitter by sympathetic nerve endings in response to stress and other stimuli; both perform important functions in neural and endocrine integration. Thus, while NE and DA are synthesized largely in the CNS and sympathetic nerve endings, E is synthesized mainly in the adrenal medulla. The pathways for the synthesis of NE and E are shown in Fig. 4.

Norepinephrine has been associated with regulation of mood, anxiety, anger, and paradoxical sleep or rapid eye movement (REM), in contrast to serotonin, which is related to slow sleep (SS). Both NE and E influence a number of physiological functions of peripheral target organs, including smooth muscles, adipose tissue, liver, heart, and myometrium, and blood pressure through two types of receptors, α and β [61]. These receptors have been named according to the potency of various agonists that activate, and that of antagonists that block, the receptor functions [62]. Among these two types of adrenergic receptors, β-receptors are most sensitive to isoproterenol and least to NE, in contrast to α receptors. Catecholamine binding to

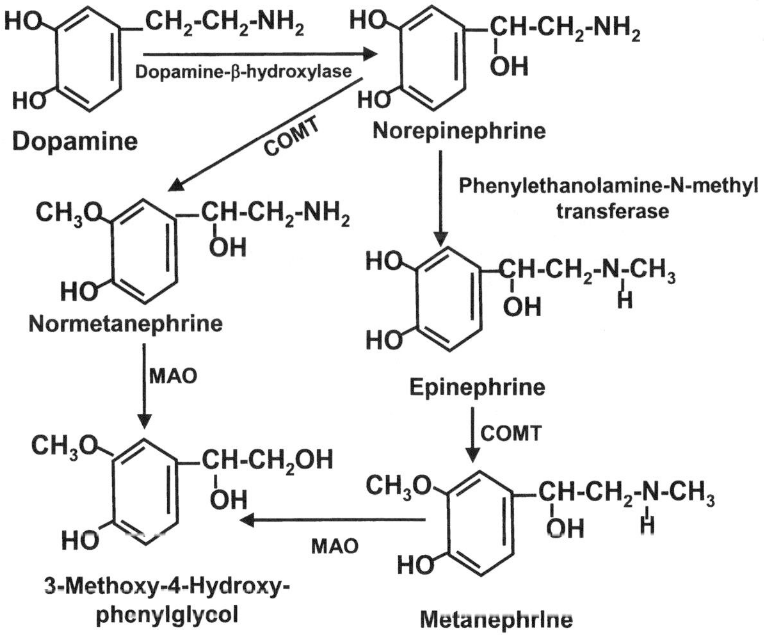

FIGURE 4 Norepinephrine and epinephrine synthesis and metabolism.

receptors activates the membrane-bound adenylate cyclase system, resulting in an increase in 3′5′-cyclic AMP concentration.

NE and E are metabolized primarily by two enzymes, COMT and MAO, respectively [63]. COMT converts NE and E to normetanephrine and metanephrine, respectively, both of which are in turn converted to vanillylmendalic acid (VMA) after oxidation by MAO. VMA is the major metabolite of NE as well as of E in the peripheral system and is measured in the urine as an index of sympathetic nervous system function and for the diagnosis of tumors such as pheochromocytomas and neuroblastomas. In the brain, the major metabolite of NE is 3-methoxy-4-hydroxylphenylglycol (MHPG), which is formed by the action of the enzymes MAO and COMT on NE. MHPG is conjugated to sulphate, which can diffuse from the brain to the general circulation. The conjugated metabolite is excreted in the urine, which is considered to reflect directly the norepinephrinergic activity in the brain. All these metabolites, metanephrine, normetanephrine, VMA, and MHPG, are secreted into blood and are also excreted in the urine as free metabolites as well as conjugated sulphates and their glucorinide derivatives, reflecting the total body activity of NE and E.

4.3 Serotonin

Since the discovery of a vasoconstrictor substance present in serum, and in the rat brain by Twarog and Page [64], serotonin has been found to play an important role in regulation of many physiological and behavioral functions, including vascular control, arterial disease, and in thrombus formation and its inhibition, sleep, control of appetite, body temperature, sexual activity, mood and pain [65,66]. Like catecholamines, serotonin is also present both in the CNS and in the peripheral systems. In the CNS, the cell bodies of the serotonin-containing neurons are localized in a series of raphe nuclei in the lower midbrain region and pons, as mentioned above. These nuclei have the highest concentration of serotonin [67], and axons from these neurons ascend primarily in the medial brain bundle and give off terminals in all brain regions, with the major proportions ascending in the hypothalamus and lowest tracts going to the cerebellum and cerebral cortex [68]. Serotonergic neurons possess specific characteristics that are different from other neurons in the way they form contacts with the other neuronal systems. For example, serotonergic neurons appear to impinge on other neurons even through nonsynaptic contacts, whereas DA- and NE-containing neurons form true synapses at the point of their contact with serotonergic fibers. Furthermore, serotonin is released from the cell bodies as well as from dendrites into the extracellular space with important functional significance, unlike other neurotransmitters, which are released exclusively from the nerve terminals. Moreover, extracellular synaptic 5-HT is also secreted into the cerebrospinal fluid in a similar manner as DA and NE. Concentration of 5-HT in CSF represents a balance between multistep processes, such as synthesis, breakdown, release, and reuptake, taking place in the presynaptic neurons as well as utilization of 5-HT by the postsynaptic neuron. A defect in one or more of these steps may contribute to a change in CSF 5-HT concentration. In a recent study, we have demonstrated a decrease in the concentration of 5-HT in CSF of patients with HIV-1 infection [42]. Although the mechanisms for this decrease are not clear, earlier reports have demonstrated low levels of 5-hydroxyindoleacetic acid (5-HIAA) in CSF of patients with depression and those who attempt or commit suicide [23,24].

Serotonin is synthesized in the brain and in enterochromaffin cells in the peripheral system from the dietary precursor aminoacid L-tryptophan. The rate-limiting enzyme for serotonin synthesis is tryptophan-5-hydroxylase, which converts tryptophan to 5-hydroxytryptophan (5HTP), which is then decarboxylated to 5-hydroxytrptamine (5-HT) by the enzyme 5-hydroxytryptophan decarboxylase in the presence of pyridoxal phosphate (vitamin B_6) as a cofactor. After synthesis, serotonin is accumulated in the storage vesicles of serotonergic neurons by a highly specific energy-dependent serotonin uptake system similar to that of norepinephrine and dopamine, operating at the level of presynaptic neuronal membrane. The storage of 5-HT in vesicles within the nerve endings can be disrupted by drugs that are serotonin releasers such as D,L-fenfluramine [69] and reserpine [70]; the latter is an antihypertensive drug. Although the uptake systems of serotonin and catecholamine can be inhibited by some of the same drugs, especially the tricyclic antidepressants, there are differences in the relative affinities of various antidepressants for the neurons that contain serotonin and catecholamine.

5-HT is metabolized to 5-hydroxyindoleacetic acid by the mitochondrial enzyme, MAO (Fig. 5). The steps involved in the breakdown of 5-HT may undergo diversification based on the internal environment of the organism. For instance, there may be an excessive breakdown of 5-HT to 5-HIAA in the presence of increased activity of MAO as found in some types of depression. The patients with depression due to increased activity of MAO respond effectively to treatment with MAO inhibitors, which results in an increased availability of 5-HT. Functions of 5-HT may also become altered in conditions of psychological distress accompanied by vitamin B_6 deficiency, a cofactor involved in 5-HT synthesis [71].

Decrease in the level of 5-hydroxyindoleacetic acid (5-HIAA), the metabolite of 5-HT, has been reported in CSF of patients with depression and those who attempt or commit suicide [23,24]. In addition to its role in

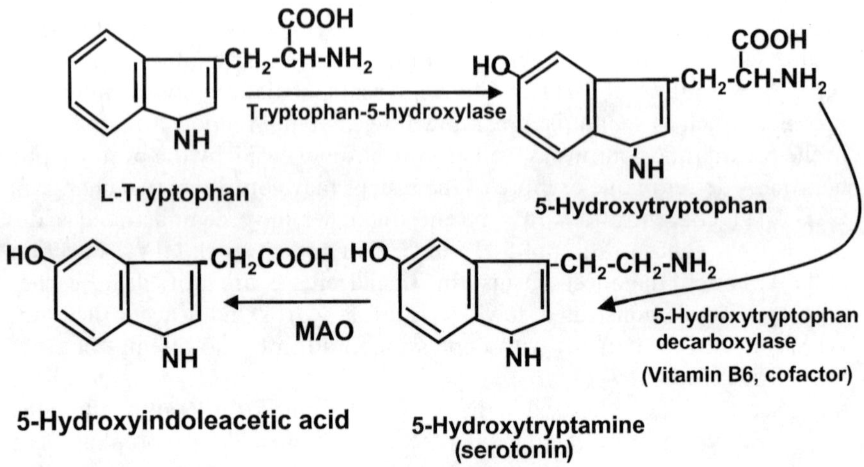

COOH
CH_2-CH-NH_2
NH
L-Tryptophan

Tryptophan-5-hydroxylase

HO
COOH
CH_2-CH-NH_2
NH
5-Hydroxytryptophan

5-Hydroxytryptophan
decarboxylase
(Vitamin B6, cofactor)

HO
CH_2-CH_2-NH_2
NH
5-Hydroxytryptamine
(serotonin)

MAO

HO
CH_2COOH
NH
5-Hydroxyindoleacetic acid

FIGURE 5 Serotonin synthesis and metabolism.

depression, dysregulation of serotonin function in the CNS has been associated with a number of other behavioral problems including aggression, obsessive compulsive disorders, schizophrenia, autism, alcoholism, sleep disorders, and suicide [24,25,72,73]. Changes in the metabolic pathways of tryptophan, the precursor amino acid of serotonin, have also been observed in certain disease conditions, such as in HIV-1 infection, where tryptophan is metabolized predominantly by the L-kynurenine pathway, producing an excessive amount of quinolininc acid instead of serotonin. Quinolininc acid, the end product of the alternate pathway of tryptophan metabolism, is a highly neurotoxic compound and has been implicated as one of the factors responsible for neurotoxicity observed in HIV-1 infection [74].

In the peripheral serotonergic system, there is a large body of evidence showing alteration in platelet 5-HT concentration, 5-HT uptake kinetics, receptor binding, and transporter in various psychiatric conditions. Thus, in a recently published study, we have reported that platelet 5-HT concentration [75] and kinetics of 5-HT uptake are altered in patients with Alzheimer's disease [75,76] and depression [77,78]. On the other hand, in schizophrenia there is an increased concentration of 5-HT in the brain, whole blood, and platelet [34,35].

Various functions of 5-HT are executed by its binding to different receptors in a similar fashion as the other neurotransmitters such as dopamine. There are multiple 5-HT receptors, which have been identified as presynaptic as well as postsynaptic receptors and include 5-HT_{1A}, 5-HT_{1B},

5-HT_{1C}, 5-HT_{1D}, 5-HT_2 and 5-HT_3, 5-HT4, and 5-HT5 [79,80]. These receptors are present in different areas of the brain and in some cases function in concert with other neurotransmitter receptors for regulating different physiological as well as behavioral functions. In the peripheral system, 5-HT_2 receptors are present in platelet and blood vessels.

The relationship between behavioral disorders and serotonin functions have been mostly derived from studies carried out with pharmacological agents such as tricyclic compounds, monoamine oxidase enzyme inhibitors, and serotonin uptake inhibitors. The effectiveness of these compounds depends on how well they regulate brain serotonin concentrations or other functions such as kinetics of 5-HT uptake, receptor numbers, receptor binding capacities, and signal transduction mechanisms. Although in animal models brain has been directly used to study the levels of tryptophan, 5-HT concentration, and its turnover rate in terms of metabolite concentration, and to study receptor binding, such studies are not possible in human during life. In human the closest window to the CNS serotonergic and catecholaminergic functions is provided by measuring their concentration in CSF. Analysis of other body fluids, including blood, urine, and to an extent saliva, also provide valuable information about functions of biogenic amines in the CNS and peripheral systems. Concentration of serotonin in human CSF is extremely low, so serotonin function in the brain has mostly been estimated in terms of its major metabolite, 5-hydroxy indoleacetic acid (5-HIAA), which has a relatively higher CSF concentration. 5-HIAA is also excreted in the urine, and its measurement in urine collected during 24 hours provides a profile of overall activity of serotonergic system in the brain and the peripheral systems.

In the peripheral system serotonin is present in various tissues including gut and blood platelets. Measurement of serotonergic activity in blood platelet has provided valuable markers for central presynaptic activity. Blood platelet is considered a peripheral model for the central presynaptic neuron because of certain common characteristics between the functions of these two cell types [81–83]. These characteristics include regulation mechanism for serotonin concentration in whole blood and platelets, uptake of 5-HT by platelet from circulation by the active energy-dependent mechanism, storage of 5-HT in storage granules, its release in response to environmental demands such as stressful situations or pharmacological agents, [^{3}H]-imipramine binding to platelet membrane, 5-HT_2 receptors present on platelet membrane, activity of platelet serotonin transporter, and platelet MAO activity. One or more of these characteristics have been used as valuable markers for diagnosis of depression and monitoring of response to pharmacotherapy [16,84–86].

5 MARKERS OF BIOGENIC AMINE FUNCTIONS

A number of markers are available for assessment of biogenic amine functions in the CNS as well as in the peripheral systems. For example, the rate limiting enzymes L-tyrosine and L-tryptophan hydroxylase can be used as markers for availability of precursors. On the other hand, precursor amino acids in blood and plasma can be used as markers. Other markers that are used during life for the diagnosis as well as treatment of a disease include measurement of biogenic amines in different body fluids such as CSF, urine, blood, and plasma. There are also markers available in the peripheral cell systems that mimic some of the functions of biogenic amines in the CNS. For example, platelets are considered markers for the central presynaptic serotonergic neurons, and some of the findings reported in platelets have been shown to reflect serotonergic functions in the CNS. However, some markers, such as the levels of biogenic amines in the CNS tissue, changes in receptor densities and their distribution, and altered binding capacities of receptors to their specific amines, can be used only in postmortem brain tissue. These markers may not provide information for treatment of the individuals during life, but they provide evidence for better understanding of the changes in the CNS mechanisms in various disease states. The methodologies described briefly in the following pages have been used for measurement of some of the markers and include concentrations of biogenic amines and their metabolites in tissues and body fluids using various analytical technologies.

6 MEASUREMENT OF BIOGENIC AMINES

Various investigations carried out to date regarding biogenic amine functions have used a number of technologies. These technologies have provided valuable resources for unraveling the mechanisms of a number of regulatory roles played by these amines and their metabolites at the physiological, emotional, and behavioral levels. While their localization and distribution has been identified by using histochemical, autoradiography, and electron microscopy, their quantitation in tissues and body fluids has been accomplished by biological, chemical, and various chromatographic techniques. Moreover, various recent technological developments have provided methods with greater sensitivity and specificity, and in addition they require a very small quantity of tissue or fluid compared to that needed in the earlier methods. These newer technologies include gas liquid chromatography–mass spectrometry, radioimmunoassay, enzymatic linked immunosorbent assay (ELISA), and high-performance liquid chromatography equipped with fluorometric (HPLC-FL), UV, or electrochemical detectors (HPLC-ECD). A short description of various assay methods used in the earlier and recent investigations is given below.

6.1 Biological Methods

The biological methods used in the earlier studies for measurement of norepinephrine and epinephrine were extremely sensitive and were used to estimate catecholamines of biological fluids and tissue extracts. Advantages of biological methods over the other methods including chemical methods were that they did not require the laborious preparations necessary for other assay methods. However, after other assay methods became available, biological methods were not used routinely for biogenic amines assays.

6.2 Histochemical Methods

Histochemical methods combined with fluorescence techniques are an important tool for the detection of monoamines (DA, NE, E, and 5-HT) in cerebral and extracerebral tissues. The technique developed by Falk et al. [11] involves visualization of monoamines in situ by condensation of these amines with formaldehyde vapors in the presence of a dried protein film, resulting in intense fluorescent products from hydroxylated phenylethylamines, indolylamines, and their respective amino acids. When tissue sections containing these compounds are exposed to formaldehyde vapors, the fluorescent derivatives formed by the reaction between the amino group of various catecholamines and formaldehyde can be distinguished by their characteristic excitation and emission spectra. While catecholamines give highly fluorescent green product with an activation wavelengh of 390–410 mμ, serotonin gives a yellow product emitting at a wavelength of 510–520 mμ. By these methods, monoamines can be detected in concentrations of 1–5 μg/g protein. These techniques have been useful for mapping the catecholaminergic pathways, especially in studies involving retrograde and anterograde degeneration after mechanical or chemical lesions. Distinguishing dopamine from epinephrine and norepinephrine in the tissue slices was difficult, but the problem was overcome by increasing the sensitivity of fluorescence by using glyoxylic acid instead of formaldehyde, which gave more intense fluorescence products; catecholamine axons could be followed in normal tissue without using pharmacological agents for increasing the tissue concentration.

This method was found to have limited use for measuring catecholamines in human brain tissue, since the processing for histochemical visualization at 45 minutes to 1 hour 30 minutes after death did not give distinguishable fluorescence. The measurable fluorescence for endogenous catecholamines could be obtained in human brain tissue only within 45 minutes of death, but not after more than 45 minutes. However, when tissue was obtained within 30 minutes of death and immediately subjected to freeze-drying, then catecholamine fluorescence could be visualized in sections prepared from various regions of brain including the thalamic and hypothalmic regions, caudate nucleus, nucleus accumbens, and some nerve terminals [87]. Another important use of the histochemical fluorescent method has been in the study of the neuronal uptake of monoamines, DA, NE, and E, in various regions of the brain. For these studies, lipid soluble monoamine analogs capable of crossing the blood–brain barrier and having characteristics for binding at the cell membrane by displacing intraneuronal monoamines have been used to provide information about uptake, binding, and release mechanisms of monoamines [88].

6.3 Electron Microscopy

Catecholamines have been localized in chromaffin granules, synaptic vesicles, and nerve endings by use of electron-microscopic technology [89]. Enzymes, such as dopamine β-hydroxylase involved in the metabolism of catecholamines, have been localized immunochemically with respect to its antigenic zone on the outer surface of the synaptic vesicle membrane and identified by electron microscopy. Synaptic vesicles are the most characteristic structural components of nerve endings and were first identified by the electron microscope [90] in mammalian brain homogenates, and later from the mitochondrial fraction of rat brain using the sucrose density gradient fractionation technique [90]. Granulated vesicles and chromaffin granules in tissue homogenates of hypothalami after fixing with 1% osmium tetraoxide for 90 minutes were observed via the electron microscope to contain 5-HT, whereas those from sympathetic nerve endings contained NE [91].

6.4 Fluorometric Techniques

Biogenic amines such as catecholamines are converted into their fluorescent derivatives by condensation either with ethylene diamine [92] or with trihydroxyindole [50], which reacts more specifically to the catechol nucleus, and the fluorescent products are measured by their characteristic fluorescence in body tissues and fluids. The fluorometric assay has been used for measuring simultaneously catecholamines and their metabolites from nervous tissue after separation on sephadex G10 column [93]. Serotonin has also been measured in blood using HPLC combined with fluorometric detection [36].

6.5 Enzymatic Radiochemical Methods

Radioenzymatic methodolgy has been used for the measurement of plasma levels of catecholamines, norepinephrine and epinephrine, the markers for sympathetic nervous system activity that regulate many physiological functions. In this method, there are four basic steps involved for the entire assay in body fluids such as plasma: (1) incubation of the plasma sample with the radioactive *S*-adenosyl-*L*-methionine-^{14}C (SAM) [94] or SAM-^{3}H methyl [95] in the presence of the enzyme COMT, to transfer the radioactive methyl group from SAM to an endogenous catecholamine acceptor molecule to form a radioactive *O*-methyl catecholamine; (2) extraction and separation of the radioactive *O*-methyl derivatives from the incubation mixture and the excess of SAM; (3) thin-layer or column chromatographic separation of the radioactive *O*-methyl catecholamines; and (4) oxidation of the susceptible radioactive vanillin, which is then measured by scintillation counting. This method is extremely sensitive and specific for catecholamine measurement in plasma and was used initially to delineate the relationship between plasma levels of norepinephrine and peripheral sympathetic nerve activity [96].

6.6 Dansylation

Catecholamines have also been measured after separation from the tissue extracts by thin-layer chromatography and converting them into highly fluorescent dansyl derivatives. Dansyl compounds are used as probes for identification of various tissue proteins. Dansyl compounds upon reaction with proteins form dansyl–amino acid derivatives that have specific fluorescent characteristics. Thus dansyl chloride would react with tyrosine or tryptophan to form highly fluorescent compounds. Since the fluorescence of dansyl–amino acid derivatives is relatively labile, radio labeling of dansyl chloride has been found to provide the most reliable and stable dansyl derivative, which has facilitated its use in metabolic studies.

6.7 Positron Emission Tomography (PET); Evaluation of Monoamine Systems

The recent development of newer technologies includes positron emission tomography (PET), which is used for investigations into various metabolic processes in the living body; there has been extensive use of PET for investigations of biogenic amine functions during life [97]. PET is an imaging method used to track the regional distribution and kinetics of chemical compounds labeled with short-lived positron-emitting isotopes in the living body [98]. PET was the first technology that was used for direct measurement of components of the dopamine system in the living human brain. For the purpose of studying catecholamine metabolism in the peripheral system such as adrenal and the heart, Christman et al. [99] labeled dopamine with short-lived ^{11}C to obtain dopamine hydrochloride-1-C^{11}. Since dopamine does not cross the blood–brain barrier, imaging studies of dopamine in the living brain have used indirect markers such as radiotracers to label dopamine receptors, dopamine transporters, precursors of dopamine, or chemical compounds that have the specificity of the enzymes that degrade synaptic dopamine. This method is accurate and useful but much too expensive for routine analysis. The other method used for the analysis of metabolic activity during life is magnetic resonance imaging (MRI), which is also sensitive but too expensive for routine analysis.

6.8 Gas–Liquid Chromatography

Measurements of monoamines and their metabolites in human body fluids is important for the diagnosis of affective disorders, pheochromocytoma, and other tumors derived from the neural crest [38]. However, it is not feasible to measure monoamines (NE, E, DA, 5-HT) or their metabolites in the central nervous system in humans during life. Although CSF is considered to represent an approximate central activity of monoamines, obtaining CSF is also an invasive procedure and is not performed routinely. Therefore most studies have examined the central monoamine activity in 24 hour urinary excretion of biogenic amines and their metabolites. Gas–liquid chromatography (GLC) has been used to measure biogenic amines in body fluids, including urine. GLC has been used to separate, characterize, and quantify urinary catecholamine metabolites after conversion into their trifluoroacetate derivatives [100]. After extraction of the derivatized compounds with ethylacetate, the samples are injected into a gas chromatogram for analysis. The peaks obtained are found to be proportional to the concentration of metabolite in the urine extract [101]. The metabolite MHPG of norepinephrine has been measured by its acetylated derivatives using GLC with an electron capture detector [102,103].

6.9 High-Performance Liquid Chromatography

In most of the technologies described above, indirect methods were used for measuring biogenic amines in the extracts of tissues or body fluids, which required long and labor-intensive purification and derivatization steps prior to their analysis. For example, fluorometric assays involved derivatization of catecholamines to their trihydroxyindoles. This derivatization was difficult to control and resulted in large variation of the values reported by different laboratories [104,105]. On the other hand, GLC has been used for measuring VMA and HVA in urine [106], and although its sensitivity and specificity could be increased by combining it with mass spectrometry for measurement of catecholamines and their metabolites, as well as indolealkylamines of the brain tissue [107], this method is expensive and time-consuming and is not suitable for routine analysis [108]. Methods such as radioenzymatic assays, where a radioactive methyl group is enzymatically transferred to the catecholamine, are specific and accurate, but they require lengthy purification and preconcentration steps before the enzymatic steps can be carried out

[109,110]. Among the recent technologies developed for measurement of biogenic amines, high-performance liquid chromatography with electrochemical detection (HPLC-ECD) has gained popularity because of its sensitivity, specificity, and ease of sample preparation and analysis. Moreover, this technology can be used for the analysis of biogenic amines, their precursor amino acids, and their metabolites in the brain and other tissues and body fluids including CSF, plasma, and urine, without any chemical transformation or derivatization steps [111–113]. An additional advantage is that a sample can be analyzed for a single analyte or for a number of compounds simultaneously. Alterations in the concentrations ranging from nanogram to picogram quantities can be detected by this method.

The use of HPLC-ECD for analysis of biogenic amines is based on the principle of reverse-phase liquid chromatography combined with detection by the redox principle. In a reverse-phase system, an analyte has a stronger affinity for ionic components in the mobile phase (elution solution) than for its binding to the matrix of the resolve column packed with octadecylsilane, C_{18} on microplate 5 μ spherical silica gel. Furthermore, the electrochemical detection is specific for substances that can be oxidized or reduced. Among the chromatographic detection systems, electrochemical detection is unique in that the substance to be detected is chemically altered by the detection process. The flow cell contains a working electrode, which is held at a constant potential relative to the mobile phase, creating a condition that is appropriate for a redox reaction. As the substance to be detected moves through the cell after its elution from the column, it passes over the working electrode and is oxidized or reduced. While the reaction occurs, the data module measures the resulting current. Thus for a specific substance in a given environment, the current varies as a function of concentration of the substance. By selecting the appropriate electrode potentials, the detection process can be utilized for the substance(s) to be analyzed.

However, there are a number of factors influencing the elution process that can affect the results. Controlling these factors is an essential part of the procedure for reliable analysis, since affinity of the analyte for the components in the mobile phase versus the column matrix depends upon (1) the concentration and combination of ionic pairing components in the mobile phase; (2) the pH of the mobile phase; (3) the flow rate of the mobile phase; and (4) the concentration of methanol in the mobile phase.

The HPLC-ECD system generally consists of (1) a programmable solvent delivery system maintained under constant high pressure; (2) an injector with an external or internal loop for injection of a known volume of extract for an assay; (3) an electrochemical detector with a glassy carbon electrode, auxillary electrodes, and an Ag/AgCL reference electrode; (4) a programmable integrator, the data module that integrates the pulses in the form of peaks, and the areas corresponding to the concentration of amines in the extract injected; (5) reverse phase chromatography columns packed with spherical silica particles of specific size and characteristics (5 μ, C_{18}) for separation of biogenic amines and/or their metabolites in the tissues or fluid extracts.

Furthermore, for the desired performance of HPLC-ECD with respect to its sensitivity and specificity for biogenic amines, it is important that stringent precautions be observed in the preparation of sample extracts, and that the equipment be maintained in order to provide reliable and reproducible results. For the preparation of a sample for analysis, whether it is a tissue or a body fluid, one of the important precautions is the addition of antioxidant for all biogenic amine analysis. Catecholamines are unstable at low concentrations at room temperature [114], so freshly obtained samples are kept chilled and antioxidants are added as preservatives. Sodium ethylenediaminetetraacetate (Na_2EDTA) and sodium metabisulphite are used as the preservatives of choice and are added prior to storing samples at −70°C. When fresh samples are analyzed, preservatives are added before extraction of samples for analysis. Similarly, serotonin and its metabolite 5-HIAA are unstable at room temperature and are sensitive to light. Samples of tissues, platelets, or plasma containing serotonin are stored at −70°C, in amber-colored eppendorf vials for protecting samples from exposure to light. Samples stored at −70°C for a period longer than 6 months to 1 year have been found to lose a significant percentage of serotonin and 5-HIAA.

The extracts of samples are prepared using different procedures for tissue, CSF, plasma, platelets, blood, and urine, since different ranges of interfering compounds are present in the tissues and body fluids. Using the Waters (Milford, MA, USA) HPLC-ECD system, we have reported simultaneous analysis of free urinary catecholamines, their metabolites, and serotonin and its metabolite, in the same sample of urine using simplified purification and extraction procedures. These metabolites included 4-hydroxy-3-methoxy-mandelic acid (HMMA), 4-hydroxy-3-methoxy-phenylglycol (MHPG), normetanephrine (NM), metanephrine (M), and 5-hydroxyindoleacetic acid (5-HIAA) [115]. In a number of separate studies, we have also reported simultaneous measurement of free urinary catecholamines [116], and metabolites, MHPG and vanillymandelic acid [117], normetanephrine and metanephrine, the intermediary metabolites of catecholamine [118]. An important feature in these measurements has been the use of internal standards and calibration of the HPLC-ECD system with an extract prepared from a catecholamine-free or metabolite-free urine sample. Prior to extraction, the pH of the urine was adjusted and HPLC-grade water was used for preparing standards and all other solutions including the mobile phase. For the assay of serotonin, HPLC-ECD was calibrated with

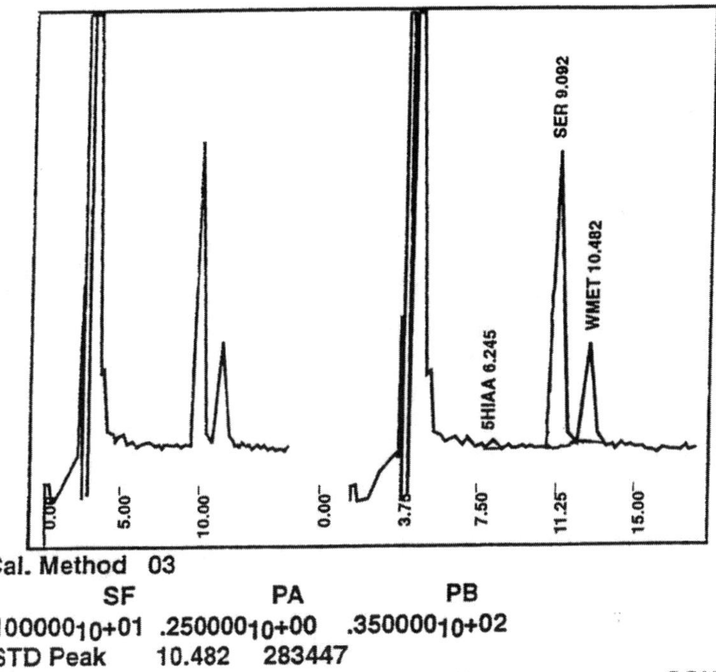

Cal. Method 03

SF	PA	PB
$.100000_{10}{+01}$	$.250000_{10}{+00}$	$.350000_{10}{+02}$

ISTD Peak 10.482 283447

No.	Name	RT	A or H	CONC
1	5HIAA	6.245	6312	3.5145
2	SER	9.092	810215	214.2852
3	NMET	10.482	283447	140.0000
	TOTAL		1099974	357.7998

FIGURE 6 A typical HPLC-ECD profile of serotonin, its metabolite 5-hydroxyindoleacetic acid (5-HIAA), and the internal standard *N*-ω-methyl-5-hydroxytryptamine (NMET) in an extract of platelet-rich plasma (PRP) as carried out in our laboratory. Peaks of 5-HIAA, serotonin (SERT), and internal standard (NMET), with their retention times, are shown with the area of each peak and the concentration of each compound. Calculations for the concentration of each analyte were carried out using the ratio method (ratio of the area *A* of the unknown to that of the internal standard) by the preprogrammed data module, and the calculation method 03 was used, as shown below the figure. Concentration is expressed as ng/mL PRP (as shown in the right column below the figure). Similar profiles are obtained for plasma and CSF (not shown here).

serotonin-free plasma, and *N*-ω-methyl-5-hydroxytryptamine (NMET) was used as an internal standard. A typical HPLC-ECD profile of the analysis of serotonin in platelet-rich plasma is shown in Fig. 6. A similar profile was obtained for plasma and CSF. We also measured the levels of dopamine [119] and serotonin [120] in CSF of HIV-infected patients by HPLC-ECD and obtained a similar profile. Our results showed a significant decrease in the concentration of both these biogenic amines in the CSF of HIV-1 infected individuals, compared to that in uninfected control subjects. Although the mechanisms related to the decrease are not clear, low levels of both these amines in CSF may be indicative of dysregulated CNS serotonergic and dopaminergic activity in HIV-1 infection.

As mentioned above, in the peripheral system serotonin circulating in blood is taken up by the platelets by the same uptake mechanism as the presynaptic neuron. Therefore it is of interest to monitor platelet serotonin in patients with depression and use this parameter as a marker for the CNS serotonergic activity. Furthermore, HPLC-ECD was also used for the analysis of catecholamines in plasma, as a peripheral marker of sympathetic nervous system activity in HIV-1 infection. It has been reported from our laboratory that release of norepinephrine in response to laboratory cold pressor challenge is blunted in HIV-1-infected individuals compared to that in the control subjects. Cold pressor is an alpha-adrenergic challenge, and it requires an individual to immerse one of his or her limbs in an ice–water mixture for two minutes. Samples of blood are collected at the baseline and at specific intervals after the limb is removed from the ice–water mixture, for investigating the pattern of catecholaminergic reactivity including the recovery after the limb has been removed. This test stimulates the release of catecholamines in blood from the peripheral nerve endings. The observation that HIV-1-infected individuals have a blunted response to this challenge is significant since it suggests that there may be peripheral neuropathy in HIV-1 infection [121]. Our investigations regarding peripheral markers of serotonin function have included measurement of the concentration of serotonin in platelet and plasma of normal subjects [122], in patients with depression [123–125], in women in relationship to aging [126], and in patients with Alzheimer's disease [75].

7 SUMMARY

Since the discovery of monoamines about half a century ago, their role in various physiological processes and behavioral regulations is being increasingly appreciated. Although an enormous amount of literature has accumulated on various aspects of their functions in health and disease, we are still at the cross roads for the precise treatment of behavioral and psychiatric disorders. Even with a number of receptors, agonists and antagonists, amine uptake inhibitors, and other drugs acting as enzyme inhibitors and activators already on the market and more in the pipelines in the drug industry, relief from various disorders still seems to be far off. Measurement of biogenic amines during life is an important aspect of assessment of their functions, and in humans the best sources of this information are sensitive and specific methods for measuring ultramicro-concentrations appearing in blood plasma, urine, or CSF in order to delineate disregulated mechanisms in various human conditions. Newer technologies developed in recent years including HPLC-ECD seem to fulfill the promise of ultrasensitivity. In this chapter an attempt has been made to provide a brief review of earlier and some of the newer methods for measurement of biogenic amines and their metabolites.

ACKNOWLEDGMENTS

This work was supported in part by the National Institute of Health (NIH) grants # DA12792, DA13550, IP50CA86966-NIC, P01MH-49548 (NIMH), and NS43982. The author is thankful to Professor Mahendra Kumar for valuable discussions.

REFERENCES

1. JH Schwartz, RE Bloom, RE Roth. Chemical basis of synaptic transmission. In: ER Kandel, JH Schwartz, eds. Principles of Neuroscience. New York: Elsevier, 1982.
2. MJ Dennis, AJ Harris, SW Kuffler. Synaptic transmission and its duplication by focally applied acetylcholine in parasympathetic neurons in the heart of frog. Proc R Soc Lond 177:509–539, 1971.
3. P Davies, AFJ Moloney. Selective loss of cholinergic neurons in Alzheimer's disease. Lancet 2:1403, 1976.
4. EA Kravitz. Acetylcholine, y-aminobutyric acid and glutamic acid: physiological and chemical studies related to their roles as neurotransmitters. In: GC Quartton, T Melne Chuck, FO Schmitt, eds. The Neurosciences. New York: Rockefeller University Press, 1967.
5. TL Perry, S Hansen, M Kloster. Hungtingtons's chorea: deficiency of gamma-aminibutyric acid in brain. N Eng J Med 288:337–342, 1973.
6. T Takahashi, M Otsuka. Regional distribution of substance P in the spinal cord and nerve roots of cat and effects of dorsal root sections. Brain Research 87:1–11, 1975.
7. J Hughes, TW Smith, HW Koteritz, LA Fothergill, BA Morgan, HR Morris. Identification of two related pentapeptides from the brain with potent opiate agonist activity. Nature 258:577–579, 1975.

8. SW Kuffler. Slow synaptic responses in autonomic ganglia and the pursuit of a peptidergic transmitter. J Exp Biol 89:257–286, 1980.
9. NA Hillarp, K Fuxe, A Dahlstrom. Demonstration and mapping of central neurons containing dopamine, noradrenaline and 5-hydroxytryptamine and their reactions to psychopharmaca. Pharmacol Rev 18:727, 1966.
10. NE Anden, A Dahlstrom, K Fuxe, K Larsson, L Olson, U Ungerstedt. Ascending monoamine neurons to the telencephalon and diencephalon. Acta Physiolo Scand 67:313, 1966.
11. B Falk, NA Hillarp, G Thieme, A Torp. Fluorescence of catecholamines and related compounds condensed with formaldehyde. J Histochem Cytochem 10:348–354, 1962.
12. SM Stahl. The human platelet: a diagnosis and research tool for the study of biogenic amines in psychiatric and neurologic disorders. Arch Gen Psychiatry 34:509–516, 1977.
13. SW Kuffler, JG Nicholls, AR Martin. A cellular approach to the function of the nervous system: In: From Neuron to Brain. 2d ed. Sunderland, MA: Sinauer Associates, 1984.
14. H Zimmerman. Vesicles recycling and transmitter release. Neuroscience 4:1773–1804, 1979.
15. TA Mellman, AM Kumar, R Kulick-Bell, M Kumar, B Nolan. Nocturnal/daytime urine noradrenergic measures and sleep in combat related PTSD. Biol Psychiatry 38:174–179, 1995.
16. HY Meltzer, MT Lowy. The serotonin hypothesis of depression. In: HY Meltzer, ed. Psychopharmacology: The Third Generation of Progress. New York: Raven Press, 1987, 513–526.
17. JJ Schildkraut, SS Kety. Biogenic amines and emotions. Science 156:21–30, 1967.
18. MM Ward, IN Mefford, SD Parker, et al. Epinephrine and norepinephrine responses in continuously collected human plasma to a series of stressors. Psychosom Med 45:471–478, 1983.
19. DG Morgan, PC May, CE Finch. Dopamine and serotonin systems in human and rodent brain. Effects of age and neurodegenerative disease. J Am Geriatric Soc 35:334–345, 1987.
20. W Koella. Serotonin and sleep. In: NN Osborne, M Hamon. Neuronal Serotonin. New York: John Wiley, 1988, pp 153–170.
21. TD Brewerton. Toward a unified theory of serotonin dysregulation in eating and related disorders. Psychoneuroendocrinology 20(6):561–590, 1995.
22. JP Zak, JA Miller, DV Sheehan, SL Fanous Balsam. The potential role of serotonin reuptake inhibitors in the treatment of obsessive compulsive disorders. J Clin Psychiatry 49(8 suppl):23–29, 1988.
23. GL Brown, MH Ebert, PF Goyer, et al. Aggression, suicide, serotonin: relationship to CSF amine metabolites. Am J Psychiatry 139:741–746, 1982.
24. M Asberg, L Traskman, P Thoren. 5HIAA in the cerebrospinal fluid; a biochemical suicide predictor? Arch Gen Psychiatry 33:1193–1197, 1976.

25. GN Pandey, SC Pandey, PG Janicak, et al. Platelet serotonin-2 receptor binding sites in depression and suicide. Biol Psychiatry 28:215–222, 1990.
26. A Carlsson, R Adolfsson, SM Aquilonius, et al. Biogenic amines in human brain in normal aging, senile dementia and chronic alcoholism. In M Goldstein, ed. Ergot Compounds and Brain Function: Neuroendocrine and Neuropsychiatric Aspects. New York: Raven Press, 1980, p 295.
27. R Adolfsson, CG Gottfries, BE Roose, et al. Changes in brain catecholamines in patients with dementia of Alzheimer's type. Br J Psychiatry 135:216, 1979.
28. H Arai, K Kosaka, T Lizuka. Changes in biogenic amines and their metabolites in postmortem brains from patients with Alzheimer's type dementia. J Neurochem 43:388, 1984.
29. J Hardy, R Adolfsson, I Alafuzoff, et al. Transmitter deficits in Alzheimer's disease. Neurochem Int 7:545, 1985.
30. B Winblad, R Adolfsson, A Carlsson, et al. Biogenic amine of patients with Alzheimer's disease. In: S Corkin, ed. Alzheimer's Disease. A Report of Progress. New York: Raven Press, 1982, p 25.
31. H Bernheimer, W Berkmayer, D Hornykiewicz, et al. Brain dopamine and the syndrome of Parkinson and Huntington: clinical, morphological and neurochemical correlations. J Neurol Sci 20:415, 1973.
32. JM Davis. Dopamine theory of schizophrenia: a two-factor theory. In: LC Wynne, RL Cromwell, S Matthyse, eds. The Nature of Schizophrenia: New Approaches to Research and Treatment. New York: John Wiley, 1978, pp 105–115.
33. HY Meltzer. The significance of serotonin for neuropsychiatric disorders. J Clin Psychiatry 52:70–72, 1991.
34. E Garelis, JC Gillinb, N Neff, et al. Elevated blood serotonin concentrations in unmedicated chronic schizophrenic patients: a preliminary study. Am J Psychiatry 132:184–186, 1975.
35. LE DeLisi, LM Neckers, DR Weinberger, RJ Wyatt. Increased whole blood serotonin concentrations in chronic schizophrenic patients. Arch Gen Psychiatry 38:647–650, 1981.
36. GM Andersson, DX Freedman, DJ Cohen, FR Volkman, EL Hoder, P McPhederan, RB Minderaa, CR Hansen, JG Young. Whole blood serotonin in autistic and normal subjects. J Clin Psychol Psychiatr 28(6): 885–900, 1987.
37. JL Haracz. The dopamine hypothesis. An overview of studies with schizophrenic patients. Schizophr Bull 8:438–445, 1982.
38. SK Wadman, D Ketting, PA Voute. Gas chromatographic determination of urinary vanilglycolic acid and vanillactic acid. Chemical parameters for the diagnosis of neurogenic tumors and the evaluation of their treatment. Clin Chim Acta 72:49–68, 1978.
39. JR Berger, M Kumar, AM Kumar, JB Fernandez, B Levine. Cerebrospinal fluid dopamine in HIV-1 infection. AIDS 8:67–71,1994.
40. AM Sardar, C Czudek, GP Reynolds. Dopamine deficits in the brain: The neurochemical basis of Parkinsonian symptoms in AIDS. Neuroreport 7:910–912, 1996.

41. A Nath, JD Geiger. Neurobiological aspects of HIV infection: neurotoxic mechanisms. Prog Neurobiolo 54:19–33, 1998.
42. AM Kumar, JR Berger, C Eisdorfer, JB Fernandez, K Goodkin, M Kumar. Cerebrospinal fluid 5-hydroxytryptamine and 5-hydroxyindoleacetic acid in Hiv-1 infection. Neuropsychobiology 44:13–18, 2001.
43. TR Kosten. Neurobiology of abused drugs. Opioids and stimulants. J Neuron and Mental Disease 178:217–227, 1990.
44. HC Fibiger, JJ Miller. An anatomical and electrophysiological investigation of the serotonergic projection from the dorsal raphe nucleus to the substantia nigra in the rat. Neuroscience 2:975–987, 1977.
45. R Kuczenski. Role of para-chlorophenylalanine on amphetamine and haloperidol-induced changes in striatal dopamine turnover. Brain Res 164: 217–225, 1979.
46. CA Dackis, MS Gold. New concepts in cocaine addiction. The dopamine depletion hypothesis. Neuroscience Biobehavioral Review 9:469–477, 1985.
47. FH Previc. Dopamine and origin of human intelligence. Brain Cognition 41:299–350, 1999.
48. O Hornykiewicz. Dopamine in the basal ganglia: Its role and therapeutic implications (including the clinical use of L-DOPA). Br Med Bull 29:172–178, 1973.
49. CE Finch. Relationship of aging changes in the basal ganglia to manifestation of Huntington's chorea. Ann Neurol 7:406, 1980.
50. S Udenfriend. Fluorescence Assay in Biology and Medicine. New York: Academic Press, 1962.
51. T Nagatsu, M Levitt, S Udenfriend. Tyrosine hydroxylase. The initial step in norepinephrine biosynthesis. J Biol Chem 239:2910–2920, 1964.
52. DM Jackson, A Westling-Danielson. Dopamine receptors: molecular biology, biochemistry and behavioral aspects. Pharmacol Ther 64:291–369, 1994.
53. DR Sibley, FJ Monsma. Molecular biology of dopamine receptors. Trends Pharmacol Sci 13:61–69, 1992.
54. RM Kessler, WO Whetshell, MS Ansari. Identification of extrastriatal D2 receptors in postmortem human brain with [^{125}I]-epidepride. Brain Res 609:237–243, 1993.
55. H Hall, G Sedvall, O Magnusson, J Kopp, C Halldin, L Forde. Distribution of D1 and D2-dopamine receptors and dopamine and its metabolite in human brain. Neuropsychopharmacology 11:245–256, 1994.
56. M Camps, R Cortes, B Gueye, A Probst, JM Palascios. Dopamine receptors in human brain: autoradiographic distribution of D2 sites. Neuroscience 28:275–290, 1989.
57. AM Murray, H Ryoo, JN Joyce. Visualization of dopamine D3-like receptors in human brain with [^{125}I]epidepride. Eur J Pharmacol 227:443–445, 1992.
58. LL Iverson. Catecholamine uptake processes. Br Med Bull 29:130–135, 1973.
59. N Kirshner. Biosynthesis of the catecholamines. In: H Blaschko, G Sayers, AD Smith, eds. Handbook of Physiology. Washington, DC: American Physiological Society, 1975, pp 341–355.

60. S Udenfriend, CR Creveling. Localization of dopamine-β-hydroxylase in brain. J Neurochem 4:350–352, 1959.
61. DM Jenkinson. Classification and properties of peripheral adrenergic receptors. Br Med Bull 29:142–147, 1973.
62. P Headqvist. Role of the alpha receptors in the control of noradrenaline release from sympathetic nerves. Acta Physiol Scand 90:156–165, 1974.
63. J Axelrod. Catechol-*O*-methyltransferase and other *O*-methyltransferases. In: H Blaschko, G Sayers, AD Smith, eds. Handbook of Physiology. Washington, DC: American Physiological Society, 1975, pp 669–676.
64. BM Twarog, IH Page. Serotonin content of some of the mammalian tissues and urine and a method for its determination. Am J Physiology 175:157–161, 1953.
65. EC Azmitia, WMP Azmitia. Awakening the sleeping giant: anatomy and plasticity of the brain serotonergic system. J Clin Psychiatry 52(suppl 12):4–16, 1991.
66. F De Clerk. Differential effects of serotonin on blood vessels and platelets in health and disease. In: A Takada, G Curzon, eds. Serotonin in the Central Nervous System and Periphery. Amsterdam/New York: Elsevier Science, 1995, pp 215–220.
67. A Dahlstrom, K Fuxe. Evidence for the existence of monoamine-containing neurons in the central nervous system. Acta Physiol Scand 62:suppl 232, 1964.
68. U Ungerstedt. Serotonergic mapping of the monoamine pathway of the rat brain. Acta Physiol Scand Suppl 367:1–48, 1971.
69. NE Rowland, J Carlton. Neurobiology of an anorectic drug, fenfluramine. Progress in Neurobiology 27:13–62, 1986.
70. FK Goodwin, MH Ebert, WE Bunney Jr. Mental effects of reserpine in man: a review. In: RI Shader, ed. Psychiatric Complications of Medical Drugs. New York: Raven Press, 1972.
71. T Baldewicz, K Goodkin, D Feaster, N Blaney, M Kumar, AM Kumar, G Shor-Posner, M Baum. Plasma pyridoxin deficiency is related to increase in psychological distress in recently bereaved homosexual men. Psychosomatic Med 60:297–308, 1998.
72. JC Ballenger, FK Goodwin, LF Major, et al. Alcohol and central serotonin metabolism. Arch Gen Psychiatry 36:224–227, 1979.
73. EF Coccaro, LJ Siever, HM Klar, et al. Serotonergic studies in patients with affective and personality disorders. Arch Gen Psychiatry 46:587–599, 1989.
74. MP Heyes, BJ Brew, A Martin, et al. Quinolininc acid in cerebrospinal and serum in HIV-1 infection: relationship to clinical and neurological status. Ann Neurol 29:202–209, 1991.
75. AM Kumar, S Sevush, M Kumar, J Ruiz, C Eisdorfer. Platelet serotonin in Alzheimer's disease. Neuropsychobiology 32:9–12, 1995.
76. AM Kumar, M Kumar, S Sevush, J Ruiz, C Eisdorfer. Serotonin uptake and its kinetics in platelets of women with Alzheimer's disease. Psychiatry Research 59:145–150, 1995.
77. PJ Goodnick, AM Kumar. Pretreatment platelet 5-HT predicts the short term response to paroxetine in major depression. Biol Psychiatry 47(4):846–847, 2000.

78. AM Kumar, PJ Goodnick. Platelet serotonin levels: a marker for treatment with SSRIs in depression. Int Soc Psychoneuroendocrinology, 30th Congress, 1999, p 52.
79. SJ Peroutka. 5-hydroxytryptamine receptors subtypes. Pharmacol Toxicol 67:373–383, 1990.
80. D Julius. Molecular biology of serotonin receptors. Ann Rev Neurosci 14:335–360, 1991.
81. LH Tecott, D Julius. A new wave of serotonin receptors. Current Opinion Neurobiology 3:310–315, 1993.
82. A Pletscher. Platelets as models: use and limitations. Experientia 44:152–155, 1988.
83. M Da Prada, AM Cesura, JM Launay, et al. Platelet as a model for neurons? Experientia 44:115–126, 1988.
84. MS Briely, SZ Langer, R Raisman, et al. Tritiated imipramine binding sites are decreased in platelets of depressed patients. Science 209:303–305, 1980.
85. HM VanPraag. Depression, suicide, and serotonin metabolism in the brain. In: RM Post, JC Ballenger, eds. Neurobiology of Mood Disorders. Vol. 1. Baltimore, MD: Willams and Wilkins, 1984, pp 601–618.
86. SM Paul, M Rehavi, P Skolnick, et al. Depressed patients have decreased binding of tritiated imipramine to platelet 'serotonin transporter'. Arch Gen Psychiatry 38:1315–1317, 1981.
87. JC De la Torre. Methods of studying cerebral monoamines and their enzymes. In: Dynamics of Brain Monoamines. New York: Plenum Press, 1972, pp 16–37.
88. A Carlsson. Modification of sympathetic function. Pharmacological depletion of catecholamine stores. Pharmacol Rev 18:541–549, 1966.
89. GK Aghajanian, EF Bloom. Electron-microscopic autoradiography of rat hypothalamus after intraventricular H-3 norepinephrine. Science 153:308–310, 1966.
90. E DeRobertis, HS Bennett. Some features of the submicroscopic morphology of synapses in frog and earthworm. J Biophys Biocem Cytolo 1:47–58, 1955.
91. JG Wood. Electron localization of amines in central nervous tissue. Nature 209:1131–1133, 1966.
92. H Weil-Malherb, AD Bone. Intracellular distribution of catecholamines in the brain. Nature 180:1050–1051, 1957.
93. NHC Westerink, J Korf. Rapid concurrent automated fluorometric assay of noradrenaline, dopamine, 3,4-dihydroxyphenylacetic acid, homovanillic acid and 3-methoxytyramine in milligrams amounts of nervous tissue after isolation on sephadex G10. J Neurochem 29:697–706, 1977.
94. K Engleman, B Portnoy. A sensitive double isotope derivative assay for norepinephrrine and epinephrine. Cir Res 26:53–57, 1970.
95. PG Passon, JD Penter. A simplified radiometric assay for norepinephrine and epinephrine. Anal Biochem 51:618–631, 1973.
96. PE Cryer. Isotope derivative measurement of plasma norepinephrine and epinephrine in man. Diabetes 25:1071–1082, 1976.

97. HN Wagner, HD Burns, RF Dannals. Imaging DA receptors in the human brain by PET. Science 221:1264–1266, 1983.
98. JS Fowler, AP Wolf. New directions in positron emission tomography. In: JA Bristol, ed. Annual Reports in Medicinal Chemistry 24:277–286, 1989.
99. DR Christmas, RM Hoyte, AP Wolf. Organic radiopharmaceuticals labeled with isotopes of short half life. I: dopamine-hydrochloride-1-^{11}C. J Nuc Med 11:474–478, 1970.
100. D Haroutune, JW Maas. An improved procedure of 3-methoxy-4-hydroxy-phenyl-ethylene glycol determination by gas–liquid chromatography. Anal Biochemistry 35:113–122, 1970.
101. FAJ Muskiet, CG Thamasson, AM Gerdin, DC Fremouw-Ottorangers, TG Nagel, BG Wolthers. Determination of catecholamines and their 3-*O*-methylated metabolites in urine by mass fragmentography with use of deuterated internal standard. Ciln Chem 25:453–460, 1979.
102. S Wilk, SE Gitlow, M Mendiowitz, MS Franklin, HK Karr, DD Clark. A quantitative method for vanillylmandelic acid (VMA) by gas liquid chromatography. Annal Biochem 12:544–551, 1965.
103. S Wilk, SE Gitlow, DD Clark, DH Paley. Determination of urinary 3-methoxy-4-hydroxyphenylethyleneglycol by gas chromatography and electron capture detection. Clin Chim Acta 16:403–408, 1967.
104. P Hjmedahl. Inter-laboratory comparison of plasma catecholamines determination using several different assays. Acta Physiol Scand 527:43–50, 1984.
105. Y Miura, V Campese V Dequattro, D Meyer. Plasma catecholamines via an improved fluorometric assay: comparison with an enzymztic method. J Lab Clin Med 89:421–428, 1977.
106. MS Roginsky, P Riederer, M Sandler. A rapid assay of 4-hydroxy-3-methoxyphenylglycol in urine. Clin Chim Acta 59:255, 1975.
107. Cattabeni F, Koslow SH, Costa E. Gas chromatography–mass spectrometry assay of four indolealkylamines of the rat pineal. Science 178:166–168, 1972.
108. CR Lake, MG Zeigler, IJ Kopin. Use of plasma noradrenaline for evaluation of sympathetic neuronal function in man. Life Sciences 18: 1315–1320, 1976.
109. JM Saavedra, M Brownstein, J Axelrod. A specific and sensitive enzymatic isotopic microassay for serotonin in tissues. J Pharmacol Exp Ther 186:508–515, 1973.
110. M DaPrada, G Zurcher. Simultaneous radioenzymatic determination of plasma and tissue adrenaline, noradrenaline and dopamine within the femtamole range. Life Sci 19:1161–1174, 1976.
111. Y Yui, C Kawai. Comparison of the sensitivity of various post-column methods for catecholamine analysis by high performance liquid chromatography. J Chromatography 206:586–592, 1981.
112. BL Lee, KS Chia, CN Ong. Measurement of urinary free catecholamines using high performance liquid chromatography with electrochemical detection. J Chromatogr 494:303–310, 1989.
113. AM Kumar, M Kumar, JB Fernandez, TA Mellman, C Eisdorfer. A

simplified HPLC-ECD technique for measurement of urinary free catecholamines. J Liquid Chromatography 14(19):3547–3557, 1991.

114. Caurethers, N Conway, P Taggart. Validity of plasma cetecholamine estimations. Lancet 2:62–67, 1970.
115. AM Kumar, JB Fernandez, K Goodkin, N Schneiderman, C Eisdorfer. An isocratic concurrent assay of free metabolites, 4-hydroxy-3-methoxy-mandelic acid, 3-methoxy-4-hydroxyphenylglycol, normetanephrine, metanephrine, and 5-hydroxyindoleacetic acid in same sample of urine extract using HPLC-ECD. J Liquid Chrom Related Technol 20(12):1931–1943, 1997.
116. AM Kumar, M Kumar, JB Fernandez, TA Mellman, C Eisdorfer. A simplified HPLC-ECD technique for measurement of urinary free catecholamines. J Liquid Chromatogr 14(19):3547–3557, 1991.
117. AM Kumar, M Kumar, JB Fernandez, N Schneiderman, C Eisdorfer. An improved assay of urinary catecholamine metabolites, 3-methoxy-4-hydroxyphenylglycol, and vanillylmandelic acid, using high performance liquid chromatography with electrochemical detection. J Liquid Chromatography 16(6):1329–1340, 1995.
118. AM Kumar, M Kumar, JB Fernandez, K Goodkin, N Schneiderman, C Eisdorfer. An isocratic HPLC-ECD assay of urinary normetanephrine and metanephrine. J Liquid Chromatography 18(11):2257–2269, 1995.
119. JR Berger, M Kumar, AM Kumar, JB Fernandez, B Levin. Cerebrospinal fluid dopamine in HIV-1 infection. AIDS 8:67–71, 1994.
120. AM Kumar, JR Berger, C Eisdorfer, JB Fernandez, K Goodkin, M Kumar. Cerebrospinal fluid 5-hydroxytryptamine and 5-hydroxyindoleacetic acid in HIV-1 infection. Neuropsychobiology 44:13–18, 2001.
121. M Kumar, R Morgan, J Szapocznik, C Eisdorfer. Norepinephrine response in early HIV-1 infection. J Acquir Immunodefic Syndr 4:782–786, 1991.
122. AM Kumar, M Kumar, D Krishnaprasad, JB Fernandez, C Eisdorfer. A modified technique for simultaneous measurement of 5-hydroxytryptamine and 5-HIAA in cerebrospinal fluid, platelet and plasma. Life Sceinces 47:1751–1759, 1990.
123. P Goodnick, AM Kumar. Pretreatment platelet 5-HT predicts the short-term response to paroxetine in major depression. Biological Psychiatry 47(9):846–847, 2000.
124. P Goodnick, C Jorge, T Hunter, AM Kumar. Nefazodone treatment in adolescent depression. An open label study of response and biochemistry. Annals Clin Psych 12(2):97–100, 2000.
125. P Goodnick, J Henry, AM Kumar. Neurochemistry and paroxetine in major depression. Biol Psychiatry 37:417–419, 1995.
126. AM Kumar, S Weiss, JB Fernandez, D Creuss, C Eisdorfer. Peripheral serotonin levels in women: role of aging and ethnicity. Gerontology 44:211–216, 1998.

8

Clinical Applications of Affinity Chromatography

William Clarke
Johns Hopkins School of Medicine, Baltimore, Maryland, U.S.A.

David S. Hage
University of Nebraska, Lincoln, Nebraska, U.S.A.

1 INTRODUCTION

Liquid chromatographic methods like reversed-phase, normal-phase, size-exclusion, and ion-exchange methods are important in modern clinical laboratories. However, a related technique that is seeing growing use is affinity chromatography. *Affinity chromatography* can be defined as a liquid chromatographic technique that utilizes "biological interactions" for the separation and analysis of specific analytes in a sample [1]. Examples of these interactions include the binding of an enzyme with an inhibitor, of hormones with receptors, or of an antibody with an antigen. Because of the highly selective nature of these binding processes, affinity chromatography is quickly becoming the method of choice for separations in fields such as pharmaceutical science and biotechnology. Similar developments are beginning to occur in clinical laboratories, thus creating a need for workers in this area to be aware of this technique.

1.1 Terms and Definitions

The key component of affinity chromatography is the immobilized binding agent, known as the *affinity ligand*, which selectively interacts with the desired analyte. This is used by immobilizing the ligand onto a solid support and placing it within a column. Once the immobilized ligand has been prepared, it can then be used for isolation or quantitation of the analyte. The ligand in affinity chromatography is the key factor that determines the success of the method. Most of the ligands for affinity chromatography are of biological origin. However, "affinity chromatography" has also been used to describe columns that contain selective ligands of a nonbiological origin. Examples of these nonbiological ligands include boronates, immobilized metal ion complexes, and synthetic dyes.

The type of ligand present in the column is often used to break down affinity methods into various subcategories. For instance, *bioaffinity chromatography* and *biospecific adsorption* are terms used to specify whether the affinity ligand is really a biological compound. Other categories that are based on the type of ligand being used include lectin, immunoaffinity, dye ligand, and immobilized metal ion affinity chromatography [2,3]. Each of these methods will be examined in more detail later in this chapter.

Another factor used to distinguish between affinity methods is the type of support present within the column. In *low-performance* (or *column*) *affinity chromatography*, the support is usually a large-diameter non-rigid gel, such as agarose, dextran, or cellulose. In contrast, *high-performance affinity chromatography (HPAC)* uses small, rigid particles based on silica or synthetic polymers that are capable of withstanding the high flow rates and/or pressures that are characteristic of HPLC systems [2,4]. Both low- and high-performance methods are used in the clinical laboratory. Low-performance affinity chromatography is generally used for sample extraction and pretreatment because it is relatively easy to set up and inexpensive. However, the better flow and pressure stability of high-performance supports make HPAC easier to incorporate into instrumental systems, giving it better speed and precision for the automated quantitation of analytes.

1.2 Principles and Separation Scheme

A typical approach for performing affinity chromatography is shown in Fig. 1. In this scheme, the sample of interest is first injected onto the affinity

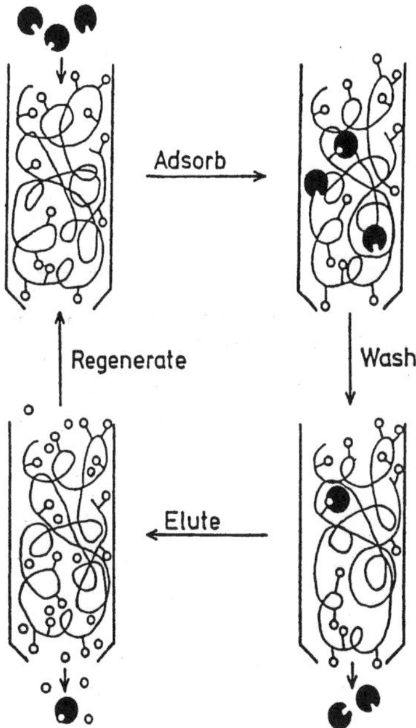

FIGURE 1 General operating scheme for affinity chromatography. (Reproduced with permission from Ref. 2.)

column under conditions in which the analyte has strong binding to the immobilized ligand. This is usually performed at a pH and ionic strength that mimic the natural environment of the ligand and analyte. Because of the specificity of the analyte–ligand interaction, other components in the sample tend to have little or no interaction with the ligand and wash quickly from the column. After these nonretained solutes have been removed, an elution buffer is applied to dissociate the analyte–ligand complex. This commonly involves changing the pH or composition of the mobile phase to decrease the strength of the interaction, or adding a competing agent to the mobile phase to displace the analyte from the ligand. As the analyte elutes, it is either detected on-line or collected for further use. Later, the system is reequilibrated with the initial application buffer and the column is allowed to regenerate prior to the next sample injection. The result is a separation that is both selective and easy to perform. These features are what make affinity chromatography so appealing for solute purification or for the quantitation of sample components in a complex mixture.

The scheme shown in Fig. 1 is known as the *on/off* or *direct detection mode* of affinity chromatography and has been the basis of numerous clinical applications. There are several reasons for the popularity of this format. For example, when this is performed by an HPLC system, the precision is generally in the range of 1 to 5%, and the run times are often as low as 5 or 6 min per sample [2,4,5]. The greater speed of these systems compared to many other ligand-based techniques (e.g., traditional immunoassays) is largely due to the better mass transfer properties and increased analyte–ligand binding rates that are produced by the supports in affinity columns. The increased precision of this approach is due to the reproducible sample volumes, flow rates, and column residence times that are possible with modern HPLC equipment. Another factor that leads to good precision is the reduced batch-to-batch variation, which is created by using the same ligand for the analysis of multiple samples and standards. In many studies it has been reported that several hundred injections can be performed on the same affinity column under properly selected elution and regeneration conditions [2,5,6].

A limitation of the direct detection format in affinity chromatography is that it requires the presence of enough analyte to allow the measurement of this compound as it elutes from the affinity column. In HPLC systems this is usually performed by on-line UV/Vis absorbance or fluorescence detectors. This requirement tends to make the direct detection mode most useful when dealing with intermediate-to-high concentration solutes. However, it is possible to employ direct detection with trace sample components if the affinity column is combined with precolumn derivatization and/or more sensitive detection schemes, like an off-line immunoassay or a suitable postcolumn reactor [5]. Another potential limitation of the direct detection mode is that samples and standards are analyzed sequentially, which limits its effectiveness in situations where high throughput is needed. This makes this format most valuable in situations where low-to-moderate numbers of samples are being processed and/or fast turnaround times per sample are desired. However, sequential analysis has the advantage of making affinity chromatography more convenient to troubleshoot than batch-mode techniques, and it is easier to determine whether the assay is operating at satisfactory levels before patient samples are tested.

In the next few sections there will be numerous examples given in which the direct detection mode is used for clinical testing with affinity ligands. This will include the use of boronates, lectins, protein A and protein G, and other miscellaneous ligands for such methods. But there are other ways in which affinity chromatography can also be used by clinical laboratories. Examples will be seen later when the topics of affinity extraction, postcolumn affinity detection, chiral separations, and studies of biomolecular interactions are discussed.

2 BORONATE AFFINITY CHROMATOGRAPHY

Affinity methods using boronic acid or boronates as ligands are one group of techniques that have been particularly successful in the clinical laboratory. This set of methods is known collectively as *boronate affinity chromatography.* At a pH above 8, most boronate derivatives form covalent bonds with compounds that contain *cis*-diol groups in their structure. Since sugars like glucose possess such groups, boronates are valuable for resolving glycoproteins from nonglycoproteins [7]. The first use of a boronate affinity column in clinical testing was for the determination of glycated hemoglobin as a means for the assessment of long-term diabetes management. This was reported by Mallia et al. in 1981 [8], where absorbance measurements at 414 nm were used to quantitate the retained and nonretained hemoglobin fractions in hemolysate samples. Similar low-performance methods have been reported or evaluated by other groups [10–13] and have been adapted for use in HPAC [9,11,14,15]. This approach has also been used as a point-of-care test for glycated hemoglobin [16].

In addition to hemoglobin, boronate columns can be used to look at other types of glycoproteins [14,17]. For example, by monitoring absorbance at 280 nm instead of 410–415 nm, the same technique used for glycated hemoglobin can be modified to determine the relative amount of all glycated proteins in a sample [14]. Boronate chromatography can also be combined with mass spectrometry to determine the extent of glycation of protein samples [18]. Alternatively, a particular type of glycoprotein can be examined by combining a boronic acid column with a detection method that is specific for the protein of interest. Examples include a boronic acid column followed by an immunoassay for the detection of glycated albumin in serum and urine [18] or glycated apolipoprotein B in serum [19], as well as the combination of boronate chromatography with latex immunoagglutination for the determination of glycated apolipoprotein A-I [20]. Additional studies have been performed using boronate affinity chromatography to isolate RNA [21].

Although hemoglobin A_{1c} is the major form of glycated hemoglobin, there are several minor species that coelute with this from boronate columns. In one study, electrospray ionization mass spectrometry was used off-line to analyze the glycated hemoglobin fraction collected from a boronate column (see Fig. 2) [22]. This hemoglobin was characterized with respect to the extent of glycation and whether this modification occurred at the α- or β-chain of hemoglobin. With this approach it was possible to demonstrate that both α- and β chains showed increased glycation as the glycated hemoglobin levels increased, but that the extent of modification of these two chains did not increase to the same degree.

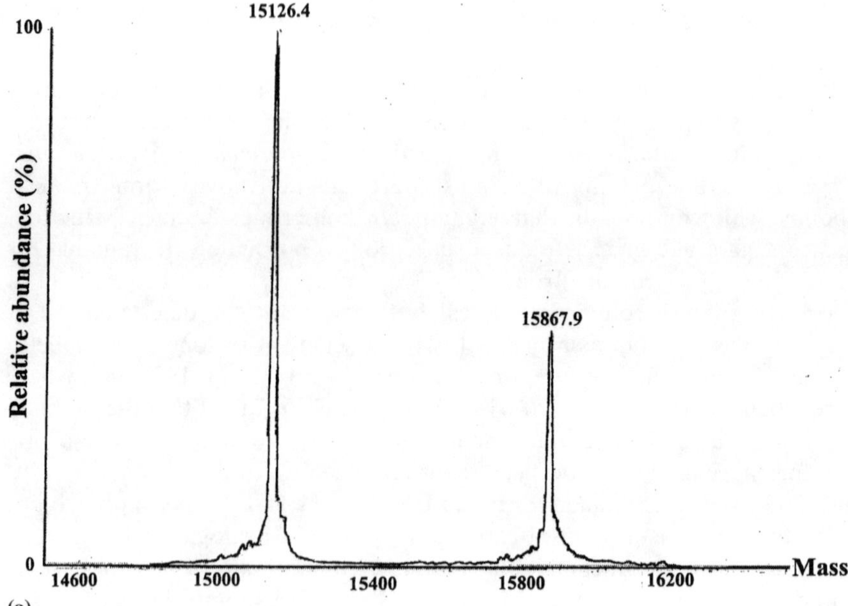

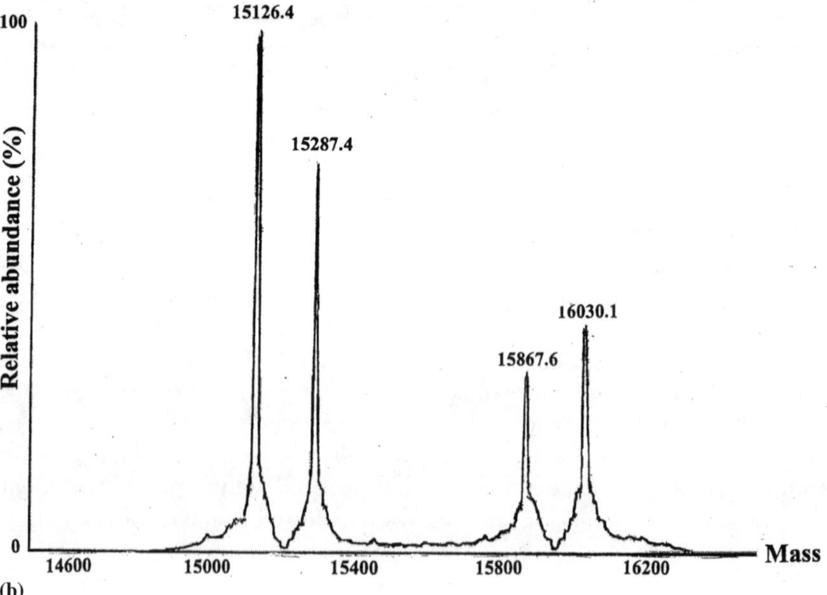

FIGURE 2 Deconvoluted electrospray mass spectra for (a) normal hemoglobin and (b) glycated hemoglobin collected from a boronate affinity column. (Reproduced with permission from Ref. 22.)

3 LECTIN AFFINITY CHROMATOGRAPHY

Lectins are another class of ligands that can be used for the detection of clinical analytes by affinity chromatography. The result is a method known as *lectin affinity chromatography*. Lectins are non-immune system proteins that are able to recognize and bind certain carbohydrate residues [23]. Two lectins often used in affinity chromatography are concanavalin A, which binds to α-D-mannose and α-D-glucose residues, and wheat germ agglutinin, which binds to D-*N*-acetylglucosamines. Other lectins that can be used are jakalin and lectins found in peas, peanuts, and soybeans. These ligands are commonly used in the isolation of polysaccharides, glycoproteins, and glycolipids [2,3].

One clinical application of lectin affinity chromatography has been in the study of prostate cancer and prostate tumor markers. Serial lectin affinity chromatography with concanavalin A and wheat germ agglutinin was used to characterize prostatic acid phosphatase (PAP) with respect to an altered asparagine-linked sugar chain structure in patients with prostatic carcinoma [24]. A similar approach using a combination of concanavalin A and phytohemagglutinin has been employed to study prostate specific antigen (PSA) [25]. These studies isolated their respective analytes from samples and divided the eluent into fractions based on their relative strength of lectin binding. The amount of PAP or PSA was quantitated by measuring the enzyme activity in the collected fractions. The results showed significant differences in both markers between patients with benign prostatic hyperplasia and those diagnosed with prostate carcinoma (see Fig. 3). These data suggest that asparagine-linked sugar chains for both PAP and PSA were significantly altered during oncogenesis and may be used to discriminate between benign prostate hyperplasia and prostate cancer.

Other glycoproteins have also been studied and quantitated by lectin affinity chromatography. For instance, low-performance columns based on concanavalin A have been used to separate apoA- and apoB-containing lipoproteins in human plasma [26], to study the microheterogeneity of serum transferrin during alcoholic liver disease [27], to examine glycoproteins produced at postsynaptic sites [28], and to characterize the carbohydrate structure of follicle-stimulating hormone and luteinizing hormone under various clinical conditions [29]. Wheat germ agglutinin columns have been used to develop and validate HPLC methods to distinguish between liver- and bone-derived isoenzymes of alkaline phosphatase [30–32]. Immobilized jakalin on low-performance supports has been employed to characterize the rabbit homolog of human MUC1 glycoprotein [33]. And the lectin *Ricinus communis* agglutinin has been coupled to a polymeric support for the high-performance affinity chromatography of glycoproteins and oligosaccharides [34].

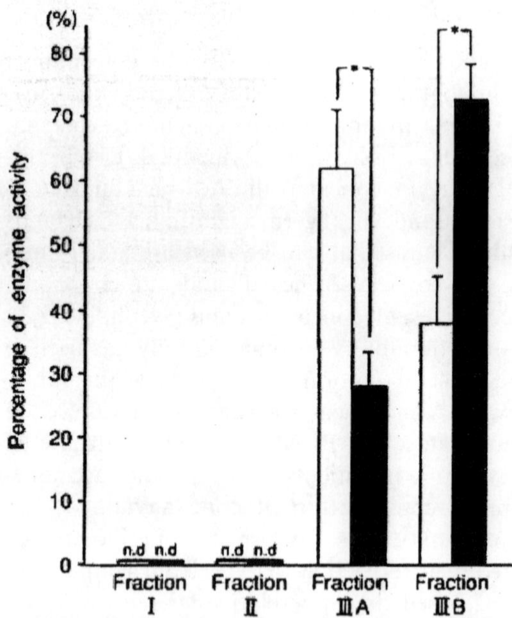

FIGURE 3 Activity of prostatic acid phosphatase (PAP) in four fractions collected from concanavalin A and phytohemagglutinin columns. The open columns represent PAP activity in patients with benign prostate hyperplasia, and the closed columns represent patients diagnosed with prostate cancer. The term "n.d." indicates that no activity was detected. (Reproduced with permission from Ref. 24.)

An interesting application of lectin affinity chromatography was demonstrated during the isolation of glycoprotein G from viral particles. Envelope glycoproteins from the outer surface of viruses play a vital role during infection, making them the focus of both structural and functional research. This has resulted in a need for chromatographic methods that maintain the activity and structure of these glycoproteins. In one study, a fish rhabdovirus known as viral haemorrhagic septicaemia virus (VHSV) was isolated and sonicated in polyethylene glycol (PEG) to liberate glycoprotein G from the viral particles. The samples were then passed over an immobilized concanavalin A column to purify glycoprotein G from the detergent-solubilized virion and virion-free PEG supernatants. The glycoprotein G was eluted using 1 M glucopyranoside and 1 M methyl-D-mannopyranoside in pH 7.5 sodium acetate buffer. The purified glycoprotein G was shown to retain its phosphatidyleserine-binding properties and was able to bind antiglycoprotein G antibodies. This suggested that purification by a concanavalin A column retained most of the native properties and conformation of this glycoprotein [35].

Lectin affinity chromatography has been combined with other analytical methods for a variety of applications. Proteomic analysis of glycoproteins has been performed by combining immobilized lectin columns with reversed-phase liquid chromatography, with the collected fractions then being analyzed by matrix-assisted laser desorption ionization (MALDI) mass spectrometry. The proteins in these fractions were identified by comparing their mass spectra to a spectral database of protein digests [36]. Other reports have used lectin affinity chromatography as the basis for affinity capture prior to the MALDI analysis of microorganisms [37]. Finally, research in the production of designer lectins has produced a synthetic ligand, based on the coupling of a triazine with 5-aminoindan, that demonstrates selective binding toward glycoproteins that parallels the behavior of concanavalin A [38].

4 PROTEIN A AND PROTEIN G AFFINITY CHROMATOGRAPHY

A third class of ligands used for affinity chromatography is antibody-binding proteins like protein A and protein G. These are bacterial cell wall proteins produced by *Staphylococcus aureus* and group G *streptococci*, respectively [39–41]. These ligands bind the constant region of many types of immunoglobulins. Protein A and protein G bind to immunoglobulins most strongly at or near neutral pH and readily dissociate from these proteins when placed into a lower pH buffer. These two ligands are different in the strength with which they bind to antibodies from different species and classes [3,39,42]. For instance, human IgG_3 binds much more strongly to protein G than to protein A, and human IgM shows no binding to protein G but does have weak interactions with protein A [3]. A recombinant protein known as protein A/G that blends the activities of these ligands is also available for use in affinity columns [3,43].

The ability of protein A and G to bind antibodies makes them valuable for the analysis of immunoglobulins, especially IgG-class antibodies, in humans. The first clinical applications of these ligands in HPLC systems were methods for the analysis of IgG in serum [44–46]. Another study used a combination of two affinity columns, one containing immobilized protein A and the other antihuman serum albumin antibodies, for the simultaneous analysis of IgG and albumin in serum for the determination of albumin/IgG ratios (see Fig. 4) [47]. An additional application of protein A and protein G has been their use as secondary ligands for the adsorption of antibodies onto supports for immunoaffinity chromatography. This method, which will be

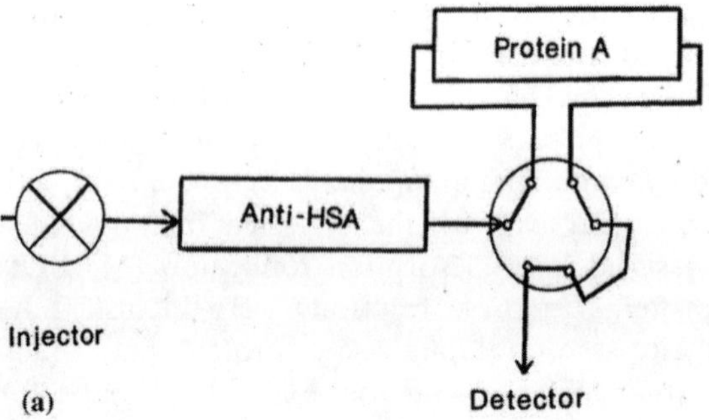

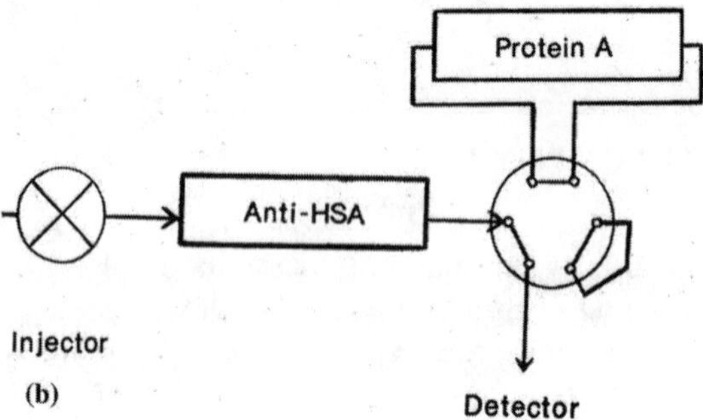

FIGURE 4 A dual-column system based on an antihuman serum albumin (anti-HSA) antibody column and a protein A column for the detection of HSA and immunoglobulin G (IgG) in serum samples. The diagram in (a) shows both columns on-line during sample application. The scheme in (b) shows the valve configuration used to elute the retained HSA with a pH 3.0 phosphate buffer, followed by a change back to the top configuration for the elution of IgG with the same buffer.

considered more in the next chapter, is advantageous when high activities or frequent regeneration is needed for antibody-based columns [5,6,48].

5 OTHER TYPES OF AFFINITY LIGANDS

The largest and most diverse group of affinity methods in clinical testing utilizes antibodies or antibody fragments as ligands. The term *immunoaffinity chromatography (IAC)* is used for chromatographic methods in which the stationary phase consists of such binding agents [5,48]. If this technique is performed as part of an HPLC system, the resulting approach is referred to as *high-performance immunoaffinity chromatography (HPIAC)* [5,48]. Owing to the wide variety of clinical applications for IAC and HPIAC, these topics will be discussed separately in the next chapter.

Besides the ligands that have already been discussed, there have been several other binding agents that have been used for the detection of clinical analytes by affinity chromatography. For instance, an immobilized heparin column has been employed for the determination of antithrombin III in human plasma [49,50]. *S*-Octylglutathione has been reported as a ligand for the separation and analysis of glutathione *S*-transferase isoenzymes in human lung and liver samples [51,52]. In addition, immobilized *p*-aminobenzamidine has been used for the separation of human plasminogen species, with the addition of an immobilized urokinase column for on-line detection [53].

Another example that demonstrates the flexibility of affinity chromatography is the use of avian ovomucoid as a receptor analog for Shiga-like toxin type 1 (Stx1) [54]. Shiga and Shiga-like toxins are important factors in the pathogenesis of infections caused by *Shigella dysenteriae* type 1 and certain serotypes of enterotoxic *Escherichia coli*. These toxins not only bind a cell surface glycolipid but also bind to the P1 antigen that is contained in avian ovomucoid. To make use of this, an ovomucoid glycoprotein fraction was prepared from pigeon egg whites and coupled to cyanogen bromide-activated Sepharose 4B. This column was then used to purify Stx1 from an *E. coli* cell lysate with almost 100% of the toxin's activity being retained during this process.

6 AFFINITY EXTRACTION

Affinity extraction refers to the use of affinity chromatography for the isolation of a specific analyte or group of analytes from a sample prior to their determination by a second method. The general operating scheme for this is the same as for other types of affinity chromatography, but it now involves combining the affinity column either off-line or on-line with some other method for the actual quantitation of analytes. Affinity extraction represents one of the most commonly employed uses of affinity chromatography in chemical analysis. This section will discuss several examples of affinity extraction, including both off-line methods and those that involve the on-line coupling of affinity columns with techniques like HPLC, gas chromatography, and capillary electrophoresis.

6.1 Off-Line Affinity Extraction

Off-line extraction is the easiest way of combining an affinity column with another analytical technique. This usually involves placing an affinity ligand onto a low-performance support (e.g., activated agarose) that is packed into a small disposable syringe or solid-phase extraction cartridge. After conditioning the affinity column with the necessary application buffer and conditioning solvents, the sample is applied, and the nonbound sample components are washed away. An elution buffer is then applied, and the analyte is collected as it is removed from the column. Occasionally this eluted fraction is analyzed directly by a second technique, but ususally the collected fraction is first dried down and reconstituted in a solvent that is more compatible with the method to be used for quantitation. If necessary, the collected solute fraction may be derivatized before it is analyzed to obtain improved detectability or more appropriate physical properties (e.g., an increase in solute volatility prior to separation and analysis by GC).

The most common ligands in affinity extraction are antibodies, as discussed in the next chapter. But these are not the only ligands used for such an approach. For example, off-line boronic acid columns have been used for the reversed-phase analysis of modified nucleosides in patients with gastrointestinal cancer [55] and in the purification of human platelet glycoclicin before analysis by anion-exchange HPLC [56]. Agarose beads derivatized with 3-aminophenylboronic acid were used to extract glycohemoglobin from a complex mixture prior to analysis [57]. Wheat germ agglutinin extraction columns have been used to extract phospholipase A2 [58] and have been combined with anion-exchange chromatography to purify and analyze angiotensinase A and aminopeptidase M in human urine and kidney samples [59]. Another lectin, concanavalin A, has been utilized to purify human paraoxonase 1 prior to analysis [60]. Additionally, sample extraction by an organomercurial agarose column followed by RPLC analysis has been used for the assessment of urinary 2-thioxothiazolidine-4-carboxylic acid, a proposed indicator of environmental exposure to carbon disulfide [61].

Another application of affinity extraction involves its use to remove specific interferences from samples. One example is the use of protein A and antimouse immunoglobulin supports for the removal of human antimouse antibodies prior to a sample's determination by an immunoassay [62]. Another illustration of this is the use of antihuman immunoglobulin immunoaffinity chromatography or protein A supports to adsorb selectively enzyme-immune complexes (i.e., macroenzymes) from patient samples [63].

Yet another unique application of affinity extraction is its use to bind to several related compounds in the same sample. Many ligands show some binding or cross-reactivity with solutes that are closely related to the desired analyte in structure. This cross-reactivity should be evaluated for each affinity extraction method by performing binding and interference studies with any solutes or metabolites that are similar to the analyte and that may be present in the samples of interest. However, this does not present a problem as long as the analyte of interest can be resolved or discriminated from cross-reacting compounds by the method used for quantitation. In many cases this can even be used to an advantage by allowing several species in the same class of compounds to be determined in a single run. For example, it will be shown in the next chapter how this approach has been used with antibody columns to look at anabolic steroids [64–66].

It is further possible to place multiple types of specific affinity ligands into the same column, as has again been reported for the analysis of anabolic steroids using immunoextraction columns [67]. In this case, a column containing up to seven different ligands was used to extract simultaneously testosterone, nortestosterone, methyl testosterone, trenbolone, zeranol, estradiol, diethylstilbestrol, and related compounds in urine. This was followed by quantitation using GC-MS. This approach allowed for extensive sample cleanup without losing any important components of the sample or prolonging the time needed for sample preparation.

An advantage of off-line extraction is that the samples collected from the extraction column can be readily derivatized or placed into a different solvent between the sample purification and quantitation steps. This is particularly important when combining affinity extraction with GC, where it is desirable to remove any water from the collected sample before injection onto the GC system and where solute derivatization is often required to improve solute volatility or detection. This advantage has been demonstrated in the analysis of anabolic steroids [68] and THC metabolites [69] by affinity extraction and GC-MS. Another advantage of off-line affinity extraction is that it is easier than on-line extraction to set up once an appropriate ligand has been obtained.

6.2 On-Line Affinity Extraction

The direct coupling of affinity extraction with other analytical methods is one other area that has been the subject of increasing research. The use of affinity extraction columns as part of HPLC systems is particularly attractive because this is compatible with automation and reduces the time needed for sample pretreatment. In addition, the high precision of HPLC pumps and injection systems provides on-line affinity extraction with better precision than off-line extraction methods, since the on-line approach has more tightly controlled sample application and elution conditions.

As with off-line affinity extractions, antibodies are generally the ligand of choice for on line extractions (see next chapter). However, other ligands, particularly boronates, have been shown to be valuable in performing on-line affinity extraction with HPLC. Examples include several methods in which

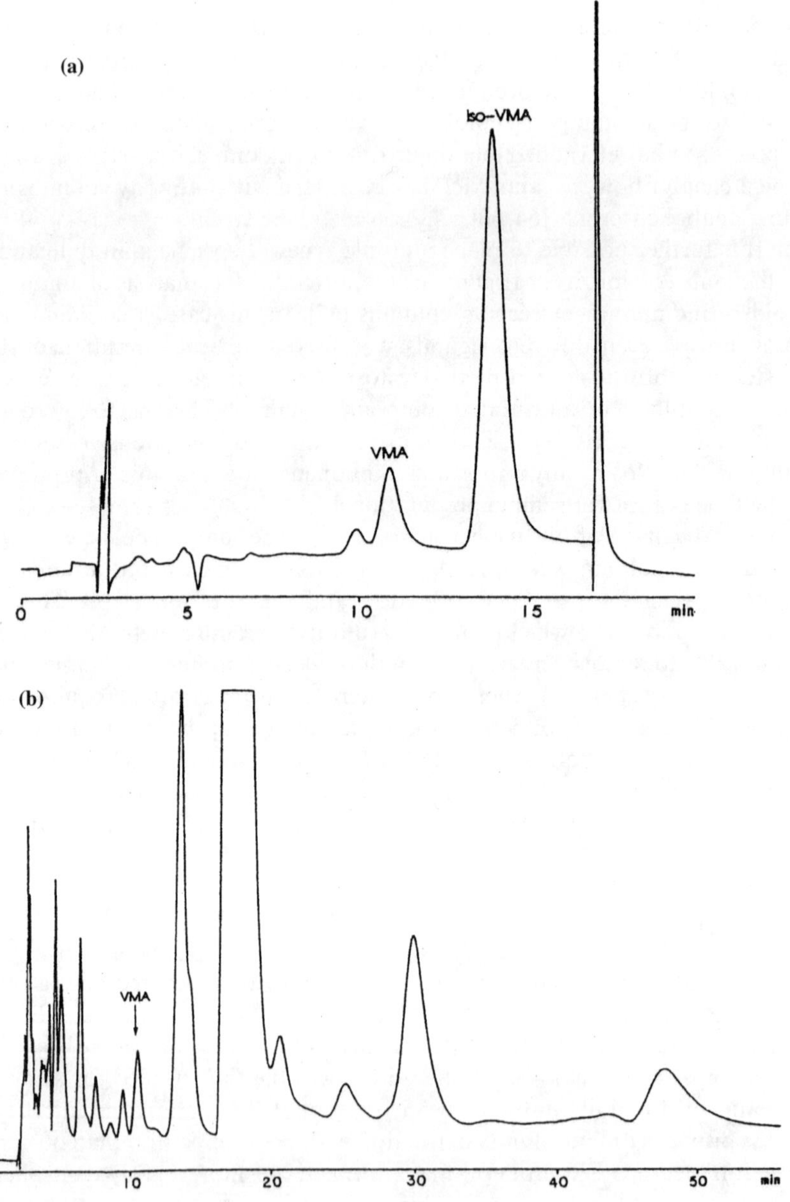

FIGURE 5 Analysis of vanilmandelic acid (VMA) in human urine using (a) a boronate extraction column on-line with an anion exchange column and (b) an anion exchange column alone. The peak labeled iso-VMA is isovanilmandelic acid, which was added to the samples as an internal standard. (Reproduced with permission from Ref. 79.)

boronate columns have been combined with HPLC columns for the analysis of catechol-related compounds like epinephrine, norepinephrine, and dopamine [70–72], dihydroxyphenylalanine [73], dihydroxyphenylacetic acid [73,74], and 5-*S*-cysteinyldopa [75]. This same approach has been adapted for profiling and quantitating ribonucleotides in urine and serum [76–78].

The analysis of vanilmandelic acid by boronate affinity extraction and anion exchange chromatography is a good example of a clinical use of on-line affinity extraction (see Fig. 5). Vanilmandelic acid is a major metabolite of epinephrine and norepinephrine and is often analyzed in urine to determine the presence of excess catecholamines in cases of neuroblastoma. When samples are passed through a boronate affinity column under acidic conditions, vanilmendelic acid (VMA) and an internal standard (i.e., isovanilmandelic acid) are retained while the remaining urine contents wash through. The analytes are eluted at a higher pH from the boronate column and passed to an on-line anion exchange column followed by electrochemical detection. This approach eliminates the need for preanalytical sample extraction and allows VMA to be measured in only 15 min [79].

7 POSTCOLUMN AFFINITY DETECTION

Another way in which affinity columns can be used is to have them monitor the elution of specific analytes from other chromatographic columns. This involves the use of a postcolumn reactor and an affinity column attached to the exit of an analytical column. Several affinity ligands have been used for this purpose. One example is the use of anion-exchange chromatography followed by an HPLC boronate column for the determination of glycated albumin in serum samples [80]. Using this approach, analysis of glycated albumin was achieved in 10 min for the analysis of samples from both diabetic and nondiabetic patients. Another example is the use of immobilized receptors for the detection of bioactive interleukin-2 as it eluted from an anti-interleukin immunoaffinity HPLC column [81].

8 AFFINITY-BASED CHIRAL SEPARATIONS

Affinity ligands are also important in the separation of chiral compounds [82]. Owing to differing pharmacological activities for enantiomeric drugs and pressure from regulatory agencies such as the U.S. Food and Drug Administration, there has been great interest in the pharmaceutical field for methods capable of discriminating between the individual chiral forms of drugs [83]. This has influenced the field of clinical chemistry, where the ability to quantitate the different forms of a chiral drug and its metabolites is seeing more use in metabolism studies and therapeutic drug monitoring. HPLC methods with chiral stationary phases have been shown to be valuable in quantitating and separating chiral compounds [82,84]. Since many of the ligands used in affinity chromatography are inherently chiral, this makes them logical choices as stationary phases for such separations.

Various naturally occurring proteins and carbohydrates, as well as derivatives of these compounds, have been used as ligands for chiral separations of clinical analytes [85–104]. Most of these separations are performed by carrying out a routine liquid–liquid or solid-phase extraction of the sample, followed by injection of the extracted contents onto a chiral affinity column. However, other approaches are possible. In some cases, a chiral column is first used to resolve the different forms of the analyte, with fractions then being collected and applied either on- or off-line to a second achiral column for further separation and quantitation [93,99]. Alternatively, an achiral column can be used first to isolate the compounds from a sample, and a chiral column can then be employed on- or off-line to resolve the different chiral forms in each peak of interest [94,95].

8.1 Protein-Based Stationary Phases

The use of proteins as chiral stationary phases has received some degree of attention. Although all proteins are chiral, only one (α_1-acid glycoprotein) has seen any significant use for the analysis of drugs in a clinical setting. α_1-Acid glycoprotein (also known as AGP, AAG, or orosomucoid) is a human serum protein that is involved in the transport of many small solutes throughout the body. AGP differs from human serum albumin in that AGP has a lower isoelectric point and contains carbohydrate residues as part of its structure. The lower isoelectric point makes AGP useful in binding cationic compounds, while the presence of carbohydrate residues may play a role in determining the steroeselectivity of AGP [84]. Many drugs and related analytes have been separated by AGP in human urine, serum, or plasma. Examples of clinical interest include bunolol [85], citalopram [86], fenoprofen [87], flurbiprofen [88], ibuprofen [87,89], ketamine [90], ketoprofen [87], methadone [91–93], norketamine [90], norverapamil [94], pindolol [95], vamicamide [96], and verapamil [94,97].

A specific example of a chiral separation based on AGP is the resolution of the *R*-(+)- and *S*-(−)-isomers of thiopentone (see Fig. 6) [98]. Thiopentone is a common anaesthetic agent that is administered as a racemate either through bolus injection or continuous intravenous infusion. Preliminary studies have shown that the *S*-(−)-isomer is more potent and more slowly eliminated than the *R*-(+)-isomer, but pharmacokinetic data for these

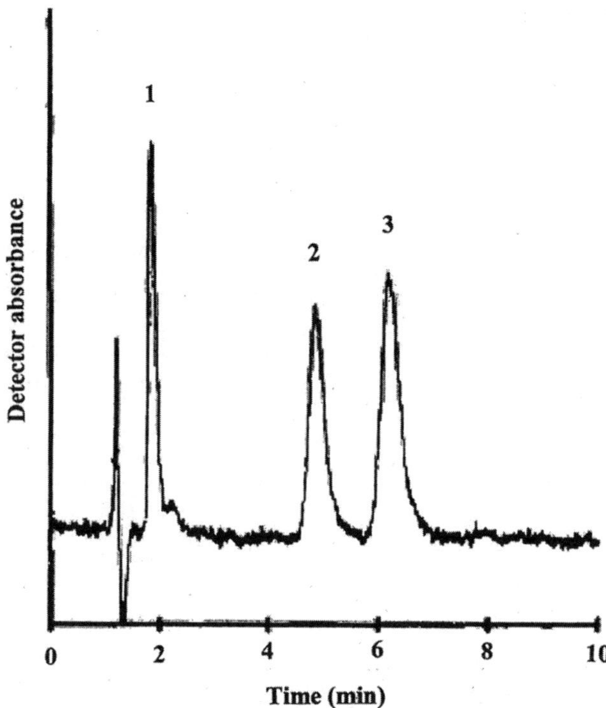

FIGURE 6 Analysis of R-(+)-thiopentone (peak 2) and S-(−)-thiopentone (peak 3) in an extract of plasma injected onto an immobilized AGP column. Ketamine (peak 1) was used as the internal standard. (Reproduced with permission from Ref. 98.)

compounds are difficult to obtain due to difficulties in differentiating between these two isomers in samples. To overcome this, an immobilized AGP column was used with UV detection at 280 nm to separate and quantitate the enantiomers of thiopentone. Using a pH 5.0 phosphate buffer as the mobile phase and a flow rate of 0.9 mL/min, the isomers were resolved in under 8 min for injections of plasma samples.

Other proteins that have received attention in clinical applications of chiral HPLC are bovine serum albumin (BSA) and ovomucoid. Ovomucoid is a glycoprotein obtained from egg whites that has shown promise in the separation of cationic solutes [84]. BSA is a member of the serum albumin family, which is involved in the transport of a wide range of small organic and inorganic compounds throughout the body, including many pharmaceutical agents [105,106]. BSA tends to bind best to neutral or anionic compounds, making this protein complementary to AGP and ovomucoid in its applications [82,84]. In clinical work, BSA has been used for the chiral separation of leucovorin in plasma [99], and ovomucoid has been used for separating the individual forms of pentazocine in serum samples [100].

8.2 Carbohydrate-Based Stationary Phases

Cyclodextrins are natural carbohydrates that can be employed as stereoselective ligands for HPLC [102–104]. These are circular polymers of α-1,4-D-glucose that are produced through the degradation of starch by the microorganism *Bacillus macerans*. The most common forms of these polymers are α-, β-, and γ-cyclodextrin, which contain six, seven, or eight glucose units, respectively [82,84]. The cone-shaped structure and hydrophobic interior cavity of cyclodextrins give them the ability to form inclusion complexes with numerous small aromatic solutes. In addition, the well-defined arrangement of hydroxyl groups about the upper and lower faces of cyclodextrins provides these agents with the ability to discriminate between various chiral compounds. Clinical applications for cyclodextrins in HPLC include methods reported for chlorpheniramine [101], citalopram,

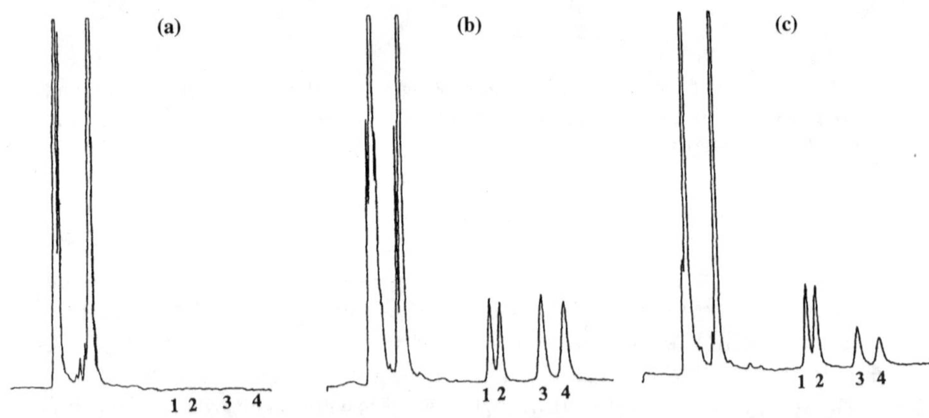

FIGURE 7 Separation of *S*- and *R*-alprenolol (peaks 1 and 2) and *S*- and *R*-propranolol (peaks 3 and 4) in a urine extract using a β-cyclodextrin column. Chromatogram (a) is blank urine sample, (b) is human urine spiked with 200 ng/mL racemic propranolol plus a fixed amount of racemic alprenolol (the internal standard), and (c) is urine from a healthy volunteer after oral administration of 80 mg of racemic propranolol. These compounds were monitored by using fluorescence detection at an excitation wavelength of 222 nm and an emission wavelength of 340 nm. (Reproduced with permission from Ref. 104.)

desmethylcitalopram, and didesmethylcitalopram [102], hexobarbital [100], and the M1 and M2 metabolites of moguisteine [103].

The most commonly used cyclodextrin, β-cyclodextrin, has been used to separate the (*R*)- and (*S*)-enantiomers of propranolol from both human plasma and urine samples, as shown in Fig. 7 [104]. Propranolol is a β-adrenergic blocking agent used in the management of patients with cardiovascular diseases. The (*S*)-enantiomer is 100-fold more active than its (*R*)-counterpart. In order to perform pharmacokinetic studies after the oral administration of racemic propranolol, a method was created to separate and quantitate the propranolol enantiomers using a β-cyclodextrin column and fluorescence detection at 340 nm. After extraction from 200 μL of plasma or urine, 50 μL of the extract was injected onto the cyclodextrin column with a polar organic mobile phase, giving a separation within 16–18 min that provided a limit of detection of 1.5 ng/mL for each enantiomer.

9 CHARACTERIZATION OF BIOLOGICAL INTERACTIONS

In addition to its application as a method for quantitating or isolating specific solutes, affinity chromatography can also be used to study the interactions in biological systems. Such an approach is known as *analytical* or *quantitative affinity chromatography*. This technique has been used to examine a variety of biological systems, including lectin/sugar, enzyme/inhibitor, protein/protein, and DNA/protein interactions [2,107]. A recent study has used affinity chromatography to study magnesium and calcium binding to human serum albumin [108]. However, most work in the clinical setting has focused on the use of affinity chromatography to study the binding of drugs or hormones to serum proteins [109–111]. In some cases, this type of protein binding is general, as in the interaction of many different drugs with melanin [112], human serum albumin, or AAG [113–122]. But on other occasions, this binding is highly specific in nature, such as the interaction of L-thyroxine with thyroxine-binding globulin or the binding of corticosteroids and sex hormones to steroid-binding globulins [123,124]. Protein binding of drugs and hormones is interesting because it plays a role in determining the final biological activity, metabolism, and elimination of these compounds. Also, the competition between drugs or between drugs and endogenous compounds (e.g., fatty acids or bilirubin) for protein binding sites can be an important source of drug–drug or drug displacement interactions [113–116,125].

Affinity chromatography has been used to examine drug–protein binding by using both immobilized drugs and immobilized proteins, but protein-based columns are more common [111]. One advantage of employing protein columns for binding studies is the ability to reuse the same ligand for multiple experiments, with up to 500 to 1000 injections per column being reported in some HPLC studies [126–128]. However, it is important when using an immobilized protein column to consider how effectively this models the behavior of the same protein in its soluble form. Fortunately, there are now many studies that show that immobilized proteins, particularly human serum albumin (HSA), can successfully be used as such models. For example, the association constants measured by equilibrium dialysis for soluble HSA with *R*- and *S*-warfarin or L-tryptophan are in close agreement with values obtained using immobilized HSA columns [129–131]. Studies have also shown that displacement phenomena and allosteric interactions found with immobilized HSA are representative of behavior observed for HSA in solution [117,131–136].

There are a number of different ways in which affinity chromatography can be used to investigate solute–protein interactions. Two techniques used for this purpose are zonal elution and frontal analysis. The next few sections will discuss each of these approaches in more detail and go over some of their applications in the study of protein binding by drugs and other solutes.

9.1 Zonal Elution

Zonal elution is the method most frequently used to study the binding of drugs and other solutes to immobilized protein columns [109,137]. The general format for these studies is to inject a small sample of the drug or analyte of interest onto a protein column in the presence of only buffer or a fixed concentration of a competing agent in the mobile phase. The injected analyte's elution time or retention factor (also known as the capacity factor, k or k') is then examined to see how this changes as a function of the competing agent concentration. An example of this type of experiment is shown in Fig. 8a [131]. Similar experiments can be used to study how changes in temperature or various solvent conditions affect solute–protein interactions [84,130,138–144] or to develop quantitative structure–retention relationships that describe these binding processes [145–147].

The most common application of zonal elution in drug–protein binding studies is its use to examine the displacement of drugs by other solutes [111,148]. Examples include displacement studies of DL-thyronine and DL-tryptophan from HSA by bilirubin or caprylate [149]; the competition of *R*/*S*-warfarin with racemic oxazepam, lorazepam, and their hemisuccinate derivatives on an HSA column [134]; the direct or allosteric competition of octanoic acid on immobilized HSA for the binding sites of *R*/*S*-warfarin, phenylbutazone, tolbutamide, *R*/*S*-oxazepam, hemisuccinate, ketoprofen A/B, and suprofen A/B [121,136]; the binding of L-thyroxine and related thyronine compounds to HSA [126,131]; and the displacement of *R*- and *S*-ibuprofen by each other at their common binding regions on HSA [150].

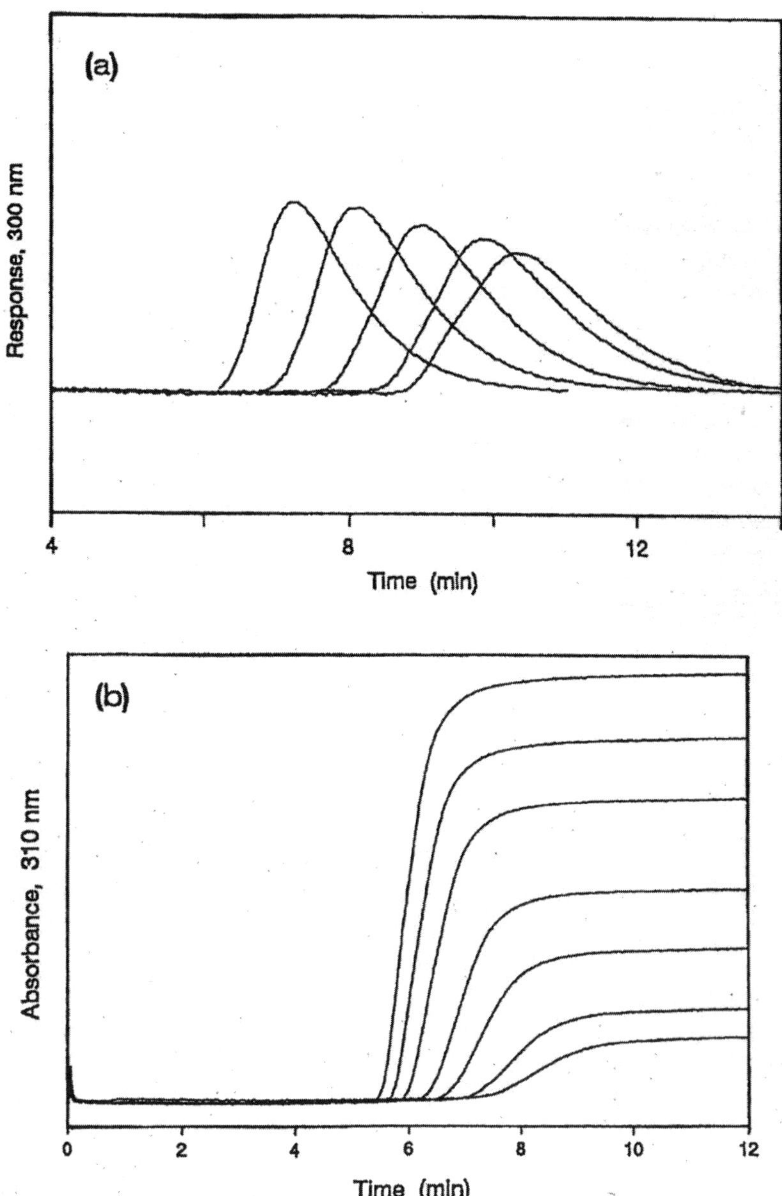

FIGURE 8 (a) Zonal elution experiment for the injection of *R*-warfarin onto an immobilized HSA column in the presence of various concentrations of a competing agent (increasing in concentration from right to left); (b) frontal analysis curves for the application of various concentrations of *R*-warfarin (increasing in concentration from right to left) on an immobilized HSA column. (Reproduced with permission from Refs. 126 and 129.)

This same technique has been used to characterize the binding sites of nonsteroidal anti-inflammatory drugs on HSA [120,151], as well as the binding of digitoxin and acetyldigitoxin [118], *cis*- and *trans*-clomiphene [119], and piroxicam [122] to HSA. In addition, this has been used to examine the displacement of nonsteroidal anti-inflammatory drugs and benzodiazepines by phenylbutazone, *R*/*S*-ibuprofen or 2,3,5-triiodobenzoic acid from serum albumin columns [152].

Besides providing qualitative information on binding and displacement effects, zonal elution can provide quantitative information on the equilibrium constants for these interactions. Many examples of this can be found in the cases already cited [117,122,126,131,136,145,147,150–153]. With proper experimental design, kinetic information on the rates of solute–protein interactions can also be obtained, provided that appropriate data are collected on the width and retention for solute peaks under various flow rate conditions. This latter case has been demonstrated for the binding of *R*/*S*-warfarin [127] and DL-tryptophan [128] to HSA columns.

9.2 Frontal Analysis

A second affinity method for studying biological interactions is *frontal analysis* [111]. In this approach, a solution containing a known concentration of analyte is continuously applied to an affinity column. As the column becomes saturated, the amount of solute eluting from the column gradually increases, forming a characteristic breakthrough curve, as shown in Fig. 8b. As long as the association and dissociation kinetics of this system are fast, the mean positions of the breakthrough curves can be related to the concentration of applied analyte, the amount of ligand in the column, and the association equilibrium constants for analyte–ligand binding.

Frontal analysis and affinity chromatography have been used to study various systems. One example is the binding of HSA to *R*- or *S*-warfarin [129,131] and D- or L-tryptophan [128,130,131,136]. This approach has further been used to determine the binding capacities of monomeric versus dimeric HSA for salicylic acid, warfarin, phenylbutazone, mefenamic acid, sulphamethizole, and sulphonylureas [154], and to examine the competition of sulphamethizole with salicylic acid for HSA binding regions [155]. Another application of this approach has been in characterizing the binding between chemically modified HSA and various site-specific probe compounds [156].

One drawback of frontal analysis is that it generally requires a larger amount of applied solute than zonal elution. However, this is largely offset by the fact that frontal analysis tends to provide binding constants that are more precise and accurate than those obtained by zonal elution methods

[111]. Another advantage of frontal analysis is that it is easy with this method to obtain information on both the binding capacity and affinity of an immobilized ligand.

10 FUTURE TRENDS AND DEVELOPMENTS

Although the application of affinity chromatography has found a wide variety of uses within clinical chemistry, there are still numerous ways this method can be improved or further adapted for work in this field. One area of ongoing research is in the creation of alternative ligands for clinical separations by affinity chromatography. For instance, ligands based on synthetic dyes, like triazine or triphenylmethane compounds, have been used for many years in enzyme and protein purification in a technique known as *dye-ligand affinity chromatography* [2,3,157–163]. However, these ligands have not yet seen any significant use in clinical labs.

Another example of a technique that is just beginning to appear in clinical labs is *immobilized metal ion affinity chromatography (IMAC)*. This uses a metal ion complexed with an immobilized chelating agent as the affinity ligand, such as iminodiacetic acid complexed with Cu^{2+}, Zn^{2+}, Ni^{2+}, Co^{2+}, or Fe^{3+}. This method separates proteins and peptides based on the interaction between amino acids like histidine, tryptophan, or cysteine with the metal ions within the immobilized chelate [164–166]. IMAC has been used in isolating human single-chain Fv antibodies [167] and fusion proteins [168]. This technique has also been employed in the separation of human blood components, including protein C and prothrombin [169]. In addition, IMAC has been used to study the interaction of human granulocyte-colony stimulating factor (rhG-CSF) with metal ions [170].

Two new affinity ligands that may soon become important in clinical testing are *aptamers* and *molecular imprints*. An aptamer is a polymer of nucleotides that has a well-defined sequence and three-dimensional structure. This is obtained by screening a random oligonucleotide library for the binding of such ligands to a given target compound; any oligonucleotides that show binding are then amplified and used in applications such as affinity chromatography [171–174]. A molecular imprint is an affinity ligand that is actually part of a support surface. This is prepared by combining the analyte of interest with a series of monomers that contain side chains capable of forming various interactions with the analyte. Polymerization is then initiated and the monomers are fixed in position about the analyte. This provides a material with binding pockets that have a known specificity [175,176] and that can be placed into a column for affinity chromatography. Both aptamers and molecular imprints are appealing candidates for affinity ligands owing to their ability to be custom designed for a given analyte, their stability over long-term use, and their moderate-to-high selectivity [171–177].

One specific application of aptamers in affinity chromatography has been their use in the isolation of a fusion protein [174]. In this study, an aptamer that could bind L-selectin was generated and immobilized within an affinity column. This column was then utilized in the purification of a recombinant human L-selectin-IgG fusion protein. This resulted in a 1500-fold purification and 83% recovery of the fusion protein in a single step. Similarly, a molecular imprint based on nortryptiline has been used as a stationary phase for the separation of structurally similar tricyclic antidepressant drugs [177]. This was based on a capillary HPLC column packed with nortriptyline-imprinted particles that were used to screen a simulated combinatorial library consisting of tricyclic antidepressants and related compounds. Using a mobile phase that contained 0.02% TFA and 0.015% TEA in acetonitrile, the retention factor for each compound was compared with its structure. This demonstrated the selectivity of the material, since compounds that shared the major structural features of nortriptyline were also the most strongly retained on the column (see Fig. 9).

Other novel ligands have appeared in recent affinity work involving clinical analytes. Immobilized synthetic oligosaccharide tumor antigens have been used to assess the response of cancer patients to carbohydrate-

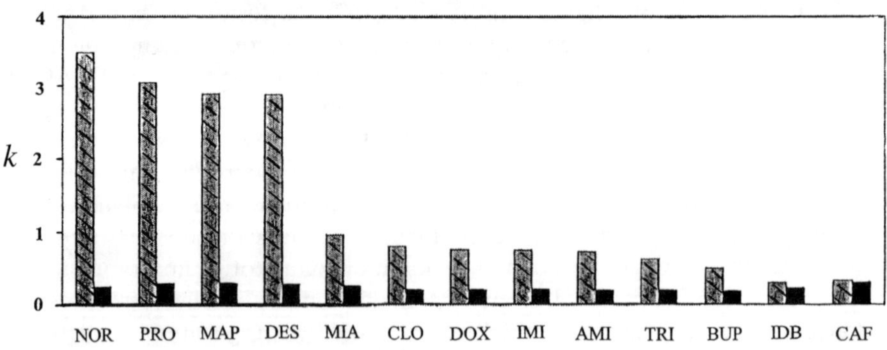

FIGURE 9 Comparison of retention factors (k) for various compounds injected onto a nortriptyline molecular imprint support (dashed boxes) and a similar support with no imprint (solid boxes). The compounds used in this study were as follows: nortriptyline (NOR), protriptyline (PRO), maprotiline (MAP), desiprimine (DES), mianserin (MIA), clomipramine (CLO), doxepin (DOX), imipramine (IMI), amitriptyline (AMI), trimipramine (TRI), bupropion (BUP), iminodibenzyl (IDB), and caffeine (CAF). (Reproduced with permission from Ref. 177.)

based vaccines [178]. There has also been a study that utilized red blood cells as an affinity stationary phase by adsorbing them to wheat germ lectin agarose gel beads [179]. This procedure was developed because previous methods of red cell adsorption were plagued with significant hemolysis, whereas immobilization to the wheat germ lectin left the majority of red cells intact. Using this cellular-based stationary phase and frontal analysis chromatography, interactions of D-glucose and cytochalasin B with the glucose transporter Glut1 were studied. Dissociation constants were measured for both analytes, and the affinity of the analytes for the Glut1 in red cells was determined to be higher than for Glut1 in cytoskeleton-depleted membrane vescicles or proteoliposomes.

Another novel affinity application involved the immobilization of liposomes containing a photosynthetic reaction center to study the binding of c-type cytochromes to this center. The liposomes were immobilized by integrating biotinyl phosphatidylethanolamine into their structures and adsorbing them to streptavidin-coupled gel beads [180]. In other work, a column with immobilized N2,N2,7-trimethyl guanosine (TMG) was used to purify TMG-binding proteins from human cells [181], and sulphate affinity chromatography was conducted to purify viral particles from crude cellular lysates [182].

Another ongoing trend in affinity chromatography is the design of improved systems and formats that will give this technique greater speed, greater selectivity, and higher sample throughput. These improvements must be made for affinity chromatography to be competitive with common clinical methods such as batch-mode immunoassays. One approach to increasing selectivity and/or increasing the number of analytes examined per assay is to combine affinity chromatography with other analytical techniques. This is already reflected in the growing popularity of using off-line affinity extraction with HPLC or GC and on-line affinity extraction with HPLC. Development of these tandem methods is expected to continue, as is further combination of on-line affinity extraction with GC, capillary electrophoresis, and mass spectrometry [183–188].

Affinity chromatography clearly possesses several attributes that make it a viable alternative to current methods of clinical analysis. It can be used for a wide variety of analyses, due to the large number of available ligands and variety of operating formats. This makes affinity chromatography a very flexible technique that can be developed for almost any analyte of clinical interest. When combined with HPLC or other methods, affinity chromatography can also be used to create both robust and reproducible analyses, which is important for the development of any clinical method. It is for these reasons that as more emphasis is placed on rapid specialized testing, affinity chromatography should play an increasing role as an important tool in clinical laboratories.

REFERENCES

1. IUPAC. Nomenclature for chromatography. http://wingate.merck.de/english/services/Chromatographie/iupac/chrnom.htm
2. DS Hage. Affinity chromatography. In: E Katz, R Eksteen, P Shoenmakers, N Miller, eds. Handbook of HPLC. New York: Marcel Dekker, 1998, Chap. 13.
3. GT Hermanson, AK Mallia, PK Smith. Immobilized affinity ligand techniques. New York: Academic Press, 1992.
4. P-O Larsson. High-performance liquid affinity chromatography. Methods Enzymol 104:212–223, 1987.
5. DS Hage. A survey of recent advances in analytical applications of immunoaffinity chromatography. J Chromatogr 715:3–28, 1998.
6. M de Frutos, FE Regnier. Tandem chromatographic-immunological analyses. Anal Chem 65:17A–25A, 1993.
7. TK Mayer, ZR Freedman. Protein glycosylation in diabetes mellitus: a review of laboratory measurements and of their clinical utility. Clin Chim Acta 127:147–184, 1983.
8. AK Mallia, GT Hermanson, RI Krohn, EK Fujimoto, PK Smith. Preparation and use of a boronic acid affinity support for the separation and quantitation of glycosylated hemoglobins. Anal Lett 14:649–661, 1981.
9. S Hjerten, JP Li. High-performance liquid chromatography of proteins on deformed nonporous agarose beads: fast boronate affinity chromatography of haemoglobin at neutral pH. J Chromatogr 500:543–553, 1990.
10. R Fluckiger, T Woodtli, W Berger. Quantitation of glycosylated hemoglobin by boronate affinity chromatography. Diabetes 33:73–76, 1984.
11. BJ Gould, PM Hall, JGH Cook. Measurement of glycosylated haemoglobins using an affinity chromatography method. Clin Chim Acta 125:41–48, 1982.
12. DC Klenk, GT Hermanson, RI Krohn, EK Fujimoto, AK Mallia, PK Smith, JD England, HM Wiedmeyer, RR Little, DE Goldstein. Determination of glycosylated hemoglobin by affinity chromatography: comparison with colorimetric and ion-exchange methods, and effects of common interferences. Clin Chem 28:2088–2094, 1982.
13. RN Johnson, JR Baker. Inaccuracy in measuring glycated albumin concentration by thiobarbituric acid colorimetry and by boronate chromatography. Clin Chem 34:1456–1459, 1988.
14. RP Singhal, SSM DeSilva. Boronate affinity chromatography. Adv Chromatogr 31:293–335, 1992.
15. N Kitagawa, LG Treat Clemens. Chromatographic study of immobilized boronate stationary phases. Anal Sci 7:195–198, 1991.
16. T Stevenson. Glycosal: the first rapid, point-of-care test for the determination

of hemoglobin A1c in patients with diabetes. Diabetes Technol Ther 1:425–431, 1999.
17. Y Li, EL Larsson, H Jungvid, I Galaev, B Mattiasson. Affinity chromatography of neoglycoproteins. Bioseparation 9:315–323, 2000.
18. AC Silver, E Lamb, WR Cattell, ABSJ Dawnay. Investigation and validation of the affinity chromatography method for measuring glycated albumin in serum and urine. Clin Chim Acta 202:11–22, 1991.
19. M Pantehini, R Bonora, F Pagani. Determination of glycated apolipoprotein B in serum by a combination of affinity chromatography and immunonephelometry. Ann Clin Biochem 31:544–549, 1994.
20. K Shishino, M Murase, H Makino, S Saheki. Glycated apolipoprotein A-I assay by combination of affinity chromatography and latex immunoagglutination. Ann Clin Biochem 37:498–506, 2000.
21. N Singh, RC Willson. Boronate affinity adsorption of RNA: possible role of conformational changes. J Chromatogr A 840:205–213, 1999.
22. KP Peterson, JG Pavlovich, D Goldstein, R Little, J England, CM Peterson. What is hemoglobin A1c? An analysis of glycated hemoglobins by electrospray ionization mass spectrometry. Clin Chem 44:1951–1958, 1998.
23. IE Liener, N Sharon, IJ Goldstein. The lectins: properties, functions and applications in biology and medicine. London: Academic Press, 1986.
24. KI Yoshida, M Honda, K Arai, Y Hosoya, H Moriguchi, S Sumi, Y Ueda, S Kitahara. Serial lectin affinity chromatography with concanavalin A and wheat germ agglutinin demonstrates altered asparagine-linked sugar-chain structures of prostatic acid phosphatase in human prostate carcinoma. J Chromatogr B 695:439–443, 1997.
25. S Sumi, K Arai, S Kitahara, K Yoshida. Serial lectin affinity chromatography demonstrates altered asparagine-linked sugar-chain structures of prostate-specific antigen in human prostate carcinoma. J Chromatogr B 727:9–14, 1999.
26. M Tavella, P Alaupovic, C Knight-Gibson, H Tournier, G Schinella, O Mercuri. Separation of ApoA- and ApoB-containing lipoproteins of human plasma by affinity chromatography on concanavalin A. Prog Lipid Res 30: 181–187, 1991.
27. T Inoue, M Yamauchi, G Toda, K Ohkawa. Microheterogeneity with concanavalin A affinity of serum transferrin in patients with alcoholic liver disease. Alcohol Clin Exp Res 20:363A–365A, 1996.
28. S Villanueva, O Steward. Glycoprotein synthesis at the synapse: fractionation of polypeptides synthesized within isolated dendritic fragments by concanavalin A affinity chromatography. Brain Res Mol Brain Res 91:137–147, 2001.
29. MJ Papandreou, C Asteria, K Pettersson, C Ronin, P Beck-Peccoz. Concanavalin A affinity chromatography of human serum gonadotropins: evidence for changes in carbohydrate structure in different clinical conditions. J Clin Endocrin Mcgab 76:1008–1013, 1993.
30. DJ Anderson, EL Branum, JF O'Brien. Liver- and bone-derived isoenzymes

of alkaline phosphatase in serum as determined by high-performance affinity chromatography. Clin Chem 36:240–246, 1990.
31. DG Gonchoroff, EL Branum, JF O'Brien. Alkaline phosphatase isoenzymes of liver and bone origin are incompletely resolved by wheat-germ-lectin affinity chromatography. Clin Chem 35:29–32, 1989.
32. DG Gonchoroff, EL Branum, SL Cedel, BL Riggs, JF O'Brien. Clinical evaluation of high-performance affinity chromatography for the separation of bone and liver alkaline phosphatase isoenzymes. Clin Chim Acta 199:43–50, 1991.
33. T Higuchi, P Xin, MS Buckley, DR Erickson, VP Bhavanandan. Characterization of the rabbit homolog of human MUC1 glycoprotein isolated from bladder by affinity chromatography on immobilized jacalin. Glycobiology 10:659–667, 2000.
34. S Cartellieri, H Helmholz, B Niemeyer. Preparation and evaluation of Ricinus communis agglutinin affinity adsorbents using polymeric supports. Anal Biochem 295:66–75, 2001.
35. L Perez, A Estepa, JM Coll. Purification of the glycoprotein G from viral haemorrhagic septicaemia virus, a fish rhabdovirus, by lectin affinity chromatography. J Virol Methods 76:1–8, 1998.
36. M Geng, X Zhang, M Bina, F Regnier. Proteomics of glycoproteins based on affinity selection of glycopeptides from tryptic digests. J Chromatogr B 752: 293–306, 2001.
37. JL Bundy, C Fenselau. Lectin and carbohydrate affinity capture surfaces for mass spectrometric analysis of microorganisms. Anal Chem 73:751–757, 2001.
38. UD Palanisamy, DJ Winzor, CR Lowe. Synthesis and evaluation of affinity adsorbents for glycoproteins: an artificial lectin. J Chromatogr B 746:265–281, 2000.
39. R Lindmark, C Biriell, J Sjoequist. Quantitation of specific IgG antibodies in rabbits by a solid-phase radioimmunoassay with ^{125}I-protein A from staphylococcus aureus. Scand J Immunol 14:409–420, 1981.
40. PL Ey, SJ Prowse, CR Jenkin. Isolation of pure IgG, IgG_{2a} and IgG_{2b} immunoglobulins from mouse serum using protein A-Sepharose. Immunochemistry 15:429–436, 1978.
41. L Bjorck, G Kronvall. Purification and some properties of streptococcal protein G, a novel IgG-binding reagent. J Immunol 133:969–974, 1984.
42. B Aakerstrom, L Bjoerck. A physiochemical study of protein G, a molecule with unique immunoglobulin G-binding properties. J Biol Chem 261:10240–10247, 1986.
43. M Eliasson, A Olsson, E Palmcrantz, K Wibers, M Inganas, B Guss, M Lindberg, M Uhlen. Chimeric IgG-binding receptors engineered from staphylococcal protein A and streptococcal protein G. J Biol Chem 263: 4323–4327, 1988.
44. S Ohlson. High performance liquid affinity chromatography (HPLAC) with protein A-silica. In: IM Chaiken, M Wilchek, I Parikh, eds. Affinity Chro-

matography and Biological Recognition. New York: Academic Press, 1983, pp 255–256.
45. SC Crowley, RR Walters. Determination of immunoglobulins in blood serum by high-performance affinity chromatography. J Chromatogr 266:157, 1983.
46. P Cassulis, MV Magasic, VA DeBari. Ligand affinity chromatographic separation of serum IgG on recombinant protein G-silica. Clin Chem 37:882–886, 1991.
47. DS Hage, RR Walters. Dual-column determination of albumin and immunoglobulin G in serum by high-performance affinity chromatography. J Chromatogr 386:37–49, 1987.
48. TM Phillips. High performance immunoaffinity chromatography. LC Mag 3:962–972, 1985.
49. AL Dawidowicz, T Rauckyte, J Rogalski. The preparation of sorbents for the analysis of human antithrombin III by means of high performance affinity chromatography. Chromatographia 37:168–172, 1993.
50. AL Dawidowicz, T Rauckyte, J Rogalski. High performance affinity chromatography for analysis of human antithrombin III. J Liq Chromatogr 17: 817–831, 1994.
51. JB Wheatley, MK Kelley, JA Mantali, COA Berry, DE Schmidt Jr. Examination of glutathione S-transferase isoenzyme profiles in human liver using high-performance affinity chromatography. J Chromatogr A 663:53–63, 1994.
52. JB Wheatley, JA Montali, DE Schmidt Jr. Coupled affinity-reversed-phase high-performance liquid chromatography systems for the measurement of glutathione S-transferase in human tissues. J Chromatogr A 676:65–79, 1994.
53. I Abe, N Ito, K Noguchi, M Kazama, KI Kasai. Immobilized urokinase column as part of a specific detection system for plasminogen species separated by high-performance affinity chromatography. J Chromatogr 565:183–195, 1991.
54. M Miyake, E Utsuno, M Noda. Binding of avian ovomucoid to shiga-like toxin type 1 and its utilization for receptor analog affinity chromatography. Anal Biochem 281:202–208, 2000.
55. K Nakano, K Shindo, T Yasaka, H Yamamoto. Reversed-phase liquid chromatographic investigation of nucleosides and bases in mucosa and modified nucleosides in urines from patients with gastrointestinal cancer. J Chromatogr 332:127–137, 1985.
56. R DeCristofaro, R Landolfi, B Bizzi, M Castagnola. Human platelet glycocalicin purification by phenylboronate affinity chromatography coupled to anion-exchange high-performance liquid chromatography. J Chromatogr 426:376–380, 1988.
57. M Fiechtner, J Ramp, B England, MA Knudson, RR Little, JD England, DE England, A Wynn. Affinity binding assay of glycohemoglobin by two-dimensional centrifugation referenced to hemoglobin $A_{1}c$. Clin Chem 38: 2372–2379, 1992.

58. J Sribar, A Copic, A Paris, NE Sherman, F Gubensek, JW Fox, I Krizaj. A high affinity acceptor for phospholipase A2 with neurotoxic activity is a calmodulin. J Biol Chem 276:12493–12496, 2001.
59. JE Scherberich, J Wiemer, C Herzig, P Fischer, W Schoeppe. Isolation and partial characterization of angiotensinase A and aminopeptidase M from urine and human kidney by lectin affinity chromatography and high-performance liquid chromatography. J Chromatogr 521:279–289, 1990.
60. RJ Brushia, TM Forte, MN Oda, BN La Du, JK Bielicki. Baculovirus-mediated expression and purification of human serum paraoxonase 1A. J Lipid Res 42:951–958, 2001.
61. LM Thienpont, GC Depourcq, HJ Nelis, AP De Leenherr. Liquid chromatographic determination of 2-thioxothiazolidine-4-carboxylic acid isolated form urine by affinity chromatography on organomercurial agarose gel. Anal Chem 62:2673–2675, 1990.
62. N Madry, B Auerbach, C Schelp. Measures to overcome HAMA interferences in immunoassays. Anticancer Res 17:2883, 1997.
63. AT Remaley, P Wilding. Macroenzymes: biochemical characterization, clinical significance, and laboratory detection. Clin Chem 35:2261–2270, 1989.
64. LA van Ginkel, H van Blitterswijk, PW Zoontjes, D van den Bosch, RW Stephany. Assay for trenbolone and its metabolite 17α-trenbolone in bovine urine based on immunoaffinity chromatographic clean-up and off-line high-performance liquid chromatography–thin layer chromatography. J Chromatogr 445:385–392, 1988.
65. LA van Ginkel, RW Stephany, HJ van Rossum, H van Blitterswijk, PW Zoontjes, RCM Hooijshuur, J Zuydendorp. Effective monitoring of residues of nortestosterone and its major metabolite in bovine urine and bile. J Chromatogr 489:95–104, 1989.
66. R Bagnati, MG Castelli, L Airoldi, MP Oriundi, A Ubaldi, R Fanelli. Analysis of diethylstilbestrol, dienestrol and hexestrol in biological samples by immunoaffinity extraction and gas chromatography-negative-ion chemical ionization mass spectrometry. J Chromatogr 527:267–278, 1990.
67. LA van Ginkel. Immunoaffinity chromatography, its applicability and limitations in multiresidue analysis of anabolizing and doping agents. J Chromatogr 564:363–384, 1991.
68. M Dubois, X Taillieu, Y Colemonts, B Lansival, J De Graeve, P Delhaut. GC-MS determination of anabolic steroids after multi-immunoaffinity purification. Analyst 123:2611–2616, 1998.
69. S Feng, MA ElSohly, S Salamone, MY Salem. Simultaneous analysis of delta9-THC and its major metabolites in urine, plasma, and meconium by GC-MS using an immunoaffinity extraction procedure. J Anal Toxicol 24:395–402, 2000.
70. PO Edlund, D Westerlund. Direct injection of plasma and urine in automated analysis of catecholamines by coupled-column liquid chromatography with post-column derivatization. J Pharm Biomed Anal 2:315–333, 1984.

71. P Ni, F Guyon, M Caude, R Rosset. Automated determination of catecholamines using on-column extraction of diphenylboronate-catecholamine complexes and high-performance liquid chromatography with electrochemical detection. J Liq Chromatogr 12:1873–1888, 1989.
72. KS Boos, B Wilmers, R Sauerbrey, E Schlimme. Development and performance of an automated HPLC-analyzer for catecholamines. Chromatographia 24:363–370, 1987.
73. PO Edlund. Determination of dihydroxyphenylalanine and dihydroxyphenylacetic acid in biological samples by coupled-column liquid chromatography with dual coulometric amperometric detection. J Pharm Biomed Anal 4:625–639, 1986.
74. L Hansson, M Glad, C Hansson. Boronic acid-silica: a new tool for the purification of catecholic compounds on-line with reversed-phase high-performance liquid chromatography. J Chromatogr 265:37–44, 1983.
75. C Hansson, B Kagedal, M Kallberg. Determination of 5-*S*-cysteinyldopa in human urine by direct injection in coupled-column high-performance liquid chromatography. J Chromatogr 420:146–151, 1987.
76. P-O Larsson, M Glad, L Hansson, O Mansson, S Ohlson, K Mosbach. High-performance liquid affinity chromatography. Adv Chromatogr 21:41–85, 1983.
77. E Hagemeier, K-S Boos, E Schlimme, K Lechtenboerger, A Kettrup. Synthesis and application of a boronic acid-substituted silica for high-performance liquid affinity chromatography. J Chromatogr 268:291–295, 1983.
78. E Hagemeier, K Kemper, K-S Boos, E Schlimme. On-line high-performance liquid affinity chromatography—high-performance liquid chromatography analysis of monomeric ribnucleoside compounds in biological fluids. J Chromatogr 282:663–669, 1983.
79. B-M Eriksson, M Wikstrom. Determination of vanilmandelic acid in urine by coupled-column liquid chromatography combining affinity to boronate and separation by anion exchange. J Chromatogr 567:1–9, 1991.
80. K Yasukawa, F Abe, N Shida, Y Koizumi, T Uchida, K Noguchi, K Shima. High-performance affinity chromatography system for the rapid, efficient assay of glycated albumin. J Chromatogr 597:271–275, 1992.
81. M Mogi, M Harada, T Adachi, K Kojima, T Nagatsu. Selective removal of β2-microglobulin from human plasma by high-performance immunoaffinity chromatography. J Chromatogr 496:194–200, 1989.
82. DW Armstrong. Optical isomer separation by liquid chromatography. Anal Chem 59:84A–91A, 1987.
83. Chiral drugs. Chem Eng News 71(Sept. 27):38–65.
84. S Allenmark. Chromatographic enantioseparation: methods and applications. New York: Ellis Horwood, 1991, Chap. 7.
85. F Li, SF Cooper, M Cote, C Ayotte. Determination of the enantiomers of bunolol in human urine by high-performance liquid chromatography on a chiral ACP stationary phase and identification of their metabolites by gas chromatography-mass spectrometry. J Chromatogr B 660:327–339, 1994.

86. D Haupt. Determination of citalopram enantiomers in human plasma by liquid chromatographic separation on a Chiral-AGP column. J Chromatogr B 685:299–305, 1996.
87. S Menzel-Soglowek, G Geisslinger, K Brune. Stereoselective high-performance liquid chromatographic determination of ketoprofen, ibuprofen and fenoprofen in plasma using a chiral α_1-acid glycoprotein column. J Chromatogr 532:295–303, 1990.
88. G Geisslinger, S Menzel-Soglowek, O Schuster, K Brune. Sterioselective high-performance liquid chromatographic determination of flurbiprofen in human plasma. J Chromatogr 573:163–167, 1992.
89. K-J Pettersson, A Olsson. Liquid chromatographic determination of the enantiomers of ibuprofen in plasma using a chiral AGP column. J Chromatogr 563:414–418,1991.
90. G Geisslinger, S Menzel-Soglowek, H-D Camp, K Brune. Stereoselective high-performance liquid chromatographic determination of the enantiomers of ketamine and norketamine in plasma. J Chromatogr 568:165–176, 1991.
91. N Schmidt, K Brune, G Geisslinger. Stereoselective determination of the enantiomers of methadone in plasma using high-performance liquid chromatography. J Chromatogr 583:195–200, 1992.
92. O Beck, LO Boreus, P LaFolie, G Jacobsson. Chiral analysis of methadone in plasma by high-performance liquid chromatography. J Chromatogr 570:198–202, 1991.
93. K Kristensen, HR Angelo, T Bloemmer. Enantioselective high-performance liquid chromatographic method for the determination of methadone in serum using an AGP and a CN column as chiral and analytical column, respectively. J Chromatogr A 666:283–287, 1994.
94. Y-Q Chu, IW Wainer. Determination of the enantiomers of verapamil and norverapamil in serum using coupled achiral-chiral high-performance liquid chromatography. J Chromatogr 497:191–200, 1989.
95. F Mangani, G Luck, C Fraudeau, E Verette. On-line column switching high-performance liquid chromatography analysis of cardiovascular drugs in serum with automated sample clean-up and zone cutting technique to perform chiral separation. J Chromatogr A 762:235–241, 1997.
96. A Suzuki, S Takagaki, H Suzuki, K Noda. Determination of the R,R- and S,S-enantiomers of vamicamide in human serum and urine by high-performance liquid chromatography on a Chiral-AGP column. J Chromatogr 617:279–284, 1993.
97. H Fieger and G Blaschke. Direct determination of the enantiomeric ratio of verapamil, its major metabolite norverapamil and gallopamil in plasma by chiral high-performance liquid chromatography. J Chromatogr 575:255–260, 1992.
98. DJ Jones, KT Nguyen, MJ McLeish, DP Crankshaw, DJ Morgan. Determination of (*R*)-(+)- and (*S*)-(−)-isomers of thiopentone in plasma by chiral high-performance liquid chromatography. J Chromatogr 675:174–179, 1996.

99. L Silan, P Jadaud, LR Whitfield, IW Wainer. Determination of low levels of the stereoisomers of leucovorin and 5-methyltetrahydrofolate in plasma using a coupled chiral–achiral high-performance liquid chromatographic system with post-chiral column peak compression. J Chromatogr 532:227–236, 1990.
100. JW Kelly, JT Stewart, CD Blanton. HPLC separation of pentazocine enantiomers in serum using an ovomucoid chiral stationary phase. Biomed Chromatogr 8:255–257, 1994.
101. J Haginaka, J Wakai. B-cyclodextrin bonded silica for direct injection analysis of drug enantiomers in serum by liquid chromatography. Anal Chem 63:997–1000, 1990.
102. B Rochat, M Amey, P Baumann. Analysis of enantiomers of citalopram and its demethylated metabolites in plasma of depressive patients using chiral reverse-phase liquid chromatography. Ther Drug Monit 17:273–279, 1995.
103. D Castoldi, A Oggioni, MI Renoldi, E Ratti, S DiGiovine, A Bernaraggi. Assay of moguisteine metabolites in human plasma and urine: conventional and chiral high-performance liquid chromatographic methods. J Chromatogr B 655:243–252, 1994.
104. C Pham-Huy, B Radenen, A Sahui-Gnassi, JR Claude. High-performance liquid chromatographic determination of (*S*)- and (*R*)-propranolol in human plasma and urine with a chiral β-cyclodextrin bonded phase. J Chromatogr B 665:125–132, 1995.
105. U Kragh-Hansen. Molecular aspects of ligand binding to serum albumin. Pharmacol. Rev 45:17–53, 1981.
106. DC Carter, JX Ho. Structure of serum albumin. Adv Protein Chem 45:153–203, 1994.
107. IM Chaiken, ed. Analytical Affinity Chromatography. Boca Raton, FL: CRC Press, 1987.
108. YC Guillaume, E Peyrin, A Berthelot. Chromatographic study of magnesium and calcium binding to immobilized human serum albumin. J Chromatogr B 728:167–174, 1999.
109. IW Wainer. Enantioselective high-performance liquid affinity chromatography as a probe of ligand–biopolymer interactions: an overview of a different use for high-performance liquid chromatographic chiral stationary phases. J Chromatogr A 666:221–234, 1994.
110. T Cserhati, K Valko. Chromatographic Determination of Molecular Interactions. Boca Raton, FL: CRC Press, 1994.
111. DS Hage, SA Tweed. Recent advances in chromatographic and electrophoretic methods for the study of drug–protein interactions. J Chromatogr B 699:499–525, 1997.
112. R Knorle, E Schniz, TJ Feurstein. Drug accumulation in melanin: an affinity chromatographic study. J Chromatogr B 714:171–179, 1998.
113. WE Lindup. Progress in drug metabolism, Vol. 10. New York: Taylor and Francis, 1987.
114. TC Kwong. Free drug measurements: methodology and clinical significance. Clin Chim Acta 151:193–216, 1985.

115. CK Svensson, MN Woodruff, JG Baxter, D Laika. Free drug concentration monitoring in clinical practice: rationale and current status. Clin Pharmacokinet 11:450–469, 1986.
116. J Barre, C Hamberger, F Didey, JC Duche, JP Tillement. Principles of methods for drug determination in biological fluids applied to therapeutic monitoring. Feuill Biol 28:47–55, 1987.
117. GA Ascoli, C Bertucci, P Salvadori. Ligand binding to a human serum albumin stationary phase: use of same-drug competition to discriminate pharmacologically relevant interactions. Biomed Chromatogr 12:248–254, 1998.
118. DS Hage, A Sengupta. Characterization of the binding of digitoxin and acetyldigitoxin to human serum albumin by high-performance affinity chromatography. J Chromatogr B 724:91–100, 1999.
119. DS Hage, A Sengupta. Studies of protein binding to nonpolar solutes by using zonal elution and high-performance affinity chromatography: interactions of *cis*- and *trans*-clomiphene with human serum albumin in the presence of beta-cyclodextrin. Anal Chem 70:4602–4609, 1998.
120. VN Russeva, ZD Zhivkova. Protein binding of some nonsteroidal anti-inflammatory drugs studied by high-performance liquid affinity chromatography. Int J Pharm 180:69–74, 1999.
121. ZD Zhivkova, VN Russeva. Stereoselective binding of ketoprofen enantiomers to human serum albumin studied by high-performance liquid affinity chromatography. J Chromatogr B 714:277–283, 1998.
122. V Russeva, Z Zhivkova, K Prodanova, R Rakovska. Protein binding of piroxicam studied by means of affinity chromatography and circular dichroism. J Pharm Pharmacol 51:49–52, 1999.
123. S Refetoff, PR Larsen. Transport, cellular uptake, and metabolism of thyroid hormone. In: LJ DeGroot, ed. Endocrinology. Philadelphia, PA: WB Saunders, 1989.
124. U Westphal. Steroid–protein interactions. New York: Springer-Verlag, 1971.
125. RH Levy, TA Moreland. Rationale for monitoring free drug levels. Clin Pharmacokinet 9:1–9, 1984.
126. B Loun, DS Hage. Characterization of thyroxine–albumin binding using high-performance affinity chromatography. 2. Comparison of the binding of thyroxine, triiodothyronine and related compounds at the warfarin and indole sites of human serum albumin. J Chromatogr B 665:303–314, 1995.
127. B Loun, DS Hage. Chiral separation mechanisms in protein based HPLC columns. 2. Kinetic studies of (*R*)- and (*S*)-warfarin binding to immobilized human serum albumin. Anal Chem 68:1218–1225, 1996.
128. J Yang, DS Hage. Role of binding capacity versus binding strength in the separation of chiral compounds in protein-based high-performance liquid chromatographic columns: interactions of D- and L-tryptophan with human serum albumin. J Chromatogr B 725:273–285, 1996.
129. B Loun, DS Hage. Chiral separation mechanisms in protein-based HPLC columns. 1. Thermodynamic studies of (*R*)- and (*S*)-warfarin binding to immobilized human serum albumin. Anal Chem 66:3814–1822, 1994.

130. J Yang, DS Hage. Characterization of the binding and chiral separation of D- and L-tryptophan on a high-performance immobilized human serum albumin column. J Chromatogr 645:241–250, 1993.
131. B Loun, DS Hage. Characterization of thyroxine-albumin binding using high-performance affinity chromatography. 1. Interactions at the warfarin and indole sites of albumin. J Chromatogr 579:225–235, 1992.
132. E Domenici, C Bertucci, P Salvadori, S Motellier, IW Wainer. Immobilized serum albumin: rapid HPLC probe of stereoselective protein-binding interactions. Chirality 2:263–268, 1990.
133. E Domenici, C Bertucci, P Salvadori, G Felix, I Cahagne, S Motellier, IW Wainer. Synthesis and chromatographic properties of an HPLC chiral stationary phase based upon human serum albumin. Chromatographia 29:170–176, 1990.
134. E Domenici, C Bertucci, P Salvadori, IW Wainer. Use of a human serum albumin-based high-performance liquid chromatography chiral stationary phase for the investigation of protein binding: detection of the allosteric interaction between warfarin and benzodiazepine binding sites. J Pharm Sci 80:164–166, 1991.
135. TAG Noctor, CD Pham, R Kaliszan, IW Wainer. Stereochemical aspects of benzodiazepine binding to human serum albumin. I. Enantioselective high-performance liquid affinity chromatographic examination of chiral and achiral binding interactions between 1,4-benzodiazepine and human serum albumin. Mol Pharmacol 42:506–511, 1992.
136. TAG Noctor, IW Wainer, DS Hage. Allosteric and competitive displacement of drugs from human serum albumin by octanoic acid, as revealed by high-performance liquid affinity chromatography, on a human serum albumin-based stationary phase. J Chromatogr 577:305–315, 1992.
137. B Sebille, R Zini, CV Madjar, N Thuaud, JP Tillement. Separation procedures used to reveal and follow drug–protein binding. J Chromatogr 531:51–77, 1990.
138. G Schill, IW Wainer, SA Barkin. Chiral separations of cationic and anionic drugs on an α1-acid glycoprotein-bonded stationary phase (Enantiopac). II. Influence of mobile phase additives and pH on chiral resolution and retention. J Chromatogr 365:73–88, 1986.
139. S Allenmark, B Bomgren, H Boren. Direct liquid chromatographic separation of enantiomers on immobilized protein stationary phases. IV. Molecular interaction forces and retention behavior in chromatography on bovine serum albumin as a stationary phase. J Chromatogr 316:617–624, 1984.
140. S Allenmark, B Bomgren, H Boren. Direct LC separation of enantiomers on immobilized protein stationary phases. III. Optical resolution of a series of *N*-aroyl D,L-amino acids by high-performance liquid chromatography on bovine serum albumin covalently bound to silica. J Chromatogr 264:63–68, 1983.
141. J Hermansson. Direct liquid chromatographic resolution of racemic drugs using α1-acid glycoprotein as the chiral stationary phase. J Chromatogr 269:71–80, 1983.

142. T Miwa, T Miyakawa, M Kayano, Y Miyake. Application of an ovomucoid-conjugated column for the optical resolution of some pharmaceutically important compounds. J Chromatogr 408:316–322, 1987.
143. S Allenmark, S Andersson, J Bojarski. Direct liquid chromatographic separation of enantiomers on immobilized protein stationary phases. VI. Optical resolution of a series of racemic barbiturates: studies of substituent and mobile phase effects. J Chromatogr 436:479–483, 1988.
144. T Fornstedt, G Zhong, Z Bensetiti, G Guiochon. Experimental and theoretical study of the adsorption behavior and mass transfer kinetics of propranolol enantiomers on cellulase protein as the selector. Anal Chem 68: 2370–2378, 1996.
145. TAG Noctor, CD Pham, R Kaliszan, IW Wainer. Stereochemical aspects of benzodiazepine to human serum albumin. I. Enantioselective high-performance liquid affinity chromatographic examination of chiral and achiral binding interaction between 1,4-benzodiazepines and human serum albumin. Mol Pharmacol 42:506–511, 1992.
146. R Kaliszan, TAG Noctor, IW Wainer. Stereochemical aspects of benzodiazepine binding to human serum albumin. II. Quantitative relationships between structure and enantioselective retention in high performance liquid affinity chromatography. Mol Pharmacol 42:512–517, 1992.
147. R Kaliszan. Retention data from affinity high-performance liquid chromatography in view of chemometrics. J Chromatogr B 715:229–244, 1998.
148. TAG Noctor, IW Wainer. The use of displacement chromatography to alter retention and enatioselectivity on a human serum albumin-based HPLC chiral stationary phase: a minireview. J Liq Chromatogr 16:783–800, 1993.
149. L Dalgaard, JJ Hansen, JL Pedersen. Resolution and binding site determination of D,L-thyronine by high-performance liquid chromatography using immobilized albumin as a chiral stationary phase. Determination of the optical purity of thyroxine in tablets. J Pharm Biomed Anal 7:361–368, 1989.
150. DS Hage, TAG Noctor, IW Wainer. Characterization of the protein binding of chiral drugs by high-performance affinity chromatography. Interaction of *R*- and *S*-ibuprofen with human serum albumin. J Chromatogr A 693:23–32, 1995.
151. S Rahim, A-F Aubry. Location of binding sites in immobilized human serum albumin for some non-steroidal anti-inflammatory drugs. J Pharm Sci 84:949–952, 1995.
152. A-F Aubry, N Markoglou, A McGann. Comparison of drug binding interactions on human, rat and rabbit serum albumin using high-performance displacement chromatography. Comp Biochem Physiol 112C:257–266, 1995.
153. Y Zhang, F Leonessa, R Clarke, IW Wainer. Development of an immobilized P-glycoprotein stationary phase for on-line liquid chromatographic determination of drug-binding affinities. J Chromatogr B 739:33–37, 2000.
154. NI Nakano, Y Shimamori, S Yamaguchi. Binding capacities of human serum

albumin monomer and dimer by continuous frontal affinity chromatography. J Chromatogr 237:225–232, 1982.
155. NI Nakano, Y Shimamori, S Yamaguchi. Mutual displacement interactions in the binding of two drugs to human serum albumin by frontal affinity chromatography. J Chromatogr 188:347–356, 1980.
156. A Chattopadhyay, T Tian, L Kortum, DS Hage. Development of tryptophan-modified human serum albumin columns for site-specific studies of drug–protein interactions by high-performance affinity chromatography. J Chromatogr B 715:183–190, 1998.
157. C Koch, L Borg, Skjodt, G Houen. Affinity chromatography of serine proteases on the triazine dye ligand Cibacron Blue F3G-A. J Chromatogr B 718:41–46, 1998.
158. K Jones. A review of biotechnology and large scale affinity chromatography. Chromatographia 21:469–480, 1991.
159. MD Scaween. Dye affinity chromatography. Anal Proceed 28:143–144, 1991.
160. YD Clonis, NE Labrou, VP Kotsira, C Mazitsos, S Melissis, G Gogolas. Biomimetic dyes as affinity chromatography tools in enzyme purification. J Chromatogr A 891:33–44, 2000.
161. J Kaminska, J Dzieciol, J Koscielak. Triazine dyes as inhibitors and affinity ligands of glycosyltransferases. Glycoconj J 16:719–723, 1999.
162. S Mori, M Nishibori, K Yamaoka, M Okamoto. One-step purification of rabbit histidine rich glycoprotein by dye-ligand affinity chromatography with metal ion requirement. Arch Biochem Biophys 383:191–196, 2000.
163. G Alberghina, S Fisichella, E Renda. Separations of G structures formed by a 27-mer guanosine-rich oligodeoxyribonucleotide by dye-ligand affinity chromatography. J Chromatogr A 840:51–58, 1999.
164. SA Lopatin, VP Varlamov. New trends in immobilized metal affinity chromatography of proteins. Appl Biochem Microbiol 31:221–227, 1995.
165. JJ Winzerling, P Berna, J Porath. How to use immobilized metal ion affinity chromatography. Methods 4:4–13, 1992.
166. J Porath. Immobilized metal ion affinity chromatography. Protein Expression Purif 3:263–281, 1992.
167. J Laroche-Traineau, G Clofent-Sanchez, X Santarelli. Three-step purification of bacterially expressed human single-chain Fv antibodies for clinical applications. J Chromatogr B 737:107–117, 2000.
168. Q Liu, P Willson, S Attoh-Poku, LA Babiuk. Bacterial expression of an immunologically reactive PCV2-ORF2 fusion protein. Protein Expr Purif 21:115–120, 2001.
169. H Wu, DF Bruley. Homologous human blood protein separation using immobilized metal affinity chromatography: protein C separation from prothrombin with application to the separation of factor IX and prothrombin. Biotechnol Prog 15:928–931, 1999.
170. M Zaveckas, B Baskeviciute, V Luksa, G Zvirblis, V Chmieliauskaite, V Bumelis, H Pesliakas. Comparative studies of recombinant human granulocyte-colony stimulating factor, its Ser-17 and (His)6-tagged forms interaction

with metal ions by means of immobilized metal ion affinity partitioning. Effect of chelated nickel and mercuric ions on extraction and refolding of proteins from inclusion bodies. J Chromatogr A 904:145–169, 2000.
171. LB McGown, MJ Joseph, JB Pitner, JB Vonk, CP Linn. The nucleic acid ligand: a new tool for molecular recognition. Anal Chem 67:663A–668A, 1995.
172. C Turek, L Gold. Systematic evolution of ligands by exponential enrichment: rna ligands to bacteriophage T4 DNA polymerase. Science 249:505–510, 1990.
173. AD Ellington, JW Szostak. In vitro selection of RNA molecules that bind specific ligands. Nature 346:818–822, 1990.
174. TS Romig, C Bell, DW Drolet. Aptamer affinity chromatography: combinatorial chemistry applied to protein purification. J Chromatogr B 731:275–284, 1999.
175. D Kriz, O Ramstrom, K Mosbach. Molecular imprinting: new possibilities for sensor technology. Anal Chem 69:345A–349A, 1997.
176. B Sellergran. Noncovalent molecular imprinting: antibody-like molecular recognition in polymeric network materials. Trends Anal Chem 16:310–319, 1997.
177. PT Vallano, VT Remcho. Affinity screening by packed capillary high-performance liquid chromatography using molecular imprinted sorbents. I. Demonstration of feasibility. J Chromatogr A 888:23–34, 2000.
178. ZG Wang, LJ Williams, XF Zhang, A Zatorski, V Kudryashov, G Ragupathi, M Spassova, W Bornmann, SF Slovin, HI Scher, PO Livingston, KO Lloyd, SJ Danishefsky. Polyclonal antibodies from patients immunized with a globo H-keyhole limpet hemocyanin vaccine: isolation, quantification, and characterization of immune responses by using totally synthetic immobilized tumor antigens. Proc Natl Acad Sci USA 97:2719–2724, 2000.
179. I Gottschalk, YM Li, P Lundahl. Chromatography on cells: analyses of solute interactions with the glucose transporter Glut1 in human red cells adsorbed on lectin-gel beads. J Chromatogr B 739:55–62, 2000.
180. Q Yang, XY Liu, M Hara, P Lundahl, J Miyake. Quantitative affinity chromatographic studies of mitochondrial cytochrome c binding to bacterial photosynthetic reaction center, reconstituted in liposome membranes and immobilized by detergent dialysis and avidin–biotin binding. Anal Biochem 280:94–102, 2000.
181. R Espuny, D Bahia, RM Barreto Cicarelli, C Codony, A Khaouja, AM Avin, R Eritja, M Bach-Elias. Preparation of N2,N2,7-trimethylguanosine affinity columns. Nucleosides Nucleotides 18:125–136, 1999.
182. CR O'Riordan, AL Lachapelle, KA Vincent, SC Wadsworth. Scaleable chromatographic purification process for recombinant adeno-associated virus (rAAV). J Gene Med 2:444–454, 2000.
183. A Farjam, JJ Vreuls, WJGM Cuppen, UAT Brinkman, GJ de Jong. Direct introduction of large-volume urine samples into an on-line immunoaffinity sample pretreatment-capillary gas chromatography system. Anal Chem 63:2481–2487, 1991.

184. TM Phillips, JJ Chmielinska. Immunoaffinity capillary electrophoresis analysis of cyclosporine in tears. Biomed Chromatogr 8:242–246, 1994.
185. NA Guzman. Biomedical applications of on-line preconcentration-capillary electrophoresis using an analyte concentrator: investigation of design options. J Liq Chromatogr 18:3751–3768, 1995.
186. LJ Cole, RT Kennedy. Selective preconcentration of capillary zone electrophoresis using protein G immunoaffinity capillary chromatography. Electrophoresis 16:549–556, 1995.
187. J Cai, J Henion. On-line immunoaffinity extraction-coupled column capillary liquid chromatography/tandem mass spectrometry: trace analysis of LSD analogs and metabolites in human urine. Anal Chem 68:72–78, 1996.
188. CS Creaser, SJ Feely, E Houghton, M Seymour, P Teale. On-line immunoaffinity chromatography–high-performance liquid chromatography–mass spectrometry for the determination of dexamethasone. Anal Commun 33:5–8, 1996.

9

Immunoaffinity Chromatography in Clinical Analysis

David S. Hage
University of Nebraska, Lincoln, Nebraska, U.S.A.
William Clarke
Johns Hopkins School of Medicine, Baltimore, Maryland, U.S.A.

1 INTRODUCTION

Immunoaffinity chromatography (IAC) is a separation and analysis technique that has a long history of use in biomedical studies. IAC is a type of affinity chromatography (described in the previous chapter) in which the specific type of stationary phase, or affinity ligand, within the column is an antibody or related binding agent. Immunoaffinity chromatography is the most common type of affinity chromatography and is rapidly gaining in use as an analytical tool for the clinical laboratory. Reasons for this interest include the specificity of antibodies in their binding to other agents and the relative ease with which antibodies can be obtained to a variety of compounds. This gives IAC high selectivity while also making it possible to adapt this approach to many analytes of clinical interest.

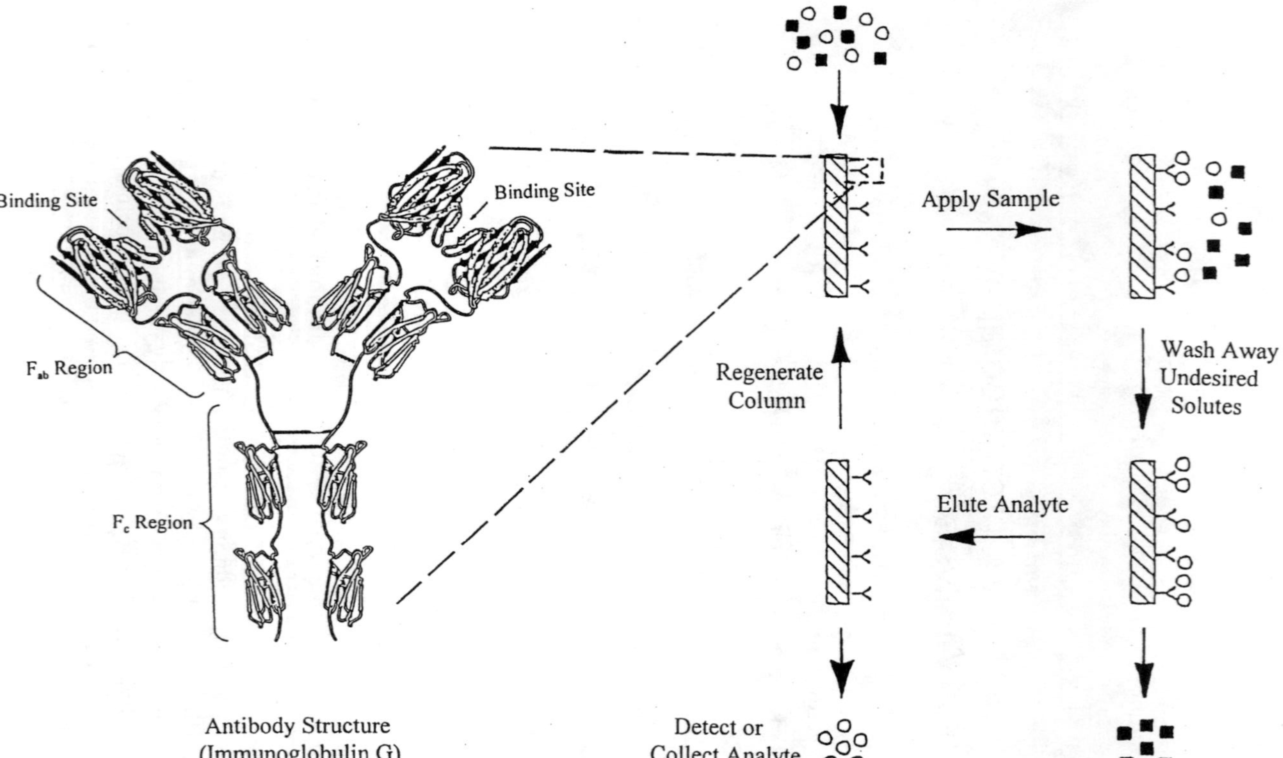

FIGURE 1 The structure of an antibody and a scheme showing how antibodies are used in immunoaffinity chromatography for the selective capture and elution of compounds. (Reproduced with permission from Ref. 19.)

The basic operation of IAC is relatively simple. As is shown in Fig. 1, a column is first prepared that contains antibodies or related ligands that are immobilized or adsorbed onto a solid-phase support. Next, a sample containing solutes that can bind to these ligands is applied to the column under conditions in which strong binding is obtained. In clinical labs, this is typically done by using an application buffer that is at or near a physiological pH (i.e., pH 7–7.4). As the sample is applied to the column, the desired solutes are bound to the ligands while other chemicals wash through nonretained. Later, a second buffer is applied to the column that causes the retained solutes to elute from the column. These compounds are then collected for further use or monitored directly by an on-line detector. If desired, the IAC column can be placed back into the initial application buffer, which allows the antibody-based ligands to regenerate. Another sample can then be applied to the column and the whole process repeated. In the next few sections of this chapter, the various components of this scheme will be examined in more detail (e.g., the antibodies, supports, and solvent conditions). Following this, the various formats in which IAC has been used for clinical testing will be considered.

1.1 Antibody Structure and Production

The item that gives IAC its selectivity is the antibody or antibody-related ligand that is used as the stationary phase. An *antibody*, also known as an *immunoglobulin*, is simply a type of glycoprotein that is produced by the body in response to a foreign agent, or *antigen*. It has been estimated that the human body can produce antibodies capable of binding between a million and a billion different foreign agents. As shown in Fig. 1, the basic structure of a typical antibody (i.e., immunoglobulin G) consists of four polypeptides that are linked by disulfide bonds to form a Y- or T-shaped structure. The amino acids in the lower, stem region generally have the same sequence within a given group of antibodies but show high variability at the two identical binding sites that are located at the upper ends of an antibody. In fact, it is this difference in amino acid composition at or near the binding sites that allows the body to produce antibodies with a variety of binding specificities and affinities to different chemical or biological substances that might enter the body.

One way to produce antibodies for a given test agent is to inject a solution of the substance (or a conjugate of this agent and a large carrier) into a laboratory animal such as a rabbit or a mouse. Samples of the animal's blood are then collected at specified intervals (typically a few weeks or months after injection) to collect any antibodies that have been produced against the foreign agent. This approach results in a heterogeneous mixture of antibodies that bind with a range of strengths and to various sites on the original injected agent or conjugate. These are known as *polyclonal antibodies*, since they are produced by different cell lines within the body. Another approach for making antibodies is to isolate single antibody-producing cells and combine them with carcinoma cells to produce new hybrid cells that are relatively easy to grow in culture. These new combined cells are referred to as *hybridomas* and their product is a single type of well-defined antibody known as a *monoclonal antibody*.

1.2 IAC Supports and Solvents

It can be seen from Fig. 1 that several components are needed along with antibodies for IAC to work. These other items include a support to hold the ligand within the column, a means for attaching the ligand to this support, and solvents to apply and elute analytes from the column. As was discussed in the last chapter for affinity chromatography, either low- or high-performance supports can be used in IAC methods. Examples of low-performance materials used for this purpose include carbohydrate-related materials and synthetic organic supports, like agarose, cellulose, acrylamide polymers, and polymethacrylate derivatives. High-performance supports used in IAC are diol-bonded silica or glass beads, azalactone supports, and glycol-coated perfusion media.

In practice, both low- and high-performance methods have found use in the clinical setting. Antibodies immobilized to low-performance supports are mainly used for analyte concentration or extraction from a sample. This is due to the low cost and relative ease with which a low-performance IAC column can be created and operated. On the other hand, the use of high-performance IAC materials, giving rise to a method known as *high-performance immunoaffinity chromatography (HPIAC)*, is more common when producing automated systems for analyte detection. This is due to the increased stability of these latter supports as well as the greater precision inherent in HPLC methods.

There are numerous techniques for immobilizing antibodies and related ligands to low- or high-performance supports. For instance, antibodies can be directly coupled to many supports by reacting the free amine groups on antibodies with materials activated by agents such as *N*,*N'*-carbonyl diimidazole, cyanogen bromide, and *N*-hydroxysuccinimide. A similar result is accomplished by using materials that have been treated to produce reactive epoxide or aldehyde groups on their surfaces. Antibodies and antibody fragments can also be immobilized through more site-selective methods. For example, the free sulfhydryl groups generated during the production of antibody Fab fragments can be used to couple these fragments to supports by using surfaces that are activated by the divinylsulfone, epoxy, iodoacetyl/bromoacetyl, maleimide, TNB-thiol, or tresyl chloride/tosyl chloride methods. Intact antibodies can undergo site-selective immobilization by oxidizing the carbohydrate residues in their stem region with periodate to produce aldehyde groups, which can then react with a hydrazide or amine-containing support [19].

Antibodies can also be placed onto IAC supports through simple noncovalent adsorption. One example is the conjugation of antibodies with biotin, which can then be used to bind these antibodies noncovalently to a support containing immobilized streptavidin. Another approach for indirect immobilization involves adsorbing the antibody to a secondary ligand such as protein A or protein G, as described in the last chapter. This makes use of the ability of both protein A and G to bind strongly to the stem region of many antibodies at a neutral pH, along with the ability of these ligands to release the adsorbed antibodies when there is a decrease in pH. This method is attractive when high antibody activity is needed and/or when it is desirable to have frequent replacement of the antibodies in the IAC column. This allows good long-term reproducibility for the column binding capacity but does require the use of much larger amounts of antibody than direct immobilization methods.

Although the selection of application buffer for an IAC column is usually straightforward (i.e., often being a neutral pH buffer), the selection of an appropriate elution solvent is not as simple. It is possible to use isocratic elution for some weak affinity antibodies, but this does not work for the high or moderate affinity antibodies that are used in the majority of IAC columns. Instead, the retained solutes must be eluted by changing the column conditions to lower the effective binding constant of the antibodies for the retained compound. This is often accomplished by using an acidic buffer (i.e., pH 1–3) and a step elution scheme. Another approach is to perform gradient elution by gradually increasing the amount of a chaotropic agent, organic modifier, or denaturing agent that is present in the mobile phase. The proper choice of elution conditions is critical in analytical applications of IAC in order to provide rapid release of analyte from the column without causing permanent harm to the immobilized antibodies or support. This item must currently be addressed on a case-by-case basis and is particularly important to consider when the same immunoaffinity column is to be employed for a large number of samples.

2 DIRECT DETECTION OF ANALYTES BY IAC

There are a variety of formats that can be used with IAC to analyze samples. The simplest approach is to use the scheme shown in Fig. 1 to capture and elute an analyte, followed by either on-line or off-line detection. This is known as the *direct detection* or *on/off mode* of IAC. In HPLC-based methods, the most common approach for directly detecting analytes as they elute from immunoaffinity columns is to use on-line UV/vis absorbance measurements. This works well for proteins, peptides, and other compounds that are present at moderate concentrations and that have relatively good chromophores in their structures. However, special methods for the detection of low-concentration analytes have also been devised. Examples include the use of precolumn derivatization to place fluorescent labels [10,11,15] or radiolabels [4] onto solutes prior to injection. Alternatively, the analytes eluted from an IAC column can be collected and later measured off-line by methods like immunoassays [8,12] or receptor assays [11].

The past fifteen years have produced many clinical reports utilizing direct analyte detection by IAC and HPIAC. Some applications that have been reported are methods for the analysis of anti-idiotypic antibodies [1,2], glucose-containing tetrasaccharide [3,4], human serum albumin [5,6], immunoglobulin G [7], immunoglobulin E [8], interferon [9,10], tumor necrosis factor-α [10], interleukins [10,11], β_2-microglobulin [12], fibrinogen [13], and transferrin [14].

One example of direct analyte detection by immunoaffinity chromatography is shown in Fig. 2. In this case, an HPIAC method was used to determine granulocyte colony stimulating factor (GCSF) in plasma, bone marrow aspirates, and cerebrospinal fluid obtained from chemotherapy patients [15]. The immunoaffinity column for this study was prepared by incubating biotinylated monoclonal anti-GCSF antibodies with avidin-coated glass beads, which were packed into a 50 $\times$ 4.6 mm ID column. The amount of GCSF in a 100 μL sample was determined by using precolumn derivatization to react GCSF with *o*-phthaldialdehyde and then measuring the fluorescence of the resulting conjugate as it eluted from the immunoaffinity column. The retained peak for GCSF appeared at approximately 7 min, or 2 min after initiation of a pH gradient for elution [15]. Similar results, with total analysis times of 5–15 min per injection, have been reported in many other clinical applications of direct detection by HPIAC [19].

In addition to the traditional on/off mode of IAC, some weak monoclonal antibodies have been used under constant elution conditions to separate carbohydrate antigens [16]. In other work, biotinylated IgY

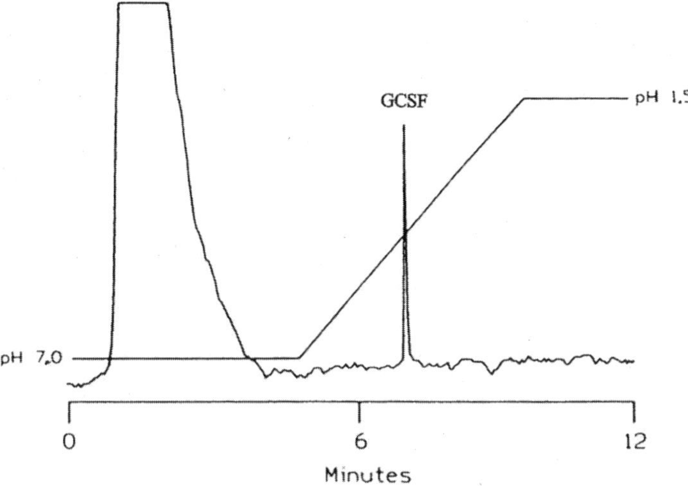

FIGURE 2 An example of the direct detection of an analyte by IAC, as illustrated by the injection and separation of granulocyte colony stimulating factor (GCSF) from the other components of a biological sample. (Adapted with permission from Ref. 15.)

has been used in conjunction with an immobilized avidin column for the purification of IgG [17]. Furthermore, it is possible to use immunoaffinity columns either separately or in combination with other affinity columns. One example given in the previous chapter was a dual-column immunoaffinity/protein A method that was developed for the analysis of HSA and IgG in serum [5]. A similar approach has been used for the simultaneous determination of cytokines in clinical samples by using fluorescent labeled samples and up to ten separate immunoaffinity columns connected in series [10].

Another variation on direct detection methods is to use other analytical techniques to monitor the nonretained fraction of immunoaffinity columns. This later approach was used in an assay that combined an HPIAC column and flow injection analysis for the determination of urinary albumin/creatinine ratios [18]. In this technique, an antialbumin column was used for the capture and detection of HSA, while a Jaffe-based colorimetric reactor was used for the quantitation of creatinine in the portion of sample that was not bound by the immunoaffinity column (see Fig. 3). The result was an automated method that allowed two separate analytes, one retained and the other nonretained, to be examined during the same sample injection.

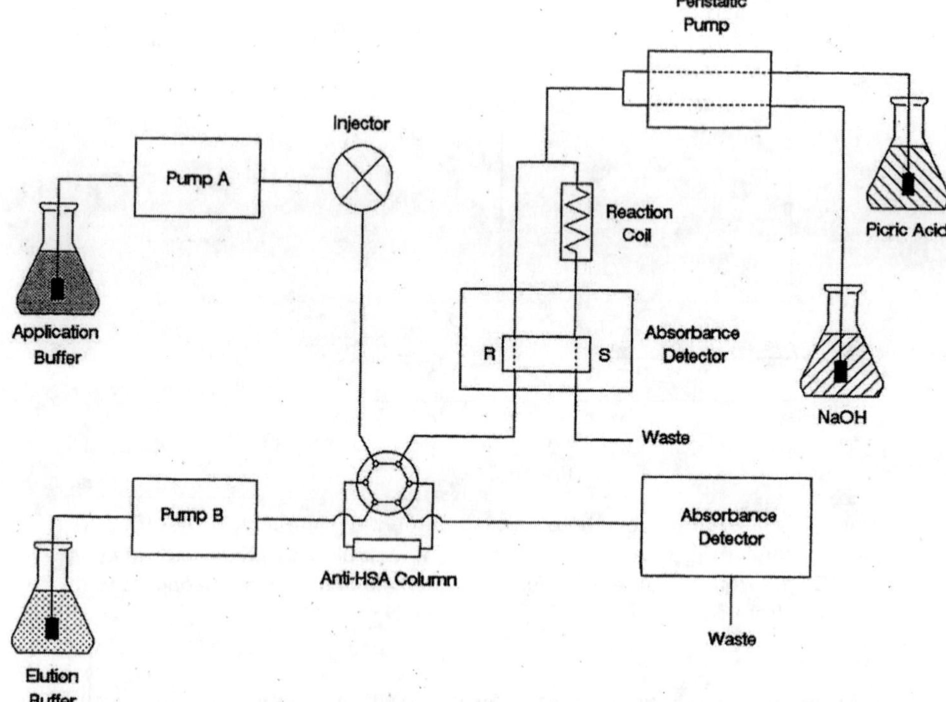

FIGURE 3 Schematic of an immunoaffinity/flow injection analysis system for the simultaneous determination of albumin and creatinine in urine. The left-hand portion of the diagram shows the chromatographic system that was used for the analysis of human serum albumin (HSA) by an anti-HSA antibody column and absorbance detector. The right-hand section shows the flow injection analysis system used for creatinine detection. In this latter portion, the nonretained sample components first passed through the reference side (R) of an absorbance detector, followed by their combination with picric acid and sodium hydroxide for the colorimetric detection of creatinine in the sample side (S) of the same detector. (Reproduced with permission from Ref. 6.)

3 CHROMATOGRAPHIC IMMUNOASSAYS

Another area of application for immunoaffinity chromatography is the use of IAC columns to perform immunoassays. The resulting technique is known as a *chromatographic* (or *flow-injection*) *immunoassay* [19]. This approach is valuable in determining trace analytes that by themselves may not produce a suitable signal for direct detection. Chromatographic immunoassays overcome this problem by using a labeled antibody or labeled analyte analog that can be used for indirect detection even at low analyte concentrations.

A large number of the labels that have been used in traditional immunoassays have also been used to perform chromatographic-based immunoassays. For instance, enzyme labels such as horeseradish peroxidase, alkaline phosphatase, and glucose oxidase have all been used in chromatographic assays. Other labels that have been employed include fluorescent tags like fluorescein, Texas red or lucifer yellow, chemiluminescent labels based on acridinium esters, and liposomes coupled with fluorescent dye molecules [20]. These labels are generally detected on-line as they elute in the nonretained or eluted peaks of the immunoaffinity column. However, fraction collection and off-line detection can be used as well.

The variety of analytes and concentrations that is present in clinical samples requires that several formats be available for the detection of these analytes by chromatographic immunoassays. These formats include procedures for conducting competitive binding immunoassays along with one-site and two-site immunometric assays. Each of these methods, along with examples of their applications, is considered in the following sections.

3.1 Competitive Binding Immunoassays

The most common way of performing a chromatographic immunoassay is to use a competitive binding format. One way of accomplishing this is to mix the sample with a fixed amount of a labeled analyte analog (i.e., the "label") and inject these simultaneously onto an immunoaffinity column that contains a relatively small amount of antibody. The ratio of bound label (B) to the maximum amount of possible bound label (B_o) is then plotted versus the concentration of the analyte. An example of a standard curve for a competitive binding immunoassay can be seen in Fig. 4, where the magnitude of the signal (B/B_o) is inversely proportional to the amount of analyte in the sample.

The particular format shown in the bottom of Fig. 4 is a *simultaneous injection competitive binding immunoassay*, in which both the sample and the label are applied at the same time to the IAC column. Up to the present, this has been the most popular approach for performing chromatographic competitive binding assays. Examples of applications in which this technique has been used include the measurement of theophylline in serum [21] and the determination of human serum albumin [22,23], immunoglobulin G [24,25], testosterone [26], or transferrin [22,27] in various aqueous or biological samples. An important advantage of this and other competitive binding

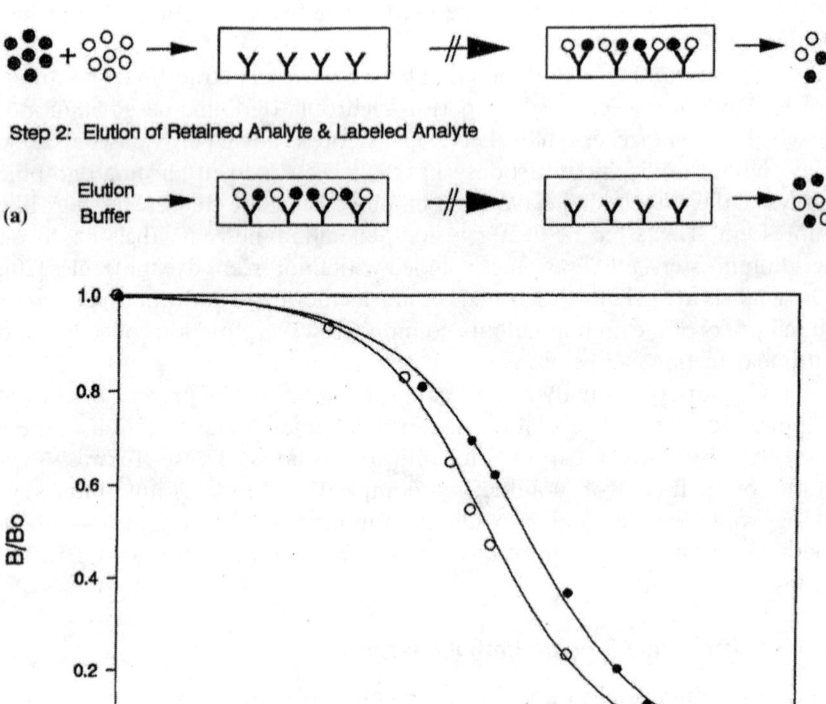

FIGURE 4 (a) Format and (b) calibration curves for a chromatographic simultaneous injection competitive binding immunoassay. The open circles in the top diagram represent the analyte and the closed circles represent a labeled analog of this compound. The calibration curves were obtained using human serum albumin as the analyte. The two curves shown in (b) were generated using two different amounts of the labeled analog. (Reproduced with permission from Ref. 23.)

formats is that they can be used equally well for small and large substances, making them applicable for a variety of analytes.

Another way of performing a chromatographic competitive binding immunoassay is to apply first only the sample to the immunoaffinity column, and then make a second separate injection of the label. This approach is known as a *sequential injection competitive binding immunoassay* and has been used for the determination of human serum albumin [22,28]. Like the simultaneous injection method, this format produces a final signal (B/B_o) that is inversely proportional to the concentration of analyte in the sample. One advantage of the sequential injection approach is that even an unlabeled preparation of analyte can be used as the "label" if it provides a sufficient signal for detection. This approach is particularly useful for complex samples that contain analytes at moderate to high concentrations. Additionally, there are no matrix interferences during detection of the label since it is never in contact with the actual sample [20].

A third type of chromatographic competitive binding technique is the *displacement immunoassay* [29]. In this technique, the immunoaffinity column is saturated with a labeled analog of the analyte of interest. The sample is then injected onto the column, with the analyte in the sample displacing label from the column. Using this format it is possible to employ a competitive binding format while also giving a signal that is directly proportional to the amount of analyte in the sample. This approach is especially useful when working with an analyte that occurs at low concentrations. Although this technique has been used in other fields, it has not yet seen any appreciable use in clinical applications [20].

3.2 Immunometric Assays

Another type of immunoassay that can be performed as part of an IAC system is an immunometric assay. The most common of these methods is the *sandwich immunoassay* or *two-site immunometric assay* [30–33]. In this technique, antibodies that bind to two different regions of the same analyte are used (see Fig. 5). One of the antibodies is immobilized to a chromatographic support and used to extract the analyte from samples. The second antibody has an easily measured label and is added in solution to the analyte either before or after sample injection. This labeled antibody places a label directly on the analyte, allowing the amount of analyte on the immunoaffinity column to be quantitated as it and the label are eluted from the column.

A specific clinical application in which a chromatographic sandwich immunoassay has been used is the determination of human chorionic gonadotropin (hCG) in serum (see Fig. 5) [34]. This method adapted an enzyme-linked immunosorbent assay for use on a perfusion chromatographic system. The sample was incubated with FITC-labeled anti-hCG antibodies and horseradish peroxidase (HRP)–labeled anti-hCG antibodies for 1 hour. A 100 μL aliquot of this mixture was then injected onto an

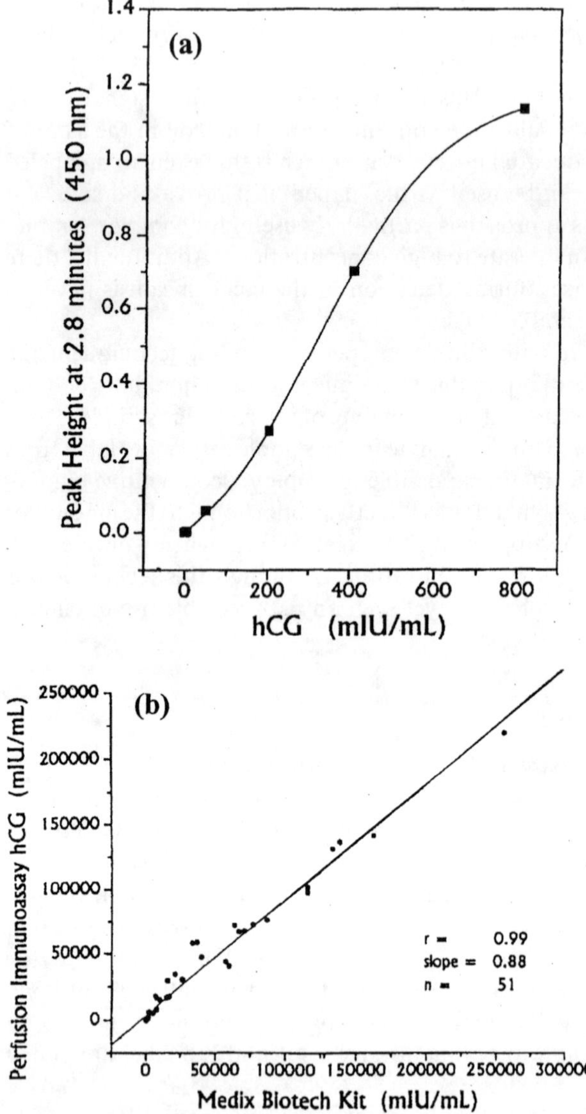

FIGURE 5 (a) Calibration curve and (b) correlation plot for the analysis of human chorionic gonadotropin (hCG) using a chromatographic-based ELISA method. (Reproduced with permission from Ref. 34.)

immunoaffinity column containing anti-FITC antibodies, resulting in retention of the sandwich immune complex within the column. The amount of retained hCG was determined by monitoring the absorbance at 450 nm that was produced by passing an *o*-phenylenediamine substrate for HRP through the column. The total analysis time was 8.6 min and the limit of detection was 5 pmol/L. Other examples of chromatographic sandwich immunoassays include those developed for parathyrin [32,33], thyroid stimulating hormone [35], various antigen-specific antibodies [31], and immunoglobulin G [30].

A chromatographic sandwich immunoassay is analogous to its traditional solid-phase counterpart in that it produces a signal for the bound label that is directly proportional to the amount of injected analyte. In addition, the fact that antibodies to two unique sites on the same analyte are used gives this technique higher selectivity than competitive binding immunoassays. The biggest disadvantage of the sandwich immunoassay is that it can only be used for analytes that are large enough to bind simultaneously two separate antibodies. Another drawback is that the use of two different antibodies makes the analysis slightly more complex to perform than a competitive binding assay.

A second type of immunometric assay that can be conducted by IAC is the *one-site immunometric assay,* as is illustrated in Fig. 6. In this technique, the sample is first incubated with a known excess of labeled antibodies of Fab fragments that are specific for the analyte of interest. After incubation, an aliquot of this mixture is injected onto an affinity column that contains an immobilized analog of the analyte. The purpose of this column is to extract any labeled antibodies or Fab fragments that have not bound to analyte in the original sample. Meanwhile, the label that is now bound to the analyte in solution passes through this column and elutes in the nonretained peak. It is the size of this peak that is then detected and used for analyte quantitation.

The one-site immunometric approach has been utilized in the analysis of such clinical analytes as thyroxine [35] and α-(difluoromethyl)ornithine [36]. This technique combines aspects of both the competitive binding immunoassay and the sandwich immunoassay. Like competitive binding immunoassays, this method can be used for both small and large molecules because only one antibody is bound to the analyte. However, like a sandwich immunoassay, the signal is directly proportional to the amount of analyte in the original sample. One drawback of this technique is that in order to provide a low background signal, relatively pure and highly active labeled antibodies or Fab fragments must be used. This makes this approach more difficult to set up than other formats and requires good label preparation methods.

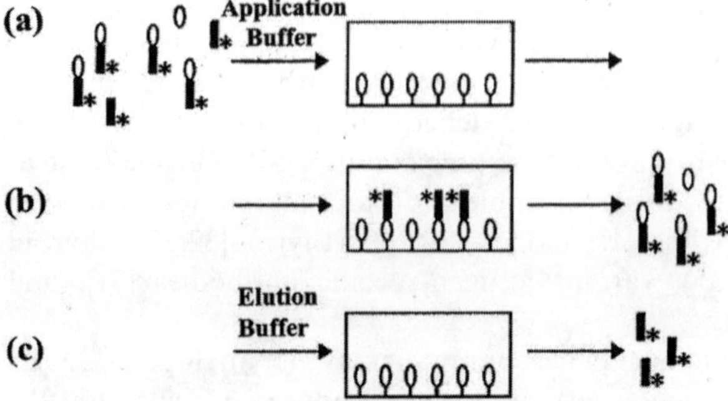

FIGURE 6 General scheme for a chromatographic one-site immunometric assay. The solid rectangles represent the labeled Fab fragments (or labeled antibodies) and the ovals represent the analyte in solution or its immobilized analog on the column. In (a), the sample and labeled Fab fragments are mixed and applied to the immobilized analog column. The Fab fragments which are already bound to analyte are then washed through this column as a nonretained peak and are detected (b). An elution buffer is later passed through the column to remove the retained Fab fragments (c). The amount of labeled Fab fragments in either the nonretained or the retained fraction is then used to determine the amount of analyte that was in the original sample.

4 IMMUNOAFFINITY EXTRACTION

As was shown in the previous chapter, affinity ligands like antibodies can also be used for sample extraction and isolation prior to the measurement of analytes by a second analytical method. The use of IAC columns for this purpose is often referred to as *immunoextraction* or *immunoaffinity extraction*. Due to the high affinity and selectivity of antibody interactions, a high degree of molecular selectivity is obtained with IAC. This means that complex biological samples can often be cleaned up in a single step with antibody columns, even if large sample volumes are required. There are two main forms of immunoextraction, off-line and on-line methods, which are both quite useful and appealing for clinical applications. Both approaches will be examined more closely in the following sections.

4.1 Off-Line Immunoaffinity Extraction

The easiest type of immunoaffinity extraction is the use of this as an off-line sample pretreatment method. This approach typically involves the use of antibodies that are immobilized onto a low-performance support packed into a small disposable syringe or solid-phase extraction cartridge. After this column has been conditioned with the necessary application buffer or other solvents, the sample is applied and any undesired sample components are washed away. The elution buffer is then passed through the column, and the retained solutes are collected as they dissociate. In some cases this eluted fraction is analyzed directly, but it may also be first dried down and reconstituted in a solvent that is more compatible with the method to be used later for analyte quantitation. If needed, the collected solute fraction can also be derivatized before it is examined by another technique. For instance, this last step can be used to improve a compound's detectability and/or volatility prior to separation and analysis by HPLC or GC [20].

Off-line immunoextraction has recently become popular in the area of drug residue analysis [20,37–40]. Examples involving human samples are the use of immunoextraction before reversed-phase liquid chromatography (RPLC) in the determination of albuterol in plasma [41], human chorionic gonadotropin in urine [42], and ochratoxin A in human serum, plasma, or milk [43]. In addition, off-line immunoextraction has been used for sample cleanup prior to analysis by gas chromatography (GC) or gas chromatography/mass spectrometry (GC/MS) in the determination of prostaglandins and thromboxanes [44–47] or alkylated DNA adducts [48,49] in human urine. Similar approaches have been used in a number of animal studies utilizing off-line immunoextraction and RPLC or GC for the detection of alkylated DNA adducts in DNA extracts from rats [50], chloramphenicol in urine and tissue samples from pigs [51], dexamethasone and flumethasone in equine urine [52,53], ivermectin and avermectin in sheep serum [54], and estrogens [55,56], nortestosterone [57], or trenbolone [58] in bovine urine and bile samples.

Like any IAC method, off-line immunoextraction requires the availability of an antibody preparation that is selective for the desired analyte or group of analytes. If such antibodies are available, then immunoextraction offers the potential of much greater specificity than traditional liquid–liquid or solid-phase extraction methods. But it should be kept in mind that most antibodies will also show some binding with compounds that are close to the desired analyte in structure. Ideally, this cross-reactivity should be evaluated for each immunoextraction support by performing binding and interference studies with any solutes or metabolites that are related to the analyte and that may be present in the samples of interest. However, even if several solutes do bind to the same IAC column, this will not present a problem as long as the analyte can be resolved or discriminated from these other compounds by the method that is to be used for quantitation. In fact, this cross-reactivity can be used as an advantage in that it may allow several structurally related compounds to be examined simultaneously by the single method.

4.2 On-Line Immunoaffinity Extraction

The direct coupling of IAC columns to other analytical systems is another area that has seen rapid growth during the past decade. The use of on-line immunoextraction with HPLC has been of particular interest [20,59]. The ease of incorporating immunoaffinity columns into an HPLC system makes this appealing as a means for automating immunoextraction methods and for reducing the amount of time required for sample pretreatment. In addition, the increased precision for on-line immunoextraction that comes with using high precision HPLC pumps and injection systems (giving more tightly controlled sample application and elution conditions) makes it an attractive alternative to off-line immunoextraction.

A large variety of analytes have already been examined by using on-line immunoextraction in HPLC (e.g., see Fig. 7). Some applications involving

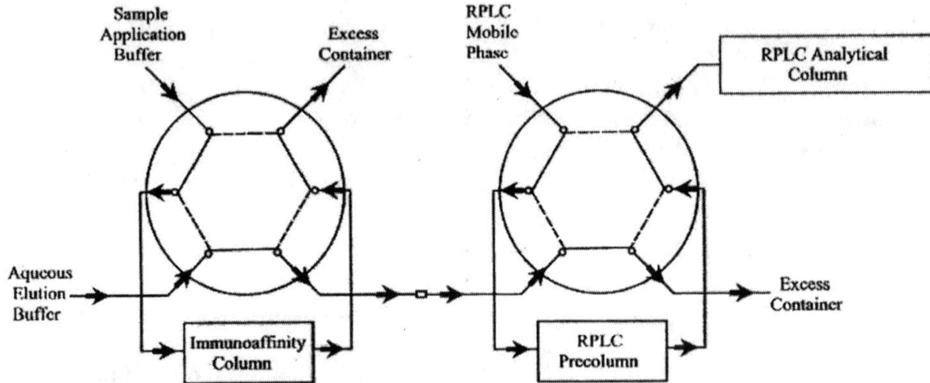

FIGURE 7 Typical scheme for the on-line coupling of immunoaffinity extraction with reversed-phase liquid chromatography (RPLC) [35]. The solid and dashed lines represent the two positions of the six-port valves shown in this scheme. When the valves are in the position represented by the solid lines, the sample is applied to the immunoaffinity column. This valve is then switched to the dashed position and an elution buffer is applied to dissociate the retained chemicals from the IAC column and reconcentrate them on a reversed-phase precolumn. When the valves are placed into their original positions (shown by the solid lines), the analytes on the precolumn come into contact with a mobile phase that has a decreased polarity, thus causing them to travel onto the larger analytical column, where they are separated and detected.

clinical samples include methods that have been reported for α_1-antitrypsin [60], cortisol [61], digoxin [62], estrogens [63,64], human epidermal growth factor [65], lysergic acid diethylamide (LSD), analogs and metabolites [66,67], phenytoin [68], propranolol [66], Δ^9-tetrahydrocannabinol [69], and transferrin [60,70]. All of these examples used immunoaffinity columns combined with standard RPLC columns. However, there have also been studies describing the use of on-line immunoextraction with size exclusion and ion-exchange chromatography [71–73].

There are several reasons why the majority of these studies use on-line immunoextraction with RPLC. One reason is the widespread use of RPLC in clinical and pharmaceutical separations. But another key reason is that the elution buffer for an immunoaffinity column is a solvent that contains little or no organic modifier, making this act as a weak mobile phase for RPLC. The consequence of this is that as a solute elutes from an IAC column to a RPLC column, it will tend to have strong retention on the reversed-phase support, leading to analyte reconcentration. This is valuable in dealing with analytes that have slow desorption from immunoaffinity columns, which might make them too dilute to analyze directly by immunoaffinity chromatography.

Though it is not as common as on-line extraction with HPLC, there have been studies investigating the use of immunoaffinity extraction coupled with GC [74]. In this method, an RPLC precolumn is used to capture and reconcentrate retained analytes as they wash off an immunoaffinity extraction column. This also removes any water from the analytes and places them into ethyl acetate, which is more compatible with GC. A portion of the RPLC eluent is then split into the injection gap of a GC system, and a temperature program is initiated for solute separation. An advantage of this approach, and also of immunoextraction with HPLC, is that dilute samples can be concentrated on the immunoaffinity column from large volumes of sample, thus providing low detection limits. The biggest drawback of coupling immunoextraction with GC is the greater complexity of this method when compared to off-line immunoextraction or on-line immunoextraction/HPLC.

A few studies have also considered the possibility of combining on-line immunoextraction with capillary electrophoresis (CE). One example is the use of immobilized Fab fragments to extract and concentrate tear samples for the CE analysis of cyclosporine and its metabolites in samples from corneal transplant patients [75]. In another study, antibodies were immobilized in microcapillary bundles or laser-drilled glass rods connected to a CE system for the on line immunoextraction and detection of immunoglobulin E in serum [76]. In addition, a capillary filled with a protein G chromatographic support has been used to adsorb antibodies for the extraction and concentration of serum insulin before quantitation by CE [77].

5 POSTCOLUMN IMMUNODETECTION

Yet another application of IAC is to employ this as a means to monitor the elution of specific analytes from other chromatographic columns. Using antibodies as ligands in these detection columns is referred to as *postcolumn immunodetection* [20,78]. The simplest approach for using postcolumn immunodetection is the direct detection mode of IAC. An example of this is a study in which size exclusion chromatography and postcolumn immunodetection were used for the analysis of acetylcholinesterase (AChE) in amniotic fluid (see Fig. 8) [79]. This report used an immunoaffinity column with immobilized anti-AChE antibodies to capture AChE as it eluted from an analytical column. A substrate solution for AChE was then passed through this column and the resulting colored product was detected on-line by an HPLC absorbance detector.

There are other formats that are also possible for postcolumn immunodetection. These include both competitive binding immunoassays [28,78,80] and sandwich immunoassays [81]. However, the one-site immunometric assay is the most common format for immunodetection and is the only approach that has been used in a clinical setting. This technique involves taking the eluent from an HPLC analytical column and combining this with a solution of labeled antibodies or Fab fragments specific for the analyte of interest. This mixture is then allowed to react in a mixing coil and passed through an immunodetection column that contains an immobilized analog of the analyte. The antibodies or Fab fragments that are already complexed with the analyte will pass through this column and into a detector, yielding a signal proportional to the amount of analyte in the original sample. The immunodetection column can later be

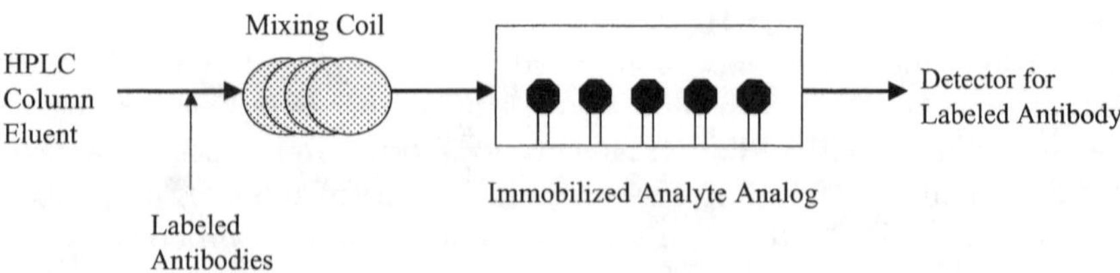

FIGURE 8 General scheme used for post-column immunodetection based on a one-site immunometric assay. The principles behind this assay are similar to those in Fig. 6.

washed with an elution buffer to dissociate the retained antibodies or Fab fragments; but generally the capacity of the immunodetection column is such that a reasonably large amount of column eluent can be analyzed before the immunodetection column must be regenerated.

One use for one-site immunometric postcolumn detection has been in the quantitation of digoxin and digoxigenin in plasma and urine samples by HPLC [82]. This was accomplished using fluorescein-labeled Fab fragments specific for digoxigenin and an immobilized digoxin support on-line with a RPLC column. This same immunodetection format has been used with a restricted-access RPLC column to monitor digoxin, digoxigenin, and related metabolites in serum samples [83].

6 FUTURE TRENDS AND DEVELOPMENTS

Immunoaffinity chromatography in its present form clearly has the potential for widespread use in the clinical laboratory. However, there is also much room for improvement and for the development of new applications for immunoaffinity supports. Some of these, such as the creation of new ligands or improved assay formats, are the same trends seen for other affinity methods (e.g., see the discussion in the previous chapter). But there also some unique advances being made with immunoaffinity supports, as will now be considered.

One example of a new development in this field is recent work in the use of rapid immunoaffinity chromatography to measure free drug fractions [84]. The free (or non-protein-bound) fraction of drugs is hypothesized to be the biologically active form in the circulation but has traditionally been difficult or time-consuming to measure. To overcome this problem, a small immunoaffinity column was developed that could extract warfarin, an anticoagulant drug, in only a few hundred milliseconds. This made it possible to remove this drug from warfarin/protein mixtures on a time scale that allowed removal of this drug's free fraction without causing significant dissociation of its protein-bound form (see Fig. 9). Similar methods are now being considered for other drugs and analytes of clinical interest.

Work is also being performed in the creation of improved supports for immunoaffinity chromatography, such as porous glass beads or monolithic supports, that might increase sample throughput [85]. In addition, efforts continue in the coupling of immunoextraction with other analytical methods like capillary electrophoresis, mass spectrometry, and biosensors [86]. Another development that is expected to impact the use of immunoaffinity supports is the creation of micromachined analytical systems. This later area may open the possibility of using arrays of antibodies for the determination

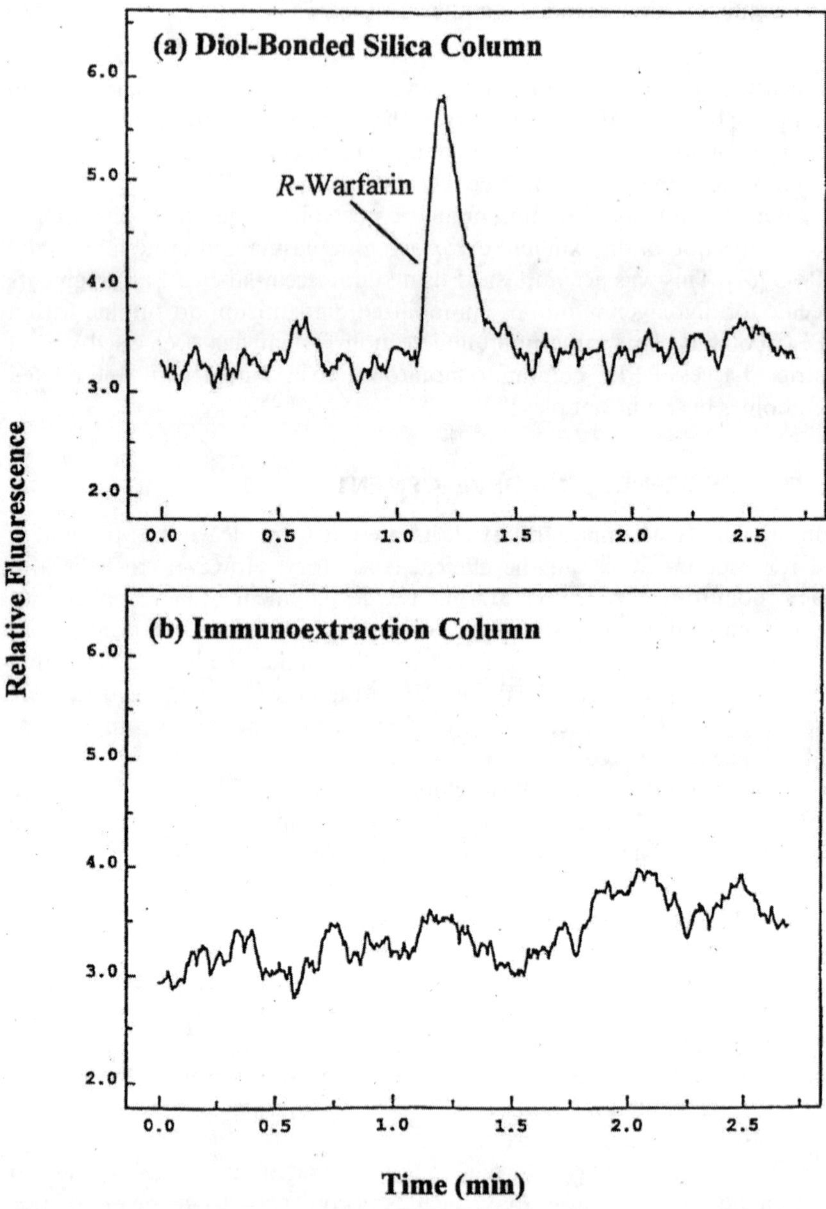

FIGURE 9 Removal of free warfarin from a sample by rapid immunoextraction. These chromatograms are for injections of warfarin onto (a) an inert diol-bonded silica column or (b) a small antiwarfarin antibody column with an allowed contact time of only 60 ms between warfarin and the columns. Quantitative extraction of the warfarin was obtained by the immunoaffinity column under these conditions, making it possible to selectively separate the free fraction of warfarin from a mixture of this drug and its transport protein, human serum albumin. (Reproduced with permission from Ref. 84.)

of a large panel of clinical analytes. As these and other applications continue to be developed, immunoaffinity chromatography and related techniques should find even greater use in clinical labs and in the analysis of biological compounds.

REFERENCES

1. Phillips TM. High-performance immunoaffinity chromatographic detection of immunoregulatory anti-idotypic antibodies in cancer patients receiving immunotherapy. Clin Chem 1988;34:1689–1692.
2. Phillips TM, Babashak JV. Isolation of anti-idiotypic antibodies by immunoaffinity chromatography on Affinichrom beads. J Chromatogr 1990;512:387–394.
3. Wang WT, Kumlien J, Ohlson S, Lundblad A, Zopf D. Analysis of a glucose-containing tetrasaccharide by high-performance liquid affinity chromatography. Anal Biochem 1989;182:48–53.
4. Zopf D, Ohlson S, Dakour J, Wang W, Lundblad A. Analysis and purification of oligosaccharides by high-performance liquid affinity chromatography. Methods Enzymol 1989;179:55–64.
5. Hage DS, Walters RR. Dual-column determination of albumin and immunoglobulin G in serum by high-performance affinity chromatography. J Chromatogr 1987;386:37–49.
6. Ruhn PF, Taylor JD, Hage DS. Determination of urinary albumin using high-performance immunoaffinity chromatography and flow injection analysis. Anal Chem 1994;66:4265–4271.
7. Phillips TM, More NS, Queen WD, Holohan TV, Kramer NC, Thompson AM. High-performance affinity chromatography: a rapid technique for the isolation and quantitation of IgG from cerebral spinal fluid. J Chromatogr 1984;317:173–179.
8. Phillips TM, More NS, Queen WD, Thompson AM. Isolation and quantitation of serum IgE levels by high-performance immunoaffinity chromatography. J Chromatogr 1985;327:205–211.
9. Phillips TM. Measurement of recombinant interferon levels by high performance immunoaffinity chromatography in body fluids of cancer patients on interferon therapy. Biomed Chromatogr 1992;6:287–290.
10. Phillips TM, Krum JM. Recycling immunoaffinity chromatography for multiple analyte analysis ion biological samples. J Chromatogr B 1998;715:55–63.
11. Phillips TM. Measurement of total and bioactive interleukin-2 in tissue samples by immunoaffinity-receptor affinity chromatography. Biomed Chromatogr 1997; 11:200–204.
12. Mogi M, Harada M, Adachi T, Kojima K, Nagatsu T. Selective removal of β_2-microglobulin from human plasma by high-performance immunoaffinity chromatography. J Chromatogr 1989;496:194–200.

13. McConnell JP, Anderson DJ. Determination of fibrinogen in plasma by high-performance immunoaffinity chromatography. J Chromatogr 1993;615:67–75.
14. Ohlson S, Gudmundsson B-M, Wikstrom P, Larsson, P-O. High-performance liquid affinity chromatography: rapid immunoanalysis of transferrin in serum. Clin Chem 1988;34:2039–2043.
15. Phillips TM. Immunoaffinity measurement of recombinant granulocyte colony stimulating factor in patients with chemotherapy-induced neutropenia. J Chromatogr B 1994;662:307–313.
16. Bergstrom M, Lundblad A, Pahlsson P, Ohlson S. Use of weak monoclonal antibodies for affinity chromatography. J Mol Recognit 1998;11:110–113.
17. Kim HO, Durance TD, Li-Chan EC. Reusability of avidin-biotinylated immunoglobulin Y columns in immunoaffinity chromatography. Anal Biochem 1999;268:383–397.
18. Eliasson M, Olsson A, Palmcrantz E, Wibers K, Inganas M, Guss B, Lindberg M, Uhlen M. Chimeric IgG-binding receptors engineered from staphylococcal protein A and streptococcal protein G. J Biol Chem 1988;263: 4323–4327.
19. Hage DS, Nelson MA. Chromatographic immunoassays. Anal Chem 2001;73:198A–205A.
20. Hage DS. A survey of recent advances in analytical applications of immunoaffinity chromatography. J Chromatogr 1998;715:3–28.
21. Locascio-Brown L, Plant AL, Chesler R, Kroll M, Ruddel M, Durst RA. Liposome-based flow-injection immunoassay for determining theophylline in serum. Clin Chem 1993;39:386–391.
22. Cassidy SA, Janis LF, Regnier FE. Kinetic chromatographic sequential addition immunoassays using protein A affinity chromatography. Anal Chem 1992;64:1973–1977.
23. Hage DS, Thomas DH, Chowdhuri AR, Clarke W. Development of a theoretical model for chromatographic-based competitive binding immunoassays with simultaneous injection of sample and label. Anal Chem 1999;71: 2965–2975.
24. De Alwis U, Wilson GS. Rapid heterogeneous competitive electrochemical immunoassay for IgG in the picomole range. Anal Chem 1987;59:2786–2789.
25. Valencia-Gonzalez MJ, Diaz Garcia ME. Flow-through fluorescent immunosensing of IgG. Ciencia 1996;4:29–40.
26. Palmer DA, Evans M, Miller JN, French MT. Rapid fluorescence flow injection immunoassay using a novel perfusion chromatographic support. Analyst 1994;119:943–947.
27. Palmer DA, Xuezhen R, Fernandez-Hernando P, Miller JN. A model on-line flow injection fluorescence immunoassay using a protein A immunoreactor and lucifer yellow. Anal Lett 1993;26:2543–2553.
28. Hage DS, Thomas DH, Beck MS. Theory of a sequential addition competitive binding immunoassay based in high-performance immunoaffinity chromatography. Anal Chem 1993;65:1622–1630.

29. Kronkvist K, Lovgren U, Svenson J, Edholm LE, Johansson G. Competitive flow injection enzyme immunoassay for steroids using a post-column reaction technique. J Immunol Methods 1997;200:145–153.
30. De Alwis WU, Wilson GS. Rapid sub-picomole electrochemical enzyme immunoassay for immunoglobulin G. Anal Chem 1985;57:2754–2756.
31. Hacker A, Hinterleitner M, Shellum C, Gubitz G. Development of an automated flow injection chemiluminescence immunoassay for human immunoglobulin G. Frenz J Anal Chem 1995;352:793–796.
32. Hage DS, Kao PC. High-performance immunoaffinity chromatography and chemiluminescent detection in the automation of a parathyroid hormone sandwich immunoassay. Anal Chem 1991;63:586–595.
33. Hage DS, Taylor B, Kao PC. Intact parathyroid hormone: performance and clinical utility of an automated assay based on high-performance immunoaffinity chromatography and chemiluminescence detection. Clin Chem 1992;38:1494–1500.
34. Johns MA, Rosengarten LK, Jackson M, Regnier FE. Enzyme-linked immunosorbent assays in a chromatographic format. J Chromatogr A 1996; 743:195–206.
35. Hage DS. Chromatographic approaches to immunoassays. J Clin Ligand Assay 1998;20:293–301.
36. Gunaratna PC, Wilson GS. Noncompetitive flow injection immunoassay for a hapten, α-(difluoromethyl)ornithine. Anal Chem 1993;65:1152–1157.
37. Van Ginkel LA. Immunoaffinity chromatography, its applicability and limitations in multi-residue analysis of anabolizing and doping agents. J Chromatogr 1991;564:363–384.
38. Haagsma N, van de Water C. Immunochemical methods in the analysis of veterinary drug residues. In: Agarwal VK, ed. Analysis of Antibiotic Drug Residues in Food Products of Animal Origin. New York: Plenum Press, 1992, pp 81–97.
39. Katz SE, Siewierski M. Drug residue analysis using immunoaffinity chromatography. J Chromatogr 1992;624:403–409.
40. Katz SE, Brady MS. High-performance immunoaffinity chromatography for drug residue analysis. J Assoc Off Anal Chem 1990;73:557–560.
41. Ong H, Adam A, Perreault S, Marleay S, Bellemare M, Du Souich P. Analysis of albuterol in human plasma based in immunoaffinity chromatographic cleanup combined with high-performance liquid chromatography with fluorimetric detection. J Chromatogr 1989;497:213–221.
42. Liu CL, Bowers LD. Immunoaffinity trapping of urinary human chorionic gonadotropin and its high-performance liquid chromatographic-mass spectrometric determination. J Chromatogr B 1996;687:213–220.
43. Zimmerli B, Dick R. Determination of ochratoxin A at the ppt level in human blood, serum, milk and some foodstuffs by high-performance liquid chromatography with enhanced fluorescence detection and immunoaffinity column cleanup: methodology and Swiss data. J Chromatogr B 1995;666:85–99.

44. Bachi A, Zuccato E, Baraldi M, Fanelli R, Chiabrando C. Measurement of urinary 8-epiprostaglandin $F_{2\alpha}$, a novel index of lipid peroxidation in vivo, by immunoaffinity extraction/gas chromatography-mass spectrometry. Basal levels in smokers and non-smokers. Free Radical Biol Med 1996;20:619–624.
45. Mackert G, Reinke M, Schweer H, Seyberth HW. Simultaneous determination of the primary prostanoids prostaglandin E_2, prostaglandin $F_{2\alpha}$ and 6-oxoprostaglandin $F_{1\alpha}$ by immunoaffinity chromatography in combination with negative ion chemical ionization gas chromatography-tandem mass spectrometry. J Chromatogr 19898;494:13–22.
46. Chiabrando C, Pinciroli V, Comoleoni A, Benigni A, Piccinelli A, Fanelli R. Quantitative profiling of 6-ketoprostaglandin $F_{1\alpha}$, 2,3-dinor-6-ketoprostaglandin $F_{1\alpha}$, thromboxane B_2 and 2,3-dinor-thromboxane B_2 in human and rat urine by immunoaffinity extraction with gas chromatography-mass spectrometry. J Chromatogr 1989;495:1–11.
47. Ishibashi M, Watanabe K, Ohyama Y, Mizugaki M, Hayashi Y, Takasaki W. Novel derivatization and immunoextraction to improve microanalysis of 11-dehydrothomboxane B_2 in human urine. J Chromatogr 1991;562:613–624.
48. Prevost V, Shuker DEG, Friesen MD, Eberle G, Rajewsky MF, Bartsch H. Immunoaffinity purification and gas chromatography–mass spectrometric quantitation of 3-alkyladenines in urine: metabolism studies and basal excretion levels in man. Carcinogenesis 1993;14:199–204.
49. Friesen MD, Garren L, Prevost V, Shuker DEG. Isolation of urinary 3-methyladenine using immunoaffinity columns prior to determination by low-resolution gas chromatography–mass spectrometry. Chem Res Toxicol 1991;4:102–106.
50. Bonfanti M, Magagnotti C, Galli A, Bagnati R, Moret M, Gariboldi P, Fanelli R, Airoldi L. Determination of O^6-butylguanine in DNA by immunoaffinity extraction/gas chromatography–mass spectrometry. Cancer Res 1990;50:6870–6875.
51. Gude T, Preiss A, Rubach K. Determination of chloramphenicol in muscle, liver, kidney and urine of pigs by means of immunoaffinity chromatography and gas chromatography with electron-capture detection. J Chromatogr B 1995;673:197–204.
52. Stanley SMR, Wilhelmi BS, Rodgers JP. Comparison of immunoaffinity chromatography combined with gas chromatography–negative ion chemical ionisation mass spectrometry and radioimmunoassay for screening dexamethasone in equine urine. J Chromatogr 1993;620:250–253.
53. Stanely SMR, Wilhelmi BS, Rodgers JP, Bertschinger H. Immunoaffinity chromatgraphy combined with gas chromatography–negative ion chemical ionisation mass spectrometry for the confirmation of flumethasone abuse in the equine. J Chromatogr 1993;614:77–86.
54. Li J, Zhang SQ. Immunoaffinity column cleanup and liquid chromatographic method for determining ivermectin in sheep serum. JAOAC Intl 1996;79:1300–1302.

55. Bagnati R, Castelli MG, Airoldi L, Oriundi MP, Ubaldi A, Fanelli R. Analysis of diethylstilbestrol, dienestrol and hexestrol in biological samples by immunoaffinity extraction and gas chromatography–negative ion chemical ionization mass spectrometry. J Chromatogr 1990;527:267–278.
56. Bagnati R, Oriundi MP, Russo V, Danese M, Berti F, Fanelli R. Determination of zeranol and b-zeranol in calf urine by immunoaffinity extraction and gas chromatography–mass spectrometry after repeated administration of zeranol. J Chromatogr 1991;564:493–502.
57. Van Ginkel LA, Stephany RW, van Rossum HJ, van Blitterswijk H, Zoontjes PW, Hooijshuur RCM, Zuydendorp J. Effective monitoring of residues of nortestosterone and its major metabolite in bovine urine and bile. J Chromatogr 1989;489:95–104.
58. Van Ginkel LA, van Blitterswijk H, Zoontjes PW, van den Bosch D, Stephany RW. Assay for trenbolone and its metabolite 17a-trenbolone in bovine urine based on immunoaffinity chromatographic clean-up and off-line high performance liquid chromatography–thin layer chromatography. J Chromatogr 1988;445:385–392.
59. de Frutos M, Regnier FE. Tandem chromatographic-immunological analyses. Anal Chem 1993;65:17A–25A.
60. Flurer CL, Novotny M. Dual microcolumn immunoaffinity liquid chromatography: an analytical application to human plasma proteins. Anal Chem 1993;65:817–821.
61. Nilsson B. Extraction and quantitation of cortisol by use of high-performance liquid affinity chromatography. J Chromatogr 1983;276:413–417.
62. Reh E. Determination of digoxin in serum by on-line immunoadsorptive clean-up high-performance liquid chromatographic separation and fluorescence-reaction detection. J Chromatogr 1988;433:119–130.
63. Farjam A, Brugman AE, Lingeman H, Brinkman UAT. On-line immunoaffinity sample pretreatment for column liquid chromatography: evaluation of desorption techniques and operating conditions using an anti-estrogen immuno-precolumn as a model system. Analyst 1991;116:891–896.
64. Farjam A, Brugman AE, Soldaat A, Timmerman P, Lingeman H, de Jong GJ, Frei RW, Brinkman UAT. Immunoaffinity precolumn for selective sample pretreatment in column liquid chromatography: immunoselective desorption. Chromatographia 1991;31:469–477.
65. Hayashi T, Sakamoto S, Wada I, Yoshida H. HPLC analysis of human epidermal growth factor using immunoaffinity precolumn. II. Determination of hEGFs in biological fluids. Chromatographia 1989;27:574–580.
66. Rule GS, Henion JD. Determination of drugs from urine by on-line immunoaffinity chromatography–high performance liquid chromatography–mass spectrometry. J Chromatogr 1992;582.103–112.
67. Cai J, Henion J. On-line immunoaffinity extraction-coupled column capillary liquid chromatography/tandem mass spectrometry: trace analysis of LSD analogs and metabolites in human urine. Anal Chem 1996;68:72–78.
68. Johansson B. Simplified quantitative determination of plasma phenytoin: on-

line precolumn high-performance liquid immunoaffinity chromatography with sample pre-purification. J Chromatogr 1986;381:107–113.
69. Kircher V, Parlar H. Determination of Δ^9-tetrahydrocannabinol from human saliva by tandem immunoaffinity chromatography–high-performance liquid chromatography. J Chromatogr B 1996;677:245–255.
70. Janis LJ, Regnier FE. Dual-column immunoassays using protein G affinity chromatography. Anal Chem 1989;61:1901–1906.
71. Riggin A, Sportsman JR, Regnier FE. Immunochromatographic analysis of proteins: identification, characterization and purity determination. J Chromatogr 1993;632:37–44.
72. Janis LJ, Regnier FE. Immunological–chromatographic analysis. J Chromatogr 1988;444:1–11.
73. Janis LJ, Grott A, Regnier FE, Smith-Gill SJ. Immunological–chromatographic analysis of lysozyme variants. J Chromatogr 1989;476:235–244.
74. Farjam A, Vreuls JJ, Cuppen WJG, Brinkman UAT, de Jong GJ. Direct introduction of large-volume urine samples into an on-line immunoaffinity sample pretreatment–capillary gas chromatography system. Anal Chem 1991;63:2481–2487.
75. Phillips TM, Chmielinska JJ. Immunoaffinity capillary electrophoresis analysis of cyclosporine in tears. Biomed Chromatogr 1994;8:242–246.
76. Guzman NA. Biomedical applications of on-line preconcentration–capillary electrophoresis using an analyte concentrator:investigation of design options. J Liq Chromatogr 1995;18:3751–3768.
77. Cole LJ, Kennedy RT. Selective preconcentration for capillary zone electrophoresis using protein G immunoaffinity capillary chromatography. Electrophoresis 1995;16:549–556.
78. Irth H, Oosterkamp AJ, Tjaden UR, van der Greef J. Strategies for on-line coupling of immunoassays to high-performance liquid chromatography. Trends Anal Chem 1995;14:355–361.
79. Vanderlaan M, Lotti R, Siek G, King D, Goldstein M. Perfusion immunoassay for acetylcholinesterase: analyte detection based on intrinsic activity. J Chromatogr A 1995;711:23–31.
80. Oosterkamp AJ, Irth H, Tjaden UR, van der Greef J. On-line coupling of liquid chromatography to biochemical assays based on fluorescent-labeled ligands. Anal Chem 1994;66:4295–4301.
81. Cho BY, Zou H, Stong R, Fisher DH, Nappier J, Krull IS. Immunochromatographic analysis of bovine growth hormone releasing factor involving reversed-phase high-performance liquid chromatography–immunodetection. J Chromatogr A 1996;743:181–194.
82. Irth H, Oosterkamp AJ, van der Welle W, Tjaden UR, van der Greef J. On-line immunochemical detection in liquid chromatography using fluorescein-labeled antibodies. J Chromatogr 1993;633:65–72.
83. Oosterkamp AJ, Irth H, Beth M, Unger KK, Tjaden UR, van der Greef J. Bioanalysis of digoxin and its metabolites using direct serum injection

combined with liquid chromatography and on-line immunochemical detection. J Chromatogr B 1994;653:55–61.
84. Clarke W, Chowdhuri AR, Hage DS. Analysis of free drug fractions by ultrafast immunoaffinity chromatography. Anal Chem 2001;73:2157–2164.
85. Schuste M, Wasserbauer E, Neubauer A, Junbauer A. High speed immunoaffinity chromatography on supports with gigapores and porous glass. Bioseparation 2000;9:259–268.
86. Willumsen B, Christian GD, Ruzicka J. Flow injection renewable surface immunoassay for real time monitoring of biospecific interactions. Anal Chem 1997; 69:3482–3489.

10

Electrospray Tandem Mass Spectrometry in the Biochemical Genetics Laboratory

Mohamed S. Rashed
King Faisal Specialist Hospital and Research Centre,
Riyadh, Saudi Arabia

1 INTRODUCTION

Biochemical genetics laboratories specialize in the diagnosis of inherited metabolic diseases (IMDs). These types of laboratories have been in existence in a number of large medical centers for over 30 years. Their clienteles are neonates or children who may present with acute onset of metabolic disturbances, failure to thrive, seizures, coma, developmental delay, dysmorphic features, and mental retardation. These laboratories started mostly as small research units in pediatric departments in medical schools. However, with the dramatic changes in the field of medical genetics in the last twenty years and the recognition of over 300 IMDs, these laboratories evolved into a well-recognized medical specialty.

In addition to the diagnosis of IMDs, the biochemical genetic laboratory may be involved in some (or all) of a number of other equally important functions such as follow-up of dietary and/or chemical treatment, biochemical reevaluation if the patient is metabolically decompensated, newborn screening, and confirmation of positive newborn screening results. These functions are achieved by either of two processes: (1) identification and quantification of certain biochemical markers in body fluids (called metabolic profiling or metabolic screening), which may reflect the enzyme deficiency; (2) enzyme assays. As enzyme assays are usually time-consuming and may require cell culturing or invasive techniques such as liver and muscle biopsies, metabolic screening is usually the first approach to diagnosing IMDs, albeit for a large number of diseases abnormal products do not appear in body fluids. With the significant development in molecular analysis techniques and the identification of the loci for many of the IMDs, a new trend is taking place where the initial metabolic screening or even newborn screening positive results are being followed directly by DNA analysis for confirmation, rather than by enzyme assays.

2 METABOLIC SCREENING AND MASS SPECTROMETRY

Metabolic screening for diagnosis, confirmation, or follow-up usually entails analysis of body fluids for metabolites of intermediary metabolism pathways such as amino acids, organic acids, acylcarnitines, acylglycines, fatty acids, purine and pyrimidines, sugars, and oligosaccharides. Traditionally, analytical techniques used for such purposes were paper chromatography, thin-layer chromatography, high-pressure liquid chromatography (HPLC), electrophoresis (EC), and gas chromatography (GC). Actually, these classical techniques were instrumental in the identification of the biochemical basis of many of the inherited disorders and in delineating many new metabolic pathways for many organic compounds. However, the need for higher levels of specificity and higher degrees of sensitivity has led to the introduction of mass spectrometry to the biochemical genetics laboratory [1].

In mass spectrometry (MS), the molecules have to be ionized to positively or negatively charged species and must be present in the gas phase at a pressure of 10^{-5} to 10^{-4} torr, where we measure mass-to-charge ratio (m/z) of the desired ions. The main hurdle in the analysis of many biochemically important molecules by traditional electron ionization (EI) and chemical ionization (CI) is their low volatility and thermal instability. Sample derivatization may lead to the formation of volatile and thermally stable compounds amenable to both gas chromatography and the mass analysis processes, e.g., trimethylsilyl esters of organic acids [1].

MS was recognized early on as an important and powerful analytical tool to the biochemical genetics laboratory. Thus in the mid 1960s Tanaka et al. used the then young technique of electron ionization gas chromatography mass spectrometry (EI-GC/MS) in the discovery of the first known organic acid catabolism disorders, namely isovaleric acidemia (IVA) [2]. With the introduction of the quadrupole mass spectrometer and the advances in capillary gas liquid chromatography, GC/MS played a pivotal role in the discovery of over 60 other genetic defects in the catabolism of organic acids [1–3]. Furthermore, the early 1970s to the late 1980s saw the discovery of important classes of IMDs, namely fatty acid oxidation defects (FOADs), steroid disorders, and peroxisomal diseases [4–13]. Here also, GC/MS particularly using the "softer" CI mode, which significantly decreased the fragmentation of ionized molecules and allowed for the development of much more sensitive assays, played an important role in the diagnosis of many patients through the analysis of plasma and urine for dicarboxylic acids, acylglycines, very-long-chain fatty acids, phytanic, and pipecolic acid [14–17]. The sensitivity, specificity, and ease of operation offered by today's inexpensive bench-top GC/MS instruments has made this technology indispensable for the biochemical genetics laboratory for some time to come.

Although derivatization and/or CI approaches to sample preparation and analysis extended the utility of GC/MS applications, the fact remained that only a small fraction of biomolecules can be analyzed by GC/MS, and that there was a need for ionization techniques that could handle polar, nonvolatile, thermally labile, and large-molecular-weight organic compounds.

3 ALTERNATIVE IONIZATION TECHNIQUES FOR MASS SPECTROMETRY

The late 1970s to the late 1980s saw the development of new "softer ionization" techniques such as thermospray ionization (TSP), fast atom bombardment (FAB), continuous flow FAB, electrospray ionization (ESI, also called ionspray), atmospheric pressure chemical ionization (APCI), and matrix-assisted laser desorption ionization (MALDI) [18–22]. The same period also saw accomplishments in interfacing the liquid chromatography system to the mass spectrometer. Thus numerous biomedical and pharmaceutical applications of liquid chromatography thermospray mass spectrometry (LC-TSP-MS) and continuous flow FAB (CF-FAB) appeared in the literature [23–26]. However, of all the soft ionization methods, only two are here to stay, ESI (including APCI) and MALDI.

ESI was and remains the most exciting technique for the field of biochemical genetics, and in combination with tandem mass spectrometry (MS/MS) it has resulted in major breakthroughs in the field.

In ESI, the sample is dissolved in an aqueous or aqueous/organic solvent where it should be partially ionized. The solution is then pumped at relatively low flow rates (5–200 μL/min) through a thin stainless steel capillary that is raised to high potential (typically 4 kV). Highly charged droplets are sprayed in a dry bath gas at near atmospheric pressure and travel down a

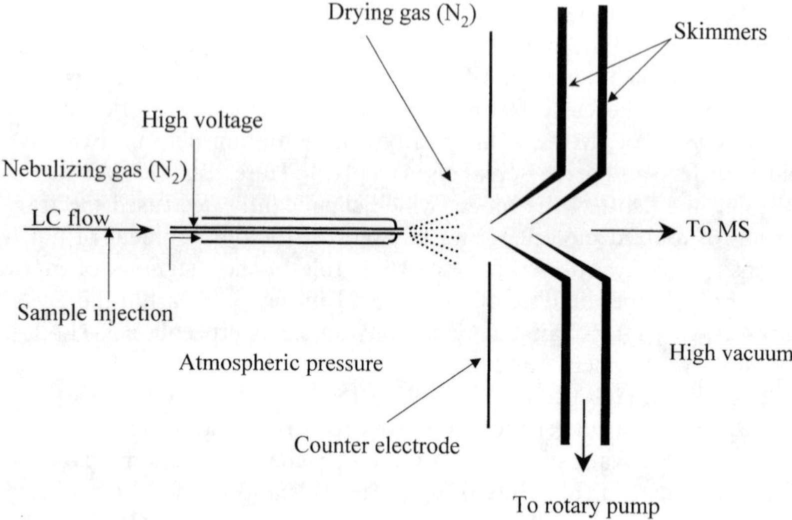

FIGURE 1 Sketch of the ES ion source.

pressure and potential gradient toward an orifice in the mass spectrometer high vacuum region (Fig. 1). These charged droplets shrink in size and are desolvated as they travel in the mildly heated ion source until the charge repulsion overcomes the surface tension forces leading to a "Coulombic explosion" and ion desorption (Fig. 2).

The ionization is "soft" in that mostly stable molecular ions are produced with little or no fragmentation. APCI is similar to ESI except that a corona discharge is used to generate reagent ions from the solvent vapor, thereby enhancing the yield of molecular ions by proton transfer. APCI has

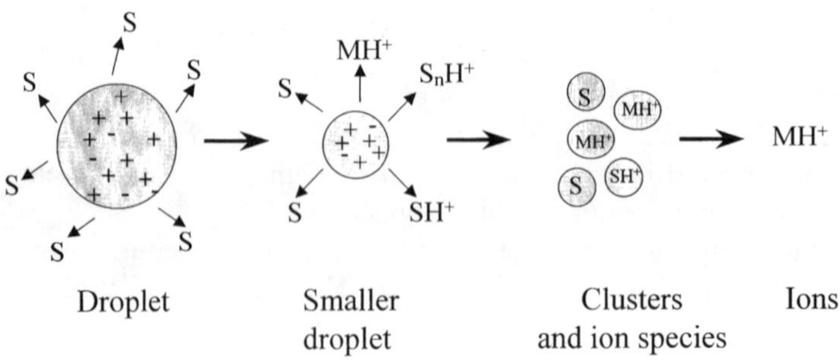

FIGURE 2 Process of ion formation in ESI.

found wide applications in the pharmaceutical industry for the analysis of drugs and their metabolites [26].

Because of the application of high potentials in ESI, larger molecules with multiple basic or acidic sites tend to give multiply charged species. Since we measure m/z, increasing the number of charges decreases m/z. This phenomenon allowed the analysis of very large molecules such as peptides and proteins on instruments with limited mass ranges and limited resolving power such as quadrupole instruments (mass range 1–4000 Da). This allowed for accurate mass measurements of peptides and proteins [27–29].

4 ELECTROSPRAY TANDEM MASS SPECTROMETRY (ESI-MS/MS)

One of the main advantages of MS/MS is that it lends itself to the qualitative and quantitative analysis of certain analytes in complex mixtures, with little or no cleanup. This advantage is due in part to the possible replacement of the separation device such as the GC column or the HPLC column with a mass analyzer (mass filter), thereby sparing all or most of the time used for resolving individual components in a mixture.

In a typical triple-quadrupole mass spectrometer fitted with an ESI ion source, the sample solution reaches the source by injecting it in a stream of liquid (mobile phase), a process called flow injection analysis (FIA). The ions produced in the ESI source at atmospheric pressure are directed in the high vacuum region of the mass spectrometer. Three scanning modes are more commonly used in MS/MS (Fig. 3A–C). In the so-called product-ion scanning an ion of interest is selected by the first mass analyzer MS1, and this ion is transmitted to the collision cell. Fragmentation is induced by a process called collision-induced dissociation (CID), whereby the transmitted ions are collided with an inert gas (typically argon or nitrogen), and low electron energy. The mass analyzer MS2 then analyzes the fragments produced. The spectrum obtained is termed the product-ion spectrum. This approach is useful for gaining structural information. In another type of MS/MS experiment, MS2 is set to transmit to the final detector a single m/z that results from fragmentation in the collision cell of ions transmitted by MS1 within a certain mass range of interest. The "precursor-ion" or "parent-ion" spectrum produced will show all the species that produced the common fragment ion. In another scanning method called "constant neutral loss," MS1 and MS2 are scanned synchronously with a fixed mass difference. The neutral loss spectrum produced will show species that underwent the loss of the common neutral fragment.

The data can be acquired in two different ways; one is termed *class-specific analysis*, for example neutral loss scanning, precursor-ion scanning,

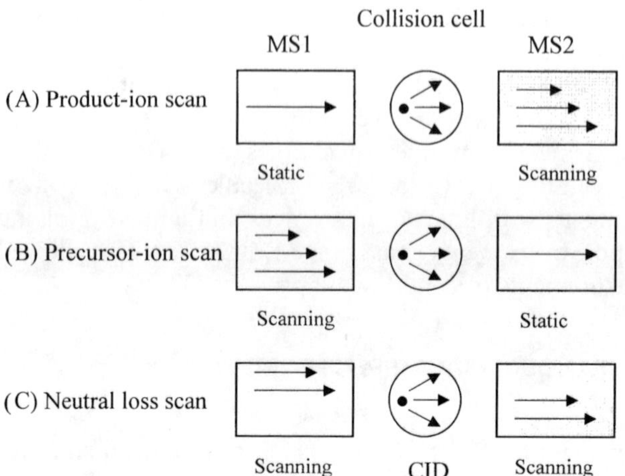

FIGURE 3 MS/MS experiments on a triple-quadrupole mass spectrometer.

and product-ion scanning. In this case mass spectra or profiles are produced. The other way is termed *target compound analysis* (selected reaction monitoring, SRM, multiple reaction monitoring, MRM) where both MS1 and MS2 are static, and certain transitions of the precursor to product are monitored. In this case a chromatogram rather than a spectrum is produced. Each of the two methods has advantages and disadvantages. SRM data allows for faster scanning speeds, higher signal-to-noise ratios, and shorter analytical time. It also allows for monitoring of target metabolites only in a certain class of compounds. On the other hand, class specific analysis is more comprehensive and may lead to the finding of new diagnostically important analytes.

Typically, these MS/MS experiments (or scan functions) would take only a few seconds each and can be done consecutively or alternately during the time the sample flows through the ion source. The process, however, is dependent on the flow rate, the scanned mass range, the scanning speed, the relative abundance of the analytes of interest, the desired quality of the data, and other parameters.

5 APPLICATIONS OF ESI-MS/MS IN BIOCHEMICAL GENETICS

Table 1 shows a list of analytes or classes of analytes that can be analyzed by ESI-MS/MS. FIA-ESI-MS/MS in certain situations allows the elimination of the separation step while retaining a high degree of specificity. For some analytes, interference from the biological matrix, other analytes, or the

TABLE 1 Biochemical Marker or Class of Markers Amenable to ESI-MS/MS Analysis

Biochemical marker
Acylcarnitines
Amino acids
Sulfur amino acids
Bile acids
Acylglycines
Pipecolic acid
Very long chain fatty acids
Organic acids
Steroids
Catecholamines and metabolites
Purines and pyrimidines
Plasmenylethanolamines
Guanidinoacetate
Hemoglobin and glycated hemoglobin
Oligosaccharides
Galactose-1-phosphate
Macromolecules

presence of isomeric or isobaric compounds require some chromatographic separation prior to mass analysis to increase sensitivity or specificity, or both.

5.1 Acylcarnitine and Amino Acid Analysis

One of the earliest breakthroughs was the ability to use automated FIA-ESI-MS/MS for the analysis of free carnitine (C0) and acylcarnitines (ACs) from plasma, blood spots, bile, urine, and amniotic fluid, in a short analytical time of 2–3 minutes, with no chromatography [30–35]. While the determination of these compounds was recognized to be of great clinical value for the diagnosis of FOADs (for reviews see Refs. 36,37), their analysis was difficult due to their polarity, zwitterionic nature, nonvolatility, relatively low concentration in body fluids, and lack of a chromophore. On the other hand, some of these properties made them ideally suited for ESI-MS/MS analysis and simplified sample preparation. Dried blood spots (DBS) were extracted with methanolic or ethanolic solutions containing isotope-labeled C0 and ACs. The separated extracts were evaporated to dryness and the residue was treated with butanolic HCl at 65°C for 15 minutes to make the butyl esters. Heat-assisted evaporation was then carried out to remove excess reagent followed by reconstitution in acetonitrile/water mixture and injection into the mass spectrometer. Slight variations to this procedure are required if plasma, serum, cerebrospinal fluid (CSF), bile, or urine sample are to be analyzed.

The ester derivatives of C0 and ACs carry a single positive charge and upon CID these derivatives yield a common fragment ion at *m/z* 85 (Fig. 4). The precursor ion spectra thereby obtained show the molecular ion species with little or no interference. The spectra (or profiles) obtained were diagnostic for a large number of FOADs, and for organic acidemias due to disorders in branched-chain amino acid (AA) catabolism disorders (see Figs. 5A–C, 6A–C, 7A–C and Table 2).

Another equally important development was the extension of the method to the analysis of neutral, acidic, and basic AAs from the same DBS, or plasma extract used for acylcarnitines. The easily ionized butyl derivatives of AAs also carry a single positive charge and therefore are ideally suited for positive-ion ESI-MS/MS analysis. These derivatives fall into two groups, the neutral/acidic group and the basic group. With the exception of six amino acids (these being asparagine, glutamine, arginine, citrulline, lysine, and ornithine), the butylated derivatives of the neutral and acidic group give a common neutral loss of 102 Da, upon CID fragmentation (Fig. 8). The basic group loses ammonia first (17 Da), followed by the neutral loss of 102 Da. Therefore, to get a more complete picture of the amino acids in the sample,

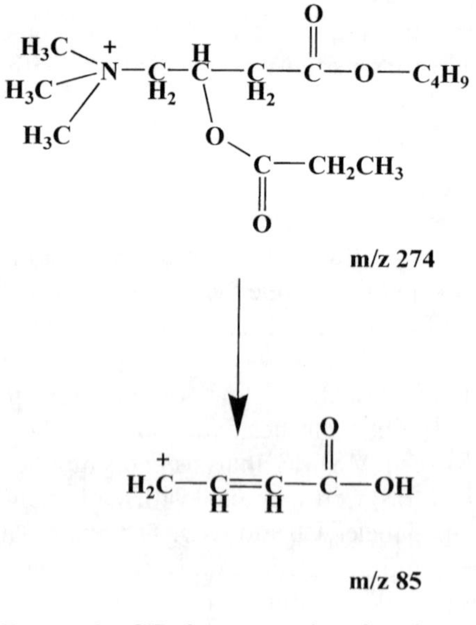

m/z 274

m/z 85

FIGURE 4 CID fragmentation for the *n*-butyl derivative of acylcarnitines as exemplified by propionylcarnitine.

two constant neutral loss scan functions were used, scanning for losses of 102 and 119 Da, generating a separate profile for each scan function.

The use of this approach allowed the determination of key biochemical markers for a number of amino acid disorders and proved valuable in the diagnosis of phenylketonuria (PKU), maple syrup urine disease (MSUD), homocystinuria (HCU), citrullinemia, tyrosinemia, and argininemia (see Figures 9A–E, 10A–C). Variations on this approach allowed for the determination of other markers such as glycine in blood and in CSF for the diagnosis of nonketotic hyperglycinemia [38].

Also, a method was developed for the detection and semiquantification of argininosuccinic acid (ASA), a key marker for the urea cycle defect argininosuccinase deficiency from the same DBS extract used for AC and AA profiling [38,39]. The tris-butyl ester of ASA obtained in the routine method for extraction and derivatization of blood spots yields a protonated molecular ion at *m/z* 459. Under CID conditions, the molecule fragments extensively with three major fragments at *m/z* 70, *m/z* 144, and *m/z* 172. We used a product-ion scan in a narrow mass range (65–150 Da) to detect two of the major fragments at *m/z* 70 and *m/z* 144. Figure 11A shows the abnormal profile obtained from the blood spot of a neonatal screening sample from a child at 24 h of life vs. Fig. 11B, which shows the profile from a normal child with no trace of ASA. The ion at *m/z* 85 appearing in both spectra is the major product ion of d_3-hexadeaconylcarnitine (d_3-palmitoylcarnitine; d_3-PC), an internal standard that is added during sample preparation. Coincidentally, the molecular ion (M^+) of d_3-PC as a butyl derivative is at *m/z* 459. Thus in the product ion scan of *m/z* 459 we get fragment ions from both ASA and d_3-PC, fortunately with no overlap in fragments produced from each species.

The routine ESI-MS/MS method of butylated DBS extracts or plasma extracts for profiling AC- and amino acids has been in use for 8 years. Numerous retrospective studies and single case reports have been published highlighting MS/MS as a powerful tool for selective screening for a large number of FOADs, organic acidemias, and aminoacidopathies [40–46]. Table 2 provides a list of diseases diagnosed by MS/MS analysis.

5.2 MS/MS-Based Newborn Screening

Almost parallel to the development of the biochemical genetics laboratory, another related activity was pursued, that is, the process of newborn screening (NBS) for a small number of genetic defects (for a review see Ref. 48). Dr. Robert Guthrie introduced NBS in 1963 as a public health measure to prevent mental retardation in PKU [48–50]. The concept stemmed from the knowledge that early detection of the disease and dietary restriction of phenylalanine prevents the cognitive deterioration of the patient. Guthrie developed

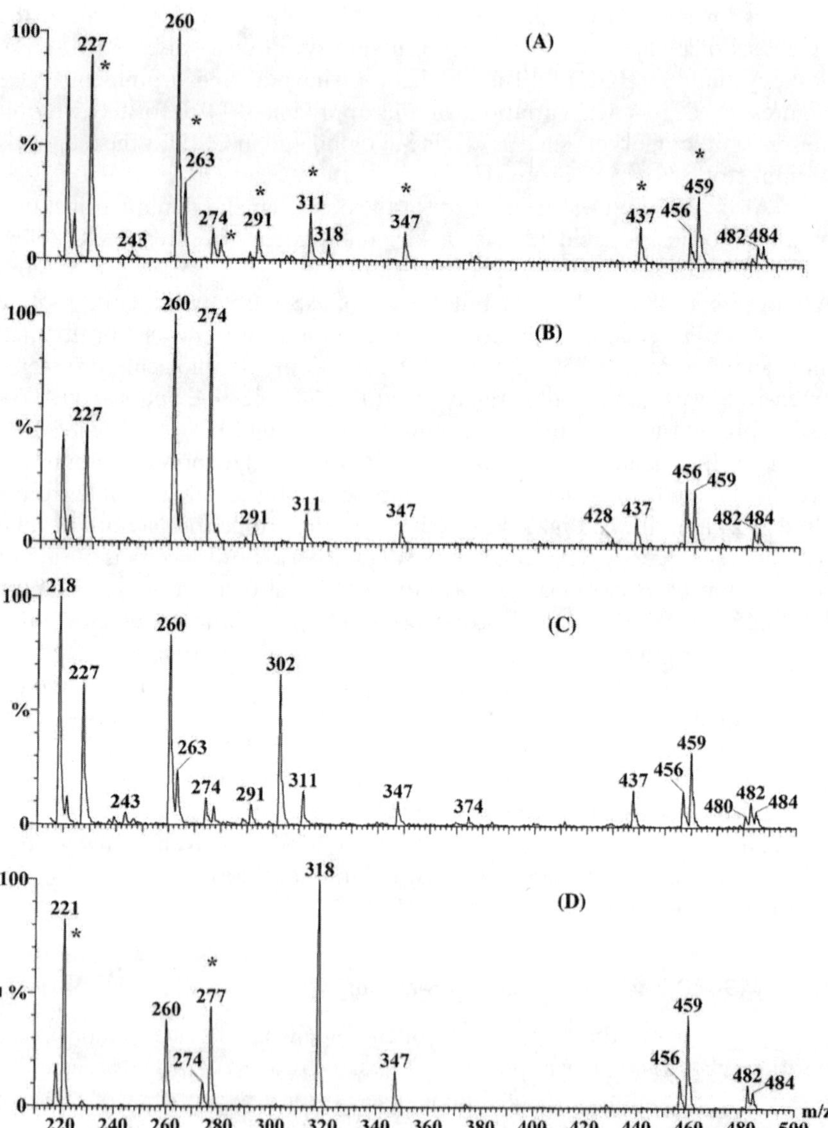
(A)
(B)
(C)
(D)
%
m/z

the bacterial inhibition test for the assay of phenylalanine in blood collected on filter paper. By the late 1960s, routine testing of neonates for PKU had spread to many countries, and many of the programs had also begun testing for other genetic defects such as galactosemia, maple syrup urine disease, and homocystinuria [50]. In the mid 1970s, a radioimmunoassay for thyroxine (T_4) was adapted to the Guthrie specimen for the identification of congenital hypothyroidism [51]. In the late 1970s, screening for congenital adrenal hyperplasia was introduced by measuring 17-hydroxprogesterone (17-OHP) by a radioimmunoassay [52]. A colorimetric method was developed and used for biotinidase screening in 1984 [53,54]. However, for each case, adding a new disease meant adding a new test using the same technology or a completely new one, "one test, one disorder."

The high sample throughput achieved by automated ESI-MS/MS, the "cleanliness" of the technique, its accuracy, reproducibility, and the ability to analyze a large number of biochemical markers in a short analytical time made it possible to develop a multidisorder test for a relatively large number of FOADs, organic acidemias, and aminoacidopathies [34,38]. This presented the biochemical genetics laboratory with a unique opportunity for the development of a comprehensive newborn screening program for a large number of IMDs where the screening system is changed from a "one test, one disorder" to "one test, many disorders" approach. An added tribute to the ESI-MS/MS approach is that it uses the same DBS specimen used for the Guthrie test, and it covers disorders that were screened for by the Guthrie method such as PKU, MSUD, and HCU, in addition to many

---

FIGURE 5 Blood spot acylcarnitine profiles obtained by ESI-MS/MS analysis with precursor-ion scanning of *m/z* 85. (A) Control; (B) propionic acidemia or methylmalonic acidemia profile; (C) isovaleric acidemia profile; (D) methylcrotonyl-CoA carboxylse profile. The peaks in the profiles are the molecular ions (M^+) of the acylcarnitine butyl esters. Their masses are as follows: free carnitine (C0, 218; $^2H_3$-isotope labeled 221; $^2H_9$-isotope labeled 227), acetyl (C2, 260; $^2H_3$-isotope labeled 263), propionyl (C3, 274; $^2H_3$-isotope labeled 277), butyryl or isobutyryl (C4, 288; $^2H_3$-isotope labeled 291), isovaleryl, or 2-methylbutyryl, (C5, 302; $^2H_9$-isotope labeled 311), 2-methyl-3-hydroxybutyryl or 3-hydroxyisovaleryl (OH-C5, 318), hexanoyl (C6, 316), octanoyl (C8, 344; $^2H_3$-isotope labeled 347), decanoyl (C10, 372), glutaryl (C4DC, 388), dodecanoyl (C12, 400), 3-methylglutaryl (C5DC, 402), tetradecanoyl (C14, 428; $^2H_9$-isotope labeled 437), hydroxy-tetradecanoyl (OH-C14, 444), hexadecanoyl (C16, 456; $^2H_3$ isotope labeled 459), hydroxy-hexadecanoyl (OH-C16, 472), octadecanoyl (C18, 484), hydroxy-octadecanoyl (OH-C18, 500).Their unsaturated analogs appear 2 Da lower in mass. * = Internal standard ions.

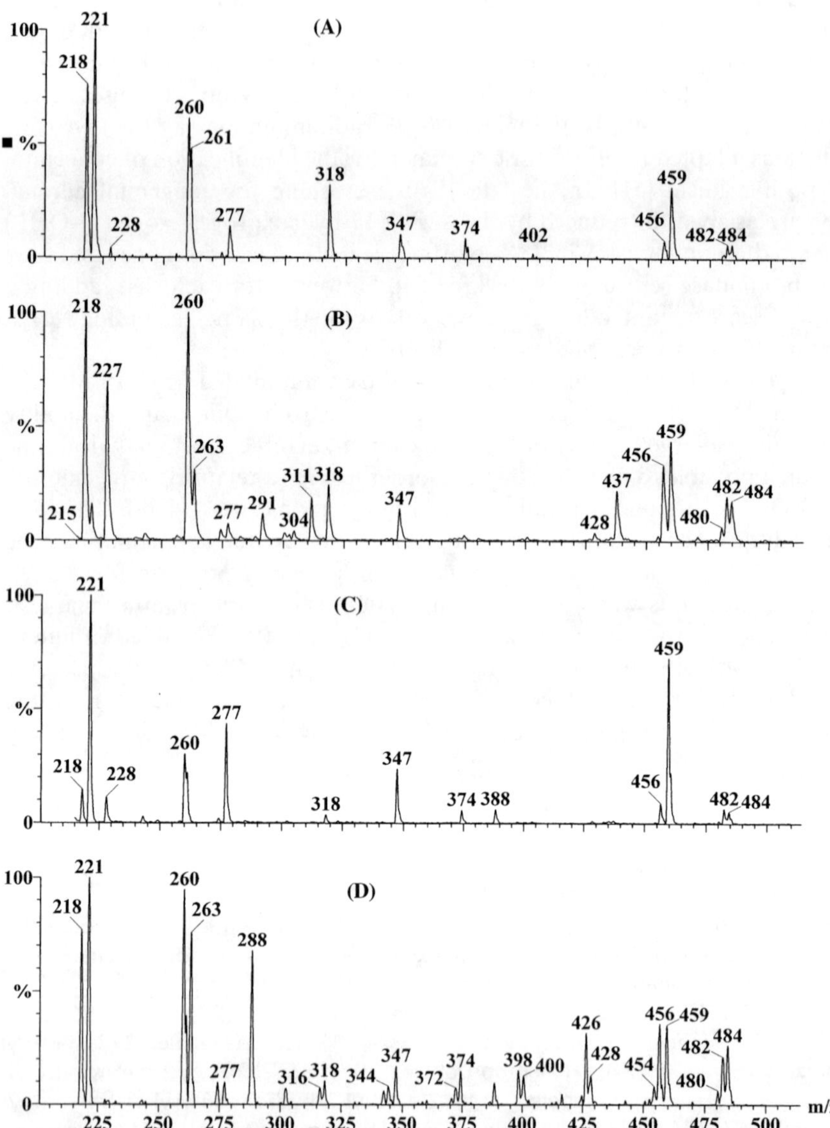

FIGURE 6 Blood spot acylcarnitine profiles obtained by ESI-MS/MS analysis with precursor-ion scanning of *m/z* 85. (A) HMG-CoA lyase profile; (B) β-ketothiolase profile; (C) glutaric acidemia type-I profile; (D) glutaric acidemia type-II profile. The masses are as indicated in the legend for Fig. 5.

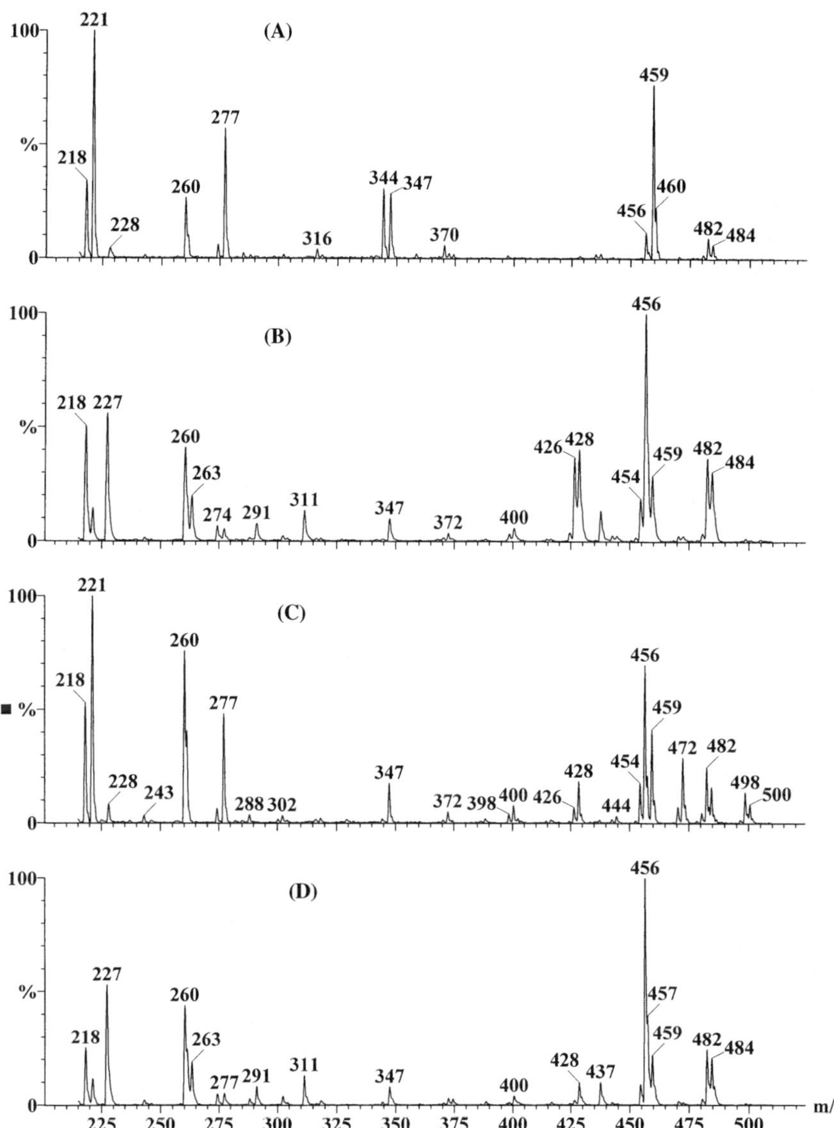

FIGURE 7 Blood spot acylcarnitine profiles obtained by ESI-MS/MS analysis with precursor-ion scanning of *m/z* 85. (A) MCAD profile; (B) VLCAD profile; (C) LCHAD or TFP profile; (D) trans or CPT-II profile. The masses are as indicated in the legend for Fig. 5.

TABLE 2 Diseases Detected by ESI-MS/MS Analysis of Blood Spots or Plasma

Fatty Acid Oxidation Defects
CTD
CPT-I
CPT-II
Translocase
MAD
SCAD
Ethylmalonic acidemia
MCAD
VLCAD
LCHAD
TFP
Dienoyl-CoA reductase deficiency
Organic acidemias
MMA (different types)
Combined methylmalonic : homocystinuria
Propionic acidemia (acute neonatal and late onset)
MCD
BKT
3-Hydroxy-3-methylglutaryl-CoA lyase deficiency
Methylcrotonyl-CoA carboxylase deficiency (isolated)
IVA
GA-I
Malonic acidemia
Aminoacidopathies
PKU (classical and biopterin dependent)
MSUD
Homocystinuria (due to CBS deficiency)
Citrullinemia (acute neonatal and mild)
Argininosuccinic acidemia (acute)
Tyrosinemia type-I
Tyrosinemia type-II
Methylenetetrahydrofolate reductase deficiency
Non-ketotic hyperglycinemia
Prolinemia type-II

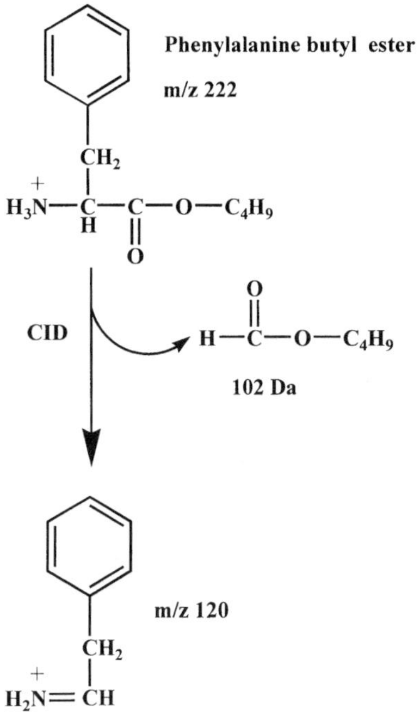

Phenylalanine butyl ester

m/z 222

$H_3N^+—CH(CH_2C_6H_5)—C(=O)—O—C_4H_9$

CID

$H—C(=O)—O—C_4H_9$

102 Da

m/z 120

$H_2N^+=CH—CH_2C_6H_5$

FIGURE 8 CID fragmentation for the protonated *n*-butyryl derivative of amino acids as exemplified by phenylalanine.

others. Furthermore, MS/MS is a highly specific approach having lower rates of false-positive results. This is because it identifies disorders based on more than one biochemical marker or ratios between metabolites. The sensitivity of the technique also is presumed to result in low false-negative results, and the ability to detect diseases on samples collected on the first day of life [40].

In the last 3 years numerous reports appeared in the literature describing prospective pilot studies on the use of ESI-MS/MS in newborn screening for IMDs. In a recent article we related our 3-year experience (1995–1998) for pilot prospective NBS using MS/MS in the Saudi population [38]. We set our cutoff values for 50 different markers at the 99.5th percentile determined from 5,000 normal newborns from the Saudi population with a birth weight ≥ 2.0 kg. We screened 27,624 blood spots from newborns and identified 20 cases yielding a frequency of 1 : 1381. No false-negative cases were identified. However, sample collection time was quite early and averaged 24.87 ± 16.4 h, which was a reason for concern. Several false-positive results were obtained. These were eliminated by repeat analysis by MS/MS on the same

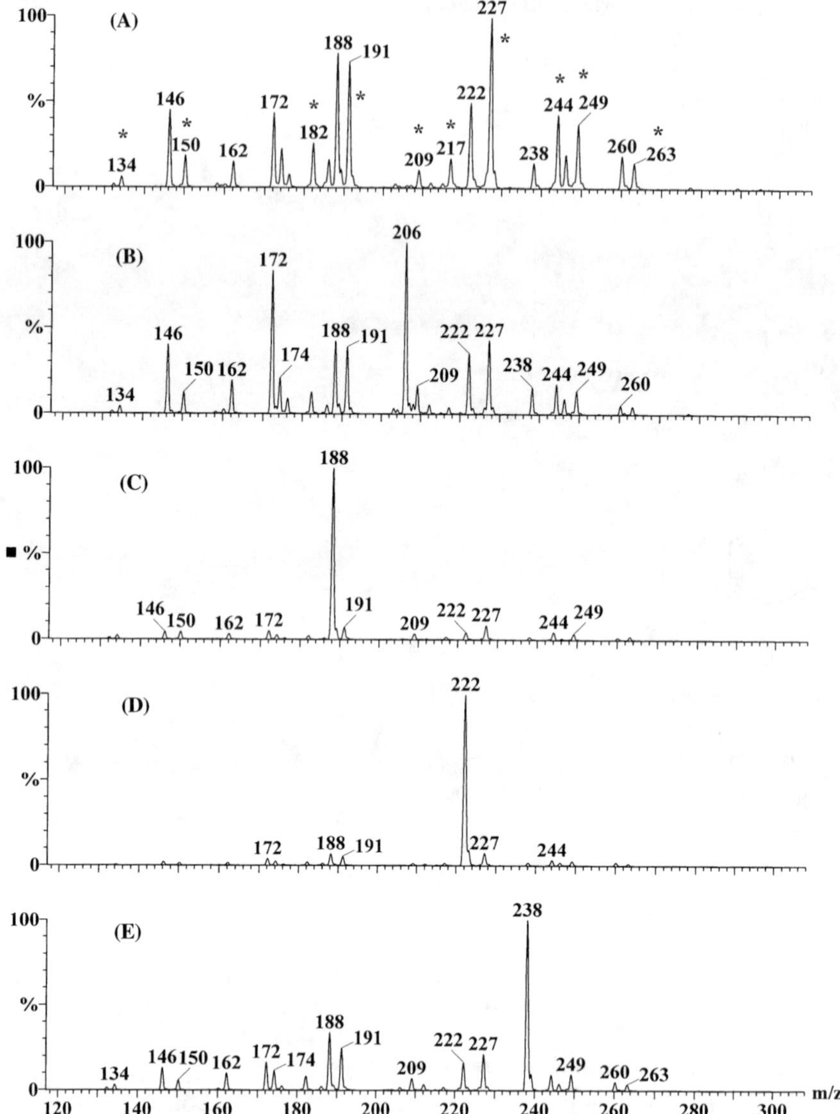

FIGURE 9 Blood spot amino acid profiles obtained by ESI-MS/MS analysis with the neutral loss scanning of 102 Da. (A) Control; (B) hypermethioninemia profile from a patient with homocystinuria; (C) MSUD profile; (D) PKU profile; (E) tyrosinemia type II profile. * = Internal standard ion. The peaks in the profiles are the protonated molecular ions (MH^+) of the amino acid butyl esters. The relevant ions are as follows: glycine (132; ^{15}N, 2-^{13}C-isotope labeled 134), alanine (146; $^{2}H_4$-isotope labeled 150), proline (172), valine (174; $^{2}H_8$-isotope labeled 182), leucine and isoleucine (188; $^{2}H_3$-isotope labeled 191), methionine (206; $^{2}H_3$-isotope labeled 209), phenylalanine (222; $^{2}H_5$-isotope labeled 227), tyrosine (238; $^{13}C_6$-ring labeled 244).

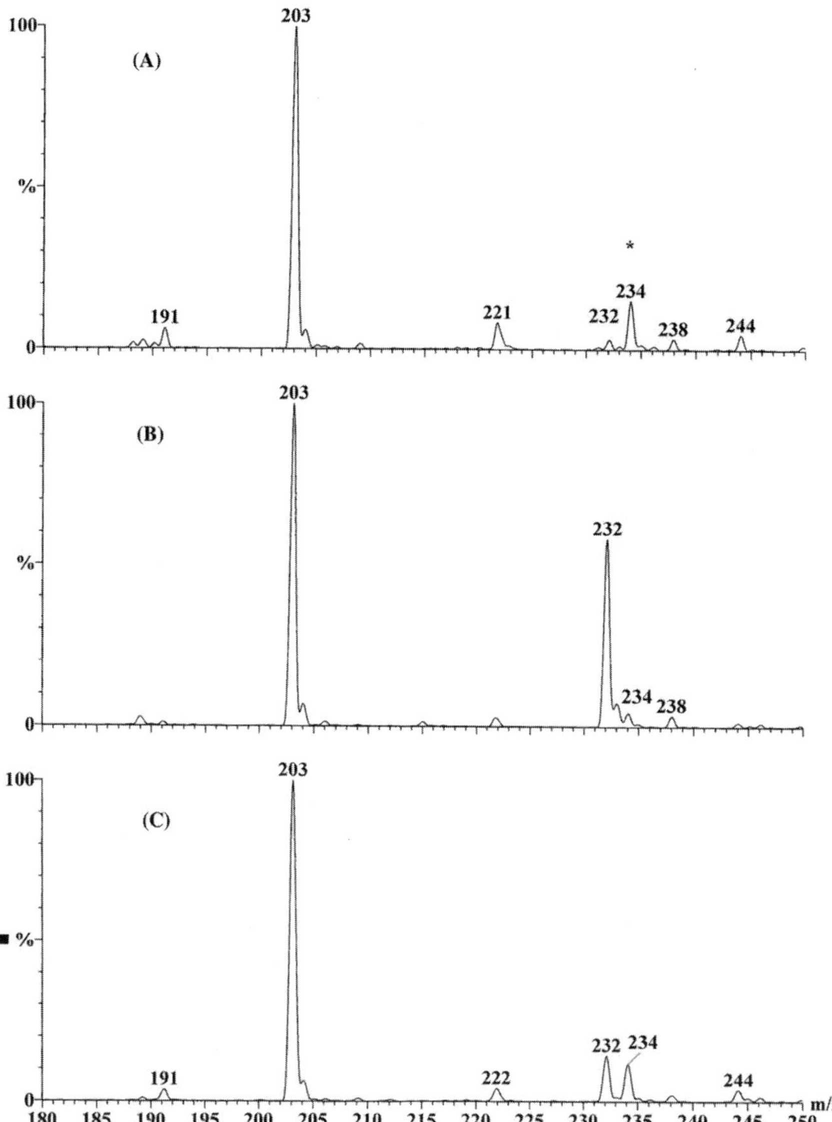

FIGURE 10 Blood spot amino acid profiles obtained by ESI-MS/MS analysis with neutral loss scanning of 119 Da. (A) Control; (B) citrullinemia profile; (C) profile from argininosuccinic acidemia patient with borderline elevated citrulline. The peaks in the profiles are the protonated molecular ions (MH^+) of the amino acid butyl esters. The relevant ions are as follows. glutamine and lysine (203), citrulline (232; $^2H_2$-isotope labeled 234).

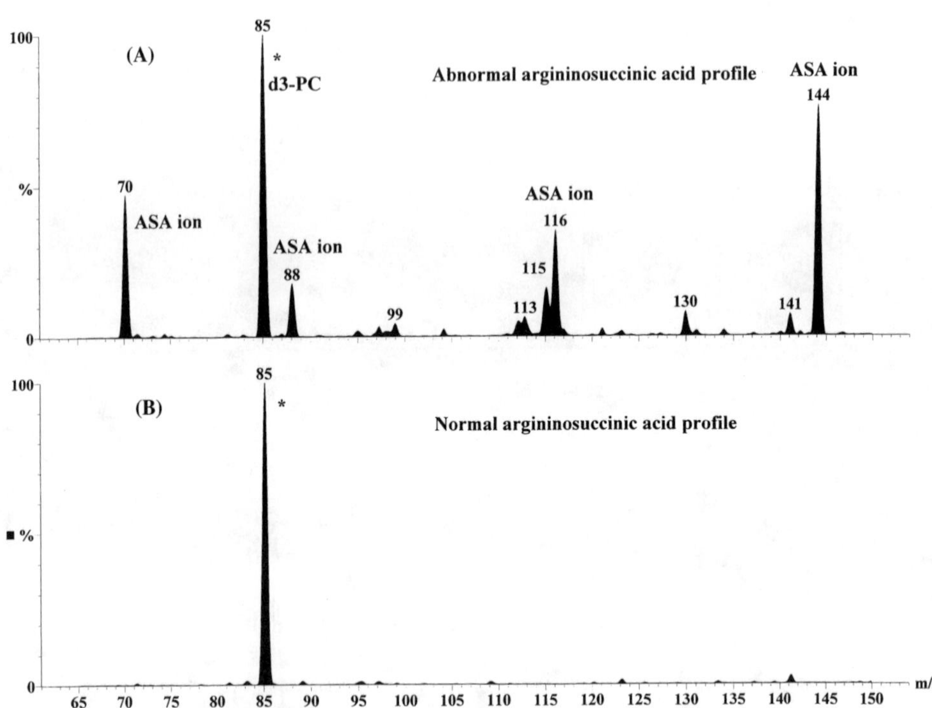

FIGURE 11 ESI-MS/MS product-ion spectrum of *m/z* 459.4 in the range of 65–150 Da. (A) Abnormal argininosuccinic acid profile from a newborn baby blood spot at 24 hours of life; (B) argininosuccinic profile from an age-matched normal newborn. * = Fragment from internal standard $^2H_3$-hexanoylcarnitine.

or a second DBS, by GC/MS analysis of urine for organic acids, and by MS/MS analysis of CSF for determination of glycine. In the study described, we have identified three cases of PKU (one of which was a biopterin-dependent PKU), two MSUD, two argininosuccinic acidemia, two citrullinemia, one non-ketotic hyperglycinemia (NKH), two glutaric acidemia type-I (GA-I), four methylmalonic acidemia (MMA), one propionic acidemia (PPA), two IVA, and two medium-chain acyl-CoA dehydrogenase deficiency (MCAD). Three metabolites contributed most to our falsely flagged samples. C3, or the C3/C2 ratio, was high in many infants, and this correlated with the early collection of blood spots. C5-carnitine could be either isovalerylcarnitine diagnostic for IVA or pivalylcarnitine. The latter interference results from the use of pivalic acid prodrug antibiotics given during pregnancy or when given to the infant. Glycine was found borderline high in several samples but was normal on repeat analysis of the same sample or a second sample.

Naylor and Chace described a 7-year study using both FAB-MS/MS and ESI-MS/MS in analyzing more than 700,000 samples from newborns from several states in the U.S.A. [55]. They prospectively diagnosed 163 cases (86 amino acid metabolism errors, 32 organic acidemias, and 45 FOADs).

Wiley et al. described a 12-month experience where they screened 137,120 blood spots using ESI-MS/MS [56]. Samples were collected between 48 and 72 h. They detected 31 babies with IMDs: 17 PKU, one tetrahydrobiopterin deficiency, three hyperphenylalaninemia, one MSUD, one tyrosinemia type II, one congenital lactic acidosis, two MCAD, one short-chain acyl-CoA dehydrogenase deficiency (SCAD), one beta-ketothiolase (BKT), two vitamin B12 deficient babies of vegetarian mothers, and one GA-I.

Shigematsu and coworkers described a pilot study where 23,000 blood spots were screened by ESI-MS/MS in Japan [57]. One PPA was detected. These authors also indicated a false positive rate in the diagnosis of IVA due to the use of pivalic-acid-containing antibiotics of 0.37%.

Liebl et al. described their one-year NBS experience using MS/MS as well as other screening tests. They screened 87,000 newborns and diagnosed 22 cases by MS/MS. Among these cases nine were PKU and six MCAD [58].

Zytkovicz et al. described a study where they screened 160,000 newborns from the New England region of the U.S.A. for 23 metabolic disorders [59]. They identified 22 babies with amino acid disorders (7 PKU, 11 hyperphenylalaninemia, one MSUD, one hypermethioninemia, one argininosuccinic acidemia, and one argininemia). They also identified 20 cases with fatty and organic acid disorders (10 MCAD, two PPA, one carnitine palmitoyltransfearse II deficiency (CPT-II), one methylcrotonyl-CoA carboxylase deficiency (MCCD), one presumptive very long-chain acyl-CoA dehydrogenase deficiency (VLCAD), and five presumptive SCAD. They set up their cutoff for different markers at 99.98th percentile so as to allow for a recall rate of 0.02%. The positive predictive value (PPV) for amino acid disorders was 8% and increased to 14% with the use of appropriate amino acid ratios. The PPV for acylcarnitine disorders was 9%. These values were better than those reported for well-established screening methods for biotinidase deficiency, galactosemia, and congenital hypothyroidism.

Lin et al. described a small pilot study of neonatal screening by ESI-MS/MS in Taiwan [60]. They screened 2,100 newborns and detected one IVA and one designated as hyperphenylalaninemia (1:1050), with a false-positive rate of 1.28%.

5.3 MS/MS-Based Selective Screening

The exciting and successful application of ESI-MS/MS in determination of AAs and ACs promoted active research toward the development of new MS/

MS methods for diagnosis of many other inherited diseases such as peroxisomal diseases, bile acids disorders, steroid diseases, and purine and pyrimidine metabolism disorders. Although the main attraction of MS/MS in flow injection analysis mode, i.e., without chromatography, remains the proven high-throughput multidisorder detection applicable to NBS, the technique has found numerous new biochemical genetics applications. The main incentive in this case was the short analytical time due to the elimination or the shortening of the chromatographic steps required for the analyte(s) of interest, the elimination of interference, the analysis of highly polar compounds, the increase in assay precision, and the simplification of sample preparation, while retaining a high level of sensitivity and specificity.

5.4 Determination of Amino Acids

Casetta and coworkers recently described a liquid chromatography ion spray MS/MS (LC-IS-MS/MS) method for the rapid determination of eighteen AAs in plasma and DBS in less than 4 min [61]. Sample preparation was quite similar to the largely accepted method of extraction in methanolic solution using $^{2}H_{4}$-alanine and $^{2}H_{5}$-phenylalanine as internal standards, evaporation, and butylation. The butyl esters were injected onto a short cyano column interfaced with the mass spectrometer. The AAs analyzed in MRM mode using specific transitions for each. The specificity offered by MS/MS analysis allowed quantification of several AAs in very short time windows despite significant coelution. The method exhibited excellent recovery, linearity, and low CVs. It also correlated well with values obtained from a standard AA analyzer. Furthermore, the method allowed for the quantification of the isobaric leucine and isoleucine, which is important for monitoring MSUD patients undergoing treatment. A particular problem was glutamine, which degraded to glutamate during the acidic esterification and thus could not be accurately measured.

Of particular interest to us was to develop a rapid method to screen sick patients who are either suspected clinically to have homocystinuria or showed an isolated elevation of methionine in their blood spots. The routine MS/MS approach to analysis of DBS does not detect homocysteine, or the disulfide homocystine. This is because of the known fact that most of homocysteine is bound to plasma protein and reduction is necessary for its release. A minor fraction is in the disulfide form either as homocystine or cysteine-homocysteine, both being highly polar compounds that are rapidly excreted in urine. We used a rapid qualitative MS/MS method to analyze urine spotted on filter paper (1/8″ punch; about 2 μL urine), followed by extraction with a mixture of 0.1 N HCl/methanol (1 : 9 v/v) for 30 min, evaporation, and derivatization to the butyl esters. The samples are then injected directly into the mass spectrometer in the FIA mode at a constant flow of 40 μL/min. A 2 min MRM scan function is carried out for three transitions, *m/z* 381→192 for homocystine bis-butyl ester (MH^+ = 381), *m/z* 367→190 for cysteine-homocysteine bis-butyl ester (MH^+ = 367), and *m/z* 335→190 for cystathionine bis-butyl ester (MH^+ = 335). A dwell time of 0.1 s is used for each transition with a cone voltage of 28 V and collision energy of 13 eV. The mass chromatogram is averaged to yield a profile that shows all three ions. So far, we have analyzed about 1,100 urine spots by this method. In over 25 homocystinuria cases due to CBS deficiency, the ratios of ions *m/z* 381 (homocystine) and of *m/z* 367 (cysteine-homocysteine) to *m/z* 335 (cystathionine) were above 1, and vice versa for controls. In each case, further confirmation was carried out by the established HPLC assay for total homocysteine in plasma [40].

Recently Magera et al. described an isotope dilution LC-MS/MS method for the determination of total homocysteine in plasma and urine [62]. In this case, sample preparation involved reduction of disulfides by dithiothreitol in presence of d_8-homocystine used as internal standard, followed by protein precipitation, centrifugation, and analysis of the supernatant by LC-MS/MS. A short cyano column was used (3 cm × 4.6 mm) at a flow rate of 1 mL/min with the LC effluent split at 1:5 yielding an analytical time of 3 min per sample. The produced homocysteine was analyzed without derivatization using SRM for the transition of *m/z* 136→90 (MH^+ to MH^+-HCOOH) and the transition *m/z* 140→94 for the internal standard ($^2H_4$-homocysteine obtained after reduction). The calibration curves constructed in the range of 2.5 to 60 μmol/L were linear with excellent intra- and interassay coefficient of variation. The method correlated well with other established assays for homocysteine and gave a high throughput of 20 samples per h.

Gempel et al. described an LC-MS/MS method for determination of total homocysteine in blood spots [63]. The method was essentially very similar to that of Magera's and showed good correlation with an established HPLC method and had the same throughput as that of the plasma method described above.

We recently described a chiral LC-ESI-MS/MS method for the determination of plasma L-pipecolic acid (L-PA), an important biochemical marker for the diagnosis of peroxisomal disorders that is usually determined as the racemate [64]. We used a narrow bore chiral macrocyclic glycopeptide teicoplanin column for the enantiomeric separation of D- and L-PA and interfaced the column directly to the ESI source of a tandem mass spectrometer. We used phenylalanine-d_5 as internal standard added to 50 μL of plasma followed by deproteinization, evaporation, and injection. The analysis was performed in the SRM mode using two transitions, *m/z* 130 to *m/z*

84 for PA, and m/z 171 to m/z 125 for phenylalanine-d_5. L-PA eluted at ~ 7 min and D-PA eluted at ~ 11 min, while phenylalanine-d_5 eluted at ~ 6 min (see Fig. 12). The turnaround time for the assay was 20 min. Linear standard curves were obtained in the range of 0.5–80 μmol/L, and the results were reproducible. At a plasma concentration of 1.0 μmol/L, the signal-to-noise ratio was 50:1. The intra- and interassay variation was 3.1–7.9% and 5.7–13.1%, respectively; at concentrations of 1, 5, and 50 μmol/L. Mean recoveries of L-PA added to plasma were 95.2% (5 μmol/L) and 102.5% (50 μmol/L). The method clearly distinguished between normal and peroxisomal disease patients.

5.5 Analysis of Acylglycines

Glycine conjugation is a detoxification pathway for some organic acids in several metabolic diseases. For some acyl-CoA groups this pathway is recognized to be quantitatively more important than conjugation with carnitine. Therefore analysis of acylglycines was found diagnostic for several diseases such as IVA, MCCD, and MCAD [15,16].

Recently, Bonafé et al. evaluated the analysis of urinary acylglycines by ESI-MS/MS. They presented a semiquantitative method for measuring acylglycines using several stable-isotope-labeled derivatives [65]. In glutaric acidemia type-II (GA-II; MAD) and MCAD acylglycines, profiles were always informative. In MCAD, hexanoylglycine, phenylpropionylglycine, and suberylglycine were the main metabolites seen. Hexanoylglycine was always high irrespective of clinical status, while phenylpropionylglycine and hydroxyoctanoylglycine were elevated during acute crises. In the neonatal form of MAD a C4-glycine (butyryl and isobutryl) and C5-glycine (methylbutyryl and isovaleryl) were the most prominent acylglycines. In IVA (five samples), PPA (47 samples), and MCCD (one sample), the typical glycine conjugates in urine were clearly elevated. However, the profiles were either inconsistent or not informative in SCAD, long-chain acyl-CoA dehydrogenase deficiency (LCAD), VLCAD, CPT-II, MMA, and GA-I.

We used the butyl esters of urinary acylglycines and ESI-MS/MS for confirmation of some of our cases such as MCCD and IVA [40]. MCCD was suspected in a 4-month-old patient based on the AC profile obtained by MS/MS analysis of the patient's blood spot that showed highly elevated ion at m/z 318 corresponding to a hydroxy-C5-carnitine. This particular ion is elevated to a different extent either by itself or in combination with other diagnostic ions in diseases related to leucine and isoleucine catabolism disorders such as MCCD, multiple carboxylase deficiency (MCD), 3-hydroxy-3-methylglutaric acidemia (HMG), BKT, and some valproate-

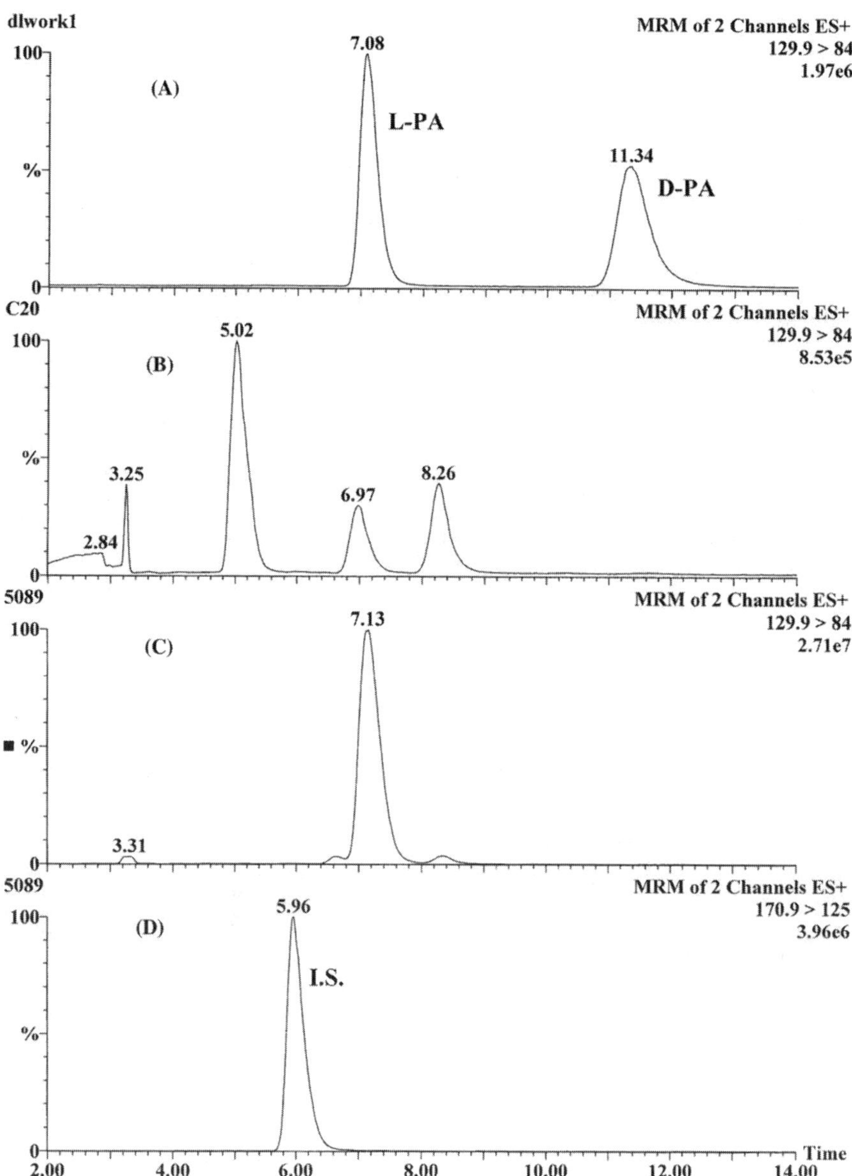

FIGURE 12 LC-MS/MS positive-ion SRM chromatograms on a Chirobiotic T column (25 cm × 2.0 mm i.d.; 5 μm). (A) Transition *m/z* 130 → *m/z* 84 for standard mixture of D- and L-PA;(B) transition *m/z* 130 → *m/z* 84 for control plasma; (C) transition *m/z* 130 → *m/z* 84 for plasma from a Zellweger patient showing highly elevated L-PA; (D) transition *m/z* 171 → *m/z* 125 for $^2H_5$-phenylalanine internal standard.

treated children. We examined the derivatized patient urine spot (~ 2 μL) by ESI-MS and observed a strong signal at *m/z* 214, possibly corresponding to a protonated C5 : 1-glycine (either methylcrotonylglycine or tiglylglycine) (data not shown). The product-ion spectrum of *m/z* 214 showed one major ion at *m/z* 83, which corresponds to the neutral loss of butylglycine ($H_2NCH_2COOC_4H_9$), thus retaining the positive charge on the molecule backbone. Urine samples from BKT patients known to excrete tiglylglycine (C5 : 1-glycine) gave the same results (data not shown). This is not the case in other butylglycines, for example isovalerylglycine (for IVA), hexanoylglycine, and phenylpropionylglycine (for MCAD). These metabolites gave a major production at *m/z* 132 corresponding to the protonated butylglycine moiety. Therefore we decided to use specific SRM functions for each metabolite (*target compound analysis*) rather than precursor-ion scans (*class-specific analysis*). Thus we used four transitions, *m/z* 214→83, 216→132, 230→132, and 264→132 for C5 : 1-glycine, C5-glycine, C6-glycine, and phenylpropionylglycine, respectively. The mass spectrometer was scanned using a cone voltage of 25 V and collision energy of 12 eV. The averaged MRM chromatograms for our MCCD patient urine yielded a spectrum dominated by a strong ion at *m/z* 214 corresponding to C5 : 1-glycine.

The methods described by Bonafé and by us lacked the specificity to pinpoint the defect, BKT or MCCD, even by two different FIA-MS/MS tests carried on different biological matrices; and GC/MS analysis in the electron impact mode was necessary for a firm diagnosis.

5.6 Analysis of Organic Acids

We introduced recently another LC-ESI-MS/MS application in selective screening for the determination of the configuration of 2-hydroxyglutaric acid (2-HG) in urine of patients with 2-hydroxyglutaric aciduria [66,40]. 2-HG is a chiral polar aliphatic dicarboxylic acid that exists in two configurations, D-2-HG and L-2-HG. The two enantiomers are intermediary metabolites normally excreted in very small amounts in mammalian urine. D-2-hydroxyglutaric aciduria and L-2-hydroxyglutaric aciduria are two distinct metabolic disorders with different phenotypes. The discovery of either defect is usually accomplished through routine urinary organic acid analysis by GC/MS. In both cases the EI mass spectra obtained are identical with standard 2-HG. The determination of the absolute configuration of 2-HG is necessary for accurate diagnosis. For this purpose we developed an enantiomeric chiral separation method using a ristocetin A glycopeptide antibiotic silica gel bonded column (25 cm × 4.6 mm). A 0.1 mL of urine was diluted with mobile phase (5 mM triethylamine acetate pH 7:

methanol; 9 : 1 v/v), filtered and injected onto the column interfaced to the mass spectrometer ion source through a T-shaped splitter at a flow rate of 0.5 mL/min with a 1 : 8.5 split ratio. We used negative-ion ESI-MS/MS in the MRM mode to monitor three transitions of MH^- to product ions (147→129, 147→85, and 147→57). The use of 100% triethylamine acetate pH 5 at a flow rate of 1 mL/min (1 : 17 split) was found later to provide better resolution of the two enantiomers.

Another application of chiral-LC-ESI-MS/MS was in the analysis of urinary D- and L-glyceric acid. Glyceric acid (2,3-dihydroxypropionic acid) is a plant and mammalian metabolite that exists in two configurations, D(+) and L(−). Both enantiomers are intermediary metabolites normally excreted in trace amounts in urine. The two enantiomers are key biochemical markers for two rare inherited metabolic diseases, with different phenotypes, D-glyceric aciduria and primary hyperoxaluria type 2 (PH2). Determination of the configuration of glyceric acid is a necessary test for the differentiation of PH2 from D-glyceric aciduria; the confirmation of these disorders and the excretion of L-glyceric is important in differentiating PH2 from primary hyperoxaluria (PH1). Similar to 2-HG, glyceric acid circulates in blood and is excreted readily in urine as a carboxylate anion and thus is ideally suited for electrospray analysis in the negative ion detection mode. To investigate this approach we optimized the conditions necessary for the detection of glyceric acid where a standard solution of the DL mixture was injected in the FIA mode. A narrow bore ristocetin glycopeptide column separated the two isomers using a triethylamine acetate buffer at pH 4.1 with 10% methanol at a flow rate of 0.3 mL/min. Figures 13A–D show the MRM profiles for the characteristic and most intense transition for glyceric acid *m/z* 105→75 for the synthetic DL mixture, a patient with D-glyceric aciduria, a patient with L-glyceric aciduria, and a control urine sample, respectively. The urine from all patients examined showed highly elevated glyceric as compared to controls [67].

Two new methods were recently reported for the determination of methylmalonic acid in plasma and urine using LC-ESI-MS/MS. This acid is an important biochemical marker for a group of inherited disorders collectively known as methylmalonic acidemias. Increased methylmalonate in plasma and urine is also found in patients with cobalamin (vitamin B_{12}) deficiency. Magera et al. described a method in which plasma is mixed with internal standard solution (d_3-methylmalonate) and extracted by anion-exchange solid phase columns. The residues were derivatized by *n*-butanolic HCl to produce the di-butyl esters. The derivatives were analyzed by LC-ESI-MS/MS using a short C18 column in a 3 minute analytical time. The LC effluent was split so that only 100 μL/min reached the ion

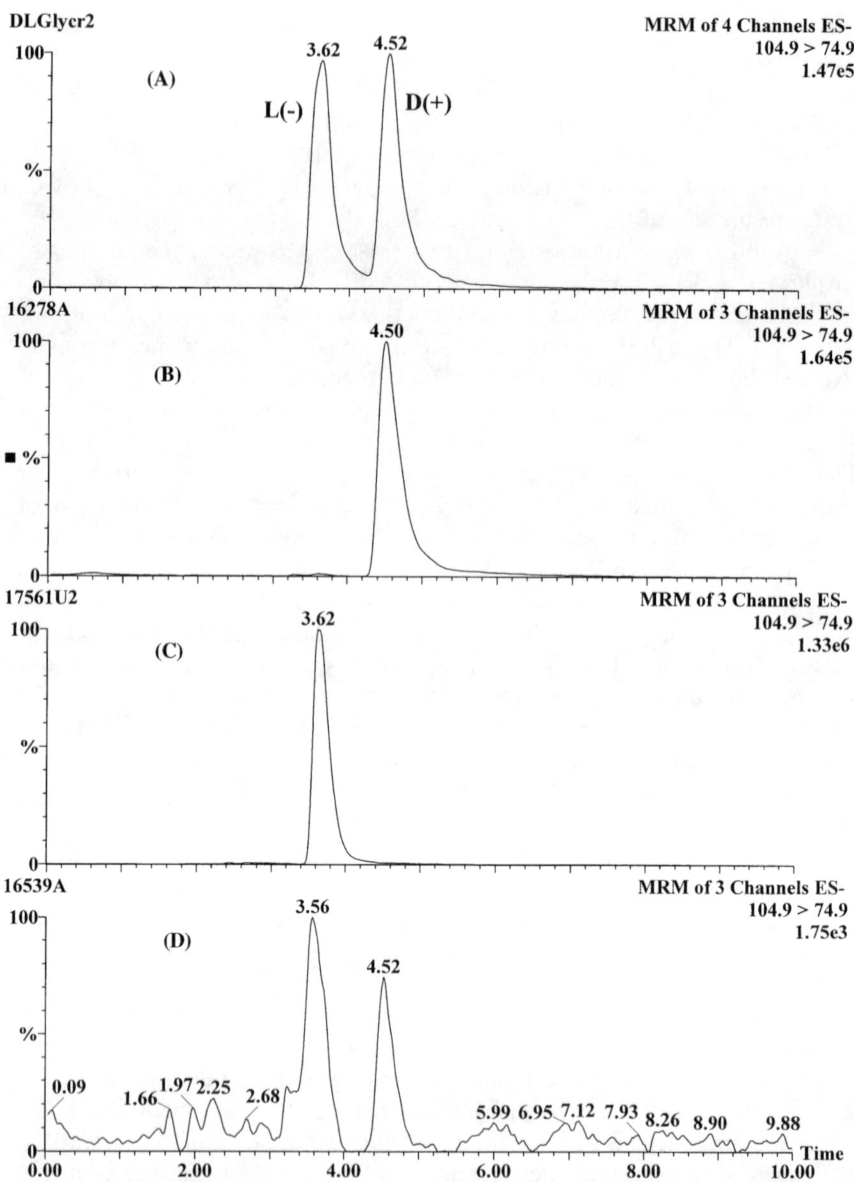

FIGURE 13 LC-MS/MS negative-ion chromatograms obtained for the SRM transition *m/z* 105→75 using a ristocetin A column (25 cm × 2.0 mm i.d.; 5 μm). (A) Standard mixture of DL-glyceric acid; (B) urine profile from a D-glyceric aciduria patient; (C) urine profile from an L-glyceric aciduria patient; (D) urine control.

source. Two positive-ion SRM transitions were used for the acid and its internal standard. Butylation of these compounds helped to minimize interference from the isomeric succinic acid, a metabolite normally present in plasma and urine. The butyl esters provided modest chromatographic separation but a significantly different CID fragmentation. This allowed for specificity in methylmalonate determination at succinate concentration up to 2000 μmol/L [68].

Kushnir et al. described a very similar method except that extraction of methylmalonic was carried out by liquid/liquid extraction using methyl-*tert*-butyl ether and a shorter analytical time of 1 minute. However, the authors indicated that 10.8% of analyzed samples needed to be reanalyzed by a method that chromatographically resolves methylmalonate and succinate [69].

Magera et al. also reported a method for determination of homovanillic (HVA) in urine by LC-ESI-MS/MS [70]. Using a stable isotope labeled internal standard ($^{13}C_6$-^{18}O-HVA), urine was extracted by an automated solid-phase extraction procedure. The underivatized extracts were dissolved in mobile phase and injected onto an amide column interfaced to the mass spectrometer. Two negative ion SRM transitions due to loss of carbon dioxide from the deprotonated parent ions were used for specific detection of HVA and $^{13}C_6$-^{18}O-HVA. The method exhibited consistent linearity and reproducibility, low inter- and intra-assay variability, and a rapid turnaround time.

5.7 Analysis of Purine and Pyrimidines

Another recent and exciting application of ESI-MS/MS was in the selective screening for inherited disorders of purine and pyrimidine metabolism. These disorders have a wide variety of clinical presentations that are often non-specific, and the severity of the cases ranges from fatal to asymptomatic. Therefore the development of a simple, rapid, and specific screening method for these diseases was highly desirable. Van Gennip and coworkers described an LC-ESI-MS/MS assay using urine-soaked filter paper [71,72]. Sample preparation was simple and essentially involved extraction of filter paper with an aqueous methanolic solution containing stable isotope labeled standards of uracil, thymine, orotic acid, thymidine, uridine, and others. The extracts were dried, reconstituted in mobile phase, and injected into a narrow-bore octadecylsilane column (250 × 2.1 mm) at a flow rate of 0.3 mL/min. Specific MRM transitions were used for a large number of analytes, and HPLC was necessary for separation of isomeric compounds. Analytical time was 15 min/sample. The mean recovery of all analytes was 87 to 102%, and CVs were from 1.6 to 13.3%. The method was evaluated with urine samples from patients with established dihydropyrimidine dehydrogenase deficiency (elevated uracil, thymine, and 5-hydroxyuracil), molybdenum cofactor deficiency (elevated xanthine), ornithine transcarbamylase deficiency (elevated orotic and uracil), purine nucleoside phosphorylase deficiency (elevated inosine and guanosine), and adenylosuccinase deficiency (elevated succinyladenosine).

The use of LC-ESI-MS/MS by the same group led to the confirmation of the first patient with β-ureidopropionase deficiency. Urine from this patient showed highly elevated *N*-carbamyl-β-alanine and *N*-carbamyl-β-aminoisobutyric acid, with normal or moderately increased levels of the pyrimidine bases and the dihydropyrimidines, respectively [73].

Struys et al. reported an isotope-dilution LC-ESI-MS/MS method for the determination of the purine derivatives *S*-adenosylmethionine (SAM) and *S*-adenosylhomocysteine (SAH) in plasma and CSF [74]. These two compounds are intermediates in the metabolic pathways of methionine/homocysteine, and variations in their concentrations are associated with several disease states. The method involved sample deproteinization by perchloric acid followed by a weak anion-exchange solid-phase extraction. Analysis was carried out on the extracts by injection onto a narrow bore C18 column directly interfaced to the ESI source of the mass spectrometer. Four positive SRM transitions of all parent M^+ ions to the same common product ion were used for SAH, its isotope-labeled standard, SAM, and its isotope-labeled standard. Isocratic elution was used and analytical time was 3 minutes. Intra- and interassay variations were low, and mean recovery for SAM and SAH was 93%.

5.8 Analysis of Bile Acids and Very-Long-Chain Fatty Acids

Another recent area for application of ESI-MS/MS analysis was the analysis of bile acids and their conjugates, which is important for the diagnosis of several bile acid metabolism disorders as well as for peroxisomal diseases. A number of inherited metabolic diseases have been characterized in which the normal conversion of cholesterol to the primary 24-carbon bile acids is disrupted. Examples include Zellweger syndrome, neonatal adrenoleukodystrophy, infantile Refsum disease, peroxisomal bifunctional protein deficiency, 3-oxoacyl-CoA thiolase deficiency, and cholestatic hepatobiliary disease [75–77].

GC/MS methods for the analysis of bile acids in urine samples are the most sophisticated and informative methods but are technically demanding and time-consuming [78]. In the 1980s, several authors demonstrated the usefulness of FAB-MS for the rapid detection of bile acids in human fluids with minimum sample preparation [79]. FAB-MS operated in the negative-ion mode showed intense ions for abnormal bile salts in urine extracts. However, the information provided was limited to the presence of the pseudomolecular ion $(M\text{-}H)^-$ of these compounds, interferences from the matrix, and lack of adequate sensitivity in some situations.

Tomer et al. using pure standards were the first to study the use of negative-ion FAB-MS/MS as an alternative approach for determination of bile salts and their conjugates [80]. Evans et al. studied the use of negative-ion LC-TSP-MS and LC-TSP-MS/MS of bile acids and their glycine conjugates and indicated more specificity by MS/MS [81]. Libert et al. using negative-ion FAB-MS/MS for the analysis of serum or urine extracts introduced the first clinical application of this methodology in patients with liver diseases and peroxisomal diseases. They reported better specificity by MS/MS and foresaw that the method might become the first line of investigation of peroxisomal disorders [82].

Warrack and DiDonato introduced the use of ion spray-MS (IS-MS) and LC-IS-MS of bile acids and their conjugates. The method demonstrated higher sensitivity than those reported for FAB-MS but no clinical applications were presented [83]. Later, Roda and coworkers used negative-ion LC-ESI-MS/MS to separate and detect a large number of free bile acids and their glycine and taurine conjugates at the picogram level in human serum and hamster bile [84].

Mills et al. recently described the use of a negative-ion ESI-MS/MS isotope dilution method for the measurement of taurotrihydroxycholanoates and glycodihydroxycholanoates in blood spots with the purpose of mass screening for cholestasis [85]. They used precursor-ion scanning for *m/z* 74 to detect glycine conjugates and precursor-ion scanning for *m/z* 80 to detect taurine conjugates. The use of deuterated internal standards allowed the construction of calibration curves for glycochenodeoxycholic acid, glycocholic acid, taurochenodeoxycholic acid, and taurocholic acids. In another study they examined the feasibility of using this method for screening for cholestatic hepatobiliary disease and extrahepatic biliary atresia [86]. They studied 218 children with cholestatic hepatobiliary disease and compared the total bile concentration (four bile acid conjugates) to 708 blood spots from normal children. Unfortunately, although significant differences in mean bile acid concentrations were found, a small but important overlap existed between the population distributions of unaffected neonates and those with cholestatic hepatobiliary disease. The separation was greater between neonates with extrahepatic biliary atresia and normal neonates, but even here the overlap was too great to make screening by this method alone a feasible option.

In a similar approach Perwaiz et al. developed an LC-ESI-MS/MS method for the determination of bile acids in human bile. The bile acids were extracted by solid-phase extraction followed by LC-MS/MS analysis in the negative-ion mode. MRM was used to monitor glycine conjugates and taurine conjugates [87]. Identification and quantification of conjugated bile acids was achieved in 5 min.

Bootsma et al. presented a method for rapid selective screening of peroxisomal diseases from plasma samples [88]. Sample preparation was quite simple. They used negative-ion ESI-MS/MS and the same internal standards used by Mills et al. [85]. However, in this case LC-MS/MS rather than FIA-MS/MS was utilized with a short narrow bore column (5 cm × 2 mm) interfaced to the ESI ion source with a flow rate of 200 μL/min. Glycine conjugates were detected by MRM using specific transitions with a mass difference of *m/z* 74 and the transitions with the mass difference of *m/z* 80 for the taurine conjugates. The C_{29} dicarboxylic bile acid was detected using the specific transition *m/z* (507→463). The method allowed measuring absolute concentrations of taurine- and glycine-conjugated cholic acid, and those of chenodeoxycholic acid.

Johnson et al. recently described a new method for the rapid and quantitative determination of unconjugated C27 bile acids in plasma and blood spots as another approach to screening for peroxisomal biogenesis defects. They prepared the acetyl dimethylaminoethyl ester derivatives of the unconjugated bile acids, which made it feasible to add a positive charge to the analytes of interest [89]. Using synthetically labeled internal standards and positive-ion ESI-MS/MS, they were able to detect and quantify di- and tri-hydroxy bile acids, which were significantly elevated in patients as compared to normal controls.

In a related approach, Johnson also described a rapid screening procedure for the diagnosis of peroxisomal disorders via the analysis of very long chain fatty acids (VLCFA) in small volumes of plasma or from DBS using the dimethylaminoethyl esters and positive-ion ESI-MS/MS analysis [90]. All the VLCFA-containing lipid species were converted to free VLCFA by heating plasma at 100°C in 10% HCl in acetonitrile for 45 min. The free acids were then converted into their corresponding acid chlorides using oxalyl chloride followed by cold esterification by *N*,*N*-dimethylethanol-amine. The derivatized acids yield strong protonated molecular ions in ESI-MS and few fragments upon CID analysis. MRM profiles were obtained for the diagnostically important acids, C20:0, C22:0, C24:0, and C26:0 using the transitions resulting from the neutral loss of 45 Da (loss of HCOOH). Using trideuterated internal standards for the four mentioned metabolites allowed quantification of these species. Calibration curves showed good correlation, but there was an overestimation of VLCFA by this method as compared to standard GC/MS methods. In plasma the diagnostic ratios of C26/C22 and C24/C22 had good precision and adequate separation between normal and affected levels. As for C20, there was significant interference, as the method measured both arachidic acid and phytanic acid as one. In blood spots the C24/C22 ratio did not differentiate between peroxisomal disease patients and control population. The author advocated the use of this method for screening purposes, and that positive results should be confirmed by GC or GC/MS.

5.9 Analysis of Steroids

The use of ESI-MS/MS in the analysis of steroids of biochemical relevance to metabolic diseases is new. The reason is that unless conjugated, most of these species are nonpolar and neutral, which makes it difficult to produce positively or negatively charged ions. In this regard, Lai et al. developed a novel method for screening for congenital adrenal hyperplasia (CAH) due to deficiency of 21-hydroxylase enzyme from neonatal DBS [91,92]. CAH is the most common inborn error of the adrenal steroid pathways and is characterized by elevation of 17-hydroxyprogetsreone (17OHP). Early diagnosis can be lifesaving, and newborn screening for CAH is part of many programs. To render 17OHP suitable for ESI analysis, Lai and coworkers extracted four 1/8″ circle blood spots deposited in microtiter plates using methanol containing 6-methylprednisolone (6MP) as internal standard. The extracts were transferred to another plate and evaporated to dryness. This was followed by the addition of a volume of Girard reagent (GirP) in ethanol with trichloroacetic as catalyst. The mixture was heated at 65°C for 50 min and evaporated to dryness and reconstituted in 50% acetonitrile in water. The carbonyl groups of 17OHP and 6MP were thus converted to water-soluble hydrazones with a permanently charged pyridine moiety (see Fig. 14A). LC-ESI-MS/MS was carried out using a short C4 column protected with a guard column. The flow rate was 50 μL/min and the mobile phase was water–acetonitrile (50:50). Both 17OHP and 6MP were doubly charged (M^{2+}) at *m*/*z* 299 and 321, respectively. Therefore two SRM functions were used, *m*/*z* 299.4→80.2 and 321.4→80.2 to provide two profiles for analyte and internal standard in a total run time of 3 min. As no buffers were used in mobile phase, 300 samples were analyzed without instrument cleaning. However, there was a need for changing the guard column after 200–300 injections. Calibration curves were linear in the range of 30–500 μg/L, and the median correlation coefficient was 0.995. The limit of detection was 10 μg/L, which was higher than standard radioimmunoassay. The interassay variation was 4.3–10%, while the intraassay variation was $< 12\%$.

Sandhoff et al. developed a method for the determination of cholesterol at the low picomole level by nano-ESI-MS/MS [93]. The method aimed at quantification of cholesterol in cells and subcellular membranes. It

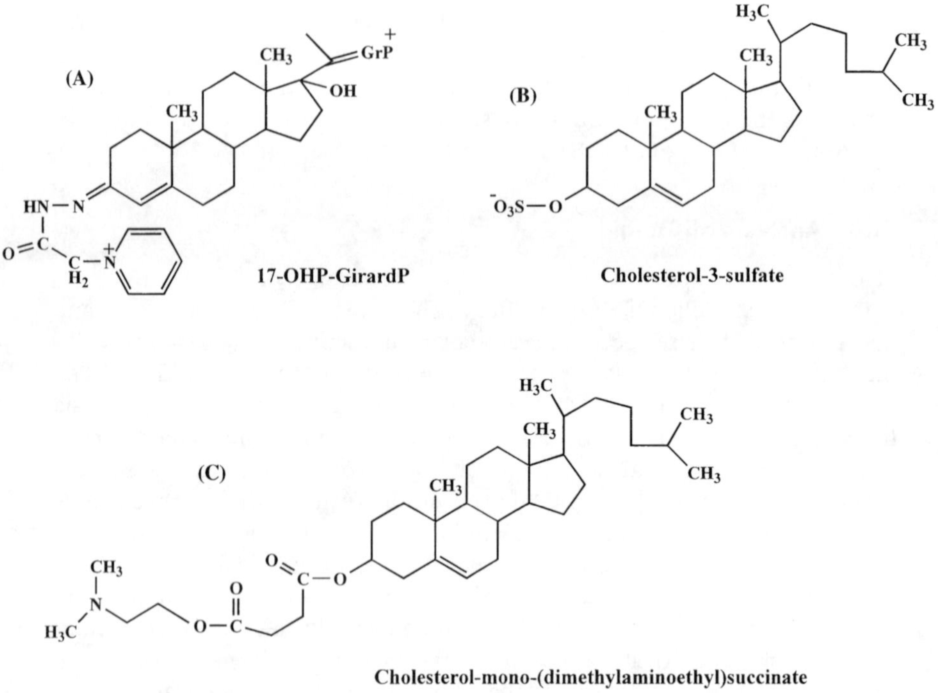

(A) 17-OHP-GirardP

(B) Cholesterol-3-sulfate

(C) Cholesterol-mono-(dimethylaminoethyl)succinate

FIGURE 14 Different types of derivatization approaches used to render steroids amenable to ESI analysis.

involved a one-step chemical derivatization to convert cholesterol to cholesterol-3-sulfate (see Fig. 14B). The sulfated cholesterol and the sulfated internal standard $^{13}C_2$-cholesterol exhibited high ionization efficiency in the negative-ion mode owing to the sulfate ester group. The analytes were monitored by precursor-ion scanning of the common product ion at m/z 97. Calibration curves were linear, and it was possible to quantify cholesterol in biological membranes to approximately 5 pmol of total free cholesterol.

A different approach was used by Johnson et al. to measure cholesterol and dehydrocholesterol in DBS and plasma as a screening method for Smith–Lemli–Optiz syndrome (SLOS) [94]. Reduced activity of 7-dehydrocholesterol reductase leads to the accumulation of 7- and 8-dehydrocholesterol and reduced total cholesterol in blood. The method involved the addition of stable-labeled cholesterol and 7-dehydrocholesterol internal standards to 5 μL plasma or plasma spotted on filter paper and the extraction of total sterols with a mixture of ethanol and sodium hydroxide. This was followed by the addition of water and hexane, and the hexane layer was separated and evaporated to dryness. The sterols in the residue were then derivatized using the synthetically prepared mono-(dimethylaminoethyl) succinate reagent dissolved in dicholoromethane and triethylamine at 70°C for 10 min (see Fig. 14C). The residual solvent was then evaporated and the residue dissolved in acetonitrile–water–formic acid mixture (50:50:0.25), and 20 μL was injected into the mass spectrometer. ESI-MS/MS was operated in the positive-ion mode using four MRM functions to detect signals for derivatized cholesterol, dehydrocholesterols, and their respective internal standards.

5.10 Analysis of Hexose Monophosphates and Oligosaccharides

Classical galactosemia is an inherited disorder in the metabolism of galactose caused by deficiency of the enzyme galactose-1-phosphate uridyl transferase. The disease leads to the accumulation of galactose and galactose-1-phosphate in blood and tissues, and if not treated it produces neonatal death or severe mental retardation, liver cirrhosis, and cataracts. The disease therefore is included in many NBS programs [47]. Jensen et al. recently developed a rapid and accurate ESI-MS/MS method to screen for the disease by measuring hexose monophosphates (HMPs) in neonatal blood spots [95]. They utilized different dissociation patterns of aldose monophosphates vs. ketose monophosphate in developing a method that favors aldose monophosphate detection. They carried out a retrospective study where they analyzed blood spots obtained 4–8 days postpartum from 12 known galactosemia patients and 2055 random controls. From each spot, one punch (3.2 mm diameter disc) was extracted in a microtiter plate using 150 μL of acetonitrile/water (50:50) for 20 min. A portion of these extracts was transferred to new microplates, sealed and centrifuged. ESI-MS/MS was carried out in the negative-ion mode, and CID conditions were set so as to favor gal-1-P detection. The predominant precursor/product ion pair *m/z* 259→79 was used to quantify total HMPs. As no internal standards were available, external calibration was carried out using control blood enriched with increasing amounts of gal-1-P. The median concentration of HMPs in the control group was 0.15 mmol/L (range 0.10–0.94 mmol/L), while in galactosemia patients HMP concentrations were 2.6–5.2 mmol/L. The method was in excellent agreement with the standard alkaline phosphatase–galactose dehydrogenase method used by some screening programs. There were two main issues with this method as a screening method for galactosemia. The first was the trend of early hospital discharge, which raised the question whether this MS/MS will be able to detect galactosemia patients at earlier postnatal age; the second was that the method is incompatible with the routine MS/MS method for amino acids and acylcarnitine analysis and thus would require a separate batch procedure.

Rozaklis et al. recently developed an ESI-MS/MS method for the determination of oligosaccharides in urine, plasma, and blood spots for the diagnosis of Pompe disease, which is a glycogen storage disease [96]. The disease is caused by a deficiency of the lysosomal enzyme, α-glucosidase, and is characterized by elevation of the urinary excretion of the glucose tetramer (Glc1-6Glc1-4Glc1-4Glc). Sample preparation for urine involved lyophilization of a volume equivalent to 1.0 μmol of creatinine and resuspension in an alkaline solution of 1-phenyl-3-methyl-5-pyrazolone (PMP) together with the internal standard methyllactose. The reaction mixture was heated for 70 min at 90°C, acidified, and diluted. The derivatized oligosaccharides were purified by solid-phase extraction. The PMP-derivatized oligosaccharides were identified by precursor-ion scanning of *m/z* 175 in the positive-ion mode and quantified by MRM analysis. Calibration curves for oligosaccharides Glc1-Glc7 were linear over the range 0.016–8 μmol/L, and all Pompe disease patients showed higher oligosaccharide concentrations than adult controls (up to 35-fold).

5.11 Prenatal Diagnosis

Another important field of application of MS/MS was in prenatal diagnosis for branched-chain catabolism disorders. Shigematsu et al. described the use of FAB-MS/MS in the prenatal diagnosis of IVA from amniotic fluid in a 32-week-old fetus [97]. Using the methyl ester derivatives and precursor-ion scans of *m/z* 99 they found elevated C5-carnitine (isovalerylcarnitine) as compared to an age-matched control. They also carried out precursor-ion scans of *m/z* 90 for acylglycines and found significantly elevated C5-glycine as compared to controls. Van Hove et al. described a retrospective prenatal study on three terminated PPA based on methylcitrate determination by GC/MS. They carried out an isotope-dilution assay for measuring C3 in the amniotic fluid using FAB-MS/MS [98]. They found that C3 was elevated by a factor of 5 as compared to 25 control amniotic fluid samples ranging in gestational age from 12 to 17.5 weeks. They proposed that AC analysis is a valid and rapid method for screening pregnancies at risk for PPA.

Shigematsu et al. carried out a larger retrospective study using ESI-MS/MS on a set of stored amniotic fluid samples in at-risk pregnancies for several organic acidemias versus controls, at early gestational age (11–20 weeks) [99]. They reported that the concentration of single AC marker is not sufficient for a valid discrimination. Instead they used concentration ratios of C3/C4 (butyryl- and isobutyrylcarnitine) for PPA and MMA, C5/C3 ratio for

IVA, C5-dioylcarnitine (glutarylcarnitine)/C3 ratio for GA-I, C5/C3 ratio for MAD. The cutoff values for the latter disease seemed not to be reliable enough for diagnosis.

In another exciting application, Nada et al. described a prenatal diagnosis study using cultured amniocytes loaded with 16,16,16-$^{2}H_3$-palmitic acid and L-carnitine [100]. After a 96 h incubation at 37°C the media were separated and $^{2}H_9$-octanoylcarnitine ($^{2}H_9$-C8) and $^{2}H_9$-isovalerylcarnitine ($^{2}H_9$-C5) were added as internal standards. The mixture was then prepared for AC analysis by FAB-MS/MS as described before. Concentrations of C8 and longer chain AC were measured relative to the concentration of the internal standard $^{2}H_9$-C8 and expressed as nmol/mg protein/96 h. In control amniocyte incubation, the AC profile showed the products of the labeled substrate at odd masses corresponding to C12, C10, C8, C6, and C4, and thus clearly distinguished from the natural metabolites, which appeared at even masses. In the amniocytes with A985G homozygosity, the profile showed significant elevation of C8 and C10 and a mild increase in C6. Analysis following the incubation of cells from the pregnancy at risk for an unspecified metabolic disorder revealed elevated amounts of long-chain AC. The most prominent species were the substrate itself and C14. The latter was undetectable in control incubations. Measurements of the enzymatic activities of the acyl-CoA dehydrogenases in both the mitochondrial membrane and the soluble fraction of the amniocyte homogenate showed a severely reduced activity of very long chain acyl-CoA dehydrogenase enzyme.

5.12 In-Vitro Loading Studies

Fibroblasts are frequently utilized to determine the impairment of fatty acid oxidation pathways and pinpoint the exact enzymatic defect. The most commonly used methods are the [^{14}C]CO_2 and [^{3}H]H2O release assays. The cells are incubated in a medium containing ^{14}C-labeled or ^{3}H-labeled fatty acids (e.g., [9,10-^{3}H]palmitic acid or [9,10-^{3}H]myristic acid), and then we quantify the released [^{14}C]CO_2 or [^{3}H]H2O. However, these global assays lack diagnostic specificity and are not sufficiently robust. Bartlett and coworkers developed methods in which the actual β-oxidation intermediates were analyzed rather than just the end products. Thus isolated mitochondria or permeabilized cells were incubated with radiolabeled palmitoyl-CoA, and then one resolves the acyl-CoA esters using radio-HPLC. Later studies led to the development of a technique in which both acyl-CoA and AC were resolved by HPLC [101].

A novel method was developed by Roe and coworkers in 1995 that involved specific AC analysis by MS/MS following incubations of intact fibroblasts or lymphopblastoid cells with deuterated long-chain fatty acids such as [17,7,18,18-$^2H_4$]linoleic acid in the presence of L-carnitine [102]. Deuterium-labeled AC and unlabeled AC (from branched AA metabolism) were detected and quantified. The profiles were often characteristic and led to immediate identification of the underlying enzyme defect. MS/MS analysis eliminated the need for purification, isolation, and separation of intermediates. This method was successfully used for the delineation of the metabolic defect in VLCAD, trifunctional protein defect (TFP), long-chain 3-hydroxyacyl-CoA dehydrogenase deficiency (LCHAD), MCAD, SCAD, MAD, carnitine:acylcarnitine translocase (Trans), and CPT-II, but not carnitine palmitoyltransfearse type-I (CPT-I) [103].

5.13 Postmortem Diagnosis

One example of postmortem diagnosis is our use of FIA-ESI-MS/MS in the analysis of bile filter paper spots obtained postmortem from infants of sudden infant death syndrome (SIDS) [32]. In this case, the bile spot extracts were prepared exactly like blood spots. Qualitative profiles of AC methyl and butyl esters were obtained by ESI-MS/MS analysis by monitoring precursor-ion scans of the common fragment at *m/z* 99 (methyl-) or *m/z* 85 (*n*-butyl). Two SIDS cases suspected of FOAD due to the appearance of fatty liver, undetectable glucose in liver tissue, or increased palmitoleic acid gave bile AC profiles strongly suggestive of LCHAD. The profiles were similar to those previously described in plasma. A third case diagnosed as GA-I by GC/MS analysis of postmortem urine gave a profile dominated by the signal corresponding to glutarylcarnitine, a pathognomonic marker for the disease. The use of both methyl and butyl esters with two different CID fragments served to increase our confidence in the nature of the diagnostic ions in the profile.

Boles and coworkers described recently a larger retrospective study where 418 cases of SIDS were examined. Several investigations were carried out on postmortem liver tissue, but bile was available in only 32 cases. The study detected 14 cases of FOAD, two MCAD, four MAD, four cases with either VLCAD or LCHAD, and four cases predicted with carnitine transport defect (CTD) [104]. Chace et al. recently reported another large study where filter-paper blood from 7058 infants from U.S. and Canadian medical examiners was analyzed by ESI-MS/MS for AAs and ACs [105]. Results on 66 specimens suggested diagnosis of metabolic disorders. The most frequently detected disorders were MCAD and VLCAD (23 and nine cases, respectively), GA-I and GA-II (three and eight cases, respectively), CPT-II/Trans (six cases), IVA/2-methylbutyryl-CoA dehydrogenase deficiencies (four cases), and LCHAD/TFP (four cases). The authors concluded that postmortem metabolic screening by MS/MS could explain deaths in infants and children, is cost-effective, and should be considered for routine use by medical examiners and pathologists in unexpected/unknown infant and child death.

6 CONCLUSIONS

The work presented in this chapter clearly demonstrates the major impact that ESI-MS/MS has made in the field of biochemical genetics. Application of ESI-MS/MS has certainly revolutionized the area of newborn screening and led to the development and implementation of broad-spectrum disease detection programs all over the world. The number of countries to adopt this approach will continue to grow in this decade. The power of the technique has also tremendously improved the ability of the biochemical genetics laboratory to diagnose old diseases and to discover new ones. New applications in the areas of prenatal, postnatal, and postmortem diagnosis will continue to appear in the literature.

ACKNOWLEDGMENTS

The author expresses his deepest gratitude to Dr. Sultan Al Sedairy, Executive Director, and Dr. Futwan Al-Muhanna, Deputy Executive Director of the Research Centre of King Faisal Specialist Hospital and Research Centre, Riyadh, for their continuous support of the Metabolic Screening Laboratory. The author is very grateful for Mr. Mohamed Al-Amoudi, Mrs. Minnie Jacob, Ms. Lujane Al-Ahaidib, Ms. Sharon Legaspi, and Mrs. Ellena Bernabe for their excellent technical assistance.

REFERENCES

1. RA Chalmers, AM Lawson, eds. Organic Acids in Man. London: Chapman and Hall, 1982.
2. K Tanaka, MA Budd, ML Efron, KJ Isselbacher. Isovaleric acidemia. A new genetic defect of leucine metabolism. Proc Natl Acad Sci USA 56:236–242, 1966.
3. L Sweetman. Organic acid analysis. In: FA Hommes, ed. Techniques in Diagnostic Human Biochemical Genetics: A Laboratory Manual. New York: Wiley-Liss, 1991, pp 143–176.
4. N Gregersen, R Lauritzen, K Rasmussen. Suberylglycine excretion in the urine from a patient with dicarboxylic aciduria. Clin Chim Acta 70:417–425, 1976.

5. S Kolvraa, N Gregersen, E Christensen, N Hobolth. In vitro fibroblast studies in a patient with C6-C10-dicarboxylic aciduria: evidence for a defect in general acyl-CoA dehydrogenase. Clin Chim Acta 126:53–67, 1982.
6. CA Stanley. Carnitine disorders. Adv Pediatr 42:209–242, 1995.
7. RJ Pollitt. Disorders of mitochondrial long-chain fatty acid oxidation. J Inher Metab Dis 18:473–490, 1995.
8. MR Seashore, P Rinaldo. Metabolic disease of the neonate and young infant. Sem Perinatol 17:318–329, 1993.
9. SJR Heales, DA Woolf, P Robinson, JV Leonard. Rapid diagnosis of medium-chain acyl-CoA dehydrogenase deficiency by measurement of cis-4-decenoic acid in plasma. J Inherit Metab Dis 14:661–667, 1991.
10. KE Niezen-Kning, TE Chapman, IE Mulder, GPA Smit, DJ Reijngoud, R Berger. Determination of medium chain acyl-CoA dehydrogenase activity in cultured skin fibroblasts using mass spectrometry. Clin Chim Acta 199:173–184, 1991.
11. RI Kelley. Diagnosis of Smith–Lemli–Opitz syndrome by gas chromatography/mass spectrometry of 7-dehydrocholesterol in plasma, amniotic fluid and cultured skin fibroblasts. Clin Chim Acta 236:45–58, 1995.
12. HW Moser, AB Moser. Measurement of saturated very long chain fatty acids in plasma. In: FA Hommes, ed. Techniques in Diagnostic Human Biochemical Genetics: A Laboratory Manual. New York: Wiley-Liss, 1991, pp 177–191.
13. RI Kelley. Quantification of pipecolic acid in plasma and urine by isotope-dilution–gas chromatography/mass spectrometry. In: FA Hommes, ed. Techniques in Diagnostic Human Biochemical Genetics: A Laboratory Manual. New York: Wiley-Liss, 1991, pp 205–218.
14. F Stellaard, HJ ten Brink, RM Kok, L van den Heuvel, C Jakobs. Stable isotope dilution analysis of very long chain fatty acids in plasma, urine and amniotic fluid by electron capture negative ion mass fragmentography. Clin Chim Acta 192:133–144, 1990.
15. P Rinaldo, JJ O'Shea, RD Welch, K Tanaka. Stable isotope dilution analysis of *n*-hexanoylglycine, 3-phenylpropionylglycine and suberylglycine in human urine using chemical ionization gas chromatography/mass spectrometry selected ion monitoring. Biomed Environ Mass Spectrom 18:471–477, 1989.
16. P Rinaldo, JJ O'Shea, PM Coates, DE Hale, CA Stanley, K Tanaka. Medium-chain acyl-CoA dehydrogenase deficiency: diagnosis by stable isotope dilution measurement of urinary *n*-hexanoylglycine and 3-phenylpropionylglycine. N Eng J Med 319:1308–1313, 1988.
17. RM Kok, L Kaster, PJM de Jong, B Poll-The, J-M Saudubray, C Jakobs. Stable isotope dilution analysis of pipecolic acid in cerebrospinal fluid, plasma, urine and amniotic fluid using electron capture negative ion mass fragmentography. Clin Chim Acta 168:143–152, 1987.
18. ML Vestal. High-performance liquid chromatography–mass spectrometry. Science 226:275–281, 1984.

19. M Barber, RS Bordoli, RD Sedgwick, AN Tyler. Fast atom bombardment of solids (FAB): a new ion source for mass spectrometry. J Chem Soc Chem Commun 789:325–327, 1981.
20. RM Caprioli, WT Moore. Continuous-flow fast atom bombardment mass spectrometry. Meth Enzymol 193:214–237, 1990.
21. CG Edmonds, RD Smith. Electrospray ionization mass spectrometry. Meth Enzymol 193:412–431, 1990.
22. M Karas, F Hillenkamp. Laser desorption ionization of proteins with molecular masses exceeding 10,000 daltons. Anal Chem 60:2299–2301, 1988.
23. AL Yergey, DJ Liberato, DS Millington. Thermospray liquid chromatography/mass spectrometry for the analysis of L-carnitine and its short-chain acyl derivatives. Anal Biochem 139:278–283, 1984.
24. DN Buchanan, J Muenzer, JG Thoene. Positive-ion thermospray liquid chromatography mass spectrometry: detection of organic acidurias. J Chromatogr 534:1–11, 1990.
25. JE Evans, A Ghosh, BA Evans. Screening techniques for the detection of inborn errors of bile acid metabolism by direct injection and micro-high performance liquid chromatography–continuous flow/fast atom bombardment mass spectrometry. Biol Mass Spectrom 22:331–337, 1993.
26. EJ Oliveira, DG Watson. Liquid chromatography–mass spectrometry in the study of the metabolism of drugs and other xenobiotics. Biomed Chromatogr 14:351–372, 2000.
27. AG Ferrige, MJ Seddon, BN Green, SA Jarvis, J Skilling. Disentangling electrospray spectra with maximum entropy. Rapid Commun Mass Spectrom 6:707–711, 1992.
28. J Yergey, D Heller, G Hansen, RJ Cotter, C Fenselau. Isotopic distribution in mass spectra of large molecules. Anal Chem 55:353–356, 1983.
29. WJ Griffiths, AP Jonsson, LIU Suya, DK Rai, Y Wang. Electrospray and tandem mass spectrometry in biochemistry. Biochem J 355:545–561, 2001.
30. MS Rashed, PT Ozand, ME Harrison, PJF Watkins, S Evans. Electrospray tandem mass spectrometry in the diagnosis of organic acidemias. Rapid Commun Mass Spectrom. 8:129–133, 1994.
31. MS Rashed, PT Ozand, MP Bucknall, D Little. Diagnosis of inborn errors of metabolism from blood spots by acylcarnitines and amino acids profiling using automated electrospray tandem mass spectrometry. Ped Res 38:324–331, 1995.
32. MS Rashed, PT Ozand, MJ Bennett, JL Barnard, DR Govindaraju, P Rinaldo. Inborn errors of metabolism diagnosed in sudden death cases by acylcarnitine analysis of postmortem bile. Clin Chem 41:1109–1114, 1995.
33. Y Shigematsu, I Hata, A Nakai, Y Kikawa, M Sudo, Y Tanaka, S Yamaguchi, C Jakobs. Prenatal diagnosis of organic acidemias based on amniotic fluid levels of acylcarnitines. Ped Res 39:680–684, 1996.
34. MS Rashed, MP Bucknall, D Little, A Awad, M Jacob, M AlAmoudi, M AlWattar, PT Ozand. Screening for inborn errors of metabolism from blood spots by electrospray tandem mass spectrometry using a microplate batch

process and a computer algorithm for automated flagging of abnormal profiles. Clin. Chem 43:129–1141, 1997.

35. AW Johnson, K Mills, and PT Clayton. The use of automated electrospray ionization tandem MS for the diagnosis of inborn errors of metabolism from dried blood spots. Biochem Soc Trans 24:932–938, 1996.
36. RJA Wanders, P Vreken, MEJ Den Boer, FA Wijburg, AH van Gennip, L Ijlst. Disorders of mitochondrial fatty acyl-CoA β-oxidation. J Inher Metab Dis 22:442–487, 1999.
37. P Rinaldo, D Matern, MJ Bennett. Fatty acid oxidation disorders. Annu Rev Physiol 64:477–502, 2002.
38. MS Rashed, Z Rahbeeni, PT Ozand. Application of electrospray tandem mass spectrometry to neonatal screening. Sem Perinatol 23:183–193, 1999.
39. MS Rashed, Z Rahbeeni, PT Ozand. Screening blood spots for argininosuccinase deficiency by electrospray tandem mass spectrometry. Southeast Asian J Trop Med Pub Health 30(suppl 2):170–173, 1999.
40. MS Rashed. Clinical applications of tandem mass spectrometry: ten years of diagnosis and screening for inherited metabolic diseases. J Chromatogr B 758:27–48, 2001.
41. P Vreken, EM van Lint, AH Bootsma, H Overmars, RJA Wanders, AH van Gennip. Quantitative plasma acylcarnitine analysis using electrospray tandem mass spectrometry for the diagnosis of organic acidemias and fatty acids oxidation defects. J Inher Metab Dis 22:302–306, 1999.
42. P Vreken, EM van Lint, AH Bootsma, H Overmars, RJA Wanders, AH van Gennip. In: PA Quant, S Eaton, eds. Current views of fatty acid oxidation and ketogenesis: from organelles to point mutations. New York: Kluwer Academic/Plenum Publishers, 1999, pp 327–337.
43. AI Al Aqeel, MS Rashed, RJA Wanders. Carnitine-acylcarnitine translocase deficiency is a treatable disease. J Inher Metab Dis 22:271–275, 1999.
44. EH Touma, MS Rashed, C Vianey-Saban, A Sakr, P Dirvy, N Gregersen, BS Andersen. A severe genotype with favourable outcome in very long chain acyl-CoA dehydrogenase deficiency. Arch Dis Child 84:58–60, 2001.
45. AI Al-Aqeel, MS Rashed, JP Ruiter, HF Husseini, MS Al-Amoudi, RJ Wanders. Carnitine palmityl transferase I deficiency in a Saudi family. Saudi Med J 22:1025–1029, 2001.
46. R Fingerhut, W Roschinger, AC Muntau, T Dame, J Kreischer, R Arnecke, R Superti-Furga, H Trosler, B Liebl, B Olgemoller, Roscher A. Hepatic carnitine palmitoyltransferase I deficiency: acylcarnitine profiles in blood spots are highly specific. Clin Chem 47:1763–1768, 2001.
47. HL Levy, S Albers. Genetic screening of newborns. Annu Rev Genomics Hum Genet 1:139–177, 2000.
48. R Guthrie. Blood screening for phenylketonuria. J Am Med Assoc 178:863, 1961.
49. R Guthrie, A Susi. A simple phenylalanine method for detecting phenylketonuria in large populations of newborn infants. Pediatrics 32:338–343, 1963.

50. R Guthrie. Screening for "inborn errors of metabolism" in the newborn infant—a multiple test program. Birth Defects IV:92–98, 1968.
51. JH Dussault, P Coulombe, C Laberge, J Letarte, H Guyda, K Khour. Preliminary report on a mass screening program for neonatal hypothyroidism. J Pediatr 86:670–674, 1975.
52. S Pang, J Hotchkiss, AL Drash, LS Levine, MI New. Microfilter paper method for 17-hydroxyprogesterone radioimmunoassay: its application for rapid screening for congenital adrenal hyperplasia. J Clin Endocrinol Metab 45:1003–1008, 1977.
53. GS Heard, JR Secor, JR McVoy, B Wolf. A screening method for biotinidase deficiency in newborns. Clin Chem 30:125–127, 1984.
54. Wolf B, Heard GS: Screening newborns for biotinidase deficiency. Pediatrics 85:512–517, 1990.
55. EW Naylor, DH Chace. Automated tandem mass spectrometry for mass newborn screening for disorders in fatty acid, organic acid, and amino acid metabolism. J. Child Neurol 14(suppl 1):S4–S8, 1999.
56. V Wiley, K Carpenter, B Wilcken. Newborn screening with tandem mass spectrometry: 12 months' experience in NSW Australia. Acta Paediatr (suppl 432) 88:48–51, 1999.
57. Y Shigematsu, I Hata, Y Kikawa, M Mayumi, Y Tanaka, M Sudo, N Kado. Modifications in electrospray tandem mass spectrometry for a neonatal-screening pilot study in Japan. J Chromatogr B 731:97–103, 1999.
58. B Liebl, R Fingerhut, W Roschinger, A Muntau, I Knerr, B Olgemoller, A Zapf, AA Roscher. Model project for updating neonatal screening in Bavaria: concept and initial results. Gesundheitswesen 62:189–195, 2000.
59. TH Zytkovicz , EF Fitzgerald, D Marsden, CA Larson, VE Shih, DM Johnson, AW Strauss, AM Comeau, RB Eaton, GF Grady. Spectrometric analysis for amino, organic, and fatty acid disorders in newborn dried blood spots: a two-year summary from the New England program. Clin Chem 47(11):1945–1955, 2001.
60. WD Lin, JY Wu, FJ Tsai, CH Tsai, SP Lin, DM Niu. A pilot study of neonatal screening by electrospray ionization tandem mass spectrometry in Taiwan. Acta Paediatr Taiwan 42:224–230, 2001.
61. B Casetta, D Tagliacozzi, B Shushan, G Federici. Development of a method for rapid determination of amino acids by liquid chromatography–tandem mass spectrometry (LC-MSMS) in plasma. Clin Chem Lab Med 38:391–401, 2000.
62. MJ Magera, JM Lacet, B Casetta, P Rinaldo. Method for the determination of total homocysteine in plasma and urine by stable isotope dilution and electrospray tandem mass spectrometry. Clin Chem 45:1517–1522, 1999.
63. K Gempel, KD Gerbitz, B Casetta, MF Bauer. Rapid determination of total homocysteine in blood spots by liquid chromatography–electrospray ionization–tandem mass spectrometry. Clin Chem 46:122–123, 2000.
64. MS Rahed, LY Al-Ahaidib, HY Aboul-Enein, M Al-Amoudi, M Jacob. Determination of L-pipecolic acid in plasma using chiral liquid chromatography electrospray tandem mass spectrometry. Clin Chem 47:2124–2130, 2001.

65. L Bonafé, H Troxler, T Kuster, CH Heizmann, NA Chaoles, AB Burlina, N Blau. Evaluation of urinary acylglycines by electrospray tandem mass spectrometry in mitochondrial energy metabolism defects and organic acidurias. Mol Genet Metab 69:302–311, 2000.
66. MS Rashed, M AlAmoudi, HY Aboul-Enein. Chiral liquid chromatography tandem mass spectrometry in the determination of the configuration of 2-hydroxyglutaric acid in urine. Biomed Chromatogr 14:317–320, 2000.
67. MS Rashed, HY Aboul-Enein, M AlAmoudi, M Jacob, LY Al-Ahiadib, A Abbad, S. Shabib, E Al-Jishi. Chiral liquid chromatography tandem mass spectrometry in the determination of the configuration of glyceric acid in urine of patients with D-glyceric and L-glyceric acidurias. Biomed Chromatogr 16:191–198, 2002.
68. MJ Magera, JK Helgeson, D Matern, P Rinaldo. Methylmalonic acid measured in plasma and urine by stable-isotope dilution and electrospray tandem mass spectrometry. Clin Chem 46:1804–1810, 2000.
69. MM Kushnir, G Komaromy-Hiller, B Shushan, FM Urry, WL Roberts. Analysis of dicarboxylic acids by tandem mass spectrometry. High-throughput quantitative measurement of methylmalonic acid in serum, plasma, and urine. Clin Chem 47:1993–2002, 2001.
70. MJ Magera, AL Stoor, JK Helgeson, D Matern, P Rinaldo. Determination of homovanillic acid in urine by stable isotope dilution and electrospray tandem mass spectrometry. Clin Chim Acta 306:35–41, 2001.
71. T Ito, AB van Kuilenburg, AH Bootsma, AJ Haasnoot, A van Cruchten, Y Wada, AH van Gennip. The application of HPLC/ESI tandem mass spectrometry on urine-soaked filter-paper strips for the screening of disorders of purine and pyrimidine metabolism. J Inher Metab Dis 23:434–437, 2000.
72. H van Lenthe, ABP van Kuilenburg, T Ito, AH Bootsma, A van Cruchten, Y Wada, AH van Gennip. Defects in pyrimidine degradation identified by HPLC-electrospray tandem mass spectrometry of urine-soaked filter paper strips. Clin Chem 46:1916–1922, 2000.
73. ABP van Kuilenburg, H van Lenthe, B Assmann, G Gohlich-Ratmann, GF Hoffmann, C Brautigam, RA Wevers, AH van Gennip. Detection of *b*-ureidopropionase deficiency with HPLC-electrospray tandem mass spectrometry and confirmation of the defect at the enzyme level. J Inher Metab Dis 24:725–732, 2001.
74. EA Struys, EEW Jansen, K De Meer, C Jakobs. Determination of *S*-adenosylmethionine and *S*-adenosylhomocysteine in plasma and cerebrospinal fluid by stable-isotope dilution tandem mass spectrometry. Clin Chem 46:1650–1656, 2000.
75. A Stiehl. Pattern of bile acids in cholestasis. In: P Gentilini, IM Arias, N McIntyre, J Rhodes, eds. Cholestasis. New York: Elsevier Science, 1994, pp 231–238.
76. I Bjorkhem. Inborn errors of metabolism with consequences for bile acid biosysnthesis. A minireview. Scand J Gastroenterol 204:68–72, 1994.

77. RJA Wanders. Peroxisomal disorders: clinical, biochemical, and molecular aspects. Neurochem Res 24:565–580, 1999.
78. KDR Setchell, A Matsui. Serum bile acid analysis. Clin Chim Acta 127:1–17, 1983.
79. AM Lawson, MJ Madigan, D Shortland, PT Clayton. Clin Chim Acta 16:221–231, 1986.
80. KB Tomer, NJ Jensen, ML Gross. Fast atom bombardment combined with tandem mass spectrometry for determination of bile salts and their conjugates. Biomed. Mass Spectrom 13:265–72, 1986.
81. JE Evans, A Ghosh, BA Evans, MR Natowicz. Screening techniques for the detection of inborn errors of bile acid metabolism by direct injection and micro–high performance liquid chromatography–continuous flow/fast atom bombardment mass spectrometry. Biol Mass Spectrom 22:331–337, 1993.
82. R Libert, D Hermans, J-P Drsye, F Van Hoof, E Sokal, E de Hoffmann. Bile acids and conjugates identified in metabolic disorders by fast atom bombardment and tandem mass spectrometry. Clin Chem 37:2102–2110, 1991.
83. BM Warrack, GC DiDonato. Ion spray liquid chromatography/mass spectrometric characterization of bile acids. Biol mass Spectrom. 22:101–111, 1993.
84. A Roda, AM Gioacchini, C Cerré, M Baraldini. High-performance liquid chromatographic-electrospray mass spectrometric analysis of bile acids in biological fluids. J Chromatogr B 665:281–294, 1995.
85. KM Mills, I Mustaq, AW Jonhnson, PD Whitfield, PT Clayton. A method for the quantitation of conjugated bile acids in dried blood spots using electrospray ionization–mass spectrometry. Ped Res 43:361–368, 1998.
86. I Mustaq, S Logan, M Morris, AW Jonhnson, AM Wade, D Kelly, PT Clayton. Br Med J 319:471–477, 1999.
87. S Perwaiz, B Tuchweber, D Mignault, T Gilat, IM Yousef. Determination of bile acids in biological fluids by liquid chromatography–electrospray tandem mass spectrometry. J Lipid Res 42:114–119, 2001.
88. AH Bootsma, H Overmars, A van Rooij, AEM van Lint, RJA Wanders, AH van Gennip, P Vreken. J Inher Metab Dis 22:307–310, 1999.
89. DW Johnson, HJ ten Brink, RC Schuit, C Jakobs. Rapid and quantitative analysis of unconjugated C27 bile acids in plasma and blood samples by tandem mass spectrometry. J Lipid Res 42:9–16, 2001.
90. DW Johnson. A rapid screening procedure for the diagnosis of peroxisomal disorders: Quantification of very long-chain fatty acids, as dimethylaminoethyl esters, in plasma and blood spots, by electrospray tandem mass spectrometry. J Inher Metab Dis 23:475–486, 2000.
91. CC Lai, CH Tsai, FJ Tsai, JY Wu, WD Lin, CC Lee. Monitoring of congenital adrenal hyperplasia by microbore HPLC-electrospray ionization tandem mass spectrometry of dried blood spots. Clin Chem 48:354–356, 2002.
92. CC Lai, CH Tsai, FJ Tsai, JY Wu, WD Lin, CC Lee. Rapid screening assay of congenital adrenal hyperplasia by measuring 17 alpha-hydroxyprogesterone

with high-performance liquid chromatography/electrospray ionization tandem mass spectrometry from dried blood spots. J Clin Lab Med 16:20–25, 2002.
93. R Sandhoff, B Brugger, D Jeckel, WD Lehmann, FT Wieland. Determination of cholesterol at the low picomole level by nanoelectrospray ionization tandem mass spectrometry. J Lipid Res 40:126–132, 1999.
94. DW Johnson, HJ ten Brink, C Jakobs. A rapid screening procedure for cholesterol and dehydrocholesterol by electrospray ionization tandem mass spectrometry. J Lipid Res 42:1699–1705, 2001.
95. UG Jensen, NJ Brandt, E Christensen, F Skovby, B Norgaard-Pedersen, H Simonsen. Neonatal screening for galactosemia by quantitative analysis of hexose monophosphates using tandem mass spectrometry: A retrospective study. Clin Chem 47:1364–1372, 2001.
96. T Rozaklis, SL Ramsay, PD Whitfield, E Ranieri, JJ Hopwood, PJ Meikle. Determination of oligosaccharides in Pompe disease by electrospray ionization tandem mass spectrometry. Clin Chem 48:131–139, 2002.
97. Y Shigematsu, Y Kikawa, M Sudo, H Kanaoka, M Fujioka, M Dan. Prenatal diagnosis of isovaleric acidemia by fast atom bomardment and tandem mass spectrometry. Clin Chim Acta 203:369–374, 1991.
98. JLK Van Hove, DH Chace, SG Kahler, DS Millington. Acylcarnitines in amniotic fluid: application to the prenatal diagnosis of propionic acidemia. J Inher Metab Dis 16:361–367, 1993.
99. Y Shigematsu, I Hata, A Nakai, Y Kikawa, M Sudo, Y Tanaka, S Yamaguchi, C Jakobs. Prenatal diagnosis of organic acidemias based on amniotic fluid levels of acylcarnitines. Ped Res 39:680–684, 1996.
100. MA Nada, C Vianey-Saban, CR Roe, J-H Ding, M Mathieu, RS Wappner, MG Bialers, JA McGlynn, G Mandon. Prenatal diagnosis of mitochondrial fatty acid oxidation defects. Prenat Diagn 16:117–124, 1996.
101. M Pourfarzam, J Schaefer, DM Turnbull, K Bartlett. Analysis of fatty acid oxidation intermediates in cultured fibroblasts to detect mitochondrial oxidation disorders. Clin Chem 40:2267–2275, 1994.
102. MA Nada, WJ Rhead, H Sprecher, H Schulz, CR Roe. Evidence for intermediate channeling in mitochondrial β-oxidation. J Biol Chem 270:530–535, 1995.
103. CR Roe, DS Roe. Recent developments in the investigation of inherited metabolic disorders using cultured human cells. Mol Genet Metab 68:243–257, 1999.
104. RG Boles, EA Buck, MG Blitzer, MS Platt, TM Cowan, SK Martin, P Rinaldo. Retrospective biochemical screening of fatty acid oxidation disorders in postmortem liver of 418 cases of sudden unexpected death in the first year of life. J Pediatr 132:924–933, 1998.
105. DH Chace, JC Diperna, BL Mitchell, B Sgroi, LF Hofman, EW Naylor. Electrospray tandem mass spectrometry for analysis of acylcarnitines in dried postmortem blood specimens collected at autopsy from infants with unexplained cause of death. Clin Chem 47:1166–1182, 2001.

11

Application of Thin-Layer Chromatography in Clinical Chemistry

Jan Bladek and Slawomir Neffe
Military University of Technology, Warsaw, Poland

1 INTRODUCTION

All chromatographic methods are suitable for examination of biological material, but only thin-layer chromatography (TLC) enables the simple, fast, cheap, and effective separation of these complex mixtures. Therefore TLC is one of the best known and thoroughly tested methods of the analysis of substances, which is significant in medical diagnosis. There are many handbooks [e.g., 1,2], book chapters [e.g., 3,4] and review publications [e.g., 5,6] that comprehensively summarized TLC applications in clinical chemistry. The TLC method has been used both as an analytical and as a preparative technique to solve problems in biochemistry, hematology, immunology, and even molecular biology. TLC is characterized by high selectivity, and it enables separation of the analyte from interfering substances. A special advantage of TLC is its versatility. It offers a great number of different sorbents in commercial form, the possibility of plate spraying with more or less specific visualizing reagents or coupling TLC with different specific detectors, as well as the ability to use a broad range (as regards polarity and selectivity of solvent) of mobile phases. Another property of TLC is that many samples can be throughput because of the possibility of performing simultaneous separation of many samples. In modern thin-layer chromatography, sample application, development, and recording of the chromatograms is realized by fast automated procedures.

2 SHORT CHARACTERISTIC OF THE METHOD

Thin-layer chromatography is a variant of liquid chromatography. A stationary phase (plate) is prepared from uniform porous materials by binding them to a support (alumina, glass, or plastic foil). When an analyzed liquid mixture is applied to the plate, the chromatogram is developed. A mobile phase (solvents or their mixtures) traverses then along the stationary phase, and analytes (solutes) move across the plate. As in other chromatographic methods, separation occurs as the consequence of different interactions of the components to be separated with both a stationary and a mobile phase. Separated substances are retained with relation to the mobile phase front, and a measure of retention is the retardation factor $R_F = z_n / z_f$, where z_n and z_f mean the distance of the nth analyte and mobile phase front travels from the start line. Unfortunately, R_F (or $R_F \times 100 = hR_F$) values are affected by a number of factors and therefore a standard analyte should be run at the same conditions as the test sample whenever possible. If it is not possible, a number (usually four) of reference compounds, whose R_F values are accurately known, can be run and then the corrected value of R_F can be estimated (cR_F).

Modern high-performance thin-layer chromatography (HPTLC) began with the introduction of high-efficiency adsorbents. HPTLC layers are fairly similar to standard TLC plates; the difference is primarily that the average size of the particles is smaller and the size distribution is tighter. The increase of efficiency results in fewer bonds broadening and hence improves resolution and makes for greater sensitivity of detection. In addition, the length of the chromatographic bed required (4–6 cm) is markedly less than that encountered in conventional TLC (10–15 cm). In general, standard HPTLC (the terms TLC and HPTLC are now used interchangeably) is sufficient for the separation of most clinical chemistry interest analytes.

The four processes taking the most important role in TLC separations of clinical interest are adsorption, partitioning, ion exchange, and affinity interactions. Adsorption takes place between the substructure of the analyte and the various adsorptive centers of the sorbents. Solutes with high adsorption capacity bind more strongly to the stationary phase, resulting in enhanced retention. Partitioning results in different solubilities of the solute molecules in the mobile phase. Solutes with higher preference for the mobile phase elute before the compounds with lower solubility. Ion exchange refers to the electrostatic forces occurring between the dissociable polar parts of the solute molecules and the ionic centers of the sorbent surface. In case of affinity TLC, separations are carried out by specific interactions of biologically related agents.

2.1 Properties of the Chromatographic Systems

The separation probability is greatly enhanced by the proper selection of a chromatographic system for particular analytes. This term refers to both stationary and mobile phases. If the stationary phase is of higher polarity than the mobile phase, the chromatographic system is called the normal phase (NP) system; if the mobile phase is of higher polarity, the system is named the reversed phase (RP) one. Representative separations of clinical interest in both NP and RP systems are presented in Table 1.

2.1.1 Stationary Phases

TLC stationary phases (silica gel, alumina, cellulose, polyamides, and other sorbents) consist of powder solids. They are chemically defined inorganic or organic materials with porous structure and relatively high specific surface area. Often modified adsorbents are recently used. During chemical modification, varied functional groups have been covalently attached to the chromatographic bed (usually silica gel material), which eliminates stripping of these groups by the mobile phase. Physical modification of adsorbent depends on their impregnation by various additives not miscible with mobile phase (additives are adsorbed on the support). Mixing the additive in the eluent used as a mobile phase can also modify the chromatographic system (dynamic modification), but the use of modified adsorbents has led to an improvement of resolution.

Among inorganic sorbents, silica gel is certainly the best TLC adsorbent. It is a polar adsorbent applied as a stationary phase and as a support to obtain physically and (especially) chemically modified sorbents. Nonmodified silica gel is widely used for routine analyses in the NP system for many biological mixtures. Alumina has a similar property but is rarely used. All aluminas tend to adsorb molecules of water from the surrounding atmosphere and thereby become deactivated. In comparison with silica gel they have somewhat larger adsorption affinity for carbon–carbon double bounds and better selectivity toward aromatic hydrocarbons and their derivatives. Alumina is also rarely used as a support material. Other inorganic sorbents such as inert silicon dioxides (diatomaceous earth or silicon dioxide) have found only limited application in clinical chemistry.

Celluloses (native or microcrystalline) are organic sorbents. They have a low specific surface area and are applied mainly in partition chromatography, especially for the separation of relatively polar compounds. Celluloses have

TABLE 1 Representative Examples of Separations in NP and RP Systems

No.	Type of analyte	Matrix	Chrom. system	Fundamental goal of analysis	Ref.
1	Biologically active amines	Urine	NP	Metabolism of the catecholamines and serotonin measurements	7
2	Bile acids	Bile	RP	Separation of unconjugated bile acids and their glycine- and taurine-amidated, 3-sulfated, 3-glucosylated and 3-glucuronidated conjugates	8
		Feces	NP	Determinations of major unconjugated bile acids (cholic, chenodeoxycholic, deoxycholic, urodeoxycholic, and lithocholic acids) in human stool specimens	9
3	Carbohydrates	Urine	NP	The lactulose determination of patients with cystic fibrosis	10
4	Vitamins	Blood	NP	A reliable procedure for the joint analysis of vitamin E (tocopherols), cholesterol, and phospholipids in the concurrent samples of human platelets in human cultured endothelial cells	11
5	Lipids	Serum	RP	Determination of cholesterol sulfate and dehydroepiandrosterone sulfate	12
		Brain	NP	Micropreparative isolation and purification of ganglosides	13
		Lung tissue	NP	Separations of lung phospholipids from matrix and their determinations	14
6	Enzymes	Urine	NP	Adenylosuccinase deficiency measurements (detection of succinyloaminoimidazolecarboxamide riboside and succinyloadenosine)	15
7	Porphyrins and their precursors	Urine, feces		Porphyria diagnosis. Evaluation of porphyrin	16
8	Alkaloids and drugs	Urine	RP	Determination of nicotine and its main metabolite cotinine for smoking and passively smoking pregnant women	17
		Blood	NP	Analysis of cyclosporine and its metabolites in peripheral blood (monitoring of transplant patients)	18
		Urine	NP	Detection of opioids, cocaine, and amphetamine	19
9	Inorganic substances	Bones, milk	NP	Separation and detection of heavy metal ions	20

primarily been used as a support material for the separation of polar substances by normal-phase liquid–liquid partition. Layers have been impregnated with buffers, chelating agents, metal ions, and other compounds. Celluloses impregnated with polyethylene imine (PEI) and polyphosphate (poly-P) or chemically modified celluloses (chemical bonded aminoethyl [AE], carboxymethyl [CM], etc.) have been recently used as ion exchange material. Polyamides (synthetic resins, mainly caprolactam or polyundecanamide) are also numbered among the organic sorbents. Polyamides show a high affinity and selectivity to the analytes that can form hydrogen bonds.

Separations of analytes in reversed-phase systems were originally carried out on silica gel or diatomaceous earth layers impregnated with a solution of paraffin, squalane, silicone oil, etc. Recently, silica gel with covalently bounded organic ligands on the surface is frequently used. Methyl (RP-2), octyl (RP-8), octadecyl (RP-18), and phenyl silicas have often been used as the nonpolar chemically bonded adsorbents. Other chemically bonded adsorbents (diol-, cyano-, and amino-modified) have hydrophilic properties. Amino layers have a polarity lower than silica gel and higher than cyano and diol layers. In this way the gap of selectivity of the extremely hydrophilic nonmodified silica gel and the nonpolar RP phases was bridged by medium polar reversed phases. The mechanisms of retention on chemically bonded reversed phases are not clearly elaborated, but R_F values for a series of solutes separated on nonpolar phases are usually reversed in comparison with the sequence on silica gel, if water constitutes a large proportion of the mobile phase. Of course, a proper separation on reversed phase is also possible if one uses entirely organic mixtures as the mobile phase. The medium polar phases can be used in both the adsorption and the reversed phase separation mode, depending on the properties of the mobile phase.

Stationary phases showing different retention characteristics from those of traditional NP and RP systems are also used in clinical applications. Chiral beads (reversed phase silica modified with Cu^{2+} and a chiral agent, e.g., β-cyclodextrin) are used for enantiomeric separation. The retention of solutes on such beads mainly depends on their steric parameters. The affinity thin-layer chromatography (ATLC) is also very significant. As in other types of affinity chromatography, the ATLC uses interactions of biologically related agents. Chiral chromatography based on binding agents of a biological origin (cyclodextrin or immobilized protein) can be considered as the affinity method.

2.1.2 Mobile Phases

Organic or inorganic solvents and even solutes of strong acids or bases, ion paring, ion-exchanger agents, etc. can be used in TLC as mobile phases. It also is worth underlining that the mobile phase is evaporated after the development of the chromatograms and it does not interfere with the solute spots during the visualization process. These two facts increase enormously the choice of eluent to be used in TLC, but organic solvents or their two-, three-, or more component mixtures are the best option. In the case of partitioning chromatography, properties of solvents are described by the Hildebrand [21] parameter of selectivity. It is a measure of the sum of disperse dipole–dipole and hydrogen-bonded interactions. In the case of adsorption chromatography, properties of solvents are defined as the elution strength. Of course, the elution strength is higher when the mobility of the solute is higher. This parameter is calculated for a given stationary system. Lists of solvents ranked according to their elution strength are called elutropic series. The best known of them is the Snyder series [22], which is linked to silica gel. It consists of 81 solvents grouped in eight classes. In laboratory practice, the Hildebrand parameter of selectivity and the elution strength are used to select the chromatographic systems to be used.

A number of mixtures with various solvent percentages can be used in clinical applications, and their compositions depend on the nature of the separated analytes. For example, four mixtures, (1) ethyl acetate–methanol–30% ammonia (85:10:15), (2) cyclohexane–toluene–diethyl amine (65:25:10), (3) ethyl acetate–chloroform (1:1), and (4) acetone, have been proposed for the separation of drugs and their metabolites on silica gel [23]. Lipids can be very well separated by a mobile phase of strong elution strength (first elution) and by a less polar mixture in a second run. Amino acids can be well separated on silica gel with very polar mixtures such as *n*-butanol–acetone–acetic acid–water (3.5:3.5:1:2) or pyridine–acetone–ammonia–water (80:60:10:35).

2.2 Sample Preparation and Application

There is a generally accepted view in laboratory practice that the TLC stage of sample preparation is not so important because TLC is also a cleanup technique. This is usually an incorrect view, especially in the case of clinical analysis, where analytes are to be found in plasma, serum or urine, sometimes whole blood, faeces, saliva, cerebrospinal fluid, gastric fluid, or body tissues. These matrices are very complex mixtures and therefore sample preparation prior to TLC analysis cannot be omitted. The process of separation from biological samples and purification of analytes is usually realized by protein precipitation, dialysis, hydrolysis, ultrafiltration, dilution, liquid–liquid, or solid-phase extraction. Originally, the most common extraction method was liquid–liquid extraction, but in recent years solid-phase extraction has became an increasingly popular method for isolation of compounds from biological matrices. Lyophilization, saponification, microwave processing and supercritical fluid extraction are less common.

The first step of separation by TLC is the application of the laboratory sample onto the plate. A line is drawn with a pencil parallel to, and 0.5–2 cm from, the bottom of the plate. The samples are spotted onto this starting line either in spots or in bands (application of bands usually results in better separation). Diverse types of capillaries were originally used for spotting of samples, but recently calibrated syringes or automated spotting equipment is used to apply precise amounts of the sample on the start line, in either spots or bands. The purpose of the separation (analytical or preparative) and the detection limit of the analytes frequently determine the sample volume. When a large volume of sample must be applied, the use of plates with preconcentrated zones or the automated spray-on technique is indispensable. The preconcentrate plate consists of two zones of different adsorbents, merging into each other but with a sharply defined boundary. One of them (the spotting zone about 3 cm long) has comparatively poor adsorptive properties, while another forms a plate for the analytical separation. Any size of spot placed on the spotting zone run in the mobile phase will become a sharp band before it gets to the analytical part of the plate. The use of the automated spray-on technique is more complicated. The sample is transferred onto the plate by means of a nitrogen or other inert gas stream. This will keep the starting zones as small as possible. Many application systems are in the market, which makes possible applying 50–500 μL of sample. Both methods improve the sensitivity of the detection and the separation efficiency.

2.3 Development Techniques

For the favorable solute separation a set of parameters (composition and characteristic of the chromatographic system, the manner of mobile phase movement, etc.) has been determined. The elution, in which the separation conditions are not changed throughout the time required for the sample separation, is defined as isocratic. In the case of a gradient elution, the mobile phase composition or its pH (rarely temperature or composition of the adsorbent) has changed during the separation. Successful separations of many complex biological mixtures by gradient elution have demonstrated the utility of this technique. Important also is the manner of mobile phase transport. This can be divided into two groups: the capillary flow (CF) and the forced flow (FF) movement methods. CF development of chromatograms is carried out with capillary forces. In case of FF development an external force is required to move the mobile phase throughout the sorbent. The most important of them are overpressure layer chromatography (OPLC) and rotator planar chromatography (RPC). OPLC is more rapid than CF TLC and offers various possibilities to improve separations. RPC is mainly used as a preparative technique.

2.3.1 Classical Techniques of Development

The one- or two-dimensional ascending or horizontal techniques are usually applied in TLC. An ascending developing chamber consists of a glass tank that has ground edges at the top to make an airtight seal with the glass lid. The development of chromatograms can be carried out in the presence (saturated chambers) or absence (nonsaturated chambers) of a gas phase in equilibrium with the mobile phase. When the mobile phase is added to the ascending chamber, the TLC plate is placed in a vertical position so that the starting line is above the level of the mobile phase. The mobile phase rises because of capillary forces. Sandwich chambers are also used in which the TLC plate is clamped between two glass plates separated by a gasket. A solvent is passed on one edge of the plate from the reservoir. The advantage of this system is that it uses less mobile phase and provides a more quickly saturated vapor system. Two-dimensional separation can be useful when particular separations are needed. The plate is run in the same direction with the first mobile phase, dried, turned 90°, and run again using a different mobile (rarely stationary) phase. In clinical analytical practice, two-dimensional developments are frequently used (Fig. 1). However, in such separation only one sample can be spotted on each plate.

2.3.2 Multiple Developments

Multiple techniques of separation involve the repeated development of the chromatogram with one or more mobile phases. Mobile phases are removed between successive developments. These techniques can be divided into three groups: (1) unidimensional; the chromatogram is developed repeatedly for the same length with the same mobile phase, (2) incremental; the same mobile phase is used, but the first development length (step) is shorter and each subsequent development is incremented, and (3) gradient multiple development. In the case of gradient elution the mobile phase composition changes for each (or for just a few) development step accompanying unidimensional or incremental development. Gradients of increasing solvent strength are preferable for complex samples spanning a wide polarity range. Decreasing solvent strength gradients are effective for simpler mixtures where a smaller separation capacity can be employed. Multiple development has some advantages over normal development. The most important of them are greater efficiency and separation capacity owing to the zone refocusing mecha-

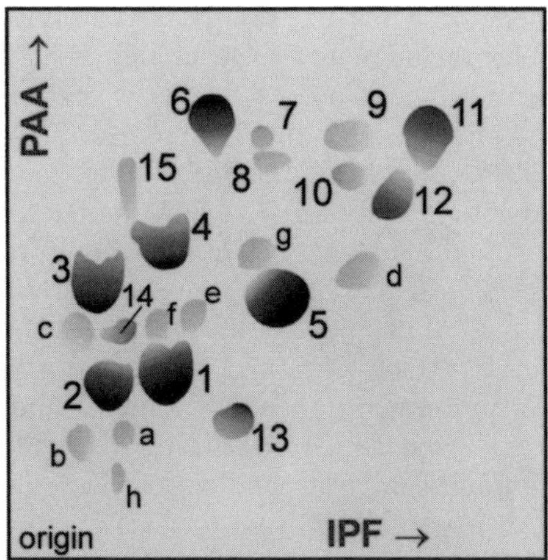

FIGURE 1 Two-dimensional chromatogram of amino acids in human urine. The sheets were developed by ascending migration with piridine–acetone–aqueous ammonium hydroxide–water (13:8.5:2.5:6, v/v; PAA) for 90 min, air-dried, turned through 90°, developed with isopropanol–formic acid–water (25:3:2, v/v; IPF). Spot notation: 1-glycine, 2-glutamine, 3-histidine, 4-serine, 5-alanine, 6-theronine, 7-trytophan, 8-tyrosine, 9-phenylalanine, 10-methionine, 11-leucine/isoleucine, 12-valine, 13-glutamic acid, 14-lysine, 15-taurine; a-arginine, b-cystine/cysteine, c-methylhistidine, d-β-aminoisobutyric acid, e-homocitrulline, f-hydroxyprolin, g-proline, h-phosphoethanolamine. (From MC Hsieh, HK Berry. Detection of metabolic diseases by thin-layer chromatography. J Planar Chromatogr 2:118–123, 1992.)

nism, optimum use of solvent selectivity, and improvement of sample detectability by scanning densitometry, owing to smaller dispersion spots or zones. Multiple chromatography is rarely required for the routine TLC separation in clinical chemistry because of isocratic separation that usually permits rapid analysis with highly reproducible R_F values. Unfortunately, many TLC screening methods must employ a multiple separation to deal with the wide range of hydrophobicity encountered among substances of interest to clinical chemistry.

2.3.3 Immunoassay Separation

The method combines the use of the solid-phase-bonded affinity ligand and conventional TLC and can be applied for both purification (preparative ATLC) and analysis (analytical ATLC) of the sample. Most ligands are of biological origin, but molecules of nonbiological origin (metal chelates, synthetic dyes) can also be used. There are many types of affinity chromatography. Bioaffinity chromatography includes any method that uses a biological molecule as the affinity ligand (nucleic acids, selectively retained DNA or RNA-binding proteins, and enzyme inhibitors and cofactors are used for enzyme binding). A special category is immunoaffinity TLC, in which the affinity ligand is an antibody or antibody-related agent (antibodies retained drugs, hormones, proteins, peptides, or viruses). In the laboratory practice the strip of a chromatographic plate in a discrete region, usually near the origin, is coated with affinity ligand. After injecting the sample on the start line, the chromatogram is developed by a mobile phase that has the proper pH value and does not wash covalently bonded affinity ligand. The solute that is complementary to the ligand is retained on the plate, while components not associated with the immunoreactive ligand migrate with the solvent front. Then the solvent that displaced solute from the strip or that promoted dissociation of the solute–ligand complex again develops the chromatogram. Sometimes the radiolabeled solute is analyzed and the latter step cannot be done.

2.3.4 Instrumentation of Developing Process

The term instrumental thin-layer chromatography is used to qualify the analytical process, in which steps occur by appropriate apparatus. Instrumental TLC is becoming more frequently used in many clinical laboratories because the instrumentation improves the separation and makes determinations and the quality assurance necessary to maintain the validity of test results.

Automated multiple (usually gradient) development (AMD) is the most representative of CF instrumentation techniques. AMD equipment enables incremental gradient multiple development in which individual steps of separation (conditioning, development, and removal of mobile phase) is automated. The number of development steps, their length, and their solvent strength gradient are programmed before separation. A typical AMD separation is shown in Fig. 2.

Progress in FF analytical separations has mainly been made as a result of overpressure instruments. In this technique the eluent is forced through the sorbent by means of a pump system using a desirable flow rate. Thanks to these, OPLC offers decreases of the development time, insignificant band spreading, and improvement of detectability. OPLC can also be performed in two-dimensional mode. RPC is another TLC technique with FF elution employing the centrifugal force of a revolving rotor to move the mobile phase and separate components of mixtures.

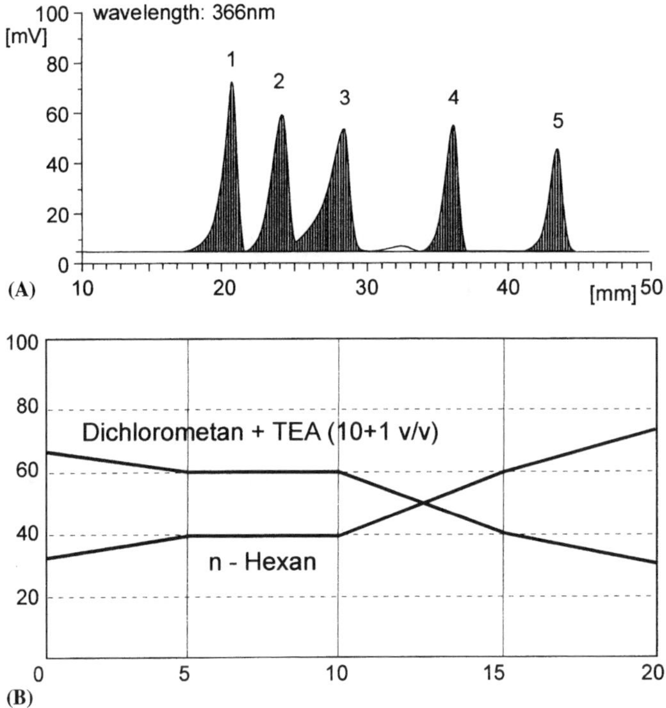

FIGURE 2 Separation of the biogenic amines (A) by use of gradient elution (B). Peaks from the left: 1-putrescine, 2-cadaverine, 3-histamine, 4-tyramine, 5-ephedrine. (From MH Vega, RF Saelzer, CE Figueroa, GG Ríos, VHM Jaramillo. Use of AMD HPTLC for analysis of biogenic amines in fish meal. J Planar Chromatogr 1:72–75, 1999.)

2.3.5 Multimodal Separation Techniques

In order to improve separation and quantitative determination of analytes, the coupled chromatographic techniques (TLC–HPLC or TLC–GC) are used. The thin-layer and column liquid chromatography coupling can be performed in the TLC–HPLC or HPLC–TLC mode. In the former case, TLC is used as a cleanup technique (indirect coupling); spots separated on TLC are scraped, and the analytes are dissolved in suitable solvents and then separated and quantified by HPLC. In the case of HPLC–TLC coupling (direct method), the suitable volume fractions of column eluent are mixed with nitrogen gas and sprayed as an aerosol onto the plate. TLC again separates analytes fractionated on column. Thanks to them, various

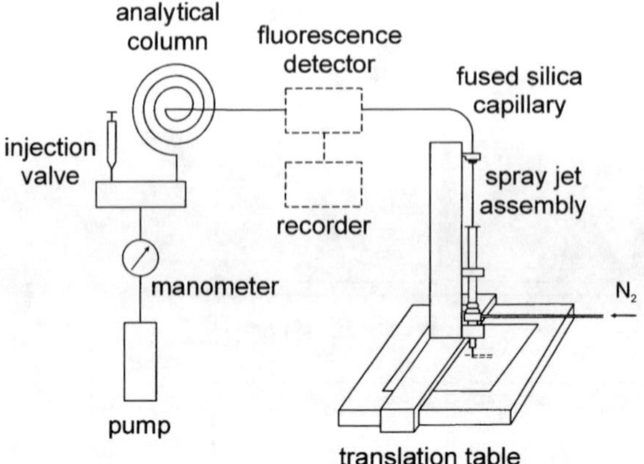

FIGURE 3 HPLC–TLC coupling. (From JW Hofstraat, S Griffioen, RJ van de Nesse, UATh Brinkman, C Gooijer, NH Velthorst. Coupling of narrow-bore column chromatography and thin layer chromatography. J Planar Chromatogr 3:220–226, 1988.)

separation mechanisms and specific visualization reactions can be used in one chromatographic process (Fig 3).

The TLC–GC coupling can be performed in direct mode, for example as the tabular TLC. The separation is carried out in a quartz tube, of which the inner surface is coated with a layer of inorganic adsorbent (usually silica gel). The tube is then driven through the scanning furnace and the separated fractions are consecutively vaporized, swept by carrier gas to the GC system. After the separation of analytes in the column, the GC detector analyses solutes. This method has been applied for analyses of phospholipids, glycolipids, bile acids, and other analytes of clinical interest. Indirect TLC–GC coupling is less complicated. As with the TLC–HPLC method, TLC is scaled down to a procedure for micropreparative separation of substances and serves as a cleanup technique.

2.4 Visualization and Quantitative Determination

The R_F value in comparison with those of the standard that had been developed under identical experimental conditions permits qualitative or semiquantitative analysis. Colored substances, which have been separated on TLC plates, can be indirectly visualized. However, the majority of solutes do not absorb light in the visible range and they must be detected by other methods. These methods usually have been divided into three groups: chemical, physical and physicochemical, and physiological–biological. Some of them can be used for quantitative determinations. Two basic techniques are available for the quantification of analytes. In the first of them (the densitometry in situ), the solutes are assayed directly on the layer. In the second one (the indirect method), analytes are removed from the layers and after the extraction assayed by other, sometimes very sophisticated, instrumental methods.

2.4.1 Chemical Methods of Visualization

In this method colorless compounds are converted into colored derivatives by use of the appropriate reagents. This can be performed before chromatographic development (the colored products are then separated) or after chromatography. Postchromatographic derivatization is the basic method of visualization especially when there is a specific reaction that can confirm the presence of particular analytes in the spot or band. The majority of chemical methods (both nonselective and selective) for derivatization of clinical chemistry interest were drawn up before 1990. Detailed information on this point can be found in the literature [24,25].

2.4.2 Physical and Physicochemical Methods of Visualization; UV Measurements

Among the physical methods of visualization, ultraviolet light (UV) measurements are used first of all. Substances containing UV chromophores can be qualitatively or quantitatively determined by UV absorption. Qualitative analysis occurs on the plates with a fluorescent (see below) indicator. The indicator placed on all surfaces of the plate emits visible light excited by UV (usually $\lambda = 254$ nm). Analyte absorbing UV prevents excitation (fluorescence of indicator is quenching) and become visible as a dark spot on the light background. Quantitative analyses can also perform on plates without fluorescent indicator by densitometer (densitometric evaluation enables rapid quantification of a lot of analytes). The separated substances are scanned with a flying spot of light or a fixed light beam in the form a rectangular slit. The absorbance of the light (UV) of the separated analytes is measured at appropriate wavelengths (maximum absorption) for the interest component. A diffusely reflected light is measured by the photosensor of the densitometer. The difference between the optical signal from the analyte-free background and that from an analyte is usually correlated with the amount of the respective fraction of calibration standard separated on the TLC plate (Fig. 4). It is obvious that a standard curve should be established on the same plate under the same conditions as for the analyte. Measurements of VIS absorption perform identically.

For the determination of very small amounts of substances, luminescence analyses have been adapted to TLC. In advance, phosphorescence

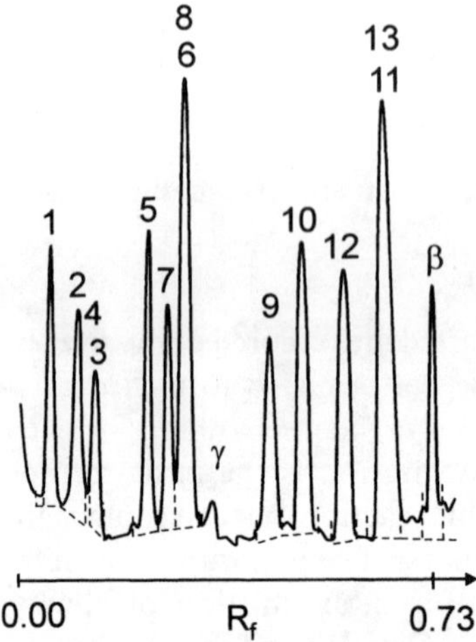

FIGURE 4 Chromatogram of benzodiazepines. Peak notation: 1-alprazolam, 2-midazolam, 3-chlordiazepoxide, 4-brotizolam, 5-carbamazepam, 6-cinolazepam, 7-nitrazepam, 8-clonazepam, 9-temazepam, 10-clobazam, 11-madazepam, 12-diazepam, 13-prazepam. (From E Hidvégi, S Perneczki, M Forstner. Data on the chromatographic behavior of solvent mixtures with similar solvent strength and selectivity. J Planar Chromatogr 6:414–419, 2000.)

analyses were used for measurements of some drugs, steroids, hydrocarbons, and other organic compounds separated on TLC plates. However, phosphorescence is rarely used in clinical chemistry, because cooling the analytical system or introducing so-called heavy atoms into the analyte is needed (an electronic oscillation in the molecule must be as small as possible). Recently, fluorescence measurements are most commonly used. They are more selective and sensitive than UV or VIS absorption measurements (the sample concentration can be 10^2–10^3 times lower). Nonfluorescing compounds do not usually interfere, so that analysis can be carried out also in the presence of such accompanying substances. Good results can also be obtained by conversion of nonfluorescing solutes into derivatives showing fluorescence. Reagents that have proved helpful in this connection include dansyl chloride (amino acids, corticosteroids, and estrogens), dansyl hydrazine (reducing sugars), and fluorescamine (amino acids, amphetamine). Quantitative determinations of analytes depend on densitometric measurements of fluorescence intensity at appropriate excitation and emission wavelengths (Fig. 5).

Isotope measurement is the method available for evaluation in the μg range, because the labeled substances generally make up only a few percent of the total isolated compound. The method is considered to be selective, since

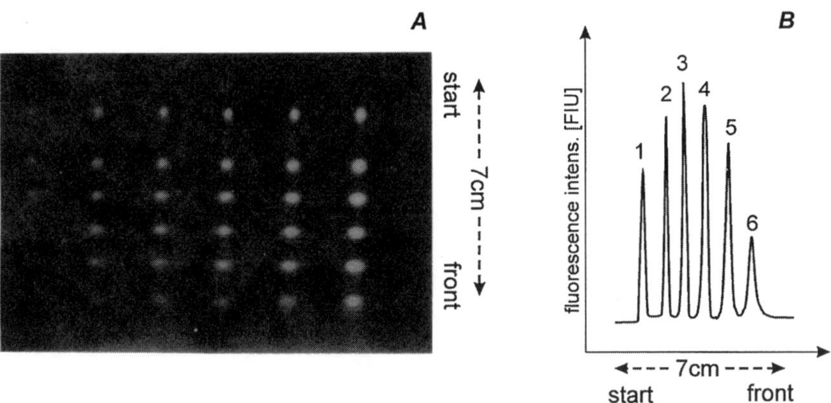

FIGURE 5 Fluorescence determinations: (A) Chromatogram of different concentrations of porphyrin standards, (B) densitogram. Peaks notation: 1-mesoporphyrin, 2-coproporphyrin, 3-pentaporphyrin, 4-hexaporphyrin, 5-heptaporphyrin, 6-uroporphyrin. (From A Junker-Buchheit, H Jork. Urinary porphyrins: ion-pair chromatography and fluorimetric determination. J Planar Chromatogr 3:214–219, 1988.)

accompanying material displays no emission and does not interfere with the evaluation. The short-lived radioisotopes ^{35}S, ^{32}P, and ^{131}I are of special interest in clinical chemistry because they do not overburden the organism. Comparatively long-lived ^{36}Cl, ^{14}C, or ^{3}H are used to facilitate analysis. Bacteria or toxins can be directly labeled (see below) prior to performing assay or metabolically labeled. The three principal methods for measuring radioactivity of labeled solutes are (1) autoradiography (an image is produced on x-ray or photographic film after exposing it to emissions from solutes), (2) in situ detection by radiation detectors, and (3) indirect zonal analysis (the plate is segmented or sectioned, adsorbent is removed from the plate, and the radioactivity is measured). All can be used for quantitative analysis. An example of radiochromatography applications is presented in Fig. 6.

2.4.3 Biological–Physiological Visualization

Biological–physiological determination depends, like affinity chromatography, on highly specific and sensitive interactions of biologically related agents. The three common versions of their method are bioautography, the overlay technique, and the direct enzymatic method.

In the case of bioautography, the test organisms are uniformly distributed in an agar or gelatin detector layer. As the solutes are separated on

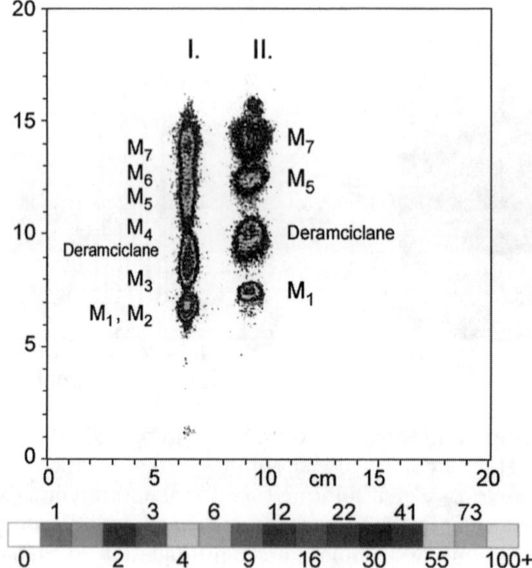

Figure 6 Digital autoradiography of two deramciclane metabolite samples obtained from nutrient matrix (I) and from liver cells (II). The different compounds found are indicated by M_1–M_7. (From K Ludányi, A Gömöry, I Klebovich, K Monostry, L Vareczkey, K Újszászy, K Vékey. Application of TLC-FAB mass spectrometry in metabolism research. J Planar Chromatogr 2:90–96, 1997.)

the plate, mobile phase residues are removed, layers are connected (sometimes through a blotting material), and the active solutes diffuse from the adsorbent into the detector layer. After some hours or even days of incubation, solutes display their inhibiting or beneficial action on the test organism. To have a result from these actions, radiolabeled substances are used or suitable reagent was either added to the detection layer or sprayed onto it after incubation (immunostaining TLC). The inhibition haloes or spots then appear as light zones on a colored background or vice versa.

Overlay detection requires pretreatment of the chromatograms with plastic. The rationale for plastic coating is to prevent flaking of silica gel from the support. For the next step the plate is overlaid with a biological agent (antibody, toxin, etc.), incubated, washed with buffer and, if necessary, overlaid with the secondary detection agents (e.g., secondary labeled antibody). Then the plate is washed again and dried for autoradiography on x-ray film or treated for secondary color development. A parallel plate is usually developed in the same chamber under identical conditions and subjected to chemical detection.

Direct enzymatic reactions depend on spraying chromatograms with enzyme solutions. After incubation time, a suitable substrate is added and the reaction, as in the chemical detection, is followed visually.

All biological–physiological methods are highly specific, and inactive accompanying substances do not interfere in determinations. The methods are used to determine hormones, antibiotic and enzyme-inhibiting action of alkaloids, mycotoxins, pesticides, etc.

2.4.4 Multimodal Techniques of Detection

Identification of the analytes by the TLC method can be (as with other kinds of chromatography) ambiguous. One commonly thinks that substances with the same R_F value are identical. This assumption is not fully justified. The same retardation factor informs only about a huge probability of identity but it does mean absolute detection. The UV or VIS absorption spectra are not selective enough, and structurally similar substances cannot generally be distinguished. Moreover, the chromophore group is needed for detection based on UV or VIS absorption. More information can decidedly be obtained by coupling TLC with modern spectroscopic methods. Several methods have been developed for coupling TLC with spectrometric techniques and their applications for clinical chemistry. These include both indirect (off-line) and direct (on-line) methods. Indirect methods are probably the most readily implemented way to couple TLC with other spectroscopic methods. Standard nondedicated spectrometers can be used, and scraping off a TLC spot and dissolving the analyte in a suitable solvent do not require special equipment. Unfortunately, indirect methods are usually time-consuming and have a potential for loss of the sample.

These days the combination of TLC-Fourier transformation infrared (FTIR) and TLC-mass spectrometry (MS) are established in clinical chemistry. Raman spectroscopy signals are very weak, and the sensitivity is too low for regular use. The quality of the FTIR or MS spectra is sufficient for discrimination between closely related substances. An advantage of these methods is detecting also nonabsorbing UV or VIS substances. Radiometry (the measurement of the radioactivity of isotopes) is also frequently used in clinical practice.

Direct coupling FTIR spectroscopy was introduced in 1989 by Glauninger et al. [26]. The principle of the method depends on scanning the plate fixed onto a computer controlled x,y-stage with an IR beam in a diffuse reflectance infrared Fourier transformation (DRIFT) unit. Of course, difficulties arise because of IR absorbance of the stationary phase. Commercially available plates (e.g., 50% silica gel and 50% magnesium tungstate) evaluate only the region between 3550 cm^{-1} and 1370 cm^{-1}, but this is enough to resolve many problems in clinical chemistry [27]. An example of an advantage resulting from direct TLC-FTIR coupling is presented in Fig. 7.

TLCMS coupling does not require a special adsorbent, but the interface is needed for introduction of the analytes separated with TLC into the MS detector. The one- or two-dimensional systems spatially resolved analyses are used (thermal evaporation of the analytes to a gas stream connected to the mass spectrometer or a direct-insertion probe into fast atom bombardment [FAB] or secondary ion [SI] sources of mass spectrometer). In this method analytes are sputtered from adsorbent by impact of the high-energy molecule stream. Liquid matrices such as glycerol or triethanolamine can be applied to the chromatogram before FAB or SI desorption. Recently matrix-assisted laser-desorption ionization (MALDI) was adapted to TLC. The method depends on pressing a previously prepared layer of matrix crystals into the TLC plate and elution of spots from the TLC plate to the MALDI layer via capillary action. A good example of MALDI applications in clinical chemistry is the paper by Isbell and co-workers [28]. They proposed a

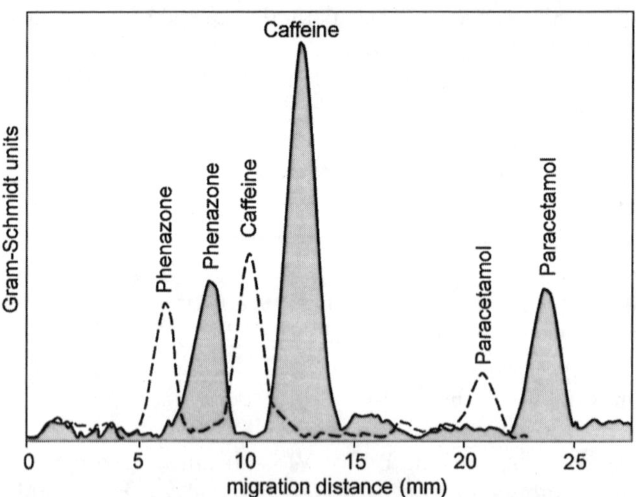

FIGURE 7 HPTLC-FTIR on-line coupling. Gram–Schmidt chromatogram of test substances on 1:1 mixed silica gel 60-magnesium tungstate (- - -) and on silica gel (—). (From GK Bauer, AM Pfeifer, HE Hauk, K-A Kovar. Development of an optimized sorbent for direct HPTLC-FTIR on-line coupling. J Planar Chromatogr 2:84–89, 1998.)

methodology for the detection of pg quantities of nucleotides directly from TLC plates without the use of radioactive labeling. The matrix-assisted laser-desorption ionization–time of flight MS (MALDI-TOFMS) method is now proposed [29,30]. It is a hybrid TLC-MALDI plate in which a silica layer and a MALDI layer are configured adjacently on a common backing material. Advantages of TLC-MS coupling are presented in Fig. 8. TLC-MS-MS

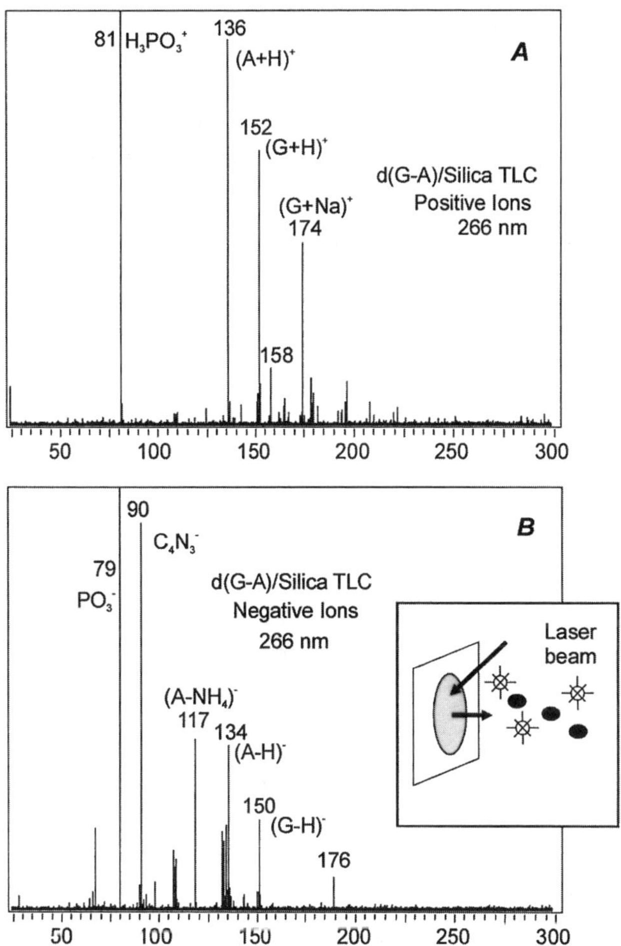

FIGURE 8 TLC MS on-line coupling. Positive (A) and negative (B) ion laser desorption mass spectra of the dinucleotide d(G-A) obtained directly from a silica gel plate. (From KL Busch. Coupling thin-layer chromatography with mass spectrometry. J Planar Chromatogr 2:72–79, 1992.)

coupling provides further advantages by giving more detailed information for particular ions.

3 APPLICATIONS

TLC is widely used in clinical chemistry for the detection of many substances, such as amino acids, carbohydrates, drugs and their metabolites, lipids, and porphyrins or prostaglandin in various biological specimens, which are significant in medical diagnosis. Research papers connected with this problem can be divided into two fundamental groups. The first one concerns investigations of the method itself, usually done with standards. In these investigations, separation techniques and visualization methods are improved, and the principles of quantitative determinations are studied. The other group involves investigation of applications. As a result of these studies, a method describing the treatment of biological sample from its collection until complete information on the analytes or their metabolism products is obtained. In such investigations the extraction and cleanup technique, the chromatographic system, and the detection methods are developed, and the efficiency of each stage for the analytical process is optimized. Both of these applications are widely represented in the literature; applications presented in this part are limited to works that illustrate the capabilities of the method as well as its practicability.

3.1 Amino Acids

TLC has proved useful for screening and quantitative analysis of amino acids in urine and blood samples. The meaning of such analyses is crucial because early estimation of free amino acids in biological fluids and tissues may prevent neurological damage and mental retardation in young infants with inborn errors of amino acid metabolism [31,32].

Almost all separations of untreated amino acids are performed in NP systems. Four stationary phases most commonly used for such separations are cellulose, silica gel, impregnated adsorbents, and ion exchangers. In the case of silica gel, two- or three-component mixtures of solvents usually carry out separation. Acetone, methanol, water, acetic acid, and formic acids, sometimes with the addition of ammonia or pyridine, are most commonly used as the components of mobile phases. The basic amino acids showed smaller R_F values than did acidic ones, because of amino group and acidic group of silica gel interaction. Impregnated silica gel, silanized, or octadecyl modified silica gels are used for some separations. Different metal ions or dodecylbenzensulfonic acid are most commonly used as impregnating agents at various concentrations. Mobile phases are then similar to those used on silica gel. Cellulose ion exchangers and ion exchange resins used for amino acid separations need eluents with an adequate ionic strength. The most widely used reagent for qualitative and quantitative assessment of nontreated amino acids is ninhydrin reagent.

A widely employed method for the separation of enantiomeric pairs of amino acids (amino acids are optically active) is chiral separation. It may be performed with β- or γ- cyclodextrin attached to dimethylpolysiloxane or on conventional stationary phases by the inclusion of chiral additive into the mobile phase.

Determinations of *N*-terminal amino acids have a dominant meaning in clinical chemistry and are applied for peptide structure researches. In such analyses, prechromatographic derivatization is usually used. Dansyl and 3-phenyl-2-thiohydantoine derivatives are most commonly investigated. Separations are carried out on silica gel or polyamide layers.

Deyl [33] presented applications of liquid chromatography for the profiling of amino acids in body fluids and tissues. The article includes many references of TLC technique. The review by Bhushan [34] summarizes the application of TLC in the analysis of amino acids and their derivatives. Breakdown of some examples of TLC applications for amino acid research is presented in Table 2.

Listing the applications of TLC for amino acid analyses, it is impossible to pass over the works presented by Yahya and coworkers [41,42]. They examined 404 urine samples to find whether inborn errors of amino acid metabolism could be the main cause of death among the children. Patients were aged between 11 days and 6 years. Of these, 41 percent had aminoacidurias. Very interesting are the works of Marklova et al. [43]. They applied two-dimensional TLC for the diagnosis of inherited metabolic diseases. Analyses of amino acids, sugars, oligosaccharides, and organic acids were carried out on silica gel and cellulose with different solvents. HPTLC and HPLC were also used for the detection of defects in the metabolism of tryptophan [44]. HPTLC was suitable at the beginning of the investigation. An HPLC method with isocratic elution and spectrometric detection was used at the next step, when pathological findings were to be confirmed and the individual metabolites quantified. The first method enables the assessment of tryptophan, 5-hydroxyindolylacetic acid, indolylacetic acid, indoly lacryloylglycine, and indolylacrylic acid and its possible precursors, namely, indolyllactic and indolylpropionic acids. The second procedure was intended for the monitoring of anthranilic, 3-hydroxyanthranilic, kynurenic and xanthurenic acids, kynurenine, 3-hydroxykynurenine and indoxyl-sulfate.

TABLE 2 Examples of TLC Application for Amino Acid Analyses

Analyte	Matrix	Chromatographic system		Visualization and quantification	Ref.
		Stationary phase	Mobile phase (v/v)		
Chain and aromatic amino acids	Serum	Polyamide	Two-dimensional: I benzene-acetic acid (9:1.7), II formic acid-water (1.5:100)	UV_{254} detection, fluorometric determination after elution	35
Phosphotyrosine, phosphoserine, phosphothreonine	Protein hydrolyzates	Cellulose	Propionic acid-1M NH_3-isopropanol (45:17.5:17.5)	Ninhydrin reagent	36
Homocysteine	Blood	Cellulose	2-mercaptoethanol	Ninhydrin reagent	37
3-phenyl-2-thiohydantoine derivatives of 16 amino acids	—	Silica gel	Three phases: I pyridine-benzene (5:40), II methanol-carbon tetrachloride (1:20), III acetone-dichloromethane (3:80)	Iodine vapors	38
Proline and hydroxyproline	Different biological samples	Impregnated silica gel	Isopropanol-water (7:3)	Ninhydrin reagent, autoradiography	39
20 main protein amino acids	—	Silica gel (single and double plates)	OPLC separation: butanol-acetonitrile-0.005 M potassium dihydrogen phosphate-acetic acid (1:5:3:1)	Ninhydrin reagent, quantification at 490 nm	40

3.2 Drugs

The most important applications of TLC are the screening of hard intoxication or drug abuse and pharmacokinetic studies. The term hard intoxication is applied to the determination of harmful effects, arising in the short time after the introduction a large dose of poison to the body. Drug abuse refers to the administration of a drug or other biologically active substance in order to produce a pharmacological effect unrelated to medical therapy. Assays of poisons and abused drugs may be required in a variety of settings, but the most urgent requests are associated with clinical toxicology. This fact has necessitated the development of rapid and sensitive methods for the detection of opiates, barbiturates, benzodiazepines, amphetamines, cannabinols, and like substances in biological samples. The common use of TLC in such analyses is due to its ability to provide necessary data. Pharmacokinetic studies of drugs include drug concentration measurements and evaluation of the efficiency of new methods of therapy. The pharmacological action of many drugs depends not on the amount taken but on concentration in the blood. Relations between these two values (dose and therapeutic concentration) depend on the drug and often have an individual character. Therefore one of the tasks of clinical chemistry is to establish relationships between the concentration of the drug in blood and its dosage. Evaluation of the efficiency of new therapy methods mainly concerns drugs newly introduced into clinical medicine. The object of investigation is then not only the concentration of the drug in blood but also its assimilation and excretion, harmfulness, metabolism, etc. Wilson [45] critically reviewed planar chromatography from the viewpoint of drug analysis in biological fluids. The capabilities of the various techniques of TLC and their advantages and disadvantages were discussed.

3.2.1 Screening of Hard Intoxication and Drug Abuse

The older diagnostic tests using TLC identification of analytes was mainly based on comparison of R_F values and on visualization with the specific color reactions. Tests for qualitative analyses of poisons and different drugs have been known for many years. The Committee for Systematic Toxicological Analysis of the International Association of Forensic Toxicologists (TIAFT) approved a test for the separation of 1600 toxicologically relevant substances [46]. The test is based on separating power measurements (database of reduced hR_F values is done) in ten standardized chromatographic systems (Table 3).

Chromogenic reagents are usually applied for the detection of drugs. Derivatization allows selective confirmation of amphetamine and primary amines (ninhydrin), barbiturates (diphenyl carbazone or mercuric sulphate),

TABLE 3 Systems for Drug Screening Approved by the Committee for Systematic Toxicological Analysis of the International Association of Forensic Toxicologists

Analytes	Chromatographic system		Separation technique
	Stationary phase	Mobile phase (v/v)	
Acidic and neutral drugs	Silica gel	Chloroform–acetone (8:2) Ethylacetate Chloroform–methanol (9:1)	Unsaturated chamber
Acidic, basic, and neutral drugs	Silica gel	Ethyl acetate–methanol–conc. ammonia (85:10:5)	
Basic drugs	Silica gel	Methanol Methanol-*n*-butanol (6:4 + 0.1 mol/L NaBr)	Saturated chamber
	Silica gel impregnated with potassium hydroxide	Methanol–conc. ammonia (100:1.5) Cyclohexane–toluene–diethylamine (75:15:10) Chloroform–methanol (9:1) Acetone	Unsaturated chamber

cocaine, antidepressants and antihistamines (iodoplatinate), narcotic analgesics (Dragendorff reagent), cannabinols (Fast blue B salt), etc. Especially useful for screening investigations are kits such as the Toxi·Prep (TP) kit, proposed by Steinberg and coworkers [47]. The method is based on TLC and involves five major steps: solid phase extraction, concentration, spotting, development of chromatograms, and detection. The kits have been found to be particularly useful for the analysis of basic and neutral drugs. Some of the newest applications of TLC for drug screening are presented in Table 4.

Performance of different techniques used to detect drug abuse in urine (based on external quality assessment) was presented in the literature. Twenty-five samples of lyophilized urine were analyzed by an average of 95 laboratories [55]. Samples contained mixtures of analytes and included replicated concentrations of morphine, methadone, amphetamine, and cocaine at 0, 1, 2, and 5 mg/L and of benzoylecgonine at 0, 0.4, 1, 2, and 4 mg/L. It turned out that some chromatographic techniques are inadequate for detecting morphine, amphetamine, cocaine, and benzoylecgonine at lower concentrations of the analytes studied. Gas–liquid chro-

TABLE 4 Screening of Hard Intoxication and Drug Abuse in Urine; Examples of Applications

Analyte	Chromatographic system		Visualization and quantification	Ref.
	Stationary phase	Mobile phase (v/v)		
Opioids, cocaine, and amphetamine	Silica gel	Two mobile phases based on chloroform, methanol, and benzene	Iodoplatinate reagent	48
Salicilates	Silica gel	Benzene–acetic acid–diethyl-ether–methanol (60:9:30:5)	Immunoassay	49
Cocaethylene and cocaine	Silica gel	Hexane:toluene:diethylamine (65:20:5)	Iodoplatinate reagent	50
Diazepam, nitrazepam, estazolam	Silica gel	Chloroform–benzene–acetone–methanol (5:5:3:1)	Densitometry at 300 or 200 nm	51
Methamphetamine	C-18	Ethyl acetate–ethanol–concentrated ammonia (36:2:2)	Fast Black K salt	52
Amphetamine and its major metabolites	Silica gel	Toluene–acetone–20% ammonia–ethanol (45:45:7:3)	HPTLC-FTIR direct coupling	53
Acidic and neutral drugs	C-18	Methanol–water (13:35)	Fast Black K salt, Iodoplatinate, Dragendorff, Marquis, and Salkovsky reagents	54
Basic, amfoteric, and quaternary drugs	Silica gel C-18	Toluene–acetone–ethanol–conc. ammonia Methanol–water–conc. HCl		

matography was least sensitive for morphine; TLC was least sensitive for the other analytes. Few significant differences in specificity were detected between techniques, although significant interference from structurally related compounds was demonstrated in assays of morphine, methadone, and amphetamine.

Jain [56] examined the utility of TLC for detection of opioids and benzodiazepines among drug addicts seeking treatment. Over a period of 5 performance years (1991–1995), 6,055 urine samples were analyzed for opioids (morphine, codeine, buprenorphine, dextropropoxyphene, pentazocine) and benzodiazepines (diazepam, nitrazepam) by TLC. Out of all the drug tests ($n = 9{,}922$) carried out, 24% of the drugs had been used during the past 72 h. Averaged across all drugs, the detection rates corresponding to 24, 48, and 72 h by TLC were 37%, 36%, and 31% respectively. A high percentage of negative TLC results were observed in these samples. Moderate sensitivity of the TLC assay procedure, low consumption of drug, short time between drug use and urine collection, and drug use history of the subjects obtained from multiple sources led to high negative results. These findings suggest that all the TLC negative results also need further confirmation by an alternative, more sensitive technique in a clinical setting.

Trying to make the drug abuse testing program more meaningful, Brzezińska et al. [57] proposed the TLC-MS method for screening of biological samples for drugs and metabolites. Several TLC systems for many compounds of toxicological interest were described. Portions of standard drug solution were applied to silica gel plates. Chromatograms were developed with six mobile phases and detected with one or more of 10 reagents. R_F values were corrected with use of four standards. Analytes were also transferred to the MS direct inlet system and evaporated at 200°C. EI mass spectra were measured at 100 eV in full-scan mode. In this way corrected R_F values and the eight most intense MS peaks for 493 drugs and their metabolites were obtained. These data were kept as a library in personal computer–based search system. The merits of the method have been confirmed in biological sample investigations. Drugs were isolated from matrices by liquid–liquid extraction at pH 3 or 9 or by SPE. Extracts were spotted on TLC plates and spots (after developing of chromatograms) were extracted with methanol-CH_2Cl_2 (1:1) and examined by MS.

3.2.2 Pharmacokinetic Studies

Drug concentration monitoring and evaluating the efficiency of new methods of therapy also require the use of relatively simple, cheap, and rapid analytical methods. From an analytical point of view, it is a more complicated process than drug screening, since quantitative analyses are needed. Investigations are mainly performed on blood, plasma, or serum samples. Example applications of TLC for such analyses are presented in Table 5.

Many investigators have confirmed the usefulness of TLC applications for pharmacokinetic studies. Forgacs and Cserhati [66] studied the interaction of eight commercial anticancer drugs with human serum albumin by charge transfer RPTLC in neutral, acidic, and basic solutions. Calculations of the relative strength of interaction and the discussion of the effect of pH and the presence of mono- and divalent cations on the strength of anticancer drugs to interact with human serum albumin are presented. Le Roux and coworkers [67] determined salbutamol concentrations by HPTLC chromatography in the sera of two sets of 10 volunteers at hourly intervals for 6 h after taking one 8 mg slow-release tablet. The influence of time lapse in processing of serum samples, i.e., centrifugation, extraction, and chromatography, was studied. A statistical significant instability of salbutamol in the sera of patients was found that was not present in standard drug-free serum samples spiked with salbutamol and used for construction of standard curves. Otsubo et al. [68] developed a rapid and sensitive method of identifying benzodiazepines and zopiclone in human serum. The drugs were developed and separated on plates for 8 to 11 min and detected by means of UV radiation and color. Each drug was accurately identified by means of the values of hR_F and the spot color in three systems. The detection limit of the benzodiazepines in serum was 0.1–0.4 μg/mL, except for cloxazolam and haloxazolam. The sensitivity was increased about tenfold over the conventional method. Authors suggested that the HPTLC system is useful for the initial detection and identification of these drugs in emergencies. Lind et al. described the measurement of urinary ifosfamide, isophosphoramide, mustard, dechloroethyl ifosfamide, and carboxyifosfamide using HPTLC with photography [69]. The technique was also used to demonstrate the large interindividual variation in the ifosfamide metabolic profile of patients receiving the drug as single-agent therapy for non-small-cell lung cancer. In addition, oral administration was shown to result in higher levels of these metabolites in the urine. Fractionation of the ifosfamide dose over several days resulted in increasing levels of metabolites in the urine, consistent with autoinduction of ifosfamide metabolism.

The pharmacokinetics and metabolism of cyclophosphamide were studied in nine pediatric patients [70]. Cyclophosphamide and its major metabolites were determined in plasma and urine using HPTLC photographic densitometry. Plasma samples were obtained from eight subjects and urine was collected from six children during a 24 h period after drug administration. Cyclophosphamide was nearly, if not completely, cleared from plasma 24 h after its administration. The plasma half-life of drug ranged from 2.15 to 8.15 h, and between 5.4 and 86.1% of the total delivered

TABLE 5 Pharmacokinetic Studies

Analyte	Matrix	Chromatographic system		Quantification	Ref.
		Stationary phase	Mobile phase (v/v)		
Sumatriptan (sulphonoamide) and its metabolite	Blood, serum	Ten TLC systems recommended by TIAFT		UV_{254}, GC-MS	58
Sinomenine	Serum	Silica gel	Chloroform-methanol (19:2)	Densitometry at 275 nm	59
Deramciclane (a new anxiolytic drug) and metabolites	Different body fluids	Silica gel	Butanol–acetic acid–water (4:1:1)	MS-MS (offline coupling)	60
Gentamycin	Plasma	Silica gel	Chloroform–methanol–20% ammonia (24:22:15)	NDB-Cl reagent, fluoro-densitometry at 436 nm	61
5-Methoxypsoralen	Serum	C-18	Methanol–water (4:1)	Fluorodensitometry at 490 nm	62
Psychosedative drugs	Serum	Silica gel	Heptane–chloroform–ethanol–ethyl acetate (10:4:4:3)	Densitometry (UV absorption)	63
Theophylline	Plasma	Silica gel	Toluene–isopropanol–acetic acid (16:2:1)	Densitometry (UV absorption)	64
Tinidazole	Serum	Silica gel	Chloroform–acetonitrile–aceticacid (60:40:2)	Densitometry (UV at 320 nm)	65

dose was recovered as unchanged drug in the urine. The major metabolites identified in plasma and urine were phosphoramide mustard and carboxyphosphamide. The study suggests that there is interpatient variability in the pharmacokinetics and metabolism of cyclophosphamide in pediatric patients. Boddy and Idle proposed [71] a method for the determination of the anticancer drug ifosfamide and its principal metabolites in urine, plasma, and cerebrospinal fluid. The urine and fluid samples were absorbed onto Amberlite XAD-2 eluting the compounds of interest with methanol. Plasma was deproteinated using cold acetonitrile and centrifuging to yield a clear supernatant liquid. The eluate and supernatant were analyzed by TLC with spot visualization. The plates were photographed for subsequent densitometric analysis. The intraassay coefficient of variation for each compound in both urine and plasma was less than 10%, and the lower limit of detection was 1 μg/mL. The method provides means for determining the full spectrum of metabolic products of ifosfamide in-patients and will allow detailed investigation of variability in metabolism and pharmacokinetics of this drug. A rapid and sensitive HPTLC assay for the measurement of nimesulide in human plasma has been evaluated by Pandya et al. [72]. Analysis was performed on plasma containing known amounts of the drug, on drug-free plasma, and on plasma containing an unknown quantity of the drug. Known amounts of extract and nimesulide (100 and 200 ng, as external standard) were spotted on silica gel plates by an autosampler. Quantification was achieved using a densitometer. The recovery of the method was 97.10 $\pm$ 2.22%. The method was applied for the determination of plasma levels and pharmacokinetic parameters of nimesulide after oral administration of two formulations (100 mg) in healthy volunteers. The authors proved that the method is a sensitive, economical, rapid, and specific assay for nimesulide in human plasma and is suitable for pharmacokinetic studies after therapeutic doses.

3.3 Carbohydrates

Carbohydrates are naturally occurring substances that contain mainly carbon, oxygen, and hydrogen. Mono-, di-, tri-, oligo-, and polysaccharides, ketose, tri-, tetra-, pento-, and hexose as well as reducing and nonreducing sugars have great importance in the life sciences. Several diseases are accompanied by the increased elimination of sugars of various groups in the urine and feces.

TLC is widely used in biomedical research and clinical laboratories for separation of carbohydrates in biological samples. Silica gel and silica gel impregnated with various inorganic ions, cellulose, polyamide, and amino-modified silica gel are the most popular stationary phases, which permits us to obtain satisfactory separations. Water is usually a mobile phase component because of the high water solubility of sugars. Owing to them the biological impurities are left on the start line. A few dozen chromogenic visualizing reagents can be applied for the visualization of carbohydrates. Reagent-free visualization proposed by Klaus et al. [73] is also very interesting. Sugars separated on amino-modified silica gel give fluorescence spots after heating at 120–150°C.

Prosek and co-workers [74] presented a review on TLC sample preparation, chromatographic systems, detection, and quantitative evaluation of carbohydrates. A breakdown of some interesting examples of TLC applications for clinical researches is presented in Table 6.

3.4 Lipids

The lipids occurring free or bound as lipoproteins comprise complex mixtures of different classes of compounds. They play a vital role in virtually all aspects of biological life. Disturbance in the lipid metabolism of the organism leads to various disorders. Analyses of lipids have therefore a great diagnostic value (Table 7).

Three kinds of adsorbents are recently used for lipid separations, namely classical silica gel, silica gel impregnated (mainly with silver—Ag TLC), and RP-modified silica gel. In case of silica gel separations, lipids with free carboxyl, keto, and hydroxyl groups give lower R_F values than those that are only fatty acid residues when petroleum ether or hexane (main components) and acetone or diethyl ether (polar modifiers) are used as mobile phases. Silica gel impregnated with silver ion is used to separate the molecular species of a single lipid class. It offers an effective means of lipid mixture fractionation into distinct fractions differing in the number of double bounds. Petroleum ether, hexane, toluene, and chloroform are then most often used as the major components of mobile phases. RP systems (usually silica gel with chemically bounded octadecyl and a polar solvent such as acetone, acetonitrile, or water) are rarely used and only for the separation of individual classes of lipids.

Various types of development (one- and two-dimensional and multiple development) are applied. Of course, the use of two-dimensional and multiple TLC is particularly valuable for the separation of complex lipid mixtures. The most popular mobile phases [91] used for the separation of lipids on silica gel are presented in Table 8.

Visualization is frequently carried out by solutions of chromogenic substances applied as spraying reagents. Aniline blue, bromophenol blue, helasol green, and alkaline blue are used to detect cholesteryl esters. Molybdic oxide and phosphomolybdic acid turned out to be very good

TABLE 6 Examples of TLC Application for Carbohydrate Analyses

Analyte	Matrix	Chromatographic system		Visualization and quantification	Ref.
		Stationary phase	Mobile phase (v/v)		
Monosaccharides	Plasma membrane	Polyamide	Ethyl acetate–acetic acid–ethanol (80 : 8 : 10)	Calorimetric, after elution with water	75
Galactose metabolites	Blood	NH_2-silica	Ethyl acetate, acetic acid, methanol, and water mixtures in different percentages	Enzyme immunostaining	76
Oligosaccharides	Gangliosides	Polyamide, NH_2-silica	Ethyl acetate, acetic acid, methanol, and water mixtures in different percentages	Enzyme immunostaining	77
Different sugars	Glycosides	Silica gel	Two-dimensional development: (I) toluene–ethyloacetone–methanol (5 : 5 : 4), (II) toluene–ethyloacetone–methanol–formic acid (5 : 5 : 4 : 0.1)	Aniline-*o*-phthalic acid reagent	78
Carbohydrates	Different biological materials	NH_2-silica	Different solvent systems	Fluorescence at 360 nm after heating at 150°C	79
Reducing sugars	Protein hydrolyzates	Silica gel	Multiple development: (I) butanol–pyridine–water (16 : 5 : 4), (II) ethyl acetate–methanol–acetic acid–water (4 : 1 : 1 : 1)	*o*-Toluidine reagent and heating at 110°C	80
Neutral sugars	Cell walls	Silica gel impregnated with phosphate buffer	Acetonitrile–amyl alcohol–water (6 : 2 : 2)	*N*-(1-Naphthyl)-ethylene diamine reagent and heating at 100°C	81
Sugars and maltodextrins	Human milk, biological fluids	Cellulose	Multiple development: (I) butanol–ethanol–water, (II) pyridine–ethyl acetate–acetic acid–water (5 : 5 : 1 : 3)	Silver nitrate, sodium thiosulfate, or Elson Morgan reagent	82
Oligosaccharides	Human milk	Anion exchanger	Ethyl acetate–acetic acid–water (5 : 5 : 4)	FAB-MS	83

TABLE 7 The Main Goal and Results of Lipid Researches by TLC

Class of lipid	Technique of measurements	The main goal and results of researches	Ref.
Lipids in plasma	TLC separation and ozonization of analyte	The elucidation of a biochemical criterion of the degree of metabolic disorders in children with insulin-dependent diabetes mellitus. It was shown that the pattern of unsaturation distribution in plasma lipid fractions might serve as a new biochemical criterion for metabolic disorders and decompensation in insulin-dependent diabetes mellitus.	84
Corneum lipid	AMD-HPTLC separation and GC determination	Investigation of the internal stratum corneum lipid in relation to depth by extraction following one, three, or five stripping. A decrease in unsaturated free fatty acids and in the unsaturated/saturated chain ratio with depth was observed. A decrease in the ratios of free fatty acids to cholesterol and free fatty acids to ceramides with depth was also observed. Results: a decrease in the ratios of free fatty acids to cholesterol and free fatty acids to ceramides with depth confirmed the diagnostic importance of this level of stratum corneum lipids in skin barrier properties.	85
Cholesterol sulphate	HPTLC separation and densitometric determination	Measurement of blood plasma and erythrocyte membrane samples of patients suffering from diabetes and Down's syndrome. It was postulated that differences in cholesterol sulphate levels might contribute to changes of erythrocyte properties in these pathological states.	86
Salivary lipids	TLC separation and densitometric determination	Research of the role of salivary lipids in oral health. A positive correlation between the body mass index and the level of saliva cholesterol concentration was found. It was shown that, in healthy adults, saliva cholesterol concentration reflects serum concentration to some extent and can be used to select individuals with high serum cholesterol levels.	87

Phospholipid a›d sph▪ngomye in	Immuno-TLC separation and determinations	Determination of pulmonary surfactant phospholipid/sphingomyelin ratio in human amniotic fluids. The method was applied to determine the surfactant phospholipid/sphingomyelin ratio in 20 μL of the amniotic fluids obtained at delivery. The amniotic fluids from women who delivered a baby suffering from respiratory distress syndrome were easily discriminated from the normal amniotic fluids.	88
Sulfatides	HPTLC separation and densitometric determination	Evaluation of sulfatides in the urine of patients with metachromatic leukodystrophy deficiency. The amount of sulfatides is expressed in relation to sphingomyelin, which copurifies with sulfatides and better reflects the level of membrane lipids in urine than commonly used parameters (creatinine, urine volume, etc.). The method is also useful as a complementary analysis for other lipidoses with high excretion of sphingolipids in urine (e.g., Fabry disease).	89
Gangliosides	Two-dimensional HPTLC and GC	Determination of gangliosides profile of cancer patients The profiles of cancer patients were compared to those of the control group, revealing a significant increase in total lipid-bound sialic acid and a specific increase in polysialogangliosides in the patients with breast cancer. An increase was noted in the ratio of gangliosides of the b-series biosynthetic pathway over those of the a-series in the cancer sera, as compared to the controls. No unusual gangliosides were found in the mixture from breast cancer patients.	90

TABLE 8 Mobile Phases for Separations of Lipids on Silica Gel

Class of lipid	Technique of separation	Solvent system (v/v)
Neutral	One-dimensional	Petroleum–ether–diethyl ester–acetic acid (80:20:1)
	Double separation in the same direction	(I) Isopropyl ether–acetic acid (94:6), (II) petroleum ether–diethyl ether–acetic acid (90:10:1)
	Two-dimensional	(I) Hexane–diethyl ether (8:2), (II) hexane–diethyl ether–methanol (7:2:1)
Phospholipids	One-dimensional	Chloroform–methanol–water (65:25:4)
	Double separation in the same direction	(I) Chloroform–methanol–water (65:25:4) (II) hexane–diethyl ether (4:1)
	Two-dimensional	(I) Chloroform–methanol–water (65:25:4), (II) *n*-butanol–acetic acid–water (6:2:2)
Glycolipids	One-dimensional	Chloroform–methanol–water (65:25:4)
	Two-dimensional	(I) Chloroform–methanol–7N ammonium hydroxide (65:30:4), (II) chloroform–methanol–acetic acid–water (170:25:25:6)
Gangliosides	One-dimensional	Chloroform–methanol–2.5 M aqueous ammonia (60:40:9)
	Two-dimensional	(I) Chloroform–methanol–0.2% aqueous $CaCl_2$ (60:35:8), (II) *n*-propanol–water–28% aqueous ammonia (75:25:5)

reagents for phospholipids. Fluorescamine, aluminum chloride, and ferric chloride can be used for analyses of fatty acids. Glycolipids are usually detected with orcinol, sulfuric acid, and α-naphthol, ganglosides with resorcinol. These reagents are suitable both for nonmodified and modified adsorbents. When radiolabeled substrates are analyzed, non-destructive reagents should be used. The I_2 vapors or fluorescamine spraying reagent does not affect the lipid and can be easily removed during the isolation processes. The most sensitive nondestructive dye for lipids also is primuline.

Many papers and review articles on TLC lipid separation are presented in the literature [e.g., 92,93]; some new and representative works on this point and on the usefulness of the proposed solutions are presented below.

3.4.1 Neutral Lipids

Modern pathophysiological researches in the human cell demand the specific analysis of neutral lipids such as cholesterol, cholosetryl esters, and triglycerides. Conventional enzymatic or calorimetric assays, while quite suitable for classical clinical chemistry, are of limited sensitivity and specificity, and the major classes of neutral lipids have to be determined separately. These results can be achieved through TLC; nanogram quantities of cholesterol, cholesteryl esters, fatty acids, and triglycerides can be detected.

Absolute specificity and high accuracy are required for the quantitation of cholesterol in small biological samples, particularly in a limited number of cells. Asmis and coworkers [94] proved that both can be achieved through TLC and phosphomolybdic acid staining, while the shortcomings of traditional spot detection are overcome by laser densitometry. The major advantage of the proposed technique is the concurrent assay of ng quantities of cholesterol, cholesteryl esters, and triglycerides. The assay is at least tenfold more sensitive than common TLC techniques and at least fourfold more sensitive than common enzymatic methods. The proposed low-cost assay is highly reproducible and may be particularly suitable for routine lipid analysis of a 10% aliquot of relatively small tissue and cell samples.

Nonesterified fatty acids (''free'' fatty acids) are usually not present in the free form. They occur in blood predominantly in association with albumins to which they are bound by electrostatic forces rather than covalent bonds. They play a significant role in the physiological control of carbohydrate metabolism. Serum lipid fatty acid compositions have been determined in diabetes, coronary disease, and artery disease, as well as renal disease. Esterified fatty acids are components of complex lipids such as glycerides and phospholipids. Sattler and coworkers presented [95] a simple, accurate, and fast procedure for quantitative analysis of fatty acids in simple lipid subclasses from different biological specimens. TLC fractionated lipid extracts of isolated plasma lipoproteins (very low-, low-, and high-density lipoproteins) into lipid subclasses on silica gel plates. Bands comigrating with lipid standards were scraped off under argon and subjected to direct in situ transesterification with BF_3/MeOH in the presence of the TLC adsorbent. Fatty acid methyl esters were subsequently extracted and quantified by capillary gas chromatography. Miwa and coworkers described [96] a study on the assay of fatty acid compositions of individual phospholipids from non-insulin-dependent diabetes mellitus patients. Eight major phospholipids were separated by a TLC method with a one-dimensional developing system without any pretreatment of the plate. The fatty acids incorporated into each phospholipid class were analyzed by an improved HPLC method with a simple elution system, advantageous with respect to resolution and analysis time. The fatty acid compositions of individual phospholipids in platelets were investigated following administration of ethyl cis-5,8,11,14, 17-eicosapentaenoate for more than 13 weeks to patients with non-insulin-dependent diabetes mellitus. The cis-5,8,11,14,17-eicosapentaenoic acid compositions of all phospholipid classes were significantly increased with decreasing platelet aggregation rates after administration. These results suggested that the method provides the complete separation of individual phospholipids in sufficient amounts to allow fatty acid analysis on the isolated phospholipid moieties.

Alvarez et al. [97] described a test for the determination of dipalmitoyl phosphatidylcholine (as free dipalmitoylglycerol) in amniotic fluid. Aliquots of amniotic fluid were hydrolyzed with *Bacilus cereus* phospholipase C, and the resulting diglycerides were analyzed by $AgNO_3$-modified HPTLC reflectance densitometry. This system provided resolution of dipalmitoyl phosphatidylcholine and palmitoylpalmitoleoylglycerol from other 1,2-diglycerides and cholesterol. The turnaround analysis time for triplicate aliquots of amniotic fluid was 40 min. Recoveries ranged between 90 and 98%. The authors suggested that the method provides a quantitative, specific, highly reproducible, and fast turnaround means of dipalmitoyl phosphatidylcholine analysis in amniotic fluid.

Multistep TLC-MS has been developed for the quantification of neutral lipids and several phospholipids extracted from mammal cells and sera [98]. The lipid classes were separated by a TLC procedure in different solvent systems. Resolved lipid bands were visualized by the lipophilic dye primulin and scanned by an automated laser-fluorescence detector. The mass of each band was determined by comparison band intensities of unknown samples with dilution curves of standards. The majority of biological lipids could be resolved and quantified with these methods. Since the detection method is nondestructive, purified lipids could then be recovered by scraping the visualized bands and extracting the lipids from the silica. MS extracted lipids were also hydrolyzed to release acyl chains and acyl chain species and then determined by GC, which confirmed the structural identities of the recovered lipids. In contrast to classical two-dimensional TLC (which allows a good resolution of some lipid species but cannot be used to analyze more than a single experimental point per plate), multistep TLC allows the direct comparative analysis of multiple samples on a single TLC plate. The method provides a good resolution for the quantification of most classes of lipid species.

3.4.2 Complex Lipids

The researches on complex lipids have seen a considerable increase in recent years. Glycolipids have been reported to be associated with differentiation, development, and organogenesis. These glycolipids have an active role in the formation of cataracts. The glycolipid biosynthesis pathway is initiated by the glucosyltransferase-catalyzed synthesis of glucosylceramide. The *n*-butyl-deoxynojisimycin is an inhibitor of this synthesis and has been shown to be an inhibitor of HIV replication in vitro. After the pioneering work of Gluck et al. [99,100] on the lecithin/sphingomyelin (*L*/*S*) ratio (the method was used for prediction of respiratory distress syndrome), TLC has become the most widely used technique for complex lipid analysis. The method has become the standard tool for resolution of ganglioside mixtures for analytical and preparative applications. The review by Muthing [101] summarizes the application of TLC in the analysis of gangliosides up to 1995. Basic general techniques and special advice are given for successful separation of glycosphingolipids. New approaches concerning continuous and multiple development, and several preparative TLC methods, are also included. Emphasis is placed on TLC immunostaining and related techniques, i.e., practical applications of carbohydrate-specific antibodies, toxins and bacteria, viruses, lectins, and eukaryotic cells.

A new and simple method for purifying phospholipids and sphingolipids by using "TLC blotting" was established by Taki and coworkers [102]. Glycosphingolipids separated by two-dimensional thin-layer chromatography (TLC) were made visible with a primuline reagent, and then bands were marked with a colored pencil. TLC blotting to a polyvinylidene difluoride membrane, together with the color marks, then transferred the glycosphingolipids that separated onto the HPTLC plate. The marked areas were scraped, after which their glycosphingolipids were extracted and monitored by TLC. By this method, 20 glycosphingolipids showing homogeneous bands on an HPTLC plate were isolated from the neutral glycosphingolipid fraction of human meconium. Moreover, 10 kinds of acidic glycosphingolipids were purified as homogeneous bands from the bovine acidic glycosphingolipid fraction. The yields of glycosphingolipids (13 different ones) ranged from 68 to 92%, the mean value was 82.3%. The authors suggested that the same procedure could also be used to purify phospholipids.

Müthing and Cacic [103] described the comparative TLC immunostaining investigation of neutral glycosphingolipids (GSLs) and gangliosides from human skeletal and heart muscle. A panel of specific polyclonal and monoclonal antibodies as well as the GM1-specific choleragenoid was used for the overlay assays, combined with preceding neuraminidase treatment of gangliosides on TLC plates. This approach proved homologies but also quantitative and qualitative differences in the expression of ganglio, globo, and neolacto series neutral GSLs and gangliosides in these two types of striated muscle tissue within the same species. The main neutral GSL in skeletal muscle was LacCer, followed by GbOse3Cer, GbOse4Cer, nLcOse-4Cer, and monohexosylceramide, whereas in heart muscle GbOse3Cer and GbOse4Cer were the predominant neutral GSLs beside small quantities of LacCer, nLcOse4Cer, and monohexosylceramide. No ganglio series neutral GSLs and no Forssman GSL were found in either muscle tissue. GM3(Neu5Ac) was the major ganglioside, comprising almost 70% in skeletal and about 50% in cardiac muscle total gangliosides. GM2 was found in skeletal muscle only, while GD3 and GM1b-type gangliosides (GM1b and GD1 alpha) were undetectable in both tissues. GM1a-core gangliosides (GM1, GD1a, GD1b, and GT1b) showed somewhat quantitative differences in each muscle; lactosamine-containing IV3Neu5Ac-nLcOse4Cer was detected in both specimens. Neutral GSLs were identified in TLC runs corresponding to e.g. 0.1 g muscle wet weight (GbOse3Cer, GbOse4Cer), and gangliosides GM3 and GM2 were elucidated in runs which corresponded to 0.2 g muscle tissue. Only 0.02 g and 0.004 g wet weight aliquots were necessary for unequivocal identification of neolacto type and GM1 core gangliosides, respectively. Muscle is known for the lowest GSL concentration from all vertebrate tissues studied up to now. Using the overlay technique, reliable GSL composition could be revealed, even from small muscle probes on a suborcinol and subresorcinol detection level.

Analyses of phospholipids and their fatty acid composition from human intestinal mucosa were performed by a method elaborated to analyze the limited amount of sample with two-dimensional TLC followed by lipid–phosphorus determination [104]. Using this method, plasmenylethanolamine was detected in human intestinal mucosa and accounted for about 7% of phospholipid in small and large intestinal mucosa. The amounts of polyunsaturated fatty acids of phosphatidylethanolamine were higher than those of other phosphoglycerides in intestinal mucosa; hence inflammation-related eicosanoids may originate from ethanolamine containing phospholipid.

Guittard et al. reported [105] results for analysis by matrix-assisted laser desorption/ionization time-of-flight mass spectrometry (MALDI-TOFMS) of native glycosphingolipids (GSLs) after development on thin layer chromatographic plates and after heat transfer of the GSLs from the plates to several types of polymer membranes. The spectral quality is better for membrane-bound analytes, in terms of sensitivity, mass resolution, and background interference. The sensitivity gain compared with liquid secondary ion mass spectrometry (LSIMS) of GSLs on thin-layer plates is 1 or 2 orders of magnitude (detection limits of 5–50 pmol vs. 1–10 nmol). Resolution and mass accuracy (0.1%) are limited by the irregular membrane surfaces, and this effect cannot be entirely compensated by delayed extraction. The best results were obtained with a polyvinylidene difluoride (PVDF) P membrane, with irradiation from a nitrogen laser. Although the Nafion membrane could not be used for molecular weight profiling, its acidic character led to sample hydrolysis at the glycosidic linkages, thus yielding a series of fragments that could be used to determine the sequence of carbohydrate residues. Structural information could also be obtained by postsource decay (PSD) experiments on mass-selected precursor ions. Samples containing both neutral and acidic components were characterized in a 1:1 combination of 2,5-dihydroxybenzoic acid and 2-amino-5-nitropyridine. GSLs that exhibited binding to antibodies in an overlay assay on the TLC plate were transferred to membranes and analyzed by MALDI-TOFMS without interference from the antibody or the salts and buffers used during the binding and visualization steps. Taking advantage of the insights into sample preparation gained from these studies, future research will extend this approach to analysis by matrix-assisted laser desorption/ionization Fourier transform ion cyclotron resonance mass spectrometry (MALDI-FTICRMS) with an external ion source.

Liu et al. compared [106] two analytical methods (TLC and IR spectroscopy) for prediction of respiratory distress syndrome (RDS) from amniotic fluid analysis. Samples of amniotic fluid were obtained by amniocentesis from 86 patients between the 28th and the 41st week of pregnancy. A small volume, 35 μL, of amniotic fluid was used to acquire infrared spectra with a commercial spectrometer. The *L*/*S* ratio was determined separately by TLC. A calibration model was established using a partial least square regression analysis, which quantitatively correlated 145 IR spectra with the TLC-based *L*/*S* ratios; the correlation coefficient was 0.95. The model was then validated using a total of 111 spectra, which also showed a high correlation coefficient ($r = 0.91$). Based upon the clinical data associated with these samples, the prediction accuracy for the presence of RDS was 67% for TLC and 83.3% for IR. The accuracy for predicting the absence of RDS was 93.4% for TLC and 96.8% for IR. The authors concluded that IR spectroscopy may become the clinical method of choice for predicting RDS from amniotic fluid analysis.

The chromatographic behavior of molecular species of sphingomyelin on HPTLC was investigated by Ramstedt et al. [107]. Sphingomyelin gave a double band pattern on HPTLC plates developed using chloroform/methanol/acetic acid/water (25:15:4:2, v/v) or chloroform/methanol/water (25:10:1.1, v/v). HPTLC analysis of acyl chain–defined sphingomyelins showed that the R_F values increased linearly with the length of the *N*-linked acyl chain. A double-banded pattern was therefore seen for natural sphingomyelins with a bimodal fatty acid composition. Racemic sphingomyelins also gave a double band pattern on HPTLC, where the lower band represented the D-erythro and the upper bands the L-threo isomer. It was showed that D-erythro-*N*-16:0-dihydrosphingomyelin migrated faster on HPTLC than D-erythro-*N*-16:0-sphingomyelin. The upper and lower band sphingomyelins from two different cell lines (human skin fibroblasts and baby hamster kidney cells) were separately scraped off the HPTLC plates, and the fatty acid and long-chain base profiles were studied using GC-MS. The lower bands contained short-chain fatty acids and most of the fatty acids in the upper bands were long. The predominant long-chain base was sphingosine, which was found in both upper and lower bands, but sphinganine was found only in the upper bands. To conclude, there are at least three possible reasons for the sphingomyelin double bands on HPTLC; acyl chain length, long-chain base composition, and stereochemistry. These reasons might sometimes overlap, and therefore HPTLC alone is insufficient for complete analysis of the molecular species of sphingomyelin.

A method of simultaneous determination of ceramides and 1,2-diacylglycerol in tissues was developed using the latroscan, which combines TLC-FID techniques [108]. Because of the relatively low amounts of these components in tissues, the fraction of nonpolar lipids, which includes ceramides and glycerides, was eluted with a chloroform/acetone mixture (3:1, v/v) through a silica acid column to eliminate the polar phospholipids. Development was carried out using three solvent systems in a four-step development technique. The relationship of the peak area ratio to the weight ratio compared with cholesteryl acetate added as an internal standard was linear. The amount of ceramides increased with incubation of rat heart homogenate and human erythrocyte membranes in the presence of sphingomyelinase (E.C. 3.1.4.12). The latroscan TLC-FID system provided a quick and reliable assessment of ceramides and 1,2-diacylglycerol.

Enhancement in separation of gangliosides on HPTLC plates has been obtained by automated multiple development chromatography [109]. A less polar mixture of the standard solvent chloroform–methanol–20 mmol aqueous $CaCl_2$ (120:85:20, v/v) was used. Lowering the water content achieved separation of two complex monosialoganglioside fractions, isolated from murine YAC 1 T lymphoma and MDAY-D2 lymphoreticular cells. Threefold chromatography in the solvent chloroform–methanol–20 mmol aqueous

$CaCl_2$ (120:85:14, v/v) resulted in TLC separation of GM1b-type gangliosides, substituted with C24 and C16 fatty acids and with Neu5Ac and Neu5Gc as well, which could not be achieved by unidirectional standard chromatography. In comparison with conventional single chromatography, the technique described allows high-resolution separation of extremely heterogeneous ganglioside mixtures and offers a convenient tool for both analytical and preparative TLC.

An efficiency assessment of a ganglioside assay procedure was carried out on human serum gangliosides from healthy subjects of different sexes and ages [110]. The analysis of the gangliosides extracted with chloroform/methanol and purified by lipid partitioning, ion exchange column chromatographic separation, and desalting procedures was performed by HPTLC followed by densitometric quantification. The yield of the procedure, expressed as radioactivity recovery, was determined by adding GM3 ganglioside, tritium labeled at the sialic acid acetyl group and at the C3 position of sphingosine, to the lyophilized serum or by associating it with the serum lipoproteins. Although the extraction and purification procedures were performed exactly as described, we found the radioactivity recovery to be variable (25–50%) and much lower than that proposed. Much of the radioactivity was found in the organic phase after lipid partitioning, while all the ganglioside purification steps led to some further loss. The recovery improved, after the introduction of some modifications to the procedure, reaching 67–79%. The analyses on 33 samples of 5 mL showed a human serum ganglioside content of about 10 nmol/mL (as corrected for the recovery) and confirmed that GM3 ganglioside is the main component of the total serum ganglioside mixture.

A human strain of influenza virus (A, H1N1) was shown to bind in an unexpected way to leukocyte and other gangliosides in comparison with avian virus (A, H4N6) as assayed on TLC plates [111]. The human strain bound only to species with about 10 or more sugars, while the avian strain bound to a wide range of gangliosides including the 5-sugar gangliosides. By use of specific lectins, antibodies, and FAB and MALDI-TOF MS, an attempt was made at preliminary identification of the sequences of leukocyte gangliosides recognized by the human strain. The virus binding pattern did not follow binding by VIM-2 monoclonal antibody and was not identical with binding by anti-sialyl Lewis x antibody. There was no binding by the virus of linear NeuAcalpha3- or NeuAcalpha6-containing gangliosides with up to seven monosaccharides per mol of ceramide. Active species were minor NeuAcalpha6-containing molecules with probably repeated HexHexNAc units and fucose branches. This investigation demonstrated marked distinctions in the recognition of gangliosides between avian and human influenza viruses. The data emphasize the importance of structural factors associated with more distant parts of the binding epitope and the complexity of carbohydrate recognition by human influenza viruses.

3.5 Other Applications

The analysis of porphyrins is important for diagnosis of the porphyries, diseases due to enzyme deficiencies in the heme biosynthetic way (porphyrins are formed as intermediates in the synthesis of heme). TLC is the most widely used technique for the routine analysis of porphyrins. Porphyrins were extracted from urine or feces and separated on silica gel with methanol, chloroform, methylene chloride, carbon tetrachloride, ethyl acetate, benzene, and toluene mixtures. The distinct porphyrin bands were observed by viewing the plate under longwave fluorescent light. Luo and Lim [112] have studied by HPTLC, HPLC, and LSI MS the porphyrin metabolisms in human porphyria cutanea tarda (PCT) and in rats treated with hexachlorobenzene (HCB). The analyses of porphyrin metabolites in the urine, feces, and liver biopsies of patients with PCT have shown that apart from uroporphyrin I and III and their expected decarboxylation intermediates and products, a complex mixture of many other porphyrins are present. The new porphyrins (meso-hydroxyuroporphyrin III, beta-hydroxypropionic acid uroporphyrin III, hydroxyacetic acid uroporphyrin III, peroxyacetic acid uroporphyrin III, beta-hydroxyproionic acid heptacarboxylic acid porphyrin III, hydroxyacetic acid hepatocarboxylic porphyrin III, and peroxyacetic acid pentacarboxylic porphyrin III) were identified.

Estimation of neopterin in urine has become part of the examination in phenylketonuria, malignities, and immunodeficiency including AIDS and HIV infection. Tomsova and Juzova [113] described a simple chromatographic method for estimation of neopterin, which does not call for HPLC or imported kits. The di- and tetra-hydroforms of neopterin are oxidized with MnO_2 and stable neopterin is obtained. The specimen is purified on a Dowex 50WX4 column and on C18 columns. Then thin-layer chromatography (silica gel 60 F_{254}) was used in a mobile phase of ethyl acetate/isopropanol/1 n NH_4OH (35:45:25, v/v). Fluorescence was assessed on a densitometer; it was linear within the 5–200 ng range. The method has also been sufficiently sensitive for the estimation of normal neopterin excretion. The authors submit the results of estimations in controls, immunopathies, and AIDS.

Maddocks and MacLachlan [114] used TLC for homocysteine detection. A new fluorescent thiol reagent, dansyl-aminophenylmercuric acetate (DAPMA), was applied to the diagnosis of homocystinuria. A disorder of homocysteine can be associated with vascular disease at an early age. DAPMA was added to urine containing metabisulphite, and the resulting fluorescent derivatives were extracted on a cyclohexyl silica column and separated by thin-layer chromatography. One hundred two coded samples were tested. The derivative of homocysteine was easily identified in samples from four children with homocystinuria but was absent from all samples from normal subjects and patients with unrelated disorders. Other thiols (cysteine, acetylcysteine, mercaptolactate, thiosulphate, and thiocyanate) were also identified in urine from healthy fasting subjects.

Steinberg et al. [115] analyzed DNA adducts by TLC (adducts occur through environmental, therapeutic, dietary, oxygen stress, and aging processes). The authors proved that a TLC technique can assess base composition and adduct formation. This requires labeling DNA by ''shot-gun'' 5′-phosphorylation of representative ^{32}P-α-deoxyribonucleotide monophosphates. Subsequent 3′-monophosphate digest ''sister exchanges'' a radioactive $^{32}PO_4^{(2-)}$ to the neighboring cold nucleotide. Separation in two-dimensional polyethyleneimine-cellulose TLC is carried out in acetic acid, $(NH_4)_2SO_4$, and $(NH_4)HSO_4$. The technique was applied to control DNA, cold substitution of dUMP, methylation, depurination, and pBR322. This technique quantifies low molecular mass adducts and DNA integrity both in vivo and in vitro.

Reddy [116] reviewed the ^{32}P-postlabeling methodology for analysis of DNA adducts derived from carcinogens containing one aromatic ring (e.g., safrole, styrene oxide, benzene metabolites, 1-nitrosoindole-3-acetonitrile) or a bulky nonaromatic moiety (e.g., mitomycin C, diaziquone). Six steps are involved: digestion of DNA to 3′-nucleotides, enrichment of adducts, 32P-labeling of adducts, separation of labeled adducts by TLC, detection, and quantitation. The first step, DNA digestion with micrococcal nuclease and spleen phosphodiesterase, is applicable to DNA modified with most carcinogens independent of their size and structure. Of the two commonly used procedures for enrichment of aromatic adducts in DNA digests, the nuclease P1 treatment is substantially more effective than butanol extraction for small aromatic and bulky nonaromatic adducts. For initial purification of these adducts from unadducted material after ^{32}P-labeling, multidirectional polyethyleneimine (PEI)–cellulose TLC using 1 M sodium phosphate, pH 6.0, as the D1 solvent is not suitable, because they are not retained on PEI-cellulose under these conditions. A higher concentration of sodium phosphate (e.g., 2.3 M) or development with D1 and D3 solvents in the same direction helps to retain adducts of safrole and of benzene metabolites. Also, transfer of adducts from multiple cutouts above the origin after D1 chromatography, as adopted for analysis of I-compounds, is potentially applicable. However, initial purification by reverse-phase TLC, followed by in situ transfer and resolution by PEI-cellulose TLC, has been found to be most effective for these adducts. Reverse-phase TLC at 4°C or in a stronger salt solution further improves the retention of some adducts (e.g., mitomycin C and diaziquone adducts). For adduct separation by PEI-cellulose TLC, salt solutions with or without urea are used.

The halopyrimidine 5-bromo-2′-deoxyuridine (BUDR) can serve as one of many indicators of tumor malignity, complementary to histologic grade. Steinberg et al. developed a TLC technique [117] that can assess tumor DNA base composition and analog BUDR incorporation, which vies with immunochemistry for BUDR. This requires postlabeling DNA by nick-translation and radioactive 5′-phosphorylation of representative ^{32}P-α-dNMPs (deoxynucleotide monophosphates). Subsequent 3′-monophosphate digest exchanges a radioactive $^{32}PO_4$ for the neighboring cold nucleotide. Separation in two-dimensional PEI-cellulose TLC is carried out in acetic acid, $(NH_4)_2SO_4$, and $(NH_4)HSO_4$. TLC of dNMPs was applied to control HeLa DNA, and HeLa cells receiving BUDR. BUDR is detected in 10(6) HeLa cells after 12–72 h incubation. Findings in HeLa DNA demonstrate normal TLC retention factors for all ^{32}P-dNMPs. This technique quantifies BUDR, which parallels the tumor S phase and serves as an indicator of the labeling index.

A rapid immunochromatographic method for qualitative and quantitative analysis of protein antigens was described [118]. The method is based on the ''sandwich'' assay format using monoclonal antibodies (Mabs) of two distinct specificities. Mabs of one specificity are covalently immobilized to a defined detection zone on a porous membrane, while Mabs of the other specificity are covalently coupled to blue latex particles, which serve as a label. The sample is mixed with the Mab-coated particles and allowed to react. The mixture is then passed along a porous membrane by capillary action past the Mabs in the detection zone. The zone binds the particles, which have antigen, bound to their surface, giving a blue color. Analysis is complete in less than 10 min, requires a minimum amount of sample (4 μL), and has a detection limit below the nanomolar range for the antigen we studied, human chorionic gonadotropin. Identification of protein phosphorylation sites is essential in order to evaluate the contribution of individual sites to the regulation of a particular protein by phosphorylation. Van der Geer and Hunter developed [119] a method for the identification of phosphorylation sites. The method is based on the digestion of ^{32}P-labeled proteins with site-specific proteases and separation of the digestion products in two dimensions on thin-layer cellulose plates using electrophoresis in the first dimension followed by chromatography. This method is very sensitive, requiring only a few hundred ^{32}P-disintegrations per minute to obtain reproducible phosphopeptide maps. Laitinen et al. showed [120] the possibility of the application of TLC to the analysis of protein–ligand affinity.

A HCHO level in cells of animal and human tissues as well as in body fluids depends on the physiological state of an organism. Rożyło et al.

[121,122] decided to find out if there are changes of HCHO level in different physiological and pathological hard tissues of teeth. Obtained results showed that there was some regularity in the level of HCHO as far as similar physiological or pathological states are concerned. This was best seen in comparison with the obtained results with mean HCHO level of the studied groups of teeth. The increase of formaldehyde level in some rare pathological cases of teeth was observed.

A TLC method for the separation of heavy metals and their complexes with dithizone, 4-(2-pyridylazo) resorcinol, and ethylenediaminetetraacetic acid was devised, and conditions for the solid-phase extraction of heavy metal ions in human bones, in placenta and milk after microwave mineralization, and in air after alkaline absorption were elaborated [123]. In the corrosively aggressive medium of the oral cavity, the use of identical dental alloys requires identification of the existing metal construction. One of the methods allowing this identification is TLC with anodic sampling [124]. Using a 4.5 V battery and suitable electrolytes, seven dental alloys for fixed and removable dentures based on cobalt were analyzed. Chromatograms of alloy samples were developed with a mixture of acetone and 2 M HCl. Scanning of the TLC spots produced chromatographic curves, and the area under the curve was proportional to the content of cobalt in the alloy studied. Regression analysis showed a very high coefficient of correlation ($r = 0.999$) between the area of the spot and the proportion of cobalt.

The interaction of 28 commercial pesticides with human and bovine serum albumin as well as with egg albumin was studied by charge-transfer reverse-phase thin-layer chromatography, and the relative strength of the interaction was calculated [125]. It turned out that only one pesticide interacted with egg albumin, whereas the majority of pesticides bound to both bovine and human serum albumins. Stepwise regression analysis proved that the hydrophobicity parameters of pesticides exert a significant impact on their capacity to bind to serum albumins. These findings support the hypothesis that the binding of pesticides to albumins may involve hydrophilic forces occurring between the corresponding apolar substructures of pesticides and amino acid side chains. No linear correlation was found between the capacities of human and bovine serum albumins to bind pesticides. TLC and twelve color reactions were proposed for identification of foreign pesticides [126]. Solubility of the agents was studied, optimal common extracting selected, and a universal method for isolation and identification of 19 new pesticides tried on the liver. The sensitivity and specificity of the method was assessed.

Analysis of mycolic acids by TLC [127] is employed by several laboratories worldwide as a method for fast identification of mycobacteria. This method was introduced in Brazil in 1992 as a routine identification technique. The method allowed earlier differentiation of *M. avium* complex-MAC (mycolic acids I, IV, and VI) from *M. simiae* (acids I, II, and IV), both with similar biochemical properties. The method also permitted to distinguish *M. fortuitum* (acids I and V) from *M. chelonae* (acids I and II) and to detect mixed mycobacterial infection cases as *M. tuberculosis* with MAC and *M. fortuitum* with MAC. Experience shows that mycolic acid TLC is an easy, reliable, fast, and inexpensive method, an important tool to put together conventional mycobacteria identification methods.

A TLC method has been developed for confirming results from the determination of aflatoxin M_1 in human urine [128]. Urine samples were cleaned on immunoaffinity columns and analyzed by means of immunochemical methods. A result higher than 5 ng/L urine was confirmed by instrumental HPTLC on silica gel layers with fluorescence detection. Chloroform/acetone/2-propanol (85:10:5, v/v) was used as a mobile phase. The chromatogram was scanned in reflectance mode at $\lambda = 366$ nm with a $\lambda = 400$ nm measuring filter. Twofold enhancement of the sensitivity of the HPTLC method was achieved by immersion of the chromatographic plate in a solution of paraffin oil in hexane. Recoveries were 75–85% in the range 20–100 ng/L urine. The limit of quantification in urine was 5 ng/L. Validation of the method was performed according to the principles used for HPTLC methods.

An autoradiographic technique for quantification of ^{10}B containing compounds used for neutron capture therapy is described [129]. Instead of applying solutions of $Cs_2B_{12}H_{11}SH$ and its oxidation products directly to solid-state nuclear track detectors, diethylaminoethyl cellulose TLC plates were utilized as sample matrices. The plates are juxtaposed with Lexan polycarbonate detectors and irradiated in a beam of thermal neutrons. The detectors are then chemically etched, and the resultant tracks counted with an optoelectronic image analyzer. Sensitivity to boron-10 in solution reaches the 1 pg/μL level, or 1 ppb. In heparinized blood samples, 100 pg ^{10}B/μL was detected. This TLC matrix method has the advantage that sample plates can be reanalyzed under different reactor conditions to optimize detector response to the boron-10 carrier material. Track etch/TLC allows quantification of the purity of boron neutron capture therapy compounds by utilizing the above method with TLC plates developed in solvent systems that resolve $Cs_2B_{12}H_{11}SH$ and its oxidative analogs. Detectors irradiated in juxtaposition to the thin-layer chromatograms are chemically etched, and the tracks are counted in the sample lane from the origin of the plate to the solvent front. A graphic depiction of the number of tracks per field yields a quantitative analysis of compound purity.

REFERENCES

1. B Fried, J Sherma. Thin-Layer Chromatography: Techniques and Applications. 4th ed. New York: Marcel Dekker, 1999.
2. JC Touchstone. Planar Chromatography in the Life Sciences. New York: John Wiley, 1990.
3. R Jain. Thin-Layer Chromatography in Clinical Chemistry. In: B Fried, J Sherma, eds. Practical Thin-Layer Chromatography: A Multidisciplinary Approach. Boca Raton, FL: CRC Press, 1996, pp 131–152.
4. J Bladek, A Zdrojewski. Clinical Chemistry: Thin-Layer (Planar) Chromatography. In: ID Wilson, C Cooke, and C Poole, eds. Encyclopedia of Separation Science. London: Academic Press, 2000, pp 2475–2484.
5. HJ Issaq, DE Jaenchen. High-performance thin-layer chromatography in clinical laboratory. Am Lab 23:44D–I, 1991.
6. DJ Anderson, F Van Lente. Clinical chemistry. Anal Chem 67:377R–524R, 1995.
7. E Dworzak, H Hauk. Quantitative determination of metabolism of catecholamines and serotonin through fluorescence quenching on thin-layer chromatography. J Chromatogr 61:162–164, 1971.
8. T Momose, M Mure, T Iida, J Goto, T Nambara. Method for the separation of unconjugates and conjugates of chenodeoxycholic acids and deoxycholic acids by two-dimensional reversed-phase thin-layer chromatography with methyl β-cyclodextrin. J Chromatogr 811:171–180, 1998.
9. M Kindel, H Ludwig-Koehn, M Lembcke. New and versatile method for determination of fecal bile acids by thin-layer chromatography and in situ spectrofluorimetry. Anal Biochem 134:135–138, 1989.
10. JA Flick, RL Schnaar, JA Perman. Thin-layer chromatographic determination of urinary excretion of lactulose, simplified and applied to cystic fibrosis patients. Clin Chem 33:1211–1215, 1987.
11. C Leray, M Andriamampandry, G Gutbier, J Cavadenti, C Klein-Soyer, C Gachet, JP Cazenave. Quantitative analysis of vitamin E, cholesterol and phospholipid fatty acids in a single aliquot of human platelets and cultured endothelial cells. J Chromatogr B: Biomed Appl 696:33–42, 1997.
12. S Serizawa, T Nagai, Y Sato. Simplified method of determination of serum cholesterol sulfate by reverse phase thin-layer chromatography. Journal of Investigative Dermatology 89:580–587, 1987.
13. J Müthing, D Heitmann. Nondestructive detection of gangliosides with lipophilic fluorochromes and their employment for preparative high-performance thin-layer chromatography. Anal Biochem 208:121–124, 1993.
14. J Krahn. Variables affecting resolution of lung phospholipids in one-dimensional thin-layer chromatography. Clin Chem 33:135–137, 1987.
15. SK Wadman, PK de Bree, M Duran, HF de Jonge. Detection of inherited adenylosuccinase deficiency by two-dimensional thin layer chromatography of urinary imidazoles. Advances in Experimental Medicine and Biology 195:21–25, 1986.

16. MJ Henderson. Thin-layer chromatography of free porphyrins for diagnosis of porphyria. Clin Chem 35:1043–1044, 1989.
17. K Tyrpień, P Bodzek, G Manka. Application of planar chromatography to the determination of cotinine in urine of active and passive smoking pregnant women. Biomed Chromatogr 15:50–55, 2001.
18. TR Roesel, BD Kahan. Thin-layer chromatographic detection of cyclosporine and its metabolites in whole blood using rhodamine B and α-cyclodextrin. Transplantation 43:274–281, 1987.
19. K Wolff, MJ Sanderson, AW Hay. A rapid horizontal TLC method for detecting drugs of abuse. Ann Clin Biochem 27:482–488, 1990.
20. I Baranowska, J Baranowski, I Norska-Borowka, C Pieszko. Separation and identification of metals in human bones, placenta and milk and in air by adsorption and ion exchange thin-layer chromatography. J Chromatogr 725:199–202, 1996.
21. JH Hildebrand, RL Scott. The Solubility of Nonelectrolytes. New York: Dover, 1964.
22. LR Snyder. Classification of the solvent properties of common liquids. J Chromatogr 92:223–230, 1974.
23. G Romano, G Carusi, G Musumarra, D Pavone, G Cruciani. Qualitative organic analysis. Part 3. Identification of drugs and their metabolites by PCA of standarized TLC data. J Planar Chromatogr Modern TLC 7:233–239, 1994.
24. H Jork, W Funk, W Fischer, H Wimmer. Thin-Layer Chromatography. Reagents and Detection Methods. Volume 1a: Physical and Chemical Detection Methods. Weinheim: WCH, 1990.
25. H Jork, W Funk, W Fischer, H Wimmer. Thin-Layer Chromatography. Reagents and Detection Methods. Volume 1b: Activation Reactions Reagent Sequences. Weinheim: WCH, 1994.
26. G Glauninger, K-A Kovar, V Hoffmann. Untersuchungen zum online DC-IR Transfer. In: G Gauglitz, ed. Software Entwicklung in der Chemie 3, Proc 3rd Workshops Comput Chem. Berlin: Springer, 1989, pp. 171–180.
27. SA Stahlmann. Ten-year report on HPTLC-FTIR online coupling. J Planar Chromatogr 1(12):5–12, 1999.
28. DT Isbell, AI Gusev, NI Taranenko, CH Chen, DM Hercules. Analysis of nucleotudes directly from TLC plates using MALDI-MS detection. Fresenius' J Anal Chem 365(7):625–630, 1999.
29. A Crecelius, MR Clench, DS Richards, J Mather, V Parr. Analysis of UK-224,671 and related substances by thin-layer chromatography–matrix-assisted laser-desorption ionization-time of flight mass spectrometry. J Planar Chromatogr TLC 13(2):76–81, 2000.
30. JT Mechl, DM Hercules. Direct TLC-MALDI coupling using a hybrid plate. Anal Chem 72(1):68–73, 2000.
31. JT Wu. Screening for inborn errors of amino acid metabolism. Ann Clin Lab Sci 21:123–126, 1991.
32. MA Edwards, S Grant, A Green. A practical approach to the investigation of amino acid disorders. Ann Clin Biochem 25:129–132, 1988.

33. Z Deyl. Profiling of amino acids in body fluids and tissues by means of liquid chromatography. J Chromatogr 379:177–250, 1986.
34. R Bhushan. Amino acids and their derivatives. J Chromatogr Sci 55:353–387, 1991.
35. L Li, G Shen, J Jiang, X Cheng, X Zhou, Ch Zhang, G Zhao. Thin-layer chromatographic determinations of the ratio of serum free branched chain amino acids to aromatic amino acids. Chinese J Med. 68:298–301, 1985.
36. E Neuffeld, HL Goren, D Boland. Thin-layer chromatography can resolve phosphotyrosine, phosphoserine and phosphothreonie in a ptotein hydrolyzate. Anal Biochem 177:138–143, 1989.
37. H Ohtake, Y Hase, K Sakemoto, T Oura, Y Wada, H Kodama. A new method to detect homocysteine in dried blood spots using thin-layer chromatography. Screening 4:17–26, 1995.
38. B Bushan, VK Mahesh, A Varma. Improved thin-layer chromatographic resolution of PTH amino acids with some new solvent systems. Biomed Chromatogr 8:69–72, 1994.
39. DC De Maglio, GK Svanberg. A thin-layer chromatographic procedure for the separation of proline and hydroxyproline from biological samples. J Liq Chromatogr 19:969–975, 1996.
40. E Tyihak, G Katay, Z Ostorics, E Mincsovics. Analysis of amino acids by personal OPLC instrument: 1. Separation of the main protein amino acids in double-layer system. J Planar Chromatogr 11:5–11, 1998.
41. NA Yahya, Z Ismail, KH Embong, SA Mohamad. High performance liquid chromatography (HPLC) method for confirming thin layer chromatography (TLC) findings in inborn errors of metabolism children in Malaysia. Southeast Asian Journal of Tropical Medicine and Public Health 26(suppl 1):130–133, 1995.
42. I Zakiah, YN Ashikin, SM Aisia, HI Ismail. Inborn errors of metabolic diseases in Malaysia: a preliminary report of maple syrup urine diseases for 1993. Southeast Asian Journal of Tropical Medicine and Public Health 26(suppl 1):134–136, 1995.
43. E Marklova. The role of TLC in the screening of inherited metabolic disease. Chromatographia 45:195–198, 1997.
44. E Marklova, H Makovickova, I Krakorova. Screening for defects in tryptophan metabolism. J Chromatogr 870:289–293, 2000.
45. ID Wilson. Thin-layer chromatography: a neglected technique. Therapeutic Drug Monitoring 18:484–92, 1996.
46. International Association of Forensic Toxicology. Report VII of the DFG Commission for Clinical-Toxicological Analysis (special issue of the TIAFT bulletin). Weinheim: VCH, 1987.
47. DM Steinberg, LJ Sokoll, KC Bowles, JH Nichols, R Roberts, SK Schultheis, CM O'Donnell. Clinical evaluation of Toxi-Prep: a semi-automated solid-phase extraction system for screening of drugs in urine. Clin Chem 43:2099–2105, 1997.

48. K Wolff, MJ Sanderson, AW Hay. A rapid horizontal TLC method for detecting drugs of abuse. Ann Clin Biochem 27:482–488, 1990.
49. RL Kincaid, MM McMullin, D Sanders, F Rieders. Sensitive, selective detection and differentiation of salicylates and metabolites in urine by a simple HPTLC method. J Anal Toxicol 15:270–271, 1991.
50. DN Bailey. Thin-layer chromatographic detection of cocaethylene in human urine. Am J Clin Path 101:342–345, 1994.
51. Simulataneous determination of diazepam, nitrazepam, estazolam in human urine by thin-layer chromatography. Chinese J Hosp Pharm 16:418–419, 1996.
52. J Klimes, P Pilarova. Thin-layer chromatography analysis of methamphetamine in urine samples. Ceska a Slovenska Farmacie 45:279–283, 1996.
53. W Pisternick, K-A Kovar, H Ensslin. High-performance thin layer chromatographic determination of *N*-ethyl-3,4-methylenedioxy-amphetamine and its major metabolites in urine and comparison with high-prformance liquid chromatography. J Chromatogr B 688:63–69, 1997.
54. I Ojanperä, R-L Ojansivu, J Nokua, E Vuori. Comprehensive TLC in 661089.8 toxicology: comparison of findings in urine and liver. J Planar Chromatogr 12:38–41, 1999.
55. JF Wilson, J Williams, G Walker, PA Toseland, BL Smith, A Richens, D Burnett. Performance of techniques used to detect drugs of abuse in urine: study based on external quality assessment. Clin Chem 37:442–447, 1991.
56. R Jain. Utility of thin layer chromatography for detection of opioids and benzodiazepines in a clinical setting. Addictive Behaviours 25:451–454, 2000.
57. H Brzezińska, P Dallakian, H Budzikiewicz. Thin-layer chromatography and mass spectrometry for screening of biological samples for drugs and metabolites. J Planar Chromatogr 12:96–108, 1999.
58. G Rochholz, B Ahrens, F König, HW Schütz, H. Schütz, H Seno. Screening and identification of sumatriptan and its metabolite by means of thin-layer chromatography, ultraviolet spectroscopy and gas chromatography/mass spectrometry. Arzneim-Forsch/Drug Res 45:941–946, 1995.
59. H Brzezinska, P Dallakian, H Budzikiewicz. Thin-layer chromatography and mass spectrometry for screening of biological samples for drugs and metabolites. J Planar Chromatogr 12:96–108, 1999.
60. K Ludanyi, A Gömöry, I Klebovich, K Monostory, L Vereczkey, K Ujszaszy, K Vekey. Application of TLC-FAB mass spectrometry in metabolism research. J Planar Chromatogr 10:90–96, 1997.
61. ChP Bhogte, VB Patravale, PV Devarajan. Fluorodensitometric evaluation of gentamycin from plasma and urine by high-performance thin-layer chromatography. J Chromatogr B 694:443–447, 1997.
62. B Mignot, Y Guillaume, S Makki, E Murret, E Cavalli, TT Truong, M Thosassin. High-performance thin-layer chromatographic determination of 5-methoxypsoralem in serum from patients. J Chromatogr B 700:283–285, 1997.
63. Y Zhang. Simultaneous determination of ten psychosedatives in serum by thin-layer chromatography. Chinese J Pharm 33:165–167, 1998.
64. A Jamshidi, M Adjvadi, S Shahmiri, A Masoumi, SW Husain, M

Mahmoodian. A new high performance thin-layer chromatography method for determination of theophylline in plasma. J Liq Crom and Rel Technol 22:1579–1587, 1999.

65. MH Guermouche, D Habel, S Guermouche. Assay of tinidazole in human serum by high-performance thin-layer chromatography—comparison with high-performance liquid chromatography. J Assoc Off Anal Chem 82:244–247, 1999.
66. E Forgacs, T Cserhati. Binding of anticancer drugs to human serum albumin studied by reversed-phase chromatography. J Chromatogr 697:265–272, 1995.
67. AM Le Roux, CA Wium, JR Joubert, PP Van Jaarsveld. Evaluation of a high-performance thin-layer chromatographic technique for the determination of salbutamol serum levels in clinical trials. J Chromatogr 581:306–309, 1992.
68. K Otsubo, H Seto, K Futagami, R Oishi. Rapid and sensitive detection of benzodiazepines and zopiclone in serum using high-performance thin-layer chromatography. J Chromatogr B 669:408–412, 1995.
69. MJ Lind, HL Roberts, N Thatcher, JR Idle. The effect of route of administration and fractionation of dose on the metabolism of ifosfamide. Cancer Chemotherapy and Pharmacology 26:105–111, 1990.
70. MJ Tasso, AV Boddy, L Price, RA Wyllie, AD Pearson, JR Idle. Pharmacokinetics and metabolism of cyclophosphamide in paediatric patients. Cancer Chemotherapy and Pharmacology 30:207–211, 1992.
71. AV Boddy, JR Idle. Combined thin-layer chromatography–photography–densitometry for the quantification of ifosfamide and its principal metabolites in urine, cerebrospinal fluid and plasma. J Chromatogr 575:137–142, 1992.
72. KK Pandya, MC Satia, IA Modi, RI Modi, BK Chakravarthy, TP Gandhi. High-performance thin-layer chromatography for the determination of nimesulide in human plasma, and its use in pharmacokinetic studies. J Pharm Pharmacol 49:773–776, 1997.
73. R Klaus, W Fischer, HE Hauck. Use of new adsorbent in the separation and detection of glucose and fructose by HPTLC. Chromatographia 28:364–366, 1989.
74. M Prosek, M Pukl, M Jamnik. Carbohydrates. Chromatogr Sci 55:439–462, 1991.
75. CH Tang, SH Li, G Gao. Separation of monosaccharides present in plasma membrane by thin-layer chromatography. Chinese J Biochem 1:15–22, 1985.
76. FG Bowling, AR Brown. Development of a protocol for new-born screening for disorders of the galactose metabolic pathway. J Inh Metab Disease 9:99–104, 1986.
77. H Higashi, Y Hirabayashi, M Ito, T Yamagata, M Matsumoto, S Ueda, S Kato. Immunostaining on thin-layer chromatograms of oligosaccharides released from gangliosides by endoglycoceramidase. J Biochem (Tokyo) 102:191–196, 1987.
78. SH Wang, L Ma. Determination of combined sugar in glycosides by 2-dimensional thin-layer chromatography. J Chinese Herb Med 20:155–157, 1989.

79. R Klaus, W Fischer, HE Hauck. Qualitative and quantitative analysis of uric acids, creatine and creatinine together with carbohydrates in biological material by HPLC. Chromatographia 32:307–316, 1991.
80. T Morcol, WH Velander. An *o*-toluidine method for detection of carbohydrates in protein hydrolyzates. Anal Biochem 195:153–159, 1991.
81. C Batisse, M-H Daurade, M Buonias. TLC separation and spectrodensitometric quantitation of cell wall neutral sugars. J Planar Chromatogr 5:131–133, 1992.
82. F Boscher-Reig, MJ Marcote, MD Minana. Separation and identification of sugars and maltodextrins by thin-layer chromatography. Application to biological fluids and human milk. Talanta 39:1493–1498, 1992.
83. S Rudloff. G Pohlentz. L Diekmann. H Egge. C Kunz. Urinary excretion of lactose and oligosaccharides in pre-term infants fed human milk or infant formula. Acta Pediat 85:598–603, 1996.
84. TI Poznyak, EV Kiseleva, TI Turkina. Distribution of the total unsaturation in lipid components of plasma as a new differential diagnostic method in clinical analysis. J Chromatogr 777(1):47–50, 1997.
85. F Bonte, A Saunois, P Pinguet, A Meybeck. Existence of a lipid gradient in the upper stratum corneum and its possible biological significance. Archives of Dermatological Research 289(2):78–82, 1997.
86. M Przybylska, M Bryszewska, U Nowicka, K Szosland, J Kędziora, RM Epand. Estimation of cholesterol sulphate in blood plasma and in erythrocyte membranes from individuals with Down's syndrome or diabetes mellitus type I. Clin Biochem 28(6):593–597, 1995.
87. S Karjalainen, L Sewon, E Soderling, B Larsson, I Johansson, O Simell, H Lapinleimu, R Seppanen. Salivary cholesterol of healthy adults in relation to serum cholesterol concentration and oral health. Journal of Dental Research 76(10):1637–1643, 1997.
88. M Iwamori, K Hirota, T Utsuki, K Momoeda, K Ono, Y Tsuchida, K Okumura, K Hanaoka. Sensitive method for the determination of pulmonary surfactant phospholipid/sphingomyelin ratio in human amniotic fluids for the diagnosis of respiratory distress syndrome by thin-layer chromatography–immunostaining. Anal Biochem 238(1):29–33, 1996.
89. L Berna, B Asfaw, E Conzelmann, B Cerny, J Ledvinowa. Determination of urinary sulfatides and other lipids by combination of reversed-phase and thin-layer chromatographies. Anal Biochem 269(2):304–311, 1999.
90. DA Wiesner, CC Sweeley. Circulating gangliosides of breast-cancer patients. International Journal of Cancer 60(3):294–299, 1995.
91. B Fried. Lipids. In: J Sherma, B Fried, eds. Handbook of Thin-Layer Chromatography. New York: Marcel Dekker, 1991, pp 593–623.
92. B Fried. Lipids. Chromatogr Sci 55:593–623, 1991.
93. JC Touchstone. Thin-layer chromatographic procedures for lipid separation. J Chromatogr B 671(1–2):169–195, 1995.
94. R Asmis, E Buhler, J Jelk, KF Gey. Concurrent quantification of cellular cholesterol, cholesteryl esters and triglycerides in small biological samples.

Reevaluation of thin layer chromatography using laser densitometry. J Chromatogr B: Biomed Appl 691(1):59–66, 1997.

95. W Sattler, H Reicher, P Ramos, U Panzenboeck, M Hayn, H Esterbauer, E Malle, GM Kostner. Preparation of fatty acid methyl esters from lipoprotein and macrophage lipid subclasses on thin-layer plates. Lipids 31(12):1302–1310, 1996.
96. H Miwa. M Yamamoto. T Futata. K Kan. T Asano. Thin-layer chromatography and high-performance liquid chromatography for the assay of fatty acid compositions of individual phospholipids in platelets from non-insulin-dependent diabetes mellitus patients: effect of eicosapentaenoic acid ethyl ester administration. J Chromatogr B: Biomed Appl 677(2):217–223, 1996.
97. JG Alvarez, B Slomovic, J Ludmir. Analysis of dipalmitoyl phosphatidylcholine in amniotic fluid by enzymatic hydrolysis and high-performance thin-layer chromatography reflectance spectrodensitometry. J Chromatogr B: Biomed Appl 665(1):79–87, 1995.
98. T White, S Bursten, D Federighi, RA Lewis, E Nudelman. High-resolution separation and quantification of neutral lipid and phospholipid species in mammalian cells and sera by multi-one-dimensional thin-layer chromatography. Anal Biochem 258(1):109–117, 1998.
99. L Gluck, MV Kulovich, RC Borer Jr. The diagnosis of the respiratory distress syndrome by amniocentesis. Am J Obstet Gynecol 109:440–442, 1972.
100. L Gluck, MV Kulovich. Lecithin-sphingomyelin ratio in amniotic fluid in normal and abnormal pregnancy. Am J Obstet Gynecol 115:539–541, 1973.
101. J Müthing. High-resolution thin-layer chromatography of gangliosides (rev). J Chromatogr 720(1–2):3–25, 1996.
102. T Taki, T Kasama, S Handa, D Ishikawa. A simple and quantitative purification of glycosphingolipids and phospholipids by thin-layer chromatography blotting. Anal Biochem 223(2):232–238, 1994.
103. J Müthing, M Cacic. Glycosphingolipid expression in human skeletal and heart muscle assessed by immunostaining thin-layer chromatography. Glycoconjugate Journal 14(1):19–28, 1997.
104. K Nakanishi, E Yasugi, H Morita, T Dohi, M Oshima. Plasmenylethanolamine in human intestinal mucosa detected by an improved method for analysis of phospholipid. Biochemistry and Molecular Biology International 33(3):457–462, 1994.
105. J Guittard, XL Hronowski, CE Costello. Direct matrix-assisted laser desorption/ionization mass spectrometric analysis of glycosphingolipids on thin layer chromatographic plates and transfer membranes. Rapid Communications in Mass Spectrometry 13(18):1838–1849, 1999.
106. KZ Liu, TC Dembiñski, HH Mantsch. Prediction of RDS from amniotic fluid analysis: a comparison of the prognostic value of TLC and infrared spectroscopy. Prenatal Diagnosis 18(12):1267–1275, 1998.
107. B Ramstedt, P Leppimaki, M Axberg, JP Slotte. Analysis of natural and synthetic sphingomyelins using high-performance thin-layer chromatography. European Journal of Biochemistry 266(3):997–1002, 1999.

108. K Okumura, K Hayashi, I Morishima, K Murase, H Matsui, Y Toki, T Ito. Simultaneous quantitation of ceramides and 1,2-diacylglycerol in tissues by Iatroscan thin-layer chromatography-flame-ionization detection. Lipids 33(5):529–532, 1998.
109. J Müthing. H Ziehr. Enhanced thin-layer chromatographic separation of GM1b-type gangliosides by automated multiple development. J Chromatogr B: Biomed Appl 687(2):357–362, 1996.
110. E Negroni, V Chigorno, G Tettamanti, S Sonnino. Evaluation of the efficiency of an assay procedure for gangliosides in human serum. Glycoconjugate Journal 13(3):347–352, 1996.
111. H Miller-Podraza, L Johansson, P Johansson, T Larsson, M Matrosovich, KA Karlsson. A strain of human influenza A virus binds to extended but not short gangliosides as assayed by thin-layer chromatography overlay. Glycobiology 10(10):975–982, 2000.
112. J Luo, CK Lim. Isolation and characterization of new porphyrin metabolites in human porphyria cutanea tarda and in rats treated with hexachlorobenzene by HPTLC, HPLC and liquid secondary ion mass spectrometry. Biomed Chromatogr 9:113–122, 1995.
113. Z Tomsova, Z Juzova. Jednoduche chromatograficke stanoveni neopterinu-markeru imunopatii. Casopis Lekaru Ceskych 128:59–61, 1989.
114. JL Maddocks, J MacLachlan. Application of new fluorescent thiol reagent to diagnosis of homocystinuria. Lancet 338:1043–1044, 1991.
115. JJ Steinberg, A Cajigas, M Brownlee. Enzymatic shot-gun 5′-phosphorylation and 3′-sister phosphate exchange: a two-dimensional thin-layer chromatographic technique to measure DNA deoxynucleotide modification. J Chromatogr 574:41–55, 1992.
116. MV Reddy. 32P-postlabelling analysis of small aromatic and of bulky non-aromatic DNA adducts. IARC Scientific Publications (Lyon) 124:25–34, 1993.
117. JJ Steinberg, GW Oliver, N Farah. P Simoni, R Winiarsky, A Cajigas. In vivo determination of 5-bromo-2′-deoxyuridine incorporation into DNA tumor tissue by a new 32P-postlabelling thin-layer chromatographic method. J Chromatogr B 694:333–341, 1997.
118. S Birnbaum, C Uden, CG Magnusson, S Nilsson. Latex-based thin-layer immunoaffinity chromatography for quantitation of protein analytes. Anal Biochem 206:168–171, 1992.
119. P van der Geer, T Hunter. Phosphopeptide mapping and phosphoamino acid analysis by electrophoresis and chromatography on thin-layer cellulose plates. Electrophoresis 15:544–554, 1994.
120. MP Laitinen, KM Sojakka, M Vuento. Thin-layer affinity chromatography in analysis of protein–ligand affinity. Anal Biochem 243:279–282, 1996.
121. TK Różyło, R Siembida. Comparative quantitative chromatographic determination of formaldehyde in different groups of physiological and pathological hard tissues of teeth. Acta Biologica Hungarica 49:413–419, 1998.
122. TK Różyło, R Siembida, A Szyszkowska, A Jambrozek-Manko. The increase of formaldehyde level in some rare pathological cases of teeth determined

with the use of quantitative TLC. Acta Biologica Hungarica 49:317–322, 1998.

123. I Baranowska, J Baranowski, I Norska-Borowka, C Pieszko. Separation and identification of metals in human bones, placenta and milk and in air by adsorption and ion exchange thin-layer chromatography. J Chromatogr 725:199–202, 1996.
124. J Zivko-Babic, V Ivankovic, J Panduric. Quantitative thin-layer chromatographic identification of dental base alloys. J Chromatogr B 710:247–253, 1998.
125. T Cserhati, E Forgacs. Charge transfer chromatographic study of the binding of commercial pesticides to various albumins. J Chromatogr 699:285–290, 1995.
126. AB Muzhanovski, AF Fartushny, AP Sukhin. The identification of new pesticide preparations in biological material. Sudebno-Meditsinskaia Ekspertiza 41:20–22, 1998.
127. CQ Leite, CW de Souza, SR Leite. Identification of mycobacteria by thin layer chromatographic analysis of mycolic acids and conventional biochemical method: four years of experience. Memorias do Instituto Oswaldo Cruz 93:801–805, 1998.
128. J Skarkova, V Ostry. An HPTLC method for confirmation of the presence of ultra-trace amounts of aflatoxin M1 in human urine. J Planar Chromatogr-Mod TLC 13:42–44, 2000.
129. JM Schremmer, DJ Noonan. Advances in analytical techniques for neutron capture therapy: thin layer chromatography matrix and track etch thin layer chromatography methods for boron-10 analysis. Medical Physics 14(5):818–824, 1987.

12

Applications of Capillary Electrophoresis in Clinical Analysis of Drugs

Jacek Bojarski and Joanna Szymura-Oleksiak
Jagiellonian University, Kraków, Poland

1 INTRODUCTION

Capillary electrophoresis (CE), one of the youngest and fastest developing separation methods, is widely applied in the analysis of pharmaceuticals. The basic principles of the method are described in Chapter 2, and the interested reader can expand his knowledge on that subject from different monographs [1,2]. Other sources can be found in the excellent review of Altria [3]. Many review articles deal with more specialized topics like the history of development [4], technical aspects [5–7], and applications [8–14]. This chapter is devoted mainly to applications of capillary electrophoresis in analysis of different drugs in body fluids and subsequent pharmacokinetic results and conclusions.

Enantioseparations of chiral drugs are also discussed, and some methods of this kind with potential applications in clinical analysis are briefly covered.

2 APPLICATION OF CAPILLARY ELECTROPHORESIS IN PHARMACOKINETICS

Pharmacokinetic studies of single- and multiple-dose kinetics, in vitro and in vivo metabolic profile, bioavailability, pharmacogenetics, and drug–drug interaction are closely associated with the measurements of concentration levels of the parent drug and its metabolites, performed predominantly in blood and other biological fluids, such as urine, bile, saliva, cerebrospinal fluid, as well as in tissue homogenates. In recent years, also intracellular and subcellular structures (biological membranes, lysosomes, microsomes, etc.) are taken into consideration. The quality of such investigations is determined not only by appropriate biochemical, physiological, and molecular biology research techniques but also by the use of appropriate, validated analytical methods that ensure true values of the determined drug concentration levels and the resultant calculated pharmacokinetic parameters.

For the above reasons, samples investigated in pharmacokinetic studies are characterized by the complexity of both the biological matrix and the analyte (parent drug, its metabolites, sometimes other drugs). Hence when characterizing the employed analytical methods, one must take into consideration both qualitative and quantitative aspects. The former are associated with the selective separation of compounds in biological fluids, while the latter include a validation procedure (accuracy, precision, linearity, limit of detection or quantitation).

Capillary electrophoresis (CE) is a relatively new technique of separation and quantitation regrouping different modes of separation, chiefly capillary zone electrophoresis (CZE) and micellar electrokinetic capillary chromatography (MEKC) [15]. In pharmacokinetic studies CE can offer numerous advantages, such as low sample volume, high separation efficiency, analysis speed, instrumentation simplicity, and reduced operating costs [16,17].

The method, when coupled with microsampling strategies, such as microdialysis (MD) or capillary ultrafiltration, offers a great potential for the rapid and accurate characterization of preclinical pharmacokinetic profiles, particularly in small and awake (not anesthetized) rodents [18].

Studies on the pharmacokinetics of gabapentin, an anticonvulsant, demonstrated that using CZE with laser-induced fluorescence detection (CZE-LIF) coupled with microdialysis sampling allows for evaluating the pharmacokinetic profile of the drug in rat's brain in vivo. These investigations were based on gabapentin level determinations in prefrontal cortex dialysates and blood plasma of the animals studied [19].

In clinical studies, the combination of MD and CE was employed in the assessment of drug distribution processes in various interstitial tissues or in solid human tumors [20].

CE has also found its use in determining the apparent equilibrium constants for molecular association; as for example in the association between small drug molecules and large entities such as polypeptides or proteins [21].

The main potential limitation of CE techniques lies in high limits of detection/quantitation, which may hinder the accurate determination of concentrations, for example in the late phase of drug elimination and in case a low drug dose is employed, owing to the small volume injected. This problem may be overcome using electrokinetic sample concentration, referred to as sample stacking, or by using classic liquid–liquid (LL) or solid phase extraction (SPE) [22,23].

The detection limit of CE techniques can be also improved by using more sensitive detection methods, such as laser-induced fluorescence (LIF) detection [24], electrochemical detection (ED), or mass spectrometry (MS) detection [1].

Recently attempts have been made at basing drug determinations in biological materials on the microchip-CE technique, which is particularly attractive for clinical applications as it may offer rapid analysis of several samples performed simultaneously [25].

A growing interest in determining the concentration values of various drugs in biological fluids using CE methods is reflected in review papers published in recent years. These reports are of a summary [26] or detailed character, describe defined groups of drugs, and take into consideration the specificity of their distribution and the possibility of measuring their concentration by CE methods to be used in pharmacokinetic studies, therapeutic drug monitoring, or pharmaceutical analysis.

Nguyen and Siegler [27] published an article on the use of the CE technique for analysis of cardiovascular drugs, such as β-blockers, acetylcholinesterase inhibitors, angiotensin-converting enzyme inhibitors, diuretics, α-adrenergic antagonists, calcium channel blockers, cardiac glycosides, hypolipidemics, vasodilators, and sodium channel blockers. The authors demonstrated that the CE technique may be employed in separation of enantiomers, in determination of drug–protein binding constants, and in assaying these drugs in various biological fluids.

Antisense oligodeoxynucleotides (ODNs) can alter gene expression by inhibiting transcription or translation via sequence-specific binding of either DNA or mRNA and become a critical tool for experimental therapeutics and research in the field of gene expression. Antisense ODN normally contains 20–25 nucleotides. The failure sequences or metabolites formed in vivo produces the family of antisense ODNs shortened by one to several nucleotides that may still possess pharmacological activity. Therefore it is important to quantify the intact oligonucleotides as well as these putative metabolic products.

Problems associated with the use of CE methods to characterize the pharmacokinetics of antisense drugs constituted the focus of the paper authored by Chen and Gallo [28]. It was found that CE has been recognized as a powerful tool for separating this type of compounds.

Iwaki et al. discussed CE methods employed to assay nicotinic acid (NiAc) and its metabolites in biological fluids [29]. In this case, CE seems to be an especially powerful technique, since metabolites of NiAc and nicotamide, which have different charges and physicochemical properties, can be analyzed simultaneously. Based on articles they analyzed, the authors showed that, using CE methods, highly sensitive detection, such as fluorescence detection, will be required to produce precise pharmacokinetic and toxicokinetic data.

Smyth and McClean [30] presented a review article on the use of CE methods in the analysis of 1,4-benzodiazepine concentration in investigations associated with the development of pharmaceuticals, therapeutic drug monitoring, and forensic toxicology. MEKC and CZE of 1,4-benzodiazepines with particular reference to biological material analysis were discussed in the text. The authors demonstrated that MEKC has been found more applicable than CZE in direct injection of biological samples with little or no sample preparation.

3 SELECTED EXAMPLES OF CE EMPLOYMENT IN PHARMACOKINETIC STUDIES

Table 1 presents the conditions of analyzing the concentration of various drugs by CE methods that have been described in the most recent publications on pharmacokinetics.

3.1 Antiviral Drugs

The thymidine analog (E)-5-(2-bromovinyl)-2′-deoxyuridine (BVdU) is a highly potent inhibitor of *Herpes simplex* virus type 1 and of *Varicella zoster* virus. BVdU is metabolized to a nonactive metabolite (E)-5-(2-bromovinyl)-uracil (BVU).

The total and unbound drug and metabolite concentrations in human plasma, as well as their urine levels, were determined using a validated CZE method with UV detection [31]. The kinetics of BVdU and BVU from plasma and urine data were estimated after oral administration of 12.5 mg and 250 mg of BVdU. The obtained results confirmed that BVdU is resynthesized from its metabolite. Inter- and intraday precision were found to be better than 10%, which qualifies the method to be useful also for drug monitoring in patients.

3.2 Anti-Inflammatory Drugs

Such properties of CZE as the possibility of using small sample size in the analysis, high separation efficiency, and high separation speed, as well as biocompatible environment, allow for conducting investigations on single biological cells. Such a technique was employed in monitoring diclofenac concentration in single human erythrocytes introduced by electrophoration. Drug concentration levels were determined by CZE with electrochemical detection. It was found that the concentration of diclofenac sodium may affect the erythrocyte membrane structure and lead the cells to lyse easily. The methodology presented in the report opens new perspectives for the assessment of drug disposition when introduced into isolated cells [32].

3.3 Antidepressant Drugs

A chip-based capillary electrophoresis/mass spectrometry (CE/MS) with electrospray ionization system was employed in quantitative determinations of imipramine and desipramine in human plasma samples [33]. The results indicated that the sensitivity of the method is not sufficient for monitoring the levels of the drugs and their metabolites in clinical settings in psychiatric patients. The LOQ values were above the therapeutic range. On the other hand, according to the authors, the method may be helpful in experimental pharmacokinetics. They suggest that improvements in mass spectrometric detection limits and the technology associated with chip-based devices may eventually help to achieve improved LOQ in the future.

3.4 Antihyperlipidemic Drugs

Dogrukol-Ak et al. [34] reported the use of the CZE method with UV detection for determination of fluvastatin in human serum. The parameters of method validation demonstrated that it can be used in bioavailability and in pharmacokinetic studies of this drug.

3.5 Anticoagulant Drugs

Coumarin in humans is metabolized by CYP2A-6 isoenzyme, the activity of which is genetically determined. The genetic variation leads to human groups with a high or low coumarin metabolism. Tegtmeier et al. [35] developed the CZE method with diode array detection allowing for quantitative determinations of the parent drug and pharmacologically important coumarin metabolites (mostly its hydroxyl derivatives) in one analytic step. This method might be used in phenotyping the process of coumarin hydroxylation as a marker of CYP2A-6 activity.

TABLE 1 Selected Capillary Electrophoresis (CE) Assays Used in Pharmacokinetic Studies

Analyte	Matrix	CE/detn. mode	Run buffer/voltage/ temperature	Sample pretreatment	LOQ	Ref.
Antiviral drugs						
BvdU and BVU	serum	CZE/UV 54 nm	50 mM disodium tetraborate adj. to pH 9.6 by 5 M NaOH, 25 kV	LLE	40 ng/mL (10 ng/mL unbound)	31
	urine		40 mM disodium tetraborate adj. to pH 10.5 by 5 M NaOH, 25 kV		170 ng/mL	
Antiinflammatory drugs						
Diclofenac	erythrocytes	CZE/EC	12.5 mM Na_2B_4O 3,13 mM NaOH, 20 kV	DSI	—	32
Antidepressant drugs						
Imipramine Desipramine	plasma	Chip-CZE/ESI-MS	10 mM ammonium acetate, 5% acetic acid (pH 3), 4 kV	LLE	5 μg/mL	33
Antihyperlipidemic drugs						
Fluvastatin	serum	CZE/UV 239 nm	10 mM borate buffer (pH 8), 28 kV, 25°C	protein precipitation	$2{\cdot}89{\cdot}10^{-6}$ M	34
Anticoagulant drugs						
Coumarin and metabolites	borate buffer	CZE/DAD 190–360 nm	0.025 M sodium borate, 1 M NaOH (pH 10) in H_2O/methanol (9:1), 30 kV, 20°C	DSI	1 μg/mL	35
Bronchodilators						
Adrenaline Salbutamol Salmeterol	serum	CZE/EC	0.06 M phosphate buffer, (pH 6.6), 12 kV	DSI	$5.0{\cdot}10^{-8}$ M $2.0{\cdot}10^{-7}$ M $2.0{\cdot}10^{-7}$ M	36

Hypoglycemic drugs						
Chlorpropamide Tolbutamide Glipizide Glibenclamide	serum	MEKC/UV	5 mM borate, 5 mM phosphate, 75 mM sodium cholate, 25 ml/L methanol (pH 8.5), 10 kV, 25°C	SPE	—	37
Immunosuppressive drugs						
Cyclosporin A	phosphate buffer	ACE/UV 200 nm	50 mM phosphate buffer (pH 8.0), 30 kV, 26°C	DSI	—	38
Protease inhibitors						
Indinavir Saquinavir Nelfinavir Amprenavir Ritonavir	serum	CZE/UV 195 nm	16 mM phosphoric acid, 0.001% HDB (pH 2.5), −30 kV, 25°C	SPE	0.02 μg/mL 0.05 μg/mL 0.05 μg/mL 0.24 μg/mL 0.12 μg/ml	39
Indinavir Nelfinavir Saquinavir	serum	NACE/UV 195 nm	1 mM formic acid, 25 mM ammonium formate; acetonitrile–methanol (80 : 20), (pH 3.5), 30 kV, 25°C	SPE		
Other drugs						
Altropane and Allylaltropane	plasma	CZE/UV 210 nm	50 mM acetate buffer, (pH 4.1), 20 kV, 20°C	SPE	0.3 μg/mL	40,41
Cimetidine	plasma	CZE/UV 214 nm	200 mM acetate buffer (pH 5.6), 0.5 mM SDS, 13 kV,	SPE	50 ng/mL	42

Abbreviations used in the table: BVdU – (*E*) – 5-(2-bromovinyl)-2′deoxyuridine, BVU – (*E*)-5-(2-bromovinyl)-uracil, ACE – affinity capillary electrophoresis, NACE – nonaqueous capillary electrophoresis, DSI – direct sample injection, LLE – liquid–liquid extraction, SPE – solid phase extraction, HDB – hexadimethrin bromide, SDS – sodium dodecylsulfate, UV – ultraviolet, EC – electrochemical, DAD – diode array detector, ESI-MS – electrospray ionization mass spectrometry.

3.6 Bronchodilators

β-Agonist drugs have been used for a long time in the treatment of bronchial asthma for their bronchodilatory effects. To monitor therapeutic use, as well as to control illegal use of β-agonists (doping agents), many methods for determination of these drugs in biological fluids have been published. Zhou et al. [36] developed the CZE method with amperometric detection (AD) to separate and determine the β-agonists adrenaline, salbutamol, and salmeterol. Based on the validation data, such as accuracy, sensitivity, and reproducibility, one may say that the proposed method most likely may be of use in biomedical analysis.

3.7 Hypoglycemic Drugs

MEKC with UV detection has been used as a quantitative method for determination of hypoglycemic agents serum concentration (chlorpropamide, tolbutamide, glipizide, gliclazide, glibenclamide). The investigators demonstrated that the analysis of gliclazide and tolbutamide pharmacokinetics employing the above method yielded results comparable to data obtained by HPLC. Both methods gave similar results that allowed for detecting overdoses of the drugs in diabetic patients [37].

3.8 Immunosuppressive Drugs

The use of the CE method for the investigation of binding properties is referred to as affinity capillary electrophoresis (ACE); it is a promising technique to monitor the molecular interaction of biomolecules. The ACE method with UV detection was employed in investigating the interactions between the enzyme cyclophilin (Cyp) and an immunosuppressive drug, cyclosporin (CsA). CsA binds and competitively inhibits this enzyme, and the Cyp-CsA complex then interacts with other proteins in a way not yet known in detail and probably suppresses cell proliferation. Using this new approach combining the mobility-shift analysis and electrophoretically mediated microanalysis, the binding constants of recombination human cyclophilin 18 (rh Cyp 18) to CsA and derivatives was estimated [38].

3.9 Protease Inhibitors

Protease inhibitors (indinavir, saquinavir, nelfinavir, amprenavir, and ritonavir) are widely used in HIV therapy. It was found that not all patients on antiretroviral therapy respond equally well. Among numerous factors that play a decisive role in the success of antiretroviral therapy are issues related to the unfavorable pharmacokinetic profile of these drugs. This is why the possibility of measuring their blood serum concentration may contribute to the optimization of their therapeutic use.

Gutleben et al. [39] developed two CE modes for determining blood serum levels of these drugs, such as CZE and nonaqueous CE (NACE). The former allows for determining all the above drugs, while NACE yielded satisfactory results in the case of indavir, nelfinavir, and saquinavir. Generally, the obtained results demonstrate the applicability of CE for the determination of protease inhibitors in serum using the versatility of the technique combined with the speed of analysis.

3.10 Other Agents

Altropane, 2β-carbomethoxy-3β-(4-fluorophenyl)-*N*-(3-iodo-*E*-allyl)nortropane, is an imaging agent that was developed recently for early detection of Parkinson's disease. Using this agent it is possible to detect the presence of dopamine-containing neurons and thus aid in detection of neuronal loss before clinical signs appear in patients with Parkinson's disease, a chronic neurodegenerative disorder.

Pharmacokinetic investigations in rats demonstrated that after iv administration, altropane is cleared very rapidly from systemic circulation with a mean residence time of about 11 min and with biological half-life time about 7–10 min. It was further shown that its metabolism is not a major factor in the disposition of altropane and the sex of the animal does not affect the pharmacological profile of this compound [41]. The results indicated that the developed CZE method with UV detection can be successfully used to study the pharmacokinetic profile of this agent [39].

Cimetidine, a histamine H_2-receptor antagonist inhibiting gastric acid excretion, was determined in human plasma by CZE mode [42]. The buffer was modified with sodium dodecylsulfate (SDS) to eliminate capillary contamination by biological macromolecules. A detection limit of 8 ng/mL was reached, and the method was successfully applied to pharmacokinetic studies in plasma samples from healthy volunteers.

4 APPLICATION OF CE IN THERAPEUTIC DRUG MONITORING

Routine therapeutic drug monitoring (TDM) in blood is mostly performed in the case of drugs with low therapeutic indices and for drugs when no ongoing assessment of therapeutic results is possible. The introduction of TDM to clinical practice allows for using drug concentration levels, pharmacokinetic criteria, and the clinical presentation as the bases for dosage optimization, detection of the patient's noncompliance, and prevention of interactions during the pharmacokinetic phase and thus increases treatment effectiveness and safety.

Three basic prerequisites underlie the introduction of TDM to pharmacotherapy, namely (1) a correlation between drug concentration and its therapeutic and toxic activity, (2) the known range of therapeutic concentration values, and (3) the availability of analytical methods that are highly sensitive, precise, and repeatable, as well as fast, simple, and economical. Attempting to meet these criteria is a continuous inspiration in the search for new, universal analytical methods.

Recently there has been increased interest in the use of CE methods for measuring concentrations of drugs in patient samples. For drug monitoring in clinical practice, essentially two methods are distinguished: CZE and MEKC. As has been mentioned, CE methods offer speed, ease, low cost of operation, and high resolution and analysis automatization. Of considerable importance is also small sample volume, which allows for collecting small blood samples employing a noninvasive method (blood drawn from the fingertip) using heparinized capillaries, which is particularly valuable in pediatric patients. The disadvantage of CE is its poor detection capability and less than desirable precision.

Detailed data on the advantages and disadvantages of CE methods and perspectives for their employment in TDM, as well as on sample preparation and method of detection, may be found in several review articles [43–45].

Table 2 presents the conditions employed in analyzing concentration values of various drugs determined by CE methods originating from recent publications on TDM.

4.1 Anthelminthic Drugs

Albendazole (ABZ) is an anthelminthic drug used for eradication of a number of a helminthic parasites. After oral administration, ABZ is usually undetectable in plasma, whereas its active metabolite albendazole sulfoxide (ABZSO) reaches its maximum concentration about 2–3 h after dosing. For optimized therapy against tissue parasites, ABZSO plasma levels should be higher than 1 μM. In view of considerable interindividual differences resulting from poor bioavailability of the parent drug, ABZSO plasma concentration levels should be monitored.

A nonaqueous capillary electrophoretic method (NACE) with UV detection for fast determination of plasma ABZ and ABZSO levels was elaborated by Procházkova et al. [46]. The limit of detection for ABZ and ABZSO was 8×10^{-7} M. The reliability of the method developed was verified via analysis of plasma samples obtained from patients treated with ABZ.

4.2 Anticancer Drugs

Doxorubicin and daunorubicin are anthracycline cytostatic drugs showing activity against leukemia and solid tumors. At the same time these agents are cardiotoxic. Therefore it is very important to monitor their plasma concentration levels during therapy. The analysis of anthracyclines in plasma is a difficult task, mainly because of the low therapeutic concentrations (10–100 nM) and the complex biological matrix. Gavenda et al. [47] developed the CE method with UV detection, sweeping preconcentration and electrokinetic injection. The results of their studies indicated that sweeping preconcentration of the sample provides a satisfactory detection limit (1×10^{-9} M). For this reason the method was employed for the determination of therapeutic levels of doxorubicin in real plasma samples.

The CZE method with laser-induced fluorescence detection for the determination of doxorubicin in human plasma was also elaborated by Hempel et al. (48). Using two different sample pretreatment modes (serum deproteinization at high drug concentrations or liquid–liquid extraction at low drug levels), the authors achieved a satisfactory limit of quantitation (2 μg/mL), which allowed for monitoring doxorubicin plasma levels for several days after administration of the drug. The results pointed to high interindividual variability in doxorubicin peak plasma levels.

4.3 Antidepressant Drugs

In view of their low therapeutic index, tricyclic antidepressants are often monitored in clinical practice. Their therapeutic concentration range lies between 100 and 250 ng/mL. Veraart and Brinkman [49] presented dialysis-solid-phase extraction sample pretreatment combined on line with nonaqueous capillary electrophoresis for the determination of tricyclic antidepressants in urine and serum. Their investigations included five tricyclic antidepressants, amitryptyline, nortryptyline, imipramine, desipramine, and maprotyline. The detection limits were in the 20–100 ng/mL range, and the day-to-day repeatability was between 2.5 and 9.5%, which is quite satisfactory and allows for employing the method in therapeutic drug monitoring.

4.4 Antiepileptic Drugs

Therapeutic monitoring of antiepileptic drugs is important to clarify and control their therapeutic and toxic effects (optimization of pharmacotherapy) and to assess the patient's compliance to therapy. In their publication, Thormann et al. [50] presented CE assays for determination of ethosuximide by MEKC via direct injection of serum, lamotrigine by CZE after protein precipitation and carbamazepine and its epoxide metabolite by MEKC after

TABLE 2 Selected Capillary Electrophoresis Assays Used in TDM

Analyte	Matrix	CE/detn. mode	Run buffer/ voltage/ temperature	Sample pretreatment	LOQ	Therapeutic range	Ref.
Anthelminthic agents							
Albendazole (ABZ) Albendazole sulfoxide (ABZSO)	plasma	NACE/UV, 280 nm	0.036 M borate buffer (pH 9.9) + methanol, *N*-methylformamide (1 : 3), 23 kV, 35°C	LLE	$8 \cdot 10^{-7}$ M	ABZSO > 1 μM	46
Anticancer drugs							
Doxorubicin Daunorubicin	plasma	MEKC/UV, 190–260 nM	100 mM phosphoric acid adj. to pH 2.5 by 50% (w/w) NaOH, 150 mM SDS, 40% (v/v) methanol, 30 kV, 25°C	LLE	1.0 mM	10–100 nM	47
Doxorubicin	plasma	CZE/LIF, 488/520 nm	100 mM phosphate buffer (pH 5.0) 60 μM spermine, acetonitrile 70% (v/v), 25 kV, 25°C	deproteination or LLE	2 μg/L		48
Antidepressant drugs							
Amitryptyline Nortryptyline Imipramine Desipramine Maprotyline	serum urine	NACE/UV, 214 nm	1 M acetic acid, 25 mM ammonium acetate in acetonitrile, 30 kV	dialysis–SPE	0.02–0.1 μg/mL	0.07–0.25 μg/mL	49
Antiepileptic drugs							
Ethosuximide	human plasma	MEKC/UV, 220 nm	phosphate–tetraborate buffer (6 mM $Na_2B_4O_7$, 10 mM Na_2HPO_4, pH 9.2), 75 mM SDS, 20 kV, 30–35°C	DSI	15 μM	280–700 μM	50
Lamotrigine	bovine plasma	CZE/UV, 210 nm	130 mM sodium acetate adj. to pH 4.5 with acetic acid, 12 kV, 27–30°C	protein precipitation	2 μM	8–24 μM	

Carbamazepine and Carbamazepine-10, 11-epoxide	human or bovine plasma	MEKC/UV, 210 nm	6 mM $Na_2B_4O_7$/10 mM Na_2HPO_4 (pH 9.2), 75 mM SDS, 5% (v/v) 2-propanol, 25 kV, 25°C	LLE	0.40 μM 0.30 μM	17–47 μM ≈ 25% of CBZ	
Zonisamide Phenobarbital Phenytoin Carbamazepine	human plasma	MEKC/DAD, 210 nm	10 mM phosphate buffer (pH 8.0) and 50 mM SDS, 30 kV, 30°C	LLE	3.0 μg/mL 2.5 μg/mL 0.8 μg/mL 0.6 μg/mL	15–40 μg/mL 10–20 μg/mL 4–10 μ/mL	52
β-Blockers							
Esmolol	serum	CZE/UV, 222 nm	50 mM phosphate buffer adj. 1 M HCL (pH 8.0), 15 kV, 25°C	deproteination and LLE	0.051 μg/mL	—	53
Purines							
Allopurinol Adenine Hypoxanthine	urine	MEKC/UV DAD, 254/300 nm	80 mM SDS in 60 mM sodium tetraborate buffer (pH 8.7), 20°C	DSI	40 μM 22 μM 18 μM		54
Hippuric acid Oxypurinol xanthine		CZE/UV DAD, 254/300 nm	60 mM sodium tetraborate buffer (pH 8.7), 20°C	DSI	16 μM 36 μM 13 μM		
2,8-Dihydroxyadenine			60 mM sodium tetraborate buffer (pH 9.7), 20°C		24 μM 10^{-8} M		
Allopurinol and oxypurinol	urine	CZE/EC	15 mM Na_2HPO_4/NaH_2PO_4 (pH 9.55), 15 kV	DSI			55
Other drugs							
Acetaminophen Salicylic acid Sulfamethoxazole Theophylline Tolbutamide Trimethoprim	plasma	MEKC/UV,	60 mM borate buffer, 200 mM SDS (pH 10.0)	DSI	5 ng/mL		56
Viagra + metabolite	serum	MEKC/UV DAD, 230 nm	10 mM phosphate buffer 30 mM SDS (pH 12.3), 25 kV, 25°C	SPE	0.6 μg/mL	0.8–4.5 μg/mL	57

Abbreviations the same as in Table 1.

solute extraction. The limit of detection for each of these methods was well below the therapeutic range. This is why the presented methods turned out to be suitable and sufficiently robust for TDM.

Zonisamide, a 1,2-benzisoxazole derivative, is a new antiepileptic drug, often employed in combination with other antiepileptics in the treatment of seizures. This results in a number of interactions, which have been reported [51]. Such interactions, as well as the nonlinear pharmacokinetics of zonisamide, produce an increase or decrease in zonisamide serum concentrations, which may lead to toxicity or ineffectiveness. In such situations, therapeutic drug monitoring is indicated.

Kataoka et al. [52] developed the MEKC method with a diode array detector for simultaneous serum determination of zonisamide and typical antiepileptic drugs such as phenobarbital, phenytoin, and carbamazepine. It was found that reproducibility of separation and quantification with MEKC for intra- and interday assays was appropriate. This method allows for a simple and efficient TDM for antiepileptic drugs, especially in patients treated with a combination therapy.

4.5 β-Blockers

Esmolol is an ultra-short-acting β_1 adrenoceptor antagonist (cardioselective) with an elimination half-life of about 9 min. It is used for rapid control of heart rate in patients with sinus tachycardia or atrial fibrillation and also for the treatment of tachycardia, especially with hypertension, during surgery and in the postoperative period when indicated. For these reasons monitoring esmolol blood levels during heart surgery requires an appropriately fast and simple concentration determination.

Malovaná et al. [53] described the CZE procedure with UV detection for determinations of esmolol in serum, plasma, and blood. The limit of drug concentration detection in serum was 0.05 μg/mL. This method was positively tested in an extensive heart surgery experiment on pigs.

4.6 Agents for Treatment of Gout

Allopurinol is used for the treatment of hyperuricemia and gout, as a potent inhibitor of xanthine oxidase, which converts hypoxanthine to xanthine, and xanthine to uric acid. Because of the extended use of allopurinol, the excretion of hypoxanthine and mainly xantines increases and can lead to the formation of xanthine stones. For these reasons it is necessary to adjust the optimum dosage of allopurinol to the lowest possible excretion level of 2,8-dihydroxyadenine (2,8-DHA). The influence of a purine-restricted diet and noncompliance are also reasons for long-term TDM of 2,8-DHA, other purine end products, and allopurinol in urine.

Wessel et al. [54] developed the MEKC method (with sodium dodecyl sulfate) and the CZE method, both with UV detection, for determinations of 2,8-hydroxyadenine, allopurinol, oxypurinol, adenine, hypoxanthine, hipuric acid, and xanthine in urine. The limit of detection for the compounds investigated was about 5 μM. These methods have been applied to the drug monitoring of one patient with DHA lithiasis for 1.5 year.

Sun et al. [55] elaborated the CZE method with end column amperometric detection for determining allopurinol and its active metabolite oxypurinol in urine. This method was demonstrated to exhibit a low detection limit of 1×10^{-8} mol/L for both compounds studied, thanks to which it can be employed in therapeutic monitoring of their concentrations.

4.7 Other Drugs

The MEKC method with UV detection and without sample pretreatment was developed by Kunkel and Watzig [56] for human plasma determinations of such drugs as acetaminophen, salicylic acid, sulfamethoxazole, theophylline, tolbutamide, and trimethoprim. SDS as micellar additive was used. The limit of detection was about 5 ng/mL. The high selectivity of this method and its cost advantages allow for its use in TDM and in pharmacokinetic studies.

The same CE mode was used for determination of citrate salt of sildenafil (Viagra) and its metabolite in human serum with the detection limits of ca. 200 ng/mL for both compounds [57]. The authors claim the method as good "with regards to simplicity, precision and sensitivity."

4.8 Chiral Drugs

Stereochemical features of drugs and their pharmacological implications were discussed elsewhere in this book (chapter by K. A. More and B. Levine). Although HPLC is now a leading analytical technique in enantioseparations of drugs in bulk products, drug forms, and biological samples, CE entered this field quite vigorously and is still expanding its scope. This trend is very well documented in a recent monograph [58] and reviews dealing with general principles [59–65] or particularly devoted to drug analysis [66–70].

Analysis of chiral drugs and their metabolites in body fluids becomes more and more popular. In one of the first reviews on this subject [71], only a few applications were mentioned, whereas in one of the latest [72], ca. 50 experimental papers are comprehensively reviewed. They report work with body fluids (mainly serum, plasma, and urine) fortified with drug enantiomers, monitoring stereoselective distribution, metabolism and elimination of chiral drugs of both therapeutic and forensic interest, as well as in vitro investigations on stereoselective drug metabolism using microsomal

TABLE 3 Recent Capillary Zone Electrophoretic Enantioseparations of Drugs and Their Metabolites in Body Fluids

Drug (metabolites)	Activity	Body fluid	Internal standard	Chiral selector	Buffer/pH (voltage)	Sample preparation	Detection	Ref.
(Albendazole sulfoxide)[a]	anthelmintic	CSF	propyphenazone	sulfated β-CD	tris/7.0 (18 kV)	LLE	UV/290 nm	73
Baclofen[a]	skeletal muscle relaxant	plasma		α-CD	borate/9.5 (9 kV)	LLE	LIF 442/500 nm	74
Carvedilol	β-blocker	plasma	carazolol	succinyl-β-CD + methyl-α-CD	phosphate/3.0 (16 kV)	LLE	LIF325/366 nm	75
		serum	(−)-propranolol	HP-β-CD	phosphate/2.5 (18 kV)	LLE	UV/200 nm	76
Ciprofibrate (glucuronide)	hypolipidemic	urine		γ-CD	phosphate/6.0 (500V/cm)	SPE	UV/230 nm	77
Disopyramide (mono-*N*-dealkyldisopyramide)	antiarrhythmic	plasma	trimethoprim	sulfated β-CD	acetate/5.0 (15kV)	LLE	UV/214 nm	78
Eticlopride	neuroleptic	serum	(−)-butaclamol	sulfated β-CD	citrate/2.9 (20 kV)	SPE	UV/220 nm	79
Fluoxetine (norfluoxetine)	antidepressant	serum, plasma		DM-β-CD + phosphated γ-CD	phosphate/2.5 (25 kV)	LLE	UV/195	80
Mianserin (desmethylmianserin, 8-hydroxymianserin)	antidepressant	plasma	propyl-norclozapine	HP-β-CD	phosphate/3.0 (30 kV)	LLE	UV/211	81
(2-, 6-, 2′-, 3′-, 4′-Hydroxymethaqualone)	hypnotic anticonvulsant	urine		HP-β-CD	phosphate/2.1 (18 kV)	LLE, SPE	UV/200–320 nm	82

Metoprolol (*O*-demethylmethoprolol, α-hydroxymetoprolol, acidic metabolite	β-blocker	urine		CM-β-CD	acetate/4.0 (10 kV)	LLE	UV/210/ 280 nm	83
Phenprocoumon	anticoagulant	urine	*S*-naproxen	α-CD	phosphate–TEA/5.4 (−24 kV)	DSI	LIF 325/ 405 nm	84
Rogletimide	antitumor	serum	*S*-(−)-amino-glutethimide	α-CD	phosphate–borate/2.5 (15 kV)	SPE	UV/220 nm	85
Salbutamol	β-blocker	urine		dermatan sulfate	tris/5.3 (24 kV)	SPE	UV/220	86
Sulpiride	neuroleptic	serum	(−)-butaclamol	sulfated β-CD	citrate/2.9 (20 kV)	SPE	UV/220 nm	79
Tramadol (*O*-demethyl tramadol glucuronide)	analgesic	urine	*S*-(+)-dimethindene	CM-β-CD + methyl-β-CD	borate/10.0 (25 kV)	DSI	LIF257/290 320 nm	87
Venflaxine (*O*-desmethylvenflaxine)	antidepressant	serum	(+)-tramadol	phosphated γ-CD	phosphate-tris (20 kV)	LLE	UV/195	88
Verapamil (norverapamil and other metabolites)	class IV antiarrhyth-mic	serum	gallopamil	CM-β-CD	phosphate/3.0 (400 V/cm)	SPE	UV/210	89

Abbreviations: CSF - cerebrospinal fluid, CD - cyclodextrin, CM - carboxymethyl, DM - dimethyl, HP - hydroxypropyl, TEA - triethylamine; other abbreviations as in Table 1.

[a] Derivatized with naphthalene-2,3-dicarboxaldehyde.

preparations. The literature is covered up to 1999, and subsequent years bring its constant expansion. The reason for this is clearly stated in the sentence: "Compared to chiral HPLC, chiral CE provides higher efficiency and is simpler, faster and cheaper, and consumes a much smaller amount of organic solvents" [72]. Here we will survey results published in the years 2000 and 2001 and those of earlier date not cited in that review. Table 3 lists several chiral drugs (and their metabolites) together with selected conditions of their determination by capillary zone electrophoresis in body fluids. In some cases, only the methods of assays in fortified body fluids are reported [74,76,79,85,86,89]; other papers deal with determinations in clinical samples [73,80,71,84,88] or report applications to the evaluation of enantioselective pharmacokinetics, metabolism, or elimination of drugs in samples from humans [75,77,78,82,83,87].

Native cyclodextrins and their derivatives are almost exclusively used as chiral selectors for these enantioseparations. The reported limits of quantitation for drug enantiomers ranged from 2 μg/mL for eticlopride to 0.01 ng/mL for fluoxetine. The methods were validated, and some of them evaluated also other drugs as possible interferents in the assays (e.g., [78]—44 drugs tested!). The CE methods are sometimes compared with the HPLC determinations of the same analytes [89].

An excellent example of the application of CE to investigations on stereoselective biotransformations of drugs is an assay of thalidomide and its hydroxylated metabolites after in vitro incubation of the drug with rat liver microsomes [90]. The chiral selector consisted of a mixture of native β-cyclodextrin and its negatively charged sulfobutyl ether; an acetate buffer of pH 4.5 was used and a diode-array detector served for detection. The method allowed the simultaneous separation of enantiomers of parent drug and its three metabolites and confirmed the stereoselectivity of the metabolism of thalidomide. Preliminary results of enantiomeric separation of noradrenaline in human cerebrospinal fluid spiked with the racemate were recently published [91]. Borate buffer (pH = 7.5) with a chiral selector of carboxymethyl-β-cyclodextrin was employed. The sample was only filtered before hydrostatic injection, and the UV detection was at 214 nm. The paper reports also electrophoretic enantioseparation of propranolol and other compounds (serotonin, dopamine, ephedrine). Stereoselective CZE was also applied for enantioanalysis of drugs such as amphetamines and their precursor (selegiline) in urine [92–95], whereas in the doping analysis racemorphan was assayed in urine using micellar electrokinetic chromatography [96]. Analysis of illicit drugs is more thoroughly discussed in another chapter (chapter by M. Bogusz).

Enantioselective capillary electrophoretic methods designed for analysis of drugs as pure active substances, in industrial bulk products, and in dosage forms may serve as a starting point for their modifications and adaptations to the determination of drug enantiomers in body fluids. Table 4 lists selected examples of such methods and their main electrophoretic conditions published recently.

The main problems with applications of these methods may lie in enhancing their sensitivity (using better detection methods) and the corroboration of methods for the preparation of samples from biological material. Other methods of enantioselective CE of different drugs were compiled in a monograph [57] and in review articles of a general nature [66–70,105–107] or

TABLE 4 Selected Examples of Recent Electrophoretic Enantioseparations of Chiral Drugs

Drug	Chiral selector	Buffer/pH (voltage)	Detection	Ref.
Chlorpheniramine	HP-β-CD	phosphate/2.5 (20 kV)	UV/210 nm	97
Flobufen	TM-β-CD	acetate/5.50[a] (20 kV)	UV/265 nm	98
Methylphenidate	DM-β-CD[b]	phosphate-TEA/3.0 (25 kV)	UV/210 nm	99
	DM-β-CD	phosphate/2.0 (28 kV)	UV/214 nm	100
Moramide	HP-β-CD[b]	phosphate-TEA/3.0 (−25 kV)	UV/210 nm	99
Naproxen	HP-β-CD + sulfated β-CD	tris-phosphate/5.5 (12 kV)	UV/210 nm	101
Propranolol	CM-β-CD + DM-β-CD	tris-phosphate/3.0 (−15 kV)	UV/214 nm	102
Terbutaline	PM-β-CD	acetate/5.0	UV/230 nm	103
Warfarin	different CDs and derivatives	phosphate-TEA/5.4 (20 kV)	UV, LIF 405/520 nm	104

Abbreviations: PM - permethylated; other abbreviations as in Table 3.
[a] Modifier - 10% ethanol.
[b] Internal standard - diphenylpyraline.

TABLE 5 Chiral Drugs Recently Enantioseparated by Capillary Electrophoresis

Drug	Reference
Alimemazine, Chlorpheniramine, Eperisone, Prenylamine, Promethazine, Tolperizone, Trimipramine[a]	110
Alprenolol, Propranolol, Sotalol, Timolol[a]	111
Alprenolol, Arterenol, Atenolol, Atropine, 4-Chloramphetamine, Chlorpheniramine, Epinephrine, Homatropine, Ketotifen, Metoprolol, Oxprenolol, Propafenone, Propranolol, Terbutaline, Terfenadine, Tetrahydrozoline, Verapamil[a]	112
Amlodipine, Nimodipine, Nisoldipine, Nitrendipine, Nivaldipine[b]	113
Amphetamines	114
Anisodamine, Bencynonate, Benzhexol, Bepridil, Bisoprolol, Carvedilol, Esmolol, Glutethimide, Glycopyrrolate, Indapamide, Lobeline, Methoxamine, Ondansetron, Phenylephrine, Pinacidil, Synephrine, Terazosin, Tramadol[a]	115
Atenolol, Bupropion, Chlorphedianol, Chlorpheniramine, Epinephrine, Isoproterenol, Ketamine, Labetalol, Methoxyphenamine, Metaproterenol, Oxprenolol, Oxyphencyclamine Pindolol, Piperoxan, Propranolol, Propafenone, Terbutaline, Tetrahydropapaveroline Trihexyphenilyl,Tetrahydrazoline, Tolperisone	116
Atenolol, Chloramphetamine, EpinephrineHomatotropine, Ketamine, Mepenzolate bromide, Metanephrine, Methoxyphenamine, Metaproterenol, Metoprolol, Piperoxan, Pindolol, Propranolol, Tolperisone Tetrahydrozoline[a]	117
Atenolol, Norephedrine, Ofloxacin, Salbutamol	118
Bepridil, Ondansetron, Pinacidil	119
Bupivacaine, Isoprenaline, Ketamine, Lobeline, Mexiletine, Propafenone, Propranolol, Verapamil	120
Bupivacaine, Mepivacaine, Orciprenaline, Prilocaine	121
Carprofen, Fenoprofen	122
Chlorpheniramine, Clorprenaline, Dimetindene, Homochlorcyclizine, Isoprenaline, Ketamine, Piperoxan, Promethazine, Terbutaline, Trimetoquinol, Verapamil[a]	123

Table 5 (Continued)

Drug	Reference
Chlorpheniramine, Methadone	124
Clenbuterol, Dobutamine, Salbutamol, Terbutaline	125
Disopyramide, Fluoxetine, Mexiletine, Praziquantel (and their metabolites)	126
Eperisone, Epinastine, Tolperisone, Trimetoquinol[a]	127
Ephedrine, Etilefrine, Metoxamine, Methylsynephrine, Norephedrine, Norepinephrine, Octopamine, Synephrine[a]	128
Ethopropazine, Promethazine, Thioridazine, Trimeprazine	129
Labetalol, Nadolol	130
Thialbarbital, Thiopental[b]	131

[a] Only those with resolution $Rs > 1.0$ are listed.
[b] Other compounds of this type were also enantioseparated.

ones devoted to particular groups of drugs [108,109] and can be found in the original papers. Some of those enantioseparations of drugs published recently are summarized in Table 5.

Enantioseparations of newly synthesized compounds with pharmacological activity may also help to find suitable experimental conditions for separations of enantiomers of drugs with structural similarity [132–134].

Enantioselective binding of drug enantiomers to proteins of biological fluids is another important field of application of electrophoretic determination quite important from the clinical point of view and its impact on pharmacokinetic results. Different methods of such analyses (affinity capillary electrophoresis, partial filling technique, zonal elution, frontal analysis, vacancy techniques, and methods with immobilized proteins) and examples of their applications were recently reviewed [21,135,136], whereas recent experimental studies on protein binding of disopyramide [137] and verapamil [138] exemplify some of the investigated problems. It was found that genetic variants of human α_1-acid glycoprotein significantly affect the enantioselectivity in binding of disopyramide, but this was not the case for verapamil. For this last drug no enantioselectivity was observed for binding with high-, low-, or oxidized low-density lipoproteins.

5 CONCLUSIONS

The material presented in this chapter clearly demonstrates the utility of CE for the determination of drugs and their metabolites in body fluids. One can expect that these bioanalytical methods will be further developed and used in TDM, pharmacokinetic studies, and the evaluation of stereochemical fates of drugs in the body. Closely related methods, not covered here, like capillary electrochromatography, can be also used for these purposes, as exemplified by the determination of enantiomers of venlafaxine and its metabolite in clinical samples [139].

The CE methods are constantly improved, and such improvements may be applied in the analysis of drugs. Some recent examples comprise the separation of enantiomers of such drugs as amphetamines [140] and adrenaline and related compounds [141] on a micromachined device, the use of miniaturized capillaries for CE separation of catecholamines [142], and a new radioactivity detector for the determination of ^{3}H-labeled drug and its metabolites in urine [143].

Different optimization methods [144] may be applied to CE separations, and among them, mathematical modeling and neural networks were recently applied to drugs of the aryl propionic acid type [145] and sulfonamides [146]. Last but not least, computer programs and internet data bases [147] can significantly assist the development of CE methods suitable for the clinical analysis of drugs.

REFERENCES

1. R Kuhn, S Hoffstetter-Kuhn. Capillary Electrophoresis: Principles and Practice. Berlin, Heidelberg: Springer-Verlag, 1993.
2. P Camilleri, ed. Capillary Electrophoresis. 2d ed. Boca Raton, FL: CRC Press, 1998.
3. KD Altria. Overview of capillary electrophoresis and capillary electrochromatography. J Chromatogr A 856:443–463, 1999.
4. HJ Issaq. A decade of capillary electrophoresis. Electrophoresis 21:1921–1939, 2000.
5. WT Kok. Capillary electrophoresis. Instrumentation and operation. Chromatographia 51(suppl):S4–S89, 2000.
6. F Steiner, M Hassel. Nonaqueous capillary electrophoresis: a versatile completion of electrophoretic separation techniques. Electrophoresis 21: 3994–4016, 2000.
7. ML Riekkola, M Jussila, SP Porras, IE Valkó. Non-aqueous capillary electrophoresis. J Chromatogr A 892:155–170, 2000.
8. LA Holland, NP Chetwyn, MD Perkins, SM Lunte. Capillary electrophoresis in pharmaceutical analysis. Pharm Res 14:372–387, 1997.

9. H Nishi. Capillary electrophoresis of drugs: Current status in the analysis of pharmaceuticals. Electrophoresis 20:3237–3258, 1999.
10. CM Boone, JCM Waterval, H Lingeman, K Ensing, WJM Underberg. Capillary electrophoresis as a versatile tool for the bioanalysis of drugs—a review. J Pharm Biomed Anal 20:831–863, 1999.
11. LG Blomberg, H Wan. Determination of enantiomeric excess by capillary electrophoresis. Electrophoresis 21:1940–1952, 2000.
12. K Altria. Application of standard methods in capillary electrophoresis for drug analysis. In: K Valkó, ed. Separation Methods in Drug Synthesis and Purification. Amsterdam: Elsevier, 2000, pp 87–105.
13. KD Altria, AB Chen, L Clohs. Capillary electrophoresis as a routine analytical tool in pharmaceutical analysis. LC GC Europe 14:736–744, 2001.
14. MD Harvey, DM Paquette, PR Banks. Clinical applications of CE. J Liq Chrom Rel Technol 24:1871–1879, 2001.
15. W Thormann, CX Zhang, A Schmutz. Capillary electrophoresis for drug analysis in body fluids. Ther Drug Monit 18:506–520, 1996.
16. D Levêque, C Gallion-Renault, H Monteil, F Jehl. Capillary electrophoresis for pharmacokinetic studies. J Chromatogr B 697:67–75, 1997.
17. SH Chen, YH Chen. Pharmacokinetic applications of capillary electrophoresis. Electrophoresis 20:3259–3268, 1999.
18. L Denoroy, L Bert, S Parrot, F Robert, B Renaud. Assessment of pharmacodynamic and pharmacokinetic characteristics of drugs using microdialysis sampling and capillary electrophoresis. Electrophoresis 19:2841–2847, 1998.
19. P Rada, S Tucci, J Pérez, L Teneud, S Chuecos, L Hernández. In vivo monitoring of gabapentin in rats: a microdialysis study coupled to capillary electrophoresis and laser-induced fluorescence detection. Electrophoresis 19:2976–2980, 1998.
20. R Mader, M Brunner, B Rizovski, C Mensik, G Steger, H Eichler, M Müller. Analysis of microdialysates from cancer patients by capillary electrophoresis. Electrophoresis 19:2981–2985, 1998.
21. KL Rundlett, DW Armstrong. Methods for the determination of binding constants by capillary electrophoresis. Electrophoresis 22:1419–1427, 2001.
22. S Pálmarsdóttir, L Mathiasson, JA Jönsson, LE Edholm. Determination of a basic drug, bambuterol, in human plasma by capillary electrophoresis using double stacking for large volume injection and supported liquid membranes for sample pretreatment. J Chromatogr B 688:127–134, 1997.
23. JR Veraart, J van Hekezen, MCE Groot, C Gooijer, H Lingeman, NH Velthorst, UAT Brinkman. On-line dialysis solid-phase extraction coupled to capillary electrophoresis. Electrophoresis 19:2944–2949, 1998.
24. F Couderc, E Caussé, C Bayle. Drug analysis by capillary electrophoresis and laser-induced fluorescence. Electrophoresis 19:2777–2790, 1998.
25. AJ Gawron, R Scott Martin, SM Lunte. Microchip electrophoretic separation systems for biomedical and pharmaceutical analysis. Eur J Pharm Sci 14:1–12, 2001.
26. Y Xu. Capillary electrophoresis. Anal Chem 71:309R–313R, 1999.

27. NT Nguyen, RW Siegler. Capillary electrophoresis of cardiovascular drugs. J Chromatogr A 735:123–150, 1996.
28. SH Chen, JM Gallo. Use of capillary electrophoresis methods to characterize the pharmacokinetics of antisense drugs. Electrophoresis 19:2861–2869, 1998.
29. M Iwaki, E Murakami, K Kakehi. Chromatographic and capillary electrophoretic methods for the analysis of nicotinic acid and its metabolites. J Chromatogr B 747:229–240, 2000.
30. FW Smyth, S McClean. A critical evaluation of the application of capillary electrophoresis to the detection and determination of 1,4-benzodiazepine tranquilizers in formulations and body materials. Electrophoresis 19:2870–2882, 1998.
31. J Olgemöller, G Hempel, J Boos, G Blaschke. Determination of (*E*)-5-(2-bromovinyl)-2′-deoxyuridine in plasma and urine by capillary electrophoresis. J Chromatogr B 726:261–268, 1999.
32. Q Dong, W Jin. Monitoring diclofenac sodium in single human erythrocytes introduced by electroporation using capillary zone electrophoresis with electrochemical detection. Electrophoresis 22:2786–2792, 2001.
33. Y Deng, H Zhang, J Henion. Chip-based quantitative capillary electrophoresis/mass spectrometry determination of drugs in human plasma. Anal Chem 73:1432–1439, 2001.
34. D Dogrukol-Ak, K Kircali, M Tunçel, HY Aboul-Enein. Validated analysis of fluvastatin in a pharmaceutical capsule formulation and serum by capillary electrophoresis. Biomed Chromatogr 15:389–392, 2001.
35. M Tegtmeier, A Mühlau, D Dücker, M Runkel, W Legrum. A new method of capillary electrophoresis for metabolites of coumarin. Pharmazie 55:94–96, 2000.
36. T Zhou, Q Hu, H Yu, Y Fang. Separation and determination of β-agonists in serum by capillary zone electrophoresis with amperometric detection. Anal Chim Acta 441:23–28, 2001.
37. R Paroni, B Comuzzi, C Arcelloni, S Brocco, S de Kreutzenberg, A Tiengo, A Ciucci, P Beck-Peccoz, S Genovese. Comparison of capillary electrophoresis with HPLC for diagnosis of factitious hypoglycemia. Clin Chem 46:1773–1780, 2000.
38. S Kiessig, H Bang, F Thunecke. Interaction of cyclophilin and cyclosporins monitored by affinity capillary electrophoresis. J Chromatogr A 853:469–477, 1999.
39. W Gutleben, ND Tuan, H Stoiber, MP Dierich, M Sarcletti, A Zemann. Capillary electrophoresis separation of protease inhibitors used in human immunodeficiency virus therapy. J Chromatogr A 922:313–320, 2001.
40. K Hettiarachchi, CE Green, S Ridge, B Wu, P Catz, MA Salem. Analysis of 2β-carbomethoxy-3β-(4-fluorophenyl)-*N*-(3-iodo-*E*-allyl)nortropane in rat plasma. I. Method development and validation by capillary electrophoresis. J Chromatogr A 895:87–100, 2000.
41. K Hettiarachchi, CE Green, S Ramanathan-Girish, B Wu, CJ Jackson, S

Ridge, MA Salem, ME Lanser. Analysis of 2β-carbomethoxy-3β-(4-fluorophenyl)-*N*-(3-iodo-*E*-allyl)nortropane in rat plasma. II. Pharmacokinetic profile in male and female Sprague–Dawley rats evaluated by capillary electrophoresis. J Chromatogr A 924:471–481, 2001.

42. JW Luo, HW Chen, QH He. Determination of cimetidine in human plasma by use of coupled-flow injection, solid-phase extraction, and capillary zone electrophoresis. Chromatographia 53:295–300, 2001.
43. LJ Brunner, JT DiPiro. Capillary electrophoresis for therapeutic drug monitoring. Electrophoresis 19:2848–2855, 1998.
44. ZK Shihabi. Therapeutic drug monitoring by capillary electrophoresis. J Chromatogr A 807:27–36, 1998.
45. ZK Shihabi. Serum drug monitoring by capillary electrophoresis. In: JR Petersen, AA Mohammad, eds. Clinical and Forensic Applications of Capillary Electrophoresis. Totowa, NJ: Humana Press, 2001, pp 355–383.
46. A Procházkova, M Chouki, R Theurillat, W Thormann. Therapeutic drug monitoring of albendazole: determination of albendazole, albendazole sulfoxide and albendazole sulfone in human plasma using nonaqueous capillary electrophoresis. Electrophoresis 21:729–736, 2000.
47. A Gavenda, J Ševčik, J Psotová, P Bednár, P Barták, P Adamovský, V Šimánek. Determination of anthracycline antibiotics doxorubicin and daunorubicin by capillary electrophoresis with UV absorption detection. Electrophoresis 22:2782–2785, 2001.
48. G Hempel, P Schulze-Westholf, S Flege, N Laubrock, J Boos. Therapeutic drug monitoring of doxorubicin in paediatric oncology using capillary electrophoresis. Electrophoresis 19:2939–2943, 1998.
49. JR Veraart, UAT Brinkman. Dialysis-solid-phase extraction combined on-line with nonaqueous capillary electrophoresis for improved detectability of tricyclic antidepressants in biological samples. J Chromatogr A 922:339–346, 2001.
50. W Thormann, R Theurillat, M Wind, R Kuldvee. Therapeutic drug monitoring of antiepileptics by capillary electrophoresis. Characterization of assays via analysis of quality control sera containing 14 analytes. J Chromatogr A 924:429–437, 2001.
51. DH Peters, EM Sorkin. Zonisamide: a review of its pharmacodynamic and pharmacokinetic properties and therapeutic potential in epilepsy. Drugs 45:760–787, 1993.
52. Y Kataoka, K Makino, R Oishi. Capillary electrophoresis for therapeutic drug monitoring of antiepileptics. Electrophoresis 19:2856–2860, 1998.
53. S Malovaná, D Gajdošová, J Benedik, J Havel. Determination of esmolol in serum by capillary zone electrophoresis and its monitoring in course of heart surgery. J Chromatogr B 760:37–43, 2001.
54. T Wessel, C Lanvers, S Fruend, G Hempel. Determination of purines including 2,8-dihydroxyadenine in urine using capillary electrophoresis. J Chromatogr A 894:157–164, 2000.
55. X Sun, W Cao, X Bai, X Yang, E Wang. Determination of allopurinol and its

active metabolite oxypurinol by capillary electrophoresis with end-column amperometric detection. Anal Chim Acta 442:121–128, 2001.
56. A Kunkel, H Watzig. Pharmacokinetic investigations with direct injection of plasma samples: possible savings using capillary electrophoresis (CE). Arch Pharm 332:175–178, 1999.
57. JJ Berzas Nevado, J Rodriguez Flores, G Castañeda Peñalvo, N Rodriguez Fariñas. Micellar electrokinetic capillary chromatography for the determination of Viagra and its metabolite (UK-103,320) in human serum. Electrophoresis 22:2004–2009, 2001.
58. B Chankvetadze. Capillary Electrophoresis in Chiral Analysis. Chichester: John Wiley, 1997.
59. K Verleysen, P Sandra. Separation of chiral compounds by capillary electrophoresis. Electrophoresis 19:2798–2833, 1998.
60. R Vespalec, P Bocek. Chiral separations in capillary electrophoresis. Electrophoresis 20:2579–2591, 1999.
61. G Gübitz, MG Schmid. Recent progress in chiral separation principles in capillary electrophoresis. Electrophoresis 21:4112–4135, 2000.
62. S Fanali. Enantioselective determination by capillary electrophoresis with cyclodextrins as chiral selectors. J Chromatogr A 875:89–122, 2000.
63. A Rizzi. Fundamental aspects of chiral separations by capillary electrophoresis. Electrophoresis 22:3079–3106, 2001.
64. TJ Ward, AB Farris III. Chiral separations using the macrocyclic antibiotics: a review. J Chromatogr 906:73–89, 2001.
65. B Chankvetadze, G Blaschke. Enantioseparations in capillary electromigration techniques: recent developments and future trends. J Chromatogr 906:309–363, 2001.
66. M Fillet, P Hubert, J Crommen. Method development strategies for the enantioseparation of drugs by capillary electrophoresis using cyclodextrins as chiral additives. Electrophoresis 19:2834–2840, 1998.
67. MR Hadley, P Camilleri, A Hutt. Enantiospecific analysis by capillary electrophoresis: Applications in drug metabolism and pharmacokinetics. Electrophoresis 21:1953–1976, 2000.
68. G Blaschke, B Chankvetadze. Enantiomer separation of drugs by capillary electromigration techniques. J Chromatogr A 875:3–25, 2000.
69. J Haginaka. Enantiomer separation of drugs by capillary electrophoresis using proteins as chiral selectors. J Chromatogr A 875:235–254, 2000.
70. A Amini. Recent developments in chiral capillary electrophoresis and applications of this technique to pharmaceutical and biomedical analysis. Electrophoresis 22:3107–3130, 2001.
71. J Bojarski, HY Aboul-Enein. Application of capillary electrophoresis for the analysis of chiral drugs in biological fluids. Electrophoresis 18:965–969, 1997.
72. S Zaugg, W Thormann. Enantioselective determination of drugs in body fluids by capillary electrophoresis. J Chromatogr A 875:27–41, 2000.
73. FO Paias, VL Lanchote, OM Takayanagui, PS Bonato. Enantioselective

analysis of albendazole sulfoxide in cerebrospinal fluid by capillary electrophoresis. Electrophoresis 22:3263–3269, 2001.

74. MT Chiang, SY Chang, CW Whang. Chiral analysis of baclofen by (-cyclodextrin-modified capillary electrophoresis and laser-induced fluorescence detection. Electrophoresis 22:123–127, 2001.
75. F Behn, S Michels, S Läer, G Blaschke. Separation of carvedilol enantiomers in very small volumes of human plasma by capillary electrophoresis with laser-induced fluorescence. J Chromatogr B 755:111–117, 2001.
76. L Clohs, KM McErlane. Development of a capillary electrophoresis assay for the determination of carvedilol enantiomers in serum using cyclodextrins. J Pharm Biomed Anal 24:545–554, 2001.
77. H Hüttemann, G Blaschke. Achiral and chiral determination of ciprofibrate and its glucuronide in human urine by capillary electrophoresis. J Chromatogr B 729:33–41, 1999.
78. VA Polisel Jabor, VL Lanchote, PL Bonato. Simultaneous determination of disopyramide and mono-*N*-dealkyldisopyramide enantiomers in human plasma by capillary electrophoresis. Electrophoresis 22:1406–1412, 2001.
79. XX Xu, JT Stewart. Chiral analysis of selected dopamine receptor antagonists in serum using capillary electrophoresis with cyclodextrin additives. J Pharm Biomed Anal 23:735–743, 2000.
80. C Desiderio, S Rudaz, MA Raggi, S Fanali. Enantiomeric separation of fluoxetine and norfluoxetine in plasma and serum samples with high detetection sensitivity capillary electrophoresis. Electrophoresis 20:3432–3438, 1999.
81. CB Eap, K Powell, P Baumann. Determination of the enantiomers of mianserin and its metabolites in plasma by capillary electrophoresis after liquid–liquid extraction and on-column sample preconcentration. J Chromatogr Sci 35:315–320, 1997.
82. F Prost, W Thormann. Enantiomeric analysis of five major monohydroxylated metabolites of methaqualone in human urine by chiral capillary electrophoresis. Electrophoresis 22:3270–3280, 2001.
83. HK Lim, PT Linh, CH Hong, KH Kim, JS Kang. Enantioselective determination of metoprolol and major metabolites in human urine by capillary electrophoresis. J Chromatogr B 755:259–264, 2001.
84. B Chankvetadze, N Burjanadze, G Blaschke. Enantioseparation of the anticoagulant drug phenprocoumon in capillary electrophoresis with UV and laser-induced fluorescence detection and application of the method to urine samples. Electrophoresis 22:3281–3285, 2001.
85. MM Hefnawy, JT Stewart. Enantioselective determination of $R(+)$ and $S(-)$ rogletimide in serum using alpha-cyclodextrin modified capillary electrophoresis and solid phase extraction. J Liq Chrom Rel Technol 25:791–804, 2000.
86. R Gotti, S Furlanetto, V Andrisano, V Cavrini, S Pinzauti. Design of experiments for capillary electrophoretic enantioresolution of salbutamol using dermatan sulfate. J Chromatogr A 875:411–422, 2000.
87. UB Soetebeer, MC Schierenberg, H Schulz, P Andresen, G Blaschke. Direct

chiral assay of tramadol and detection of the phase II metabolite *O*-demethyl tramadol glucuronide in human urine using capillary electrophoresis with laser-induced native fluorescence detection. J Chromatogr B 765:3–13, 2001.

88. S Rudaz, C Stelle, AE Balant-Gorgia, S Fanali, JL Veuthey. Simultaneous stereoselective analysis of venlafaxine and *O*-desmethylvenlafaxine enantiomers in clinical samples by capillary electrophoresis using charged cyclodextrins. J Pharm Biomed Anal 23:107–115, 2000.
89. E Brandšteterová, G Endresz, G Blaschke. Chiral separation of verapamil and some of its metabolites by HPLC and CE. Pharmazie 56:536–541, 2001.
90. M Meyring, C Mühlenbrock, G Blaschke. Investigation of the stereoselective in vitro biotransformation of thalidomide using dual cyclodextrin sysytem in capillary electrophoresis. Electrophoresis 21:3270–3279, 2000.
91. W Maruszak, M Trojanowicz, M Margasińska, H Engelhardt. Application of carboxymethyl-β-cyclodextrin as a chiral selector in capillary electrophoresis for enantiomer separation of selected neurotransmitters. J Chromatogr A 926:327–336, 2001.
92. J Ševcik, K Lemr, B Smysl, D Jirovsky, P Hradil. One-run chiral separation of methamphetamine and its related metabolites by capillary electrophoresis. J Liq Chrom Rel Technol 21:2473–2484, 1998.
93. YJ Heo, YS Whang, MK In, KJ Lee. Determination of enantiomeric amphetamines as metabolites of illicit amphetamines and selegiline in urine by capillary electrophoresis using modified β-cyclodextrin. J Chromatogr B 741:221–230, 2000.
94. S Chinaka, S Tanaka, N. Takayama, K Komai, T Ohshima, K Ueda. Simultaneous chiral analysis of methamphetamine and related compounds by capillary electrophoresis. J Chromatogr B 749:111–118, 2000.
95. EM Kim, HS Chung, KJ Lee, HJ Kim. Determination of enantiomeric metabolites of *l*-deprenyl, *d*-methamphetamine, and racemic methamphetamine in urine by capillary electrophoresis: comparison of deprenyl use and methamine use. J Anal Toxicol 24:238–244, 2000.
96. E Ban, S Choi, JA Lee, DS Seok, YS Yoo. Cyclodextrin-mediated micellar electrokinetic chromatography and matrix-assisted laser desorption/ionization time-of-flight mass spectrometry for the enantiomer separation of racemorphan in human urine. J Chromatogr A 853:439–447, 1999.
97. SW Sun, YR Lin. Optimization of capillary electrophoretic separation of chlorpheniramine enantiomers by a Plackett–Burman design. Determination of enantiomeric purity of dexchlorpheniramine. J Liq Chrom. Rel Technol 24:2051–2066. 2001.
98. A Lambert, JM Ballon, A Nicolas. Enantioseparation of flobufen with cyclodextrins studied by capillary electrophoresis and NMR. Pharm Res 18:886–893, 2001.
99. OM Denk, DG Watson, GG Skellern. Chiral analysis of methylphenidate and dextromoramide by capillary electrophoresis. J Chromatogr B 761:61–68, 2001.
100. WX Huang, Q Gao, M Harris, SD Fazio, RV Vivilecchia. Separation of

ritalin racemate and its by-product racemates by capillary electrophoresis. Electrophoresis 22:3226–3231, 2001.
101. H Lu, Z Ruan, J Kang, Q Ou. Simultaneous chiral separation and determination of the optical purity of naproxen and methyl naproxen by capillary electrophoresis with dual-cyclodextrin system as chiral selector. Anal Lett 34:1657–1668, 2001.
102. KA Assi, BJ Clark, KD Altria. Enantiomeric purity determination of propranolol by capillary electrophoresis using dual cyclodextrins and a polyacrylamide-coated capillary. Electrophoresis 20:2723–2725, 1999.
103. C Garcia-Ruiz, ML Marina. Fast enantiomeric separation of basic drugs by electrokinetic chromatography. Application to the quantitation of terbutaline in a pharmaceutical preparation. Electrophoresis 22:3191–3197, 2001.
104. B Chankvetadze, N Burjanadze, J Crommen, G Blaschke. Enantioseparation of warfarin by capillary electrophoresis with UV and LIF detection using single and dual cyclodextrin type chiral selectors. Chromatographia (suppl) 53:S296–S301, 2001.
105. K Otsuka, S Terabe. Enantiomer separation of drugs by micellar elektrokinetic chromatography using chiral surfactants. J Chromatogr A 875:163–178, 2000.
106. HH Yarabe, E Billiot, IM Warner. Enantiomeric separations by use of polymeric surfactant electrokinetic chromatography. J Chromatogr 875:179–206, 2000.
107. H Nishi, Y Kuwahara. Enantiomer separation by capillary electrophoresis utilizing noncyclic mono-, oligo- and polysaccharides as chiral selectors. J Biochem Biophys Methods 48:89–102, 2001.
108. J Bojarski. Chiral barbiturates—synthesis, chromatographic resolutions and biological activity. In: HY Aboul-Enein, IW Wainer, eds. The Impact of Stereochemistry on Drug Development and Use. New York: John Wiley, 1997, pp 201–234.
109. J Bojarski, HY Aboul-Enein. Recent chromatographic and electrophoretic enantioseparations of cardiovascular drugs. Biomed Chromatogr 13:197–208, 1999.
110. H Matsunaga, J Haginaka. Separation of basic drug enantiomers by capillary electrophoresis using ovoglycoprotein as a chiral selector: Comparison of chiral resolution ability of ovoglycoprotein and completely deglycosylated ovoglycoprotein. Electrophoresis 22:3251–3256, 2001.
111. MG Schmid, O Lecnik, U Sitte, G Gübitz. Application of ligand-exchange capillary electrophoresis to the chiral separation of α-hydroxy acids and β-blockers. J Chromatogr A 875:307–314, 2000.
112. M Tacker, P Glukhovskiy, H Cai, G Vigh. Nonaqueous capillary electrophoretic separation of basic enantiomers using heptakis(2,3-dimethyl-6-sulfato)-β-cyclodextrin. Electrophoresis 20:2794–2798, 1999.
113. T Christians, U Holzgrabe. Enantioseparation of dihydropyridine derivatives by means of neutral and negatively charged β-cyclodextrin derivatives using capillary electrophoresis. Electrophoresis 21:3609–3617, 2000.

114. S Cherkaoui, S Rudaz, E Varesio, JL Veuthey. On-line capillary electrophoresis–electrospray mass spectrometry for the stereoselective analysis of drugs and metabolites. Electrophoresis 22:3308–3315, 2001.
115. X Ren, Y Dong, J Liu, A Huang, H Liu, Y Sun, Z Sun. Separation of chiral basic drugs with sulfobutyl-β-cyclodextrin in capillary electrophoresis. Chromatographia 50:363–368, 1999.
116. W Zhu, G Vigh. Enantiomer separations by nonaqueous capillary electrophoresis using octakis(2,3-diacetyl-6-sulfato)-γ-cyclodextrin. J Chromatogr A 892:499–507, 2000.
117. DK Maynard, G Vigh. Heptakis(2-*O*-methyl-3,6-di-*O*-sulfo)-β-cyclodextrin: a single-isomer, 14-sulfated β-cyclodextrin for use as a chiral resolving agent in capillary electrophoresis. Electrophoresis 22:3152–3162, 2001.
118. W Wang, J Lu, X Fu, Y Chen. Carboxymethyl-β-cyclodextrin for chiral separation of basic drugs by capillary electrophoresis. Anal Lett 34:569–578, 2001.
119. X Ren, A Huang, T Wang, Y Sun, Z Sun. Enantiomeric separation of three chiral drugs by nonaqueous capillary electrophoresis with triethylamine as additive. Chromatographia 50:625–628, 1999.
120. G Li, X Lin, C Zhu, A Hao, Y Guan. New derivative of β-cyclodextrin as chiral selectors for the capillary electrophoretic separation of chiral drugs. Anal Chim Acta 421:27–34, 2000.
121. A Amini, U Paulsen-Sörman, D Westerlund. Dependence of chiral separations on the amount of cyclodextrins as selectors, employing the partial filling technique in capillary zone electrophoresis. Chromatographia 51:226–230, 2000.
122. 17. TJ Ward, AB Farris III, K Woodling. Synergistic chiral separations using the glycopeptides ristocetin A and vancomycin. J Biochem Biophys Methods 48:163–174, 2001.
123. H Matsunaga, J Haginaka. Separation of basic drug enantiomers by capillary zone electrophoresis using glucuronyl glucosyl β-cyclodextrin as a chiral selector. Electrophoresis 22:3382–3388, 2001.
124. M Wind, P Hoffmann, H Wagner, W Thormann. Chiral capillary electrophoresis as predictor for separation of drug enantiomers in continuous flow zone electrophoresis. J Chromatogr A 895:51–65, 2000.
125. HY Aboul-Enein, MD Efstatiade, GE Baiulescu. Cyclodextrins as chiral selectors in capillary electrophoresis: a comparative study for the enantiomeric separation of some beta-agonists. Electrophoresis 20:2686–2690, 1999
126. PS Bonato, VAP Jabor, FO Paias, VL Lanchote. Chiral capillary electrophoretic separation of selected drugs and metabolites using sulfated β-cyclodextrin. J Liq Chrom Rel Technol 24:1115–1131, 2001.
127. T Tsukamoto, T Ushio, J Haginaka. Chiral resolution of basic drugs by capillary electrophoresis with new glycosaminoglycans. J Chromatogr A 864:163–171, 1999.
128. MG Schmid, M Laffranchini, D Dreveny, G Gübitz. Chiral separation of sympathomimetics by ligand exchange capillary electrophoresis. Electrophoresis 20:2458–2461, 1999.

129. CE Lin, KH Chen. Enantioseparation of phenothiazines in capillary zone electrophoresis using cyclodextrins as chiral selectors. J. Chromatogr A 930:155–163, 2001.
130. SL Tamisier-Karolak, MA Stenger, A Bommart. Enantioseparation of β-blockers with two chiral centers by capillary electrophoresis using sulfated β-cyclodextrins. Electrophoresis 20:2656–2663, 1999.
131. 12. U Schmitt, J Bojarski, U Holzgrabe. Enantioseparation of chiral thiobarbiturates using cyclodextrin-modified capillary electrophoresis. Electrophoresis 22:3237–3242, 2001.
132. O Lecnik, MG Schmid, CK Kappe, G. Gübitz. Chiral separation of pharmacologically active dihydropyrimidinones with carboxymethyl-β-cyclodextrin. Electrophoresis 22:3198–3202, 2001.
133. B Proksa, R Cižmáriková. Separation of β-adrenolytics derived from 4-hydroxyacetophenone by capillary electrophoresis in the presence of cyclodextrins. Anal Chim Acta 434:75–79, 2001.
134. S Fanali, C Cartoni, C Desiderio. Chiral separation of newly synthesized arylpropionic acids by capillary electrophoresis using cyclodextrins or a glycopeptide antibiotic as chiral selectors. Chromatographia 54:87–92, 2001.
135. DS Hage. Chromatographic and electrophoretic studies of protein binding to chiral solutes. J Chromatogr A 906:459–481, 2001.
136. Y Tanaka, S Terabe. Recent advances in enantiomer separation by affinity capillary electrophoresis using proteins and peptides. J Biochem Biophys Methods 48:103–116, 2001.
137. Y Kuroda, Y Kita, A Shibukawa, T Nakagawa. Role of biantennary glycans and genetic variants of human α_1-acid glycoprotein in enantioselective binding of basic drugs as studied by high performance frontal analysis/capillary electrophoresis. Pharm Res 18:389–393, 2001.
138. NAL Mohamed, Y Kuroda, A Shibukawa, T Nakagawa, S El Gizawy, HF Askal, ME El Kommos. Enantioselective binding analysis of verapamil to plasma lipoproteins by capillary electrophoresis–frontal analysis. J Chromatogr A 875:447–453, 2000.
139. S Fanali, S Rudaz, JL Veuthey, C Desiderio. Use of vancomycin silica stationary phase in packed capillary electrochromatography II. Enantiomer separation of venlafaxine and *O*-desmethylvenlafaxine in human plasma. J Chromatogr 919:195–203, 2001.
140. SR Wallenborg, IS Lurie, DW Arnold, CG Bailey. On-chip chiral and achiral separation of amphetamine and related compounds labeled with 4-fluoro-4-nitrobenzofurazane. Electrophoresis 21:3257–3263, 2000.
141. MA Schwarz, PC Hauser. Rapid chiral on-chip separation with simplified amperometric detection. J Chromatogr A 928:225–232, 2001.
142. LA Woods, TP Roddy, TL Paxton, AG Ewing. Electrophoresis in nanometer inner diameter capillaries with electrochemical detection. Anal Chem 73:3687–3690, 2001.
143. KO Boernsen, JM Floeckher, GJM Bruin. Use of a microplate scintillation

counter as a radioactivity detector for miniaturized separation techniques in drug metabolism. Anal Chem 72:3956–3959, 2000.
144. AM Siouffi, R Phan-Tan-Luu. Optimization methods in chromatography and capillary electrophoresis. J Chromatogr A 892:75–106, 2000.
145. JP Wolbach, DK Lloyd, IW Wainer. Approaches to quantitative structure–enantioselectivity relationship modeling of chiral separations using capillary electrophoresis. J Chromatogr A 914:299–314, 2001.
146. M Javali-Heravi, Z Garkani-Nejad. Prediction of electrophoretic mobilities of sulfonamides in capillary zone electrophoresis using artificial neural networks. J Chromatogr A 927:211–218, 2001.
147. J. Reijenga, HK Lee. Software and internet resources for capillary electrophoresis and micellar electrokinetic capillary chromatography. J Chromatogr A 916:25–30, 2001.

Index